OPTICAL NETWORKS AND TECHNOLOGIES

IFIP – The International Federation for Information Processing

IFIP was founded in 1960 under the auspices of UNESCO, following the First World Computer Congress held in Paris the previous year. An umbrella organization for societies working in information processing, IFIP's aim is two-fold: to support information processing within its member countries and to encourage technology transfer to developing nations. As its mission statement clearly states,

> *IFIP's mission is to be the leading, truly international, apolitical organization which encourages and assists in the development, exploitation and application of information technology for the benefit of all people.*

IFIP is a non-profitmaking organization, run almost solely by 2500 volunteers. It operates through a number of technical committees, which organize events and publications. IFIP's events range from an international congress to local seminars, but the most important are:

- The IFIP World Computer Congress, held every second year;
- Open conferences;
- Working conferences.

The flagship event is the IFIP World Computer Congress, at which both invited and contributed papers are presented. Contributed papers are rigorously refereed and the rejection rate is high.

As with the Congress, participation in the open conferences is open to all and papers may be invited or submitted. Again, submitted papers are stringently refereed.

The working conferences are structured differently. They are usually run by a working group and attendance is small and by invitation only. Their purpose is to create an atmosphere conducive to innovation and development. Refereeing is less rigorous and papers are subjected to extensive group discussion.

Publications arising from IFIP events vary. The papers presented at the IFIP World Computer Congress and at open conferences are published as conference proceedings, while the results of the working conferences are often published as collections of selected and edited papers.

Any national society whose primary activity is in information may apply to become a full member of IFIP, although full membership is restricted to one society per country. Full members are entitled to vote at the annual General Assembly, National societies preferring a less committed involvement may apply for associate or corresponding membership. Associate members enjoy the same benefits as full members, but without voting rights. Corresponding members are not represented in IFIP bodies. Affiliated membership is open to non-national societies, and individual and honorary membership schemes are also offered.

OPTICAL NETWORKS AND TECHNOLOGIES

IFIP TC6 / WG6.10 First Optical Networks & Technologies Conference (OpNeTec), October 18-20, 2004, Pisa, Italy

Edited by

KEN-ICHI KITAYAMA
Department of Electronics and Information Systems
Osaka University, Japan

FRANCESCO MASETTI-PLACCI
Alcatel
Vimercate, Italy

GIANCARLO PRATI
Consorzio Nazionale Interuniversitario per le Telecomunicazioni – CNIT, Pisa, Italy
Scuola Superiore Sant'Anna, Pisa, Italy

 Springer

Ken-Ichi Kitayama
Osaka University
Osaka, JAPAN

Francesco Masetti-Placci
Alcatel Italia
Vimercate, ITALY

Giancarlo Prati
Consorzio Nazionale
Interuniversitario per le
Telecomunicazioni -
CNIT
Pisa, ITALY

Library of Congress Cataloging-in-Publication Data

A C.I.P. Catalogue record for this book is available from the Library of Congress.

Optical Networks and Technologies / Edited by Ken-Ichi Kitayama, Francesco Masetti-Placci, Giancarlo Prati.
    p.cm. —(The International Federation for Information Processing)
    Includes bibliographical references and Index.

ISBN 978-1-4419-3583-0   e-ISBN 978-0-387-23178-5   Printed on acid-free paper.

Printed in the United States of America.

9 8 7 6 5 4 3 2 1
springeronline.com

Contents

viii

Preface

There has continuously been a massive growth of Internet traffic for these years despite the "bubble burst" in year 2000. As the telecom market is gradually picking up, it would be a consensus in telecom and data-com industries that the CAPEX (Capital Expenditures) to rebuild the network infrastructure to cope with this traffic growth would be imminent, while the OPEX (Operational Expenditures) has to be within a tight constraint. Therefore, the newly built 21st-century network has to fully evolve from voice-oriented legacy networks, not only by increasing the transmission capacity of WDM links but also by introducing switching technologies in *optical domain* to provide full-connectivity to support a wide variety of services.

This book stems from the technical contributions presented at the Optical Networks and Technology Conference (OpNeTec), inaugurated this year 2004 in Pisa, Italy, and collects innovations of optical network technologies toward the 21st century network. High-quality recent research results on optical networks and related technologies are presented, including IP over WDM integration, burst and packet switchings, control and managements, operation, metro- and access networks, and components and devices in the perspective of network application. An effort has been made throughout the conference, hopefully reflected at least partially in this book, to bring together researchers, scientists, and engineers working both academia and industries to discuss the relative impact of networks on technologies and vice versa, with a vision of the future. Too often the photonic communication field is approached as it were a mature field where systems and technologies have their own lives. Photonics is still in its infancy, playing the correlation and reciprocal influence of technology and system/network solutions a key role, deserving more attention and consideration on both sides.

Ken-ichi Kitayama
Francesco Masetti-Placci
Giancarlo Prati

Acknowledgments

The editors wish to express their sincere thanks to the members of the International Program Committee of the First Optical Networks & Technologies Conference (OpNeTec), which was held in October 18-20, 2004 in Pisa, Italy, whose cooperation was essential to the organization of the conference and to the publication of this book.

The conference would not have been possible without the support of the Italian National Consortium for Telecommunications (CNIT) and the work of the Organizing Committee.

The editors would also like to gratefully acknowledged the following organizations and institution hereafter:

Agilent Technologies
Alcatel
Anritsu
IFIP-Technical Committee on Communication Systems (TC6)
Marconi Communications
Siemens
Telecom Lab Italia

Committees

General Chairman
Giancarlo Prati, Scuola Superiore Sant'Anna & CNIT

Technical Program Co-Chairs
K. Kitayama (Co-chair), Osaka University, Japan
F. Masetti-Placci (Co-chair), Alcatel CIT, France

International Program Committee
S. Araki, NEC Corporation, Japan
K. Asatani, Kogakuin University, Japan
K. Blow, Aston University, U.K.
A. Bonati, Alcatel Italy
A. Bononi, University of Parma, Italy
A. Cantoni, W. Australian Telecomm. Research Inst., Australia
R. Castelli, Alcatel, Italy
D. Chiaroni, Alcatel CIT, France
W. Chujo, Communication Research Lab., Thailand
S. Dixit, Nokia Research Center, USA
A. Fumagalli, University of Texas at Dallas, USA
P. Franco, Pirelli Labs, Italy
P. Gambini, Agilent Technologies, Italy
R. Gangopadhyay, IIT Kharagpur, India
P. Ghiggino, Marconi, U.K.
C. Glingener, Marconi, Germany
E. Guarene, Telecom Italia Lab, Italy

L. Jereb, Budapest Univ. of Tech. & Econ., Hungary
L. Kazovsky, Stanford University, USA
S. Bae Lee, KIST, Korea
H. T. Muftah, University of Ottawa, Canada
G. Morthier, Ghent University, Belgium
F. Neri, Politecnico di Torino, Italy
H. Onaka, Fujitsu Lab, Japan
H. Perros, North Carolina State University, USA
F. Russo, University of Pisa, Italy
R. Sabella, Ericsson, Italy
S. Saracino, Siemens CNX, Italy
K. Stubkjaer, Tech.. University of Denmark, Denmark
L. Wosinska, Royal Institute of Technology, Sweden

Organizing Committee

A. Bogoni, CNIT, Italy
P. Castoldi, Scuola Superiore Sant'Anna & CNIT, Italy
E. Ciaramella, Scuola Superiore Sant'Anna & CNIT, Italy
S. Cinquini, Telecom Italia, Italy
F. Di Pasquale, Scuola Superiore Sant'Anna & CNIT, Italy
K. Ennser, CNIT, Italy
S. Giordano, University of Pisa, Italy
L. Poti', CNIT, Italy

CNIT Secretariat

M. E. Razzoli, CNIT, Italy
A. Letta, Scuola Superiore Sant'Anna, Italy

Publications

K. Ennser, CNIT, Italy

PERSPECTIVES ON OPTICAL NETWORKS AND TECHNOLOGIES

DEVELOPMENTS IN OPTICAL SEAMLESS NETWORKS
Invited paper

Andrea Spaccapietra[1],Giovanni Razzetta[2]
[1]*VicePresident Optical Core Networks, Marconi Corporation,*
New Century Park, Coventry, West Midlands, CV3 1HJ,United Kingdom,
andrea.spaccapietra@marconi.com
[2]*Photonics System Design Manager, Marconi Communications,*
Via A. Negrone 1A, 16153 Genova, Italy
giovanni.razzetta@marconi.com

Abstract: This paper give a view on key technologies that are emerging as the enabler for evolving Core Transport Network towards the delivery of the customer experience expected by the service users community. Optical technologies will be dealt with first, explaining how they are fitting in the medium and long term evolution of commercial optical transmission systems. Hardware and software technologies involved in the shift toward data centric services are addressed, identifying the path to a full integrated transport and switching core network, with the ultimate objective of maximizing the user benefits and reducing cost.

1.   INTRODUCTION

National incumbents, large fixed line second operators and some key Mobile operators have spent significant time and Capital building large, highly reliable, resilient, carrier-class Networks to support Voice and leased lines services. This had been a stable business model for well over a decade, with the largest challenge being how to scale.

As we know the existing model is being challenged. It is anticipated that the revenue our customers will generate from traditional voice services will be flat at best, with growth expected to come from the provision of new broadband data services (i.e. triple play: voice, data and video).

Residential broadband services typically consume 10 times the bandwidth of narrowband users, but are offered at no more than twice the existing narrowband subscription rates. Equally the increasing requirement for Enterprise businesses to store, protect and retrieve information and records is doubling data traffic in the Wide area every two years.

Broadband services dramatically increase capacity demands on the network but do not return a proportional increases in revenue. Therefore network operators need to substantially reduce network operational costs, provide capacity at much lower cost per bit and deliver new revenue generating services.

In the following the enabling technologies for achieving such objectives are illustrated and placed in the context of new generation core transport network, delivering seamless services to network operators and their customers.

1.1 Optical seamless network

Optical Seamless network is a key component in building service infrastructures that deliver a delighting customer experience, innovative products, rapid time to market for new services and transform the cost base for the network. An example can be found in Figure 1.

The goal of the service infrastructure is to provide a "simple and complete" communications service to customers, regardless of time or place.

The pillars of this network vision are:

- an high performance, integrated, cost effective transport infrastructure, evolving from current transport networks;
- new platforms for services based on the delivery of content and applications, supporting both multi-media and mobile services;
- OSS increasingly becoming part of the service, and ultimately converging with the network intelligence components;
- standards to define the architecture components.

The optical seamless network is a rationalized optical transport enabling an ultra broadband data network. Access networks will be converging onto a multi-service platform.

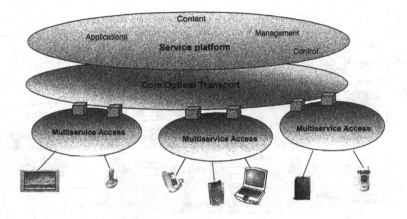

Figure 1- Service network architecture

2. THE REQUIREMENTS FOR THE OPTICAL SEAMLESS NETWORK

After positioning the Optical Seamless network in the bigger picture of overall service network, we need to identify the requirements that service objectives set to the transport network and how they can be best addressed using emerging technologies. Requirements can be better understood if the transport network is logically divided into a data plane and a control plane. From the technological point of view the data plane can be further divided into three technology layers: the optical plane, the electrical plane and the data plane in strict sense (i.e. packet/cell based).

Key requirements and related technologies for the optical plane enabling the seamless operation can be identified in transparent bypass of wavelengths (OXC, ROADM, EDFA and Raman amplifiers) and compatibility with existing fibre plant (dispersion resistant modulation schemes for combined 40Gbit/s/10Gbit/s transport).

Emerging technologies for the electrical plane ready for a massive roll-out in the electrical plane are integration of SDH and wavelength (ODU) switching, standards compliant mapping of data into transport frames (GFP/LCAS/VCAT).

The data plane will be initially provided by separated plug-in cards and ultimately a protocol agnostic transport switching can allow further integration with the electrical plane.

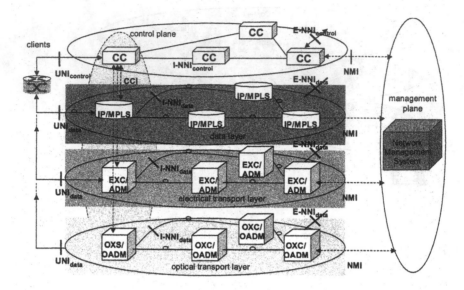

Figure 1 – Network planes

The control of the network will progressively migrate from centralized operation support systems or network management system to a distributed control plane (the same way large data networks are managed today), although functions like Alarm Supervision, Performance Monitoring and Inventory will remain under the TMN domain.

A common control scheme for data and transport simplifies operational processes and ease the convergence of data and transport.

2.1 Optical plane

After the massive investment in the "Holy Grail" of optical technologies during the years of the "Telecom bubble", we are now in a situation where components suppliers are bringing selected optical technologies to the marketplace for being usefully utilized in the design of optical system. The economical benefit of the so called "optical bypass " is now recognized in the transmission community and a number of component and system technologies are available. Some of them are well established, such as dynamic power management, dispersion management, error correction, some of them are being developed and will become common in the near future, such as electrical distortion compensation (EDC), innovative modulation formats and sophisticated methods for optical performance monitoring & fault localization. All of these ingredients are preparing current transmission infrastructure for scaling towards high channel bit-rates (i.e. 40G), that has already been experimented in the field, both in Europe and US.

All the above needs to be achieved through a "platform concept", in order to deliver inexpensive unlimited bandwidth with the maximum flexibility. For example a single common chassis type that has multi function slots that can accommodate the relevant transponders, amplifiers or optical switching unit.

2.2 Electrical plane

If we look at today SDH layer of a Core transmission network we can see that we have a potential like for like element replacement of 16:1 or greater. The high capacity OCS nodes can consolidate multiple 1st generation ADM Elements into one element. This results in considerable infrastructure savings with the associated benefits this brings. This allows significant improvements to be made at all levels of a core implementation. It also allows simplification of potential co-located elements which are interconnecting to other Access rings bringing operational and real estate benefits.

Moving to the layer above SDH we can see savings gained through integration between layers of the core Network. The recently standardized Optical Transport Network frame structure (OTN/ODU) is now integrated in optical cross-connects, that now can handle wavelength bandwidth granularity.

Along with the evolution toward larger granularity and massive integration, the capability of handling data streams by simply mapping them in transport structure is now being deployed. Bandwidth optimisation through aggregation and dynamic bandwidth allocation enhance the competitiveness of the solutions with respect to pure data networks. This identifies a definite trend for transport networks to encompass layer 2 and even layer 3 functionalities, but with the carrier class availability performance that transport network only can deliver.

This provides a base architecture that can adapt to accommodate any service mix.

2.3 Control Plane

The "Seamless' attribute of new generation transport network is associated with an increased level of intelligence that is inherent within the network elements and is realised via the implementation of an ASTN/GMPLS control plane, which simplifies network operation and optimises network resources.

Via the neighbour discovery function newly added network elements and nodes are automatically recognised by the network. The additional capacity and route diversity becomes automatically part of the network resource pool and can immediately be used. Long and cumbersome manual configuration processes belong to the past. The end user can requested new services directly via the standardised user-network-interface (UNI). The network automatically finds the best and least expensive route through the mesh and protects it according to the selected grade of service. New services are provisioned at a fraction of the time and cost compared to today's manual processes. Additional flexibility can be offered to the end-user, optical virtual private networks (OVPN) become a compelling reality.

Mesh based restoration schemes increase the resilience of the network against any type of failure. Therefore new survivability schemes can be offered in addition to the well-known SDH/OTN protection mechanisms and guaranteeing comparable switching time.

2.4 The convergence of data and transport

The reality of an optical layer capable of dynamic provisioning and restoration of optical circuits offers the opportunity for an architecture where a reconfigurable SDH/OTN network delivers connectivity to the nodes of a packet backbone. The reconfigurable optical layer can be shared among other service networks such as ATM, Frame Relay or leased lines.

User to Network Interface between IP and transport systems is standardised and interoperability between the two has been demonstrated and tested. IP over transport network approach is key to guarantee the scalability of the switched optical backbones and the cost-effectiveness of this approach has been demonstrated by network modelling studies.

The final convergence step will be represented by a unique Core Node where the Core Data functionality (IP/MPLS) is fully integrated inside the Transport Node. As mentioned before the availability of protocol agnostic switching fabrics will surely enable this evolution

3. CONCLUSIONS

Optical Seamless network developments are based around four key elements which combine to deliver against the requirements outlined
  - Integration
  - Flexibility
  - Intelligence
  - Standardisation.

The consolidated of network elements by increasing switching capacity and vertical layer integration, simplifies the network and improve the efficiency of the core network, enabling simpler operation due to fewer elements.

The flexibility of a single platform, Multi-application, next generation solution also simplifies the core network, significantly reducing the operational expenditure due to reduced Maintenance, powers, space, spares and training. Modular & Scalable platforms enables the network to scale with traffic requirements whilst also ensuring low first in cost.

Through implementation of intelligent switching and software mechanisms, the implementation of products with "change aware" hardware enables Dynamic TMN capabilities to the network allowing on the fly provisioning of services and restoration. This dynamic network environment can be utilised for creating competitive advantage by introducing new enhanced differentiated dynamic services.

Open interfaces conformant to the standards developed by the telecom industry ensure inter-operability between different layer networks and between different vendors equipment in the same layer.

By the evolution path outline, significant and measurable benefits of the optical seamless network are realised.

ACKNOWLEDGMENTS

The authors wish to thank the technology strategy team of Marconi for the useful discussions and for their encouragement and support.

REFERENCES

[1] Stefan Bodamer, Jan Späth and Christoph Glingener: "Impact of traffic behaviour on the planning of multi-layer transport networks." OFC 2004
[2] Stefan Bodamer, Jan Späth, Ken Guild, Christoph Glingener: "Is dynamic optical switching really efficient in Multi-Layer Transport Networks?." ECOC 2003
[3] Jan Späth: "Impact of traffic behaviours on the performance of dynamic WDM transport networks." ECOC 2002
[4] L. Blair et al.: "Impact of switch node architecture upon capacity efficiency in Williams North American network." ECOC 2002.
[5] Sudipta Sengupta, Vijay Kumar, Debanjan Saha: "Switched Optical Backbone for Cost-Effective Scalable Core IP Networks." IEEE Communications Magazine, June 2003
[6] Agostino Damele, Andrea Spaccapietra: "Telecom Network Architecture: Multi-Layer Switching Solution." FITCE 2002
[7] Joerg-Peter Elbers: "High-capacity DWDM/ETDM transmission." OFC 2002
[8] L.M. Gleeson, M.F. Stephens, P. Harper, A.R. Pratt, W. Forysiak, D.S. Govan, B.K. Nayar, I.D. Phillips, B. Charbonnier, M.D. Baggott, H.S. Sidhu,

I.E.Tilford, P.M. Greig: "40×10.7 Gb/s meshed ULH network with remotely managed all-optical cross connects and add-drop multiplexing." ECOC2003

[9] Stefan Herbst, Heinrich Lücken, Cornelius Fürst, Silvia Merialdo, Jörg-Peter Elbers, Christoph Glingener: "Routing criterion for XPM limited transmission in transparent optical networks." ECOE 2003.

[10] Silvia Merialdo, Jörg-Peter Elbers, Cornelius Fürst, Stefan Herbst, Christoph Glingener: "Path tolerant dispersion management for transport optical networks." ECOC 2003

[11] Helmut Griesser, Joerg–Peter Elbers, Christoph Glingener: "A generalised concatenated error correcting code for optical fibre transmission." ECOC 2003

[12] Cornelius Fürst, Roman Hartung, Jörg-Peter Elbers, Christoph Glingener: "Impact of spectral hole burning and Raman effect in transparent optical networks." ECOC 2003

[13] G. L. Jones, W. Forysiak, J. H. B. Nijhof : "Economic benefits of all-optical cross connects and multi-haul DWDM systems for European national networks." OFC 2004

[14] A. R. Pratt, P. Harper, S. B. Alleston, P. Bontemps, B. Charbonnier, W. Forysiak, L. Gleeson, D. S. Govan, G. L. Jones,

[15] D. Nesset, J. H. B. Nijhof, I. D. Phillips, M. F. C. Stephens, A. P. Walsh, T. Widdowson and N. J. Doran: "5,745 km DWDM transcontinental field trial using 10 Gbit/s dispersion managed solitons and dynamic gain equalization." OFC 2003

[16] A. R. Pratt, B. Charbonnier, P. Harper, D. Nesset, B. K. Nayar and N. J. Doran: "40 x 10.7 Gbit/s DWDM transmission over a meshed ULH network with dynamically re-configurable optical cross connects." OFC 2003

CINEMA-CLASS DIGITAL CONTENT DISTRIBUTION VIA OPTICAL NETWORKS
Invited paper

Tetsuro Fujii, Kazuhiro Shirakawa, Mitsuru Nomura, and Takahiro Yamaguchi
NTT Network Innovation Laboratories
1-1 Hikarinooka Yokosuka-shi,Knanagawa 239-0847 Japan
fujii.tetsuro@lab.ntt.co.jp

Abstract: To transmit and display high quality movies via optical networks, a
new Super High Definition (SHD) digital cinema distribution system
with the resolution of 8-million pixel is developed. Its image quality
is four times of HDTV in resolution, and enables us to replace
conventional 35mm films. This system is based on JPEG 2000
coding technology and transmits high quality digital cinema over
high-speed IP networks. All digital cinema data are continuously
transmitted at up to 500 Mbps. This system opens the door to the next
generation of cinema-class digital content distribution over optical
networks.

1.   INTRODUCTION

The growth of broadband networks has stimulated the development of
applications that use high quality image communications. To satisfy professional
users in industry, i.e. printing, medicine, and image archiving, a precision color
imaging system is required to achieve the digital images of excellent quality
beyond HDTV. An image category, called Super High Definition (SHD) images
[1,2] is defined to have a resolution of at least 2000 pixels vertically with 24-bit
color separation. The SHD images surpass the quality of 35-mm films in terms of
spatial resolution. In our first study on the SHD images and their applications, we
developed a high quality still image system with 28.3 inches LCD display of
2560x2048 pixel resolution. This image system features GbE as high speed

12

network interface and can transmit 2000 scanning line class still images within one second.

At the same time, we have developed a new platform for high quality digital cinema with 8 million pixels, called SHD digital cinema. This SHD digital cinema scales the heights now occupied by 35mm film. It offers large venue support, large screen projection to fully realize the promise of digital cinema, scalable integrated media production, and film-less cinema distribution via broadband networks. The new SHD digital cinema format is defined as 2000 or more scan lines, progressively scanned, running at 24 frames per second. To evaluate this new format, we have developed real-time DECODER and a projector capable of handling SHD digital cinema with an effective resolution of 3840 x 2048 pixels (square sample). SHD digital cinema contains roughly four times the picture information of HDTV 1080p/24. A comparison is made in Figure 1. SHD digital cinema features RGB color encoding and 30 bits per pixel, for a much more film-like visual richness. Motion picture people in Hollywood count up the cinema resolution from the point of horizontal pixel. Therefore, they call our system as "4K Digital Cinema".

To transmit the movie contents using optical networks, an exceptionally high performance decoder and an imaging system are required to process the movies in real-time. This is because the total bit rate of an SHD digital cinema can equal 5.6 Gbps (3840 x 2048 pixels, 24 fps and 30-bit color), and the movie should be compressed to 10:1 - 20:1 in order to transmit them by wide-area IP networks. Eventually a special combination of a real-time DECODER and a projection device is required to show SHD digital cinema. In this paper, we introduce an SHD digital cinema distribution system.

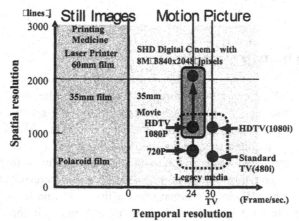

Figure 1. Resolution and frame rate of SHD images

2.   SHD DIGITAL CINEMA SYSTEM

We have developed a prototype digital cinema system that can store, transmit and display SHD digital cinema of 8 million (3840x2048) pixel resolution using JPEG2000[3] coding algorithm. The SHD digital cinema distribution system is shown in figure 2 and 3. This is the third generation of our SHD digital cinema distribution system[4,5,6]. The transmission from the server to the real-time DECODER is done over GbE (Gigabit Ethernet). It consists of three main devices, a video server, a real-time DECODER, and a LCD projector. We assume that the movie data have been compressed and stored in advance. The real-time DECODER decompresses the video streams transmitted from the server using parallel JPEG2000 processors, and outputs the digital video data to an LCD projector with 3840x2048 pixel resolution and RGB 30 bits with 24 fps.

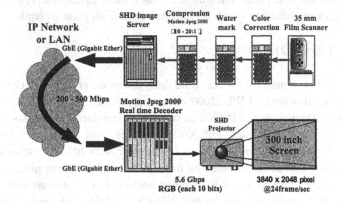

Figure 2. System conciguration of prototype SHD digital cinema distribution system

Figure 3. Photograph of SHD digital cinema distribution system

2.1 Image Coding Algorithm:

The SHD digital cinema system uses the JPEG2000 algorithm to construct a real-time DECODER. From the point of communication traffic and storage cost, inter-frame coding such as MPEG-2 is better at compressing movie data. However, we selected JPEG2000 for the following five reasons. (1) There is no international standard to compress video for RGB 30 bits or more. (2) Other sets of compressed data with lower resolution can be generated easily by using the facility for embedded scheme of layered coding algorithm. (3) Intra-frame coding schemes remain important because of their support of video editing. (4) It is much easier to implement a parallel processing decoder by using compact JPEG2000 CODEC chips. (5) The decoder is robust against data error. Error recovery is achieved simply by discarding the corrupted image frame. The reasons (3) to (5) are common for JPEG and JPEG2000, but only JPEG2000 satisfies reasons (1) and (2) for excellent image reproduction which is requested by motion picture people

2.2 Real-time DECODER

The DECODER can perform the real-time decompression at a speed of 500M pixels per second, using parallel JPEG2000 processing elements. As a JPEG2000 processor, Analog Devices Inc. ADV202 is selected. The decoder consists of 2 circuit blocks, a PC/LINUX part with GbE interface, and newly developed JPEG2000 decoder boards shown in figure 4. 4 chips of ADV202 are installed in each board. Total 4 boards are installed on the PCI-X-bus in order to process 24 frames of 4K x 2K pixel up to 36-bit RGB color images (4 :4 :4) in a second. At the same time, This board supports 10 bits YCbCr (4 :2 :2) mode and transfer function from YCbCr to RGB is installed. This kind of flexibility is obtained by the FPGA based circuit design.

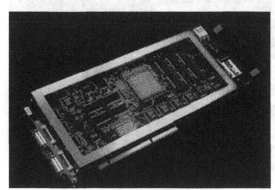

Figure 4. Photograph of JPEG2000 decoder board

The PC part receives the coded streams of 200M to 500Mbps, then transfers them to JPEG2000 decoder boards. The PC part consists of dual CPUs running with LINUX. A control program runs as an application that consists of two threads that share the PC's main memory as a large data buffer. One thread reads the data received from GbE-NIC, and the other reformats and forwards them to each of four decoder boards. The buffer size is only limited by the main memory size of the PC, usually set to 200MB.

2.3   LCD Projector

The prototype projector itself was developed by JVC using D-ILA technology. The high picture quality of D-ILA is derived from the advanced LCOS (Liquid Crystal on Silicon) technology and a high-precision optical system. The major features of D-ILA are high brightness, high resolution, high contrast ratio, analog gradation and high-speed response. The SHD LCD projector uses three pieces of 3840x2048 pixel reflective D-ILA panels for each RGB 10-bit colors, and its size is 1.7 inches. The effective brightness exceeds 5000 ANSI lumens by using a 1600W xenon lamp, which is bright enough to show images on as large as 300-inch diagonal screens. The refresh rate of the projector is chosen to 96Hz. This high refresh rate thoroughly eliminate flicker, and is compatible to 24fps movie. Every frame of decoder output is simply displayed for times in the projector without any interpolation between adjacent frames.

2.4   PC Video Server

The PC server consists of LINUX running on dual CPUs (Pentium III, 1.44GHz), an IDE-RAID (200GB x 6, RAID0, striping mode), and a GbE NIC. The movie films are digitized to a large set of still images, and compressed in advance. The original movie data are divided into 960x512 pixel image tiles, and they have 30 bits RGB color components. A data transfer command reads the data from the RAID and writes them periodically to the GbE NIC. Like the DECODER control program, a large size shared buffer is used in the server in order to enhance the maximum transmission rate by averaging the disk read speed. Contrary to an ordinal streaming system that runs by its own clock, the data rate of the server is precisely controlled by the DECODER via acknowledge signal. This is because the decoder generates the master-clock of the movie system of 24Hz to yield smooth replay without the lack or duplicate of frame

3.   EXPERIMENTAL RESULTS

The quality of SHD movies and performance of the transmission system are evaluated by using long movie data that are digitized and compressed from a variety of actual movies.

3.1   Digital Cinema Data Acquitision

To evaluate the SHD movie system, a lot of movie sequences are required to compare with the conventional film movies, So we provided various types of high quality movie data sequences with less blurs, scratches, nor grain noises. Some of the sequences are digitized from original negative films, such as "Circle of Love" provided by ARRI and short test sequences provided by Hollywood studios. Others are from inter-positive (IP) films, such as "Tomorrow's Memory" shot by NTT, and dupe-negative (DN) films, such as "Tomb Raider" of full Hollywood movie. Usually, we use an IMAGICA's "Imager XE" film scanner to digitize these films. This scanner can yield image files of 4096x3012 pixel as maximum size with RGB 10 bit log Cineon format and it takes 10 seconds per frame for scanning. This is a very time consuming process of data acquisition. Recently, Thmoson Grass Valley announced the development of high speed film digitizer. This kind of high speed film digitizer is indispensable to make large variety of digitized cinema contents to form a cinema archiving with reasonable costs in the near future.

Master image file is obtained from this 10 bit log Cineon format image files through the color correction process. As master image file format, we use RGB 30bit TIFF file for each frame. In the case of "Tomb Raider" of 101 minutes movie, the size of master file is 4.5 Tera Bytes from 144,000 frames.

As far as sound material, we don't apply any compression technologies and we use uncompressed sequence. Usually, master sound can be obtained as the TASCAM DA98 tapes. The 6-channel sound tracks are extracted from this tape. The digital sound data are stored in the Linux/PC server as WAV files and transmitted as IP stream with image data. As the highest sound quality, 24 bit per channel with 96 KHz sampling is available for the play back of orchestra and musical.

3.2   Network Transmission

We selected TCP for the connection protocol from server to real-time DECODER. TCP is adequate for the stable connected transmission and best method to share the bandwidth of IP router based network. Within the multiplex cinema, it is very easy to use. But for large RTT (Round trip time) network, it is very difficult to extract its full performance. In order to verify the performance, we had an experiment with a long distance high-speed IP network environment called Internet 2, on 29th October, 2002. In this case, we used the second generation of our SHD digital cinema distribution system[5,6].

The network configuration is shown in figure 5. We set up the server in the Electronic Visualization Laboratory (EVL) at the University of Illinois, Chicago (UIC). The real-time DECODER and the projector were installed at the Robert Zemeckis Center of the School of Cinema-Television at the University of Southern California (USC), Los Angeles. The distance between the server and the decoder was more than 3000 km. There were six router hops between them. The RTT (Round Trip Time) of the network was measured to be 59ms. The target transmission rate is 300 Mbps. Many power users were sharing Internet2 while our experiments.

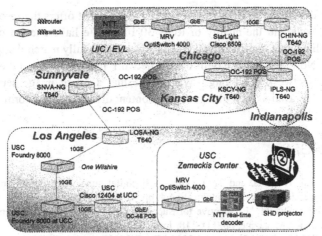

Figure 5. Network configuration of long distance transmission with Internet2

To overcome the long distance, we applied (1) a large TCP window, 4 MByte, (2) multiple TCP connections, and (3) a shaping control function. The TCP window size is the amount of data able to be sent without acknowledgement. There are theoretical limitations to TCP-window-based flow control. The configuration guideline is "window size > Required TCP throughput x RTT ". The TCP window size of the system was extended to 4 MByte from its initial 64 KByte value. Measured throughput was increased from 8 Mbps to 50 Mbps. Theoretically, extension of window size is enough, but it is impossible to extract real high speed.

To improve the performance, we increase the number of TCP connections between the server and the real-time CODEC. The server application divided the movie data into equal segments and sequentially wrote them to multiple TCP sockets. As the number of TCP connections increased, it was confirmed that throughput went up. Finally, using 64 TCP connections, we could get 200 Mbps. But the stream's bit rate could not be raised even when the number of connections was increased more than 64.

We used an application traffic monitor to observe the traffic pattern to a resolution of 1 ms. We found that the nature of data transmission was very bursty. To suppress the burstiness, a shaping control function for the data transmission was built into the socket writing process of the server application. As a result, transmitted movie data traffic reach 300 Mbps. We could succeed in transmitting SHD movie over 3000 Km at 300 Mbps with TCP/IP protocol. To implement adequate functions, TCP can be applied to long distance transmission.

4.   CONCLUSIONS

We have developed an SHD digital cinema system that offers the resolution of 3840x2048 pixels and the quality of RGB 30 bits. We have exerted concerted efforts in realizing a complete digital cinema system that fully match the quality of 35mm film. Now the technology of optical network has matured and the new era of broadband network is coming. We believe that a lot of image service will appear using this kind of high quality digital content distribution platform.

REFERENCES

[1] S. Ono et al, N. Ohta, and T. Aoyama, "All-digital super high definition images", Signal Processing: Image Communication 4, pp. 429-444 1992.
[2] S. Ono and J. Suzuki, "Perspective for Super-High-Definition Image Systems," IEEE Communication Magazine, pp. 114-118, Jun. 1996
[3] ISO/IEC JTC1/SC29/WG1 15444-1:2000 "JPEG 2000 image coding system -- Part 1", 2000
[4] T. Fujii, M. Nomura, et al. "Super High Definition Digital Movie System", SPIE VCIP'99, Vol. 3653, pp. 1412-1419, Jan. 1999
[5] T. Fujii et al. "IP Transmission System for Digital Cinema Using 2048 scanning line resolution", IEEE Globecom 2002, GEN-01-2, Nov. 2002
[6] T. Fujii, et al. "Super High Definition Digital Cinema Delivery System with 8 Million Pixel Resolution", RICHMEDIA 2003, pp.79-88, Lausanne, Switzerland, October 2003

NEXT GENERATION NETWORKS –
A VISION OF NETWORK EVOLUTION

Howard Green[1], Pierpaolo Ghiggino[1]
[1]Marconi Communications, Stoneleigh House, New Century Park, Coventry, CV3 1HJ, UK

Abstract: This article presents a view of the needs and developments for the "Next
              Generation Networks". It starts from a market and service context following
              the burst of the Internet bubble and sketches the likely evolution of services
              by end user type. It is centered, however on a vision of network evolving to
              architectures necessary to support the needs of operators with special
              emphasis on the European environment.

1.   MARKET & SERVICE CONTEXT

1.1   Network traffic

Total service revenues are still increasing, but revenue segmentation in service
categories shows a move towards broadband access and mobile (see Fig. 1, 2)

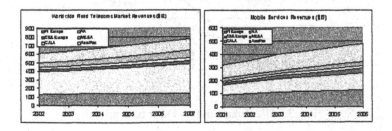

Figure 1 & 2 - Global fixed and mobile revenues by region (Source Gartner)

20

The dominating revenue stream remains voice traffic. On a worldwide basis, fixed network voice revenues are largely flat. Growth is mostly in new markets, offset by decline in mature areas (most pronounced in Western Europe). Prices continue to fall where competition exists. Mobile call volumes and revenues are continuing to grow quite rapidly (although prices are falling, and there is some evidence that revenue growth is slowing slightly). Mobile telephony revenues are set to exceed fixed line revenues in the near future[1]. In many less developed markets, mobile telephony is supplying the basic voice communications need, as fixed line infrastructure is not available.

In general, fixed line network operators in the developed world are expecting a slow long-term decline in revenues from these traditional services. They are developing defensive strategies (pricing tactics against mobile services, for example) but nonetheless expect these revenues to decline (3% per annum is a European operator expectation).

1.2   Service evolution

1.2.1   Residential Customers

Residential broadband is a major success as a consumer technology. Worldwide, the dominant method of provision is DSL. In the UK and US, cable modem has also been a major success (see fig. 3 for the UK scenario).

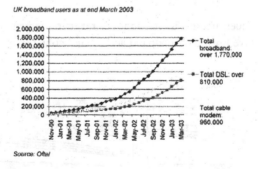

Fig. 3 – UK Broadband end user base growth

---

[1] A major source of recent fixed line revenue has been calls to mobile networks. These prices are now being reduced by regulators.

It is expected that residential broadband services will be a key growth area over the next 5 years or so, growing at a faster rate than business services, to become one of the most important revenue segments.

Broadband service users typically consume 10 times the bandwidth of narrowband users per person, while revenues are simply based on subscription (no more than twice a typical narrowband subscription). This causes a decrease in revenue per bit carried, leading to cost pressure within service provider networks.

Increase in revenue is likely to be generated not just by higher penetration of broadband for fast Internet access, but also by additional services. For this to be profitable, however, such new services need to be made available without significant additional infrastructure investment.

Such services are likely to be based on specific content and application needs, such as (personal interest) video-on-demand and online gaming via specialized game consoles (e.g. PS-2). Whilst the business models have yet to evolve, there are two likely scenarios. Firstly, the ISP could become the focus (via their portal) from which end-users purchase such content and applications direct [the ISP as "media aggregator"]. Secondly, such purchases could be made from the content / application owners, with a proportion of the revenue going to the service provider for such things as portal placement, localized caching and (later) session-based services providing increased capacity and/or QoS. (The two models can be thought of as a retail scenario and a wholesale scenario.)

Within a few years, typical residential customers will use some sort of home media server. Via in-home networking technologies (typically 802.11 WLAN), these will connect using xDSL transport to such services in an "always-on" manner.

An increasing amount of voice traffic will be handled by mobile networks. What remains on the fixed networks could be decreased further (in PSTN terms) by VoIP, because of voice-enabled game software, instant messaging packages & multimedia contact centres.

An important opportunity for the fixed network operator is to capture parts of the mobile revenue stream by facilitating fixed/mobile convergence. For example:
- Services allowing terminal mobility from the mobile to the fixed network – e.g. use of a WiFi enabled mobile terminal on fixed network tariffs at home using packet voice over DSL.
- Integration of fixed and mobile directory services, and the use of "presence" applications to keep in touch with people.

An increasing percentage of people are working from home, at least on a part-time basis. Secure IP VPNs through the Internet will become the predominant way to connect such teleworkers to their corporate networks.

1.2.2 Small and Medium Enterprises (SMEs)

Similar to the situation with residential broadband, business broadband over DSL is expected to show healthy year-by-year growth.

As DSL products were initially developed for residential customers and priced for the mass market, these services can prove very attractive to SMEs. On the downside, there is a significant risk of substitution ("cannibalization") of existing leased line revenues.

Service performance and guaranteed SLAs are the main means of differentiation with leased line services, as customers are willing to pay a premium if their specific needs are met with distinct products. However, the delivery mechanism is not as important to SMEs as the service provided for a given cost.

It is likely then that data VPN services will grow rapidly where they are offered over xDSL broadband access. Such provision could then result in service providers offering seamless service evolution from xDSL technologies towards higher capacity fiber based solutions.

It can be expected that, where SMEs operate over multiple sites, service provider provision of data VPN services could become much more attractive than the renting of leased lines and self-provisioning of services on them.

Coupled to this, customers will expect to get a multitude of services delivered at a single interface (with appropriate quality. This requires technical mechanisms for service separation (e.g. VLAN tags, MPLS labels, ATM VCs), session admission control and flow policing so as to provide appropriate guarantees of quality and security.

VoIP/Centrex will be among the interesting business options for such customers based on advanced broadband access.

1.2.3 Large Corporate Customers

Large corporate customers are examining their WAN costs as part of reviews of IT infrastructure, and will expect to continue to get more bandwidth and lower prices. Nonetheless, there are opportunities to increase overall communications revenue, in particular by convincingly demonstrating that other costs can be avoided using communications. Such opportunities are likely to arise in respect of supply chain integration, outsourcing, travel avoidance, and reduction in fixed office costs. There is also an increased focus on network security and disaster recovery, leading to requirements for network-wide encryption and authentication, and to the need for geographically remote storage backup.

Centralized servers organized in hub-and-spoke configuration are a common model. The dominating technologies used to build such WANs are FR and ATM, as well as SDH-based leased line networks, but growth in data VPN services is expected to eat into this.

There is a spectrum of these data VPN services, ranging from "multiplexed Ethernet private line" (multiple "Ethernet virtual circuits" from a large site switched to different destinations in the network on the basis of VLAN tag), through "complex L2" VPN technologies (such as Virtual Private LAN Service (VPLS) – formerly "draft Lassere-Vkompella") to L3 VPN technologies based on IP/MPLS routing (e.g. RFC 2547) to VPNs based on IPSec. From the service perspective, these offerings can be differentiated by scalability, security and QOS characteristics (where large numbers of endpoints are involved, L3 routing is necessary, and delay-sensitive intersite services still require leased line-like QOS guarantees). Not all of these services will find favour in the marketplace. Moreover, many operators are finding these services more complex to provide and administer than traditional leased lines, whilst service prices are usually lower.

Home worker and Telecommuters now frequently use the Internet to access company servers by means of secure tunneling. Utilization of wireless LANs is currently quite low, but likely to expand within both the corporate environment, and via the use of Wireless hot spots.

ISDN PBX is the predominant infrastructure for corporate voice. There is no compelling short-term reason to replace this infrastructure, however as replacement opportunities arise, packet voice will be the new technology of choice. VoIP PBXs are already taking substantial market share. An increasing amount of these packet voice service opportunities will be outsourced to service providers. There will also be business development of PC instant messaging products to provide "virtual presence" and interface via VoIP with mobile networks.

The physical service interface is likely to be fast or Gigabit Ethernet, allowing for flow separation by VLAN tags or MPLS labels, and providing a "fractional rate" service[2].

2.   IMPACT ON OPERATORS

After the "bursting of the Internet bubble", network operators are under significant business pressure. As consequences, operators have two major objectives:
1. Need for substantial reduction of ongoing network operational costs, and incremental capacity at much lower cost per bit.
2. Need for new service revenues. There is unlikely to be any single "killer application", and so this likely requires many new services to be deployed.

[2] That is, a "subscribed bit rate" less than physical line rate, policed by the network, and variable in relatively small increments by service provisioning.

Operators are at different stages of emergence from corporate reconstruction and debt reduction programs. Moreover, they have different views of the appropriateness of their basic network infrastructure. Nonetheless, they are increasingly beginning to consider new programs of capital investment to address the objectives above (OPEX reduction and multi-service flexibility). These investment programs will have major consequences both in network and in OSS architecture for operators.

3.   NETWORK CONSEQUENCES

Operators are discovering a real business need to make convergence happen. Reducing the number of edge platforms deployed in high volume in their networks, and reducing the number of service-specific overlay networks are major opportunities for increased efficiency and cost reduction. They need to find new service revenues from added value on top of basic broadband and new services to small business. These new services need to be delivered with appropriate quality, and will need usage-based charging.

3.1   Operational Support Systems (OSS) consequences

In mature organizations there are opportunities to rationalize operational support systems (which are often based on individual service needs rather than common business processes). The rationalizations may be linked with network evolution but can also take place independently. For example, a flexible billing system operating across many services and several networks can be introduced independently of the evolution of the networks. A common fault management process can be introduced in a similar way. There are also opportunities to introduce new processes and systems where there may be nothing today: for example post processing of usage records can help to improve fraud detection and reduce revenue 'leaks'. Introduction of auto-discovery algorithms and centralized inventory may assist in validating network equipment usage and result in better utilization of capacity. Where there are demonstrable short to medium term returns on investment operators will invest in OSS in order to:

- Improve operational efficiency - minimize data entry, end user control, error reduction, supply chain integration, fast fulfillment, speed of fault location and repair.
- Improve capital investment utilization – accurate inventory, efficient utilization of installed capacity, timely investment in new capacity
- Increase/secure revenue – flexibility of service definition, bundling and tariffing, improved customer retention, revenue assurance

4. A NETWORK VISION

Future communication networks will provide a significantly broader range of services. As profit margins will continue to be low, these networks will need to be realized by a minimum number of technologies[3].

IP will be the near-universal technology of the service layer. MPLS will eventually form a common virtual-circuit switching backbone layer, running over an SDH/OTN transport core (with integrated DWDM). Specialized high capacity traffic streams (e.g. SAN) will be mapped with minimum overhead onto core transport. For high capacity business services, the access and metro networks will have a traffic aggregation and transport role, connecting customer IP routers to IP service switches at the "edge of the core". For residential broadband and SMEs, a multi-service access node will manage the continuing diversity of "last mile" technologies (including copper, wireless and some fibre). (See Fig. 4 & 5)

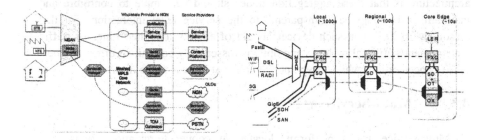

Fig. 4 & 5 – Next Generation Network structure and transport views

4.1 Customer premises Equipment

As the network simplifies, most of the required network terminals will become standardized consumer products. Currently we already have standardized ADSL and cable modems, and integration with "Internet gateway" switches and firewalls. We believe that this is likely to extend into the business market. Of particular significance is the Ethernet in the First Mile (EFM) initiative, which may well create a market for both copper and fibre termination of fractional Ethernet (and possibly also E1 & Ethernet together over fibre). There will remain a market for low cost Ethernet over SDH terminals, and for access C/DWDM.

---

[3] For a broader description of some key aspects of such technologies see H Green, P Ghiggino: "An overview of key technologies for the next generation networks" – Submitted to OpNeTec 2004

4.2 Edge Transport

Access and metro transport will be based on Ethernet L2 aggregation techniques, using frame labels such as VLAN tags or MPLS for "permanent virtual circuits" with fine bandwidth granularity. Both in the access and regional transport area, these frame streams may be carried either as native Ethernet (with OAM additions such as those proposed by EFM) or GFP encoded to SDH virtual concatenation groups. SDH will continue to be used for high-quality synchronous transport (E1s for PBX interconnect, mobile backhaul etc.) especially where timing/clock distribution requirements are important, and is a natural choice where ring topology protection is useful.

While transport devices will evolve the capability to cross-connect labeled frame streams, we do not believe that this will imply the need for a full IP/MPLS control plane. There will be a move towards more automatic provisioning of these label switched paths, most likely based on ASTN concepts. A vital feature of this architecture is that these aggregation nodes should be simple to configure and maintain, and should be "transparent" to the service level interaction. For this reason, we believe that such devices should offer only a limited range of traffic management/statistical multiplexing options (essentially committed effort and best effort concentration) which can be policed once at ingress to the network.

4.3 Multi-Service Access

Multi-service access platforms, located at copper termination points (and perhaps also in street cabinets) are a vital part of the next generation network. They will support a graceful evolution from POTS to packet voice (using "combo" line cards and integrated packet voice gateways), and allow a range of value-added services to be deployed on top of residential broadband access. They will gradually replace many different service-specific devices at the edge of the network, allowing significant simplification and operational cost saving, whilst allowing continued diversity of "last mile" options.

MSANs can be seen as a new generation of DSLAMs. As such, they need to maintain the ATM aggregation capabilities of DSLAMs on the line side. In our view, ATM remains an effective way of managing controlled multiplexing of services on relatively low-speed lines. On the aggregate side, we expect MSANs to offer Ethernet transport (native and NG-SDH).

Current generation DSLAMs transport ATM VCs per customer back into the network, where they are terminated at a B-RAS. The B-RAS forms the point of service termination, calling upon external servers (e.g. RADIUS) for the so-called "triple-A" (Authorization, Authentication, & Accounting) functions, and then forward the virtual circuits to ISP edge routers using IP Layer 2 Tunneling protocol

(L2TP). This architecture does not scale well, especially when added value services require multiple classes of service per customer. Hence we believe that this B-RAS functionality will be distributed to the MSANs, which will terminate multiple ATM VCs per customer, and directly map to L2TP inside provisioned MPLS tunnels for different QOS classes (or use "draft Martini" MPLS label stacking). The MSAN will continue to use centralized servers for "triple-A"[4].

4.4   MPLS Core

In current large data networks, there are often three kinds of network elements: edge routers, core routers, and L2 switches (usually ATM).

These architectures are rightly felt to be complicated to configure and difficult to maintain. The router elements have no concept of differentiated service quality, and hence there is required to be an underlying network of virtual circuits separating traffic.

Because of resilience and control plane scaling issues, routers are often arranged in complex hierarchies, implying many packet hops and long latency.

For these reasons most operators believe that their current generation core packet networks will require major change to make them suitable as the core of a large multi-service network offering service guarantees. We believe that there will be two kinds of network element remaining:

1.   Large (~1Tb/s) "edge of core" MPLS switch/routers, arranged in a "sparse mesh"[5] topology, forming a universal network core.
2.   IP service switches for high-value services (e.g. IPSec VPN) only accessed as needed.

   [This assumes a "wholesale" model of services, where the core network provider is connecting customers to multiple service providers, whose switches are accessed through the core. Where the core network operator is providing retail services directly (e.g. RFC 2547 VPNs, internet access) there may be a need for IP services nodes "before the core"]

MPLS core switches will also need to interwork with remaining ATM-based network services (which, like TDM, will have a long "old age").

---

[4] Whilst this does imply some distribution of complexity, it remains service agnostic. Thus new services can be added without impact on high-volume network elements

[5] That is, high-usage routes are all directly connected. Some lower-usage routes may utilise transit nodes. Where required simply by geography, transit is probably better provided at the transport layer.

4.5 Core transport

It has been suggested that a next-generation network does not need a core transport layer, since it can be formed simply from "IP over light" – that is, IP routers connected together using high capacity optical (perhaps DWDM) interfaces. We think by contrast that there continues to be an important role for a core transport network at VC4/ODU granularities, for the following reasons:

1. There will still be a significant quantity of "leased line" business inter-site services. Even if these are presented as "Ethernet Private Line", there is no need to use expensive router capacity to provide these services. Cross-connection in the transport layer is a lower-cost option and easier to administer (especially when using ASTN-based automated provisioning, in which explicitly routed paths are downloaded from the network planning process).

2. The mesh core topology envisaged above requires the interconnection of large numbers of MPLS nodes at bit rates in the range ~100Mb/s to 20Gb/s. This requires a "virtual port" technology allowing single high capacity physical interfaces to be routed to different destinations. The most plausible options are STM-64/256 canalized at STM1, and GigE/10GigE using MPLS or VLAN tags. In either case, there is need of a transport layer to "fan out" these interfaces, and to arrange for transit where required (in the Ethernet case, the transport network must map tags to VC4 or ODU1 granularity VCGs).

3. Large networks of routers take a considerable time to respond to faults. Transport-based protection can handle most common faults in <50msec timescales, avoiding any impact on the router layer.

This transport layer will be built from large SDH/OTN cross-connects with integrated WDM and a sophisticated ASTN control plane.

4.6 Control Plane Scalability

Many of the key challenges to next generation network design lie in this area. As we have argued, the network will require underlying resource reservation mechanisms to ensure appropriate quality in supplying a wide range of services. This requires a concept of "virtual circuit". VCs are likely to be implemented in the network using several base technologies (at least MPLS, ATM & Ethernet VLAN tagging) which will need to interwork.

It is very easy to design network architectures involving meshes of virtual circuits which require impossibly large numbers of circuits, or "automatic discovery" architectures sharing vast amounts of information. In general, our view is that the network continues to need to be partitioned both horizontally (hierarchical layers) and vertically (access/core), but that complexity can be reduced by using similar mechanisms in each domain. Individual service types will in general use "virtual networks" consisting of provisioned tunnels. Thus, for example, most services will need to use IP routing (Label Edge Router functionality) as they enter the network core. To give an example, Internet access packets destined for a single ISP will be unpacked from access LSPs (at least 1 per MSAN) and repacked to a single LSP.

4.7   Bandwidth management

For a multi-service network providing service guarantees, it is not enough just to reserve resources for appropriate traffic aggregates. It is also necessary to operate admission control for services requiring the guarantees, so as to ensure that the aggregates are not over-filled. This issue is already recognized for packet voice, but in a multi-service network is applicable to all guaranteed services. Admission control requires a count of resource used throughout the network, to decide whether a new service instance can be admitted.

This counting of resource for admission control can in principle be done at least three ways.
1.   Directly by reserving the resource by signalling through the set of network nodes (using RSVP, for example)
2.   In the application, for example using circuit and route control in a softswitch
3.   Using a separate logical layer of "bandwidth managers"

We believe it is important that next generation networks allows common mechanisms to support these functions for all services needing guarantees, and thus we think this function should be separated from individual applications. For this reason we generally support the creation of a "bandwidth management" function with a standardized application interface. The bandwidth management layer will also need to communicate across operator boundaries, and hence standardization is essential. There is work going on (for example, in ETSI Tiphon and the Multi-Switching Forum) on architectures and protocols for such bandwidth managers.

4.8   Service Platforms and Gateways

The network described above is capable of carrying the full range of services. Of course such a network will still need a variety of platforms to support the logic of particular service types, and to perform specialized data plane functions (e.g. hardware support for IPSec encryption). Our view is that these platforms will be subtended as "server pools" from the network core. This allows them to be used only as required, and gives great flexibility for centralization of resource, network-wide redundancy etc.

Similarly gateways to the TDM network and to other operators are accessed through the core and can be placed wherever convenient, supporting a view in which operators can use building space more flexibly and concentrate resource at fewer locations.

5.   CONCLUSION

We have presented a view of the development of next generation networks. It is being driven by:

- Flat to declining revenues from traditional fixed line services
- Rapid growth in bandwidth demand, predominantly from data services
- Reducing revenues per bit carried
- Business need for major operating cost reduction

A set of emerging technologies does allow operators to address these business problems.

- Emergence of Internet Protocol as an increasingly ubiquitous service interface
- A wide variety of broadband access technologies (xDSL, optical Ethernet, Wireless) supporting a common IP service framework
- Data-enabled transport, including "Next Generation SDH" and Ethernet developing for the WAN, utilising a control plane for fast restoration and implementation of provisioning.
- Maturing MPLS standards and implementations supporting packet Quality of Service

The resulting network architecture is likely to be characterised by:

- Multi-service access – a single family of platforms supporting the full range of services at the network edge, with traditional access and transport coming together
- Aggregation of packet flows in data-aware transport devices
- A densely interconnected network core of large MPLS nodes, supported by intelligent transport providing resilience, flexibility and fan-out.

Service nodes and network gateways located as "server pools" at convenient locations accessed through the network core.

AN OVERVIEW OF KEY TECHNOLOGIES FOR THE NEXT GENERATION NETWORKS

Howard Green[1], Pierpaolo Ghiggino[1]
[1]Marconi Communications, Stoneleigh House, New Century Park, Coventry, CV3 1HJ, UK

Abstract: This article presents a view of some of the key new technologies that are likely to play a significant role in the deployment of the next generation network architectures necessary to support the needs of operators.

1.   OPTICAL TRANSPORT NETWORKS (OTN)

Most large operator networks today have a substantial transport layer of SDH/SONET devices, with fibre bottlenecks and shortages being avoided by the use of WDM point-point systems. As networks evolve to carry mostly data, there is scope to reduce network complexity by "de-layering". Nonetheless, we believe that there continues to be a clear role for a transport layer[1], providing flexible protected transport between network service nodes, carrying customer leased line services, and supporting the scalability of higher network layers. In the short term, many operators have substantial overcapacity in their core transport networks. Nonetheless, this capacity will be exhausted. In a three to five year time frame, transport networks will need increased capacity. Moreover, operators will wish to use this opportunity to simplify their networks and reduce operational costs.

In this context, we believe that there will be a substantial integration of the SDH and WDM sub-layers in the transport network. The ITU OTN standards (e.g. G.872, G.709 "Digital Wrapper", G.798, G.959.1) provide for a consistent hierarchy of bearers, multiplexing and trail overhead, allowing the transport of all kinds of signal formats up to and including services (e.g. Fibre Channel, DVI

---

[1] In particular in the regional and core parts of networks. There is a tendency towards the "collapsing" of transport, aggregation and switching at the edge of the network .

video, low latency GigE and 10GigE) which are currently transported as "wavelength services". This extension of standard transport principles will also allow the deployment of 40Gb/s line rates in a consistent fashion when demand requires and the technology is cost-effective. It allows for standardized inter-vendor and inter-operator connectivity, and supports consistently a standard set of features including dedicated and shared protection and virtual concatenation.

Thus we believe that the dominant switching paradigm in transport networks for the foreseeable future will be "opaque" O-E-O switching. There will be a role for transparency, in particular to provide "optical bypass" of intervening traffic nodes to save the cost of transponders. WDM technology will be progressively integrated into OTN-capable transport platforms, starting with coloured interfaces and later allowing integration of filters, "managed optical patch panel" etc. In this context we believe that WDM will increasingly be thought of as a network and interface technology, rather than a network layer in its own right.

2. AUTOMATED SWITCHED TRANSPORT NETWORKS (ASTN)

Today there is a significant difference in operating data networks and transport networks. While data networks tend to use distributed protocols for switching and routing, transport networks are centrally managed.

Part of this difference just reflects the differing roles of these network layers.

At the network edge in the service layer, network elements must deal with many fine-grain service requests from very large numbers of users. Thus by necessity control must be distributed and respond in real time. By contrast, at the network core in the transport layer, long term planning is unavoidable. Fibre must be laid and capacity hired. Thus trails are bound to be of longer lifetime and higher capacity.

However, these network layers have historically been designed quite separately. They have different control technologies, operating procedures and user interfaces.

However we believe that these elements (bandwidth planning, customer contracts allowing variation of capacity, network optimization) can and will be integrated together into operator systems. These systems will automatically drive fast provisioning in the network. Standardization between vendors, between operators and between technologies will allow much simpler systems, which will translate into long term cost savings for operators.

In the short term, there is already widespread recognition of the value of signalling-based restoration in core transport networks. This can serve either to reduce "Out of hours" maintenance costs as a supplement to standard protection (replacing a backup path in the event of a failure, so as to allow extended time to repair) or to avoid deploying new capacity by using shared instead of dedicated protection.

A likely next step is the widespread use of "explicit routing" to implement provisioning decisions made by external systems. This (in conjunction with the maturing of standards for UNI and inter-vendor NNI) will allow a common process across a multi-vendor network, and increasingly support a "flow-through" management philosophy.

The biggest potential OPEX saving is where there are most trails to provision – relatively low bandwidth circuits based on VC12-nv, VC4-nv, often supporting fractional Ethernet. Thus we expect to see this control-plane based provisioning gradually extend towards the network edge.

Clearly the extension to use distributed discovery and routing protocols is critically dependent on resolving the relationship with the network management systems.

3. "NEXT-GENERATION SDH"

Next-Generation SDH is not a precisely defined term. It refers to the new elements of flexibility in the SDH standards aimed to support data interfaces, standard data encapsulation and functionality supporting data aggregation. It is usually taken to include:

- Virtual Concatenation at VC12, VC3 and VC4 – building of "VCx-nv" Virtual Concatenation Groups (VCGs)
- LCAS (G.7042) allowing "hitless" variation of the size of a VCG.
- More highly integrated line-side xWDM I/F (and pluggable optical modules)
- Data interfaces (predominantly FastE and GigE, but also Fibre Channel, ESCON/FICON, and ATM). This trend includes increasing use of the "Ethernet in the First Mile" (EFM or IEEE802.3ah) standards for optical Ethernet with OAM.
- GFP (G.7041) encoding of these data interfaces into VCGs, both using delay-sensitive "transparent mode" and (more commonly) "frame mode".
- "Layer 2" data aggregation capability. Here there is considerable variation. Most commonly supported is the ability to create "Permanent Virtual Circuits" over Ethernet based on the use of VLAN tags to distinguish customer frame streams. This may or may not include "QinQ" – the ability to stack VLAN tags so as to preserve customer VLAN information. Also commonly included is some use of "MPLS draft Martini" based on provisioned MPLS labels. The traditional enterprise

meaning of "Ethernet switching" (MAC address based learning bridges) is, in our view, of limited value in public networks (since MAC addresses cannot be summarized and spanning tree implies an acyclic network).

- Traffic management. Here again there are many variants. Policing of "fractional Ethernet" (subscribed bit rate < physical line rate) is fundamental. Concentration of "best effort" traffic is also appropriate; it saves hauling bits back to core routers just to throw them away. We do not believe that "frame relay-like" statistical multiplexing is appropriate in a transport device, since it greatly complicates the network and the service being offered may not be easily explained to a customer.

This collection of functionality provides attractive options for operators to optimize their transport networks in a mostly data environment. At the same time, it allows the preservation of the "carrier class" service qualities which are a major brand value for traditional operators, by providing guaranteed service and transport quality fast protection. In the short term, it will be provided by add-in cards extending the installed base. In our view, transport nodes in the medium term will be natively able to provide "permanent virtual circuit cross-connection" based on Ethernet and MPLS tags (and interworking with ATM). We are using the term "Frame Cross-Connect" (FXC) to describe this function.

These new data-oriented functions also imply new responsibilities for transport network management, which will need to represent end to end data services across transport paths (facilitating surveillance, and adding test capability to aid fault location and repair actions).

There has also been discussion of more extensive integration of IP functionality with transport ("Layer 3"). We are not currently persuaded of the value of this integration. Tying router functionality together with transport reduces the opportunity to scale both technologies independently. Moreover, it greatly increases the complexity of the device, and makes it substantially more complex to manage. Finally, it can get in the way of the Layer 3 interaction between the customer router and an IP service switch in the network, making the service less predictable (in particular concerning QOS). Our view is that network transport should be "the router's best friend" (providing managed aggregation and fast protection) and not a bad router.

4. ETHERNET IN THE TRANSPORT NETWORK

4.1 Transport

Just as IP is becoming the dominant interface at the service level, we expect the same to be true for Ethernet at the physical level. Thus there is a clear case for many network services to be presented to customers as Ethernet. Some have argued from this premise that Ethernet should become the universal transport mechanism, "in order to avoid format conversions", and that this will lead to a substantial reduction in network cost. In our view, the comparison is not so clear. If we compare a "best effort" unprotected Ethernet transport network with statistical gain to a protected NG-SDH network offering guaranteed subscribed bit rates, then no doubt Ethernet is cheaper. However, we believe that operators do need to offer service guarantees in order to sustain their revenues.

In general, WAN transport does not have the same requirements as LAN. There are new requirements for OAM, guaranteed packet throughput, and latency in the WAN, and some LAN technologies (e.g. MAC address "learning bridges") do not scale appropriately. Therefore not all of the functions for WAN transport and aggregation can benefit from the high volumes of enterprise Ethernet components.

We believe that "native Ethernet" and "Ethernet over SDH" transport will co-exist in the network, and have somewhat complementary roles.

In the access region, initiatives such as Ethernet in the First Mile (EFM) do allow for cost-effective Ethernet transport. We believe that these interfaces (over fibre and copper) will take a significant share of business access. At the same time, where TDM circuits still need to be carried, or protected access is required, Ethernet over SDH (GFP encoded to a Virtual Concatenation Group) is an effective solution.

Further into the network, requirements for protection and OAM are critical. Whilst there is certainly a role for "native" Gigabit Ethernet transport, we believe that the predominant mode of Ethernet transport will be GFP encoding over VC4-nv or ODU bearers. Ethernet (GigE and 10GigE) will be widely used as a cost-effective way of interconnecting network equipment inside buildings.

4.2 Aggregation

A key advantage of Ethernet as a physical interface is that the cost is not significantly affected by the physical line rate. Thus service providers have the opportunity to fit network termination equipment with substantial "headroom", and hence a long life without further visits to customer sites. However, most customers will not wish to pay for a full FastE or GigE through the network. Therefore there is a key role in access of policing these interfaces to the customer subscribed bit rate ("fractional Ethernet"). This subscribed rate needs to be flexible with relatively fine granularity.

A second advantage of high bandwidth Ethernet customer interfaces is the ability to multiplex several frame streams cost-effectively over a single broad interface. These "multiplexed Ethernet Private Line" services are commonly distinguished using VLAN tags. Transport nodes will need to preserve these "virtual circuit" labels across the network, so as to avoid using expensive IP routers to support basic private line services. They will also need to be able to interwork multiple virtual circuit technologies. So there is a need to map VLAN tagged frames to MPLS "draft Martini", and also to ATM VCs.

Hence transport nodes will need to evolve to support these "frame" technologies. At the same time, we do not see a good justification for a full IP/MPLS control plane, which would greatly complicate such node and especially its management. These services remain provisioned virtual circuits.

Since the services offered over these bearers are predominantly "new generation private line", the normal use requires guarantees of subscribed bit rate ("committed effort"). Transport nodes will also need to offer fair concentration of best effort traffic (allowing controlled overbooking of bearers for services such as Internet access).

4.3   Interfacing to Routers

As networks come to be mostly packet based, and the cost per bit of transport falls, there is a new opportunity for network optimization.

Current router networks use several levels of hierarchical concentration towards the core to reduce the total number of bits carried. We believe that (just as occurred in voice network digitalization in the 1980s) this router hierarchy will significantly flatten, and that there is a key opportunity to use transport nodes to reduce transit traffic through routers. Moreover, since transport node interfaces are a small fraction of the price of "carrier class" router cards of the same bandwidth, there is a role for the transit network in ensuring that router line cards are well filled.

Both on the access and the core side, a "de-layered" network of routers needs many more ports than it has physical line card slots. Therefore there is need for a flexible mechanism supporting multiple "virtual ports" on a single physical port, to support the core network scale which will be required. Moreover, the size of these "virtual routes" across the core will range widely (from 10s of Mbits/s to many Gbit/s).

There are several possible mechanisms to support this requirement. The most likely seem to be:

1. Channelized line cards (e.g. STM-1 channels in STM-64). Currently using POS ("Packet over SONET") encoding – likely to move to GFP.
2. IP over ATM line cards (at rates up to STM-64).
3. GFP in (for example) STM-64 with use of the multiplexing field in the linear extension header to define frame streams, with transport nodes routing individual multiplex values to different VCGs (probably at VC4 granularity).
4. "Tagged Ethernet". Multiple frame streams (distinguished by VLAN tag or MPLS tunnel label) inside GigE or 10GigE.

The ATM option above will almost certainly be required, given the need to interwork with current large ATM networks, but probably does imply extra cost. The "GFP multiplexing" option is in some technical respects the most elegant, but relies upon the existence of appropriate transport layer functionality which is currently unavailable. Therefore, another option likely to be successful is "Tagged Ethernet", with an intelligent core transport node directing the different tagged streams to different VCGs (most likely VC4 granularity) and encoding with GFP.

[This implies functionality from the router, both in respect of per-LSP queuing and traffic management, and in respect of multiple routing adjacencies per physical interface.]

5.   MULTI-SERVICE ACCESS NODES

Operators are discovering a real business need to make convergence happen. Reducing the number of edge platforms deployed in high volume in their networks is a major opportunity for OPEX reduction. They need to find new service revenues from added value on top of basic broadband and new services to small business. At the same time, they see the need to begin to plan their evolution to packet voice (even where no compelling short term reason for implementation exists).

In this context, "first generation" DSLAMs are being replaced by much more flexible multi-service Access nodes (MSANs), capable of terminating a wide variety of access technologies.

MSANs continue to support the DSLAM functionality, and hence require flexible ATM aggregation. They increasingly require to terminate Ethernet bearers over copper, fibre and even fixed wireless, and can also support Ethernet backhaul on the network side. They will support "combo" POTS/DSL line cards as well as packet voice gateways, allowing a graceful evolution from old to new voice networks without on-site intervention.

A key enabler for new service revenues is the ability to support "session-based services" – that is, usage-based charging for quality guarantees for particular content streams. A first step in this direction is the integration of the so called Local Access Concentrator (LAC) and (part of) Broadband Remote Access Server (B-RAS) functions [terminating PPP sessions from customers, initiating authentication, authorization and accounting (AAA) functions for the user (interacting with centralized network servers for these functions), and forwarding the traffic in appropriate "tunnels"]. There will also be a need for "media firewall" (flow identification and policing) functions for packet-based access to multiple services.

The predominant trend in current networks is for deployment of MSANs at copper distribution frames (collocated with existing switch remotes and DSLAMs). Many operators are also considering the possibility of street electronics, as an evolution bringing fibre nearer to the customer and allowing higher bandwidth (e.g. based on VDSL). This can be accommodated within an MSAN architecture, using small remote line shelves and aggregator nodes.

Finally, there is obviously scope for bringing together MSAN with edge transport functions (e.g. VC12 granularity SDH). The likely outcome is an architecture allowing for "remote line shelves" subtended from aggregation nodes.

6. MPLS AND ATM

Most profitable data services today are carried over ATM networks. Moreover, the most important new source of bandwidth in operator networks is residential broadband, which is aggregated using ATM. Hence "rumours of the death of ATM have been greatly exaggerated". Most operators have substantial and profitable ATM networks which they will not abandon quickly, and which they will require to evolve without service disruption. Whilst best effort IP traffic has huge volume, most such traffic does not support profitable services. Operators recognize that obtaining such profitability requires service delivery with appropriate and dependable quality. In the context of real attempts at network convergence including voice and video, differentiated QOS is completely essential.

[Note that Quality of Service is not completely assured by Class of Service technologies such as DiffServ. While DiffServ can ensure that some packets are treated with priority, it cannot ensure that there is sufficient capacity in the network to provide appropriate quality for the delivery of any particular service flow. This requires in addition the reservation of resources and admission control]

MPLS is intended to solve this problem by integrating a hierarchical "virtual circuit" technology with IP forwarding, so as to obtain a scalable IP control plane supporting QOS. As currently standardized, MPLS does not fulfill that promise completely. In particular, it does not support QOS-aware path selection beyond the boundaries of a single Internal Gateway Protocol (IGP) area, and it does not have adequate OAM capabilities. Scalable QOS support requires either changes to existing gateway protocols (IS/IS, OSPF) or the adoption of a scalable QOS-aware alternative (e.g. PNNI, which already addresses these problems for ATM networks).

We believe these problems can be solved, but they are not solved yet. Current generation routers are not appropriate and are probably not extensible to this functionality. We think, therefore, that MPLS is the most likely medium term solution to a converged core network, but that the necessary preconditions are not yet in place. When MPLS is deployed for that purpose, it will need to interwork effectively with the ATM installed base.

7.   WIRELESS ACCESS TECHNOLOGIES

Wireless will be the predominant home networking technology, used not just for telephony but also for residential LANs and for distribution of multi-media.

Wireless LAN (IEEE 802.11x) technology will be widely deployed at "hot spots". There will be data solutions using roaming between WiFi and GPRS/UMTS, and wireless connected (e.g. Bluetooth) voice solutions allowing roaming between the fixed and mobile networks. Since WiFi will be deployed by many small access providers (shopping malls etc.) there will be sharing of backhaul infrastructure between many networks and technologies. Extensions to 802.11 allowing extended point-point connectivity, as well as new technologies such as IEEE 802.16, will be used as "fixed wireless", and will be integrated into MSANs.

In total, these new technologies will eat into the revenues expected of UMTS. However, this will remain the network with the largest coverage used for high volume mobile telephony.

8.   MULTIMEDIA SOFTSWITCHES

Packet voice technology has been "emerging" for a long time in public networks. Whilst there has been substantial deployment of IP PBXs, most operators have not made major investments in public network systems. There are several reasons for this:

1.   Existing voice systems are stable, dependable and do not require major new investment

2.   Base fixed voice demand is not growing. There has been growth (in particular because of dial IP access) but this has been addressed where

necessary by specialized bypass solutions, and is now disappearing because of broadband substitution.

3.   Whilst basic packet voice technology is not in doubt, there are still major concerns about quality, security and scalability for the public network, and the evolution issues are complex.

In our view, this situation is now changing. Operators are beginning to have real obsolescence concerns about their voice infrastructure, and they recognize that any replacement strategy will take some time to deploy. Moreover, they are now convinced that they cannot survive on their traditional revenues, and therefore need a flexible service architecture.

The challenge in designing a public network packet voice solution is to balance several issues:

a.   Internet-derived technologies such as SIP give great flexibility for rapid deployment of new services. Operators must be able to support new kinds of services and new media if they are to invest in a whole new service infrastructure.

b.   A totally "transparent" IP network gives operators no opportunity to provide appropriate quality and dependability for a public network offering, and no way of charging for their services.

c.   A valuable service depends on universal inter-operability with fixed and mobile networks, and hence must interface to current standards and feature sets, and support current regulatory and government requirements.

Once again, we believe that the technology is now becoming sufficiently mature to address these concerns. State of the art softswitch systems consist of a set of components connected together by standardized interfaces. They offer:

- controlled access to resource (and other networks) based on "media firewalls", which police and translate between multiple IP address spaces and also manage QOS for media streams on the boundaries of an operator IP domain
- a full range of gateway protocols including MGCP, H.323 and H.248.
- controlled access to SIP servers and devices by way of SIP servers, gateways, and SIP-T tunneling of PSTN protocols.
- Full interoperability with existing network features and protocols
- Flexible service intelligence based on "new generation IN"-in particular, based on Parlay
- Service interworking with 2.5G and 3G mobile networks using CAMEL & Parlay
- High dependability and network wide fault tolerance,
- Flexibility in size and location, allowing deployment at sites convenient to the operator (typically a small number of centralized locations)

9.   OSS STANDARDS AND PROCESSES

As a result of network, technology and service evolution most operators run a number of separate business addressing different service needs. Organizational responsibilities are partitioned in order to give focus and enable effective decision-making. Even within an individual business there are typically several types of network and these may be partitioned in different ways organizationally and will use equipment from multiple vendors. The operational support systems reflect the organizational boundaries, resulting in multiple management systems that overlap in functionality, and have been interconnected in an ad hoc manner. In recognition of the inefficiencies inherent in this pragmatic method of developing OSSs, many operators are attempting to move to an approach based on operational process modelling, on the basis that organizations can change but the underlying business processes are stable.

In addition there are a number of industry initiatives aimed at better standardization of the techniques and terminology used to define operational support systems. There are independent software vendors who are developing components for the next generation of OSS including platforms to support business to business interfaces, web based user interfaces, and distributed applications, as well as the applications themselves.

We believe that there will be a general trend towards deployment of next generation OSS architecture supporting a component-based approach. But there are significant organizational barriers to be overcome and significant expenditure to fund. Operators will take advantage of these developments depending on their own internal priorities and they will fall into two broad categories. Firstly, the large majority of OSS changes will be those that can show a positive business benefit in less than a year. They will be targeted at operational cost saving largely in an existing network context. [Major OSS changes de-coupled from network infrastructure changes often result in the need for complex migration strategies to ensure continuity of service using legacy management interfaces]. Secondly, there will be strategic investments that will be combined with significant changes to networks. These will involve organizational change with a view to major restructuring of the business cost base and adding flexibility to deliver new services. Here again there may be the issue of migration of crucial service specific data and functionality if the new network platform has to deliver existing services.

In the telecommunications industry two standards cultures exist, the ITU-T and the IETF. Associated with these there are two fundamentally different approaches to management interfaces at the equipment level. These are the ISO protocol based Q3/Qx from the ITU-T and the IP protocol based SNMP from the IETF. IP equipment often also uses proprietary command line interfaces (CLI) and web browser style interfaces (HTTP based) while many different proprietary interfaces exist on traditional telco equipment. Most network operators have implementations that use both types of interface as they have both transport (ITU-T) and packet (IETF) based networks. If we assume that IP/MPLS is the de facto protocol for the core of future networks then it is likely that the de facto management interface will be SNMP. Because of the security shortcomings of SNMP v1 and v2 a migration to SNMP v3 seems probable (and may be mandated by some influential customers).

There are drives towards process and architecture commonality championed by the TeleManagement Forum's NGOSS initiative. NGOSS (New Generation Operations Systems and Software) is a set of guidelines for the industry to build software in a more structured and complementary way than has been done historically.

- a "loosely coupled" distributed component architecture
- along with functioning application components
- the components interact through a common information bus
- the components can be programmed through the use of a process management tool to control the business processes of the service provider using the functionality provided by the components

Due to the cost of change, all these interface styles are likely to persist for a long time. Thus there will be an ongoing requirement for multiple mediation functions in an integrated management system. This is another reason why significant change in the OSS of an established operator generally needs to be coupled with significant change in the equipment layer when the consideration of legacy interfaces can be minimized.

9.1 OSS Technology evolution

A number of technology threads will have significant influence on OSS implementations.

The general development of computing and storage technology will allow more complex functions to be performed in near real time and will allow more functionality to be deployed on the elements themselves. [Many of the complexities of existing element management applications arise from constraints on processing power in equipment controllers and DCN bandwidth].

The development of more robust and functional distributed application environments, web services and better information modelling techniques supported by tools is enabling faster implementation and more automation of operational processes. For example, the combination of Java-based mobile code frameworks and environments supporting a "component lifecycle" model allow standard ways of deploying new functionality and retiring old.

10. CONCLUSIONS

The emergence of improved processing frameworks and information modeling tools is having a significant impact on the time to develop new OSS applications, but to date no one has found a simple way to migrate from existing to new systems. All such processes require some form of encapsulation concept and the implementation of an adaptation function, which may be able to be retired when the legacy system is replaced.

Telecommunication networks and services will need to evolve on order to allow for a more efficient delivery of existing and new broadband services, whilst facing a significant decrease in revenue per transported and switched bits.

As broadband new services require more bandwidth and a more complex delivery, so network complexity and global capacity is growing fast while revenues per bit are falling.

Operator face a difficult task requiring on one hand to converge their networks in order to achieve better cost efficiency, whilst at the same time making sure that considerable network scalability is permitted in order to cope with the growing demand.

Several technologies are however available to meet this challenge and allow for an evolutionary network migration. This article provides a brief overview of the most significant, which are most likely to be employed in the making of the next generation networks

PART A1:

OPTICAL PACKET SWITCHING/ OPTICAL BURST SWITCHING

GUARANTEEING SEAMLESS END-TO-END QOS IN OBS NETWORKS

Maurizio Casoni,[1] Maria Luisa Merani[1], Alessio Giorgetti[2], Luca Valcarenghi[2], Piero Castoldi[2]

[1]*Department of Information Engineering, University of Modena and Reggio Emilia - Via Vignolese, 905 - Modena - Italy*

[2]*Sant'Anna School of University Studies and Doctoral Research, Center of Excellence for Communications Networks Engineering (CEIRC) - via Cisanello, 145 - Pisa - Italy*

Abstract:

In this paper the authors propose a method to guarantee end-to-end QoS to multiple traffic classes in optical burst switched (OBS) scenario, even in case of network congestion. The OBS network utilizes a core node architecture with no fiber delay lines and a limited set of wavelength converters. Traffic class performance differentiation is achieved by allowing high class traffic to utilize more node functionalities than low class traffic. To improve the likelihood of finding a route, even upon network congestion, bursts are allowed either to be deflected from their default route or to choose from a set of preplanned end-to-end paths. Network performance evaluation is focused on burst blocking probability and end-to-end TCP throughput. Performance is determined under the assumption of exponentially distributed burst interarrival times and arbitrarily distributed burst durations. Numerical results show that the proposed approach is able to guarantee different end-to-end TCP throughput performance to each traffic class. Moreover the proposed burst routing policies allows to decrease, with respect to a shortest path routing policy, the burst blocking probability. Thus the end-to-end TCP throughput is seamlessly guaranteed even in case of network congestion.

1. INTRODUCTION

Optical networking ultimate goal is the development of a full optical Internet, where signals carried within the network never leave the optical domain [1]. A first important step in this direction is to have optical networks transparent at least for data, with the control part converted and processed in electronics. In Optical Burst Switching (OBS) [2,3] the key idea is to dynamically set up an optical path, i.e. a lightpath, whenever a *large* data flow is identified and needs to traverse the network. In OBS data never leave the optical domain but for each data burst assembled at the network edge, a reservation request is sent as a separate control packet, well in advance, and processed within the electronic domain. The control packet, carrying relevant forwarding and routing information, precedes each burst by a basic offset time. The offset time is set to accommodate the non-zero electronic processing time of the control packets inside the network nodes. OBS network nodes can be classified as either edge or core routers. The main task of edge nodes is the burst assembly function: as they represent the border between "traditional" electrical LAN/MAN IP networks and a high speed optical transport network, they must collect incoming IP datagrams and assembly them into bursts according to suitable algorithms. Core routers, on the other hand, deal with data bursts and the related control packets; they have to set up internal optical paths on the fly for switching bursts and take them hop-by-hop to their final destination. The control packet offset time allows the core switches to be buffer-less, thus avoiding the utilization of optical memories, e.g. fibre delay lines, required on the other hand by optical packet switching [4].

In a previous paper [5] some of the authors have evaluated, in terms of burst loss and delay, the performance of OBS networks that utilize the Just Enough Time (JET) reservation mechanism and mechanisms for service differentiation. In this paper the authors propose a scheme for differentiating the blocking probability experienced by three different burst classes and for making the end-to-end throughput unaffected by network congestion. The considered OBS network is built of core nodes equipped with a limited set of full range optical wavelength converters. Burst class differentiation is implemented by allowing different class bursts to utilize different node functionalities. In addition selected class bursts are allowed either to be deflected from their default shortest path route or to choose among a set of preplanned end-to-end paths. Thus during the optical path routing different burst classes experience a different blocking probability. Furthermore the round trip time experienced by the bursts can be either unpredictable, in case of deflection routing, or increase of a bounded amount, in case of alternate end-to-end re-routing.

The metrics utilized to evaluate the performance of the proposed scheme are the burst blocking probability, i.e. the probability that a burst is blocked because the request of bandwidth at the traversed optical node cannot be fulfilled, and the end-to-end throughput, obtained from the burst blocking probability and the round trip time (RTT) through a widely-accepted TCP model. Numerical results shows that assigning to different classes the right of utilizing different node functionalities allows to differentiate burst classes blocking probability. In addition the possibility for the optical paths to be rerouted allows bursts to overcome network congestion. Moreover while deflection routing unpredictably affects the RTT, network congestion avoidance through alternate end-to-end re-routing allows to limit the burst blocking probability and only slightly affects RTT and therefore does not downgrade the end-to-end TCP throughput.

The rest of the paper is organized as follows. In Section 2 the core network with the related routing algorithms is shown. Section 3 provides the framework for the end-to-end performance evaluation when a transport protocol like TCP is employed. Numerical results are then collected in Section 4 while concluding remarks are in Section 5.

2. OBS NETWORK MODEL

This work assumes that OBS nodes support JET and that traffic is mapped into three (but in general N_{QoScl}) QoS classes. Each class is characterized by a different statistical traffic description and features different QoS requirements determined by specifying an upper bound of burst blocking probability and/or end-to-end delay. Traffic incoming into edge nodes is supposed to be $M/Pareto$. If, for instance, $\overline{x}_1$, $\overline{x}_2$ and $\overline{x}_3$ are the mean duration times of the ON periods due to aggregation of short, medium and long-sized datagrams, respectively, on each incoming link the offered load ρ is the sum of three distinct contributions:

$$\rho = \sum_{i=1}^{3} \rho_i = \sum_{i=1}^{3} p_i \lambda \overline{x}_i, \tag{1}$$

where p_1, p_2 and p_3 represent the occurrence probabilities of the three classes and λ is the mean arrival rate.

The core optical routers are assumed to be equipped with $M \times M$ optical interfaces capable of supporting N wavelengths each and with a limited pool of wavelength converters wc. The optical nodes are buffer-less, i.e., no fiber delay lines (FDLs) are present in order to resolve contention for an output fiber, output wavelength (Figure 1). It is however reasonable to assume the use of a set of input FDLs [6], whose exclusive task is to re-align the OBS data burst and its control packet, so as to guarantee a minimal offset time at any intermediate node, as depicted in Figure 1.

Bursts are created according to three class of service, class 1 carrying time-sensitive data, class 2 and 3 loss-sensitive data. The bursts default route is the shortest hop path from source to destination. However routing can be modified in order to better meet the performance required by each class; in particular no time-sensitive bursts are allowed to be re-routed along longer paths to decrease their blocking probability. Two alternative re-route schemes, deflection routing at intermediate nodes and end-to-end re-routing, are compared for traffic class

2 and 3 in terms of achievable burst blocking probability and TCP throughput. Deflection simply deflects bursts toward another output fiber if not any wavelength is available in the desired output fiber; while to minimize the blocking probability end-to-end re-route can be performed as presented in [7] by avoiding the contention among different optical paths.

The overall strategy for service differentiation is as follows: class 1 bursts are given the highest priority through an additional extra-offset and the use of wavelength converters in the core nodes; class 2 bursts have medium priority by just using the converters; class 3 bursts have low priority since they have no extra offset and cannot exploit wavelength converters; on the other hand, class 2 and 3 can use alternative sub-optimal variable delay paths through deflection or end-to-end re-routing.

In this paper the network considered covers most European countries (Figure 2). London, Oslo and Stockolm are sources of information flows and operate as edge routers, whereas Madrid, Rome and Athens are the possible destinations. All the other nodes work as core routers. Moreover, no flow is supposed to enter or leave the network at intermediate steps.

Section 4 will show the performance, in terms of overall edge-to-edge burst blocking probability for the three classes of bursts in the OBS network, when each node implements the JET reservation mechanism with a limited set of wavelength converters available.

3. END TO END PERFORMANCE

The throughput is here studied on end-to-end basis when TCP is assumed as transport layer protocol. In fact, in wide area data networks like Internet congestion control mechanisms have a fundamental role for the global functioning. TCP is a reliable window-based acknowledgment-clocked flow control protocol, thought to avoid to overload the network and to react to a possible congestion at network level. In the system under investigation TCP Reno is assumed to be employed by hosts. TCP Reno is modeled following the approach detailed in [8] where the throughput (bit/s) is approximated by:

$$Thr_{TCP} = \frac{MSS}{RTT\sqrt{\frac{2bp}{3}} + T_0 min\left(1, 3\sqrt{\frac{3bp}{8}}\right)p(1 + 32p^2)} \quad (2)$$

being MSS the maximum segment size expressed in bits, RTT the round trip time, p the segment loss probability, T_0 the time out and b the number of packets acknowledged by ACKs.

The performance of TCP over OBS networks have been studied in some previous works [9] [10] but it can still be considered an open issue and thus a challenge for the research community. Here we want to add some thoughts to the worldwide discussion. Let us assume, as classified in [9], to have only slow

TCP sources which emit at most one segment during the interval $(0, T_{max})$. It also means that at most one segment for each connection is contained in a burst generated by edge nodes and injected into the OBS network. Therefore, even if approximated, for this type of sources the segment loss probability p can be assumed equal to the burst blocking probability.

Now, considering the reference network topology (Figure 2) an average link length of 800 Km can be assumed and the number of hops, N_{hops}, is in the range [3 − 5]. Since the light propagation speed in the fibers is roughly 70% the speed of the light in the vacuum, the propagation delay for each hop, T_{hop}, is 4 ms. Therefore the one-way delay edge-to-edge, from entering the ingress edge to leaving the egress edge, $T_{e2e1way}$ in the OBS network is:

$$T_{e2e1way} = T_{assembly} + N_{hops} \times T_{hop} + T_{disass} \tag{3}$$

which can be bounded to 30 ms. If in addition the network has a symmetric behavior, RTT is approximately 60 ms. Actually, RTT has also to consider the delays given by the access networks before entering the OBS network: this means that 60 ms can be considered as a kind of lower bound.

In the next Section the behavior of the throughput of TCP will be shown referring to the above assumptions for p and RTT.

4. NUMERICAL RESULTS

The numerical results are obtained for the reference network, in which the performance figures are the burst loss and the end-to-end TCP throughput computed by exploiting the aforementioned analytical model of TCP. The performance of a complete burst switching network have been determined by simulation through an *ad-hoc* event-driven C++ object oriented simulator. Regarding core routers, it is set $wc = 20$, $M = 2$ with $N = 16$ wavelengths per fiber. The following values for p_1, p_2 and p_3 are considered, $(0.5, 0.2, 0.3)$. Incoming traffic into edge nodes is supposed to be $M/Pareto$ with $\alpha_{on} = 1.2$, $\bar{x}_1 = 218$ bytes, $\bar{x}_2 = 576$ bytes $\bar{x}_3 = 1500$ kbytes. The extra offset for class 1 bursts is set at $6\mu s$. It is worth reminding that only class 1 and 2 bursts exploit a set of 20 wavelength converters; however, class 2 and 3 can be deflected in case of unavailability of wavelengths on the outgoing fibre or end-to-end re-routed along an alternate path. Figure 3 shows the total burst blocking probability for the three burst classes having Oslo as source and Rome as destination, when deflection routing is utilized for traffic class 2 and 3. In order to have loss values in the range of 10^{-3} for class 1, 1% for class 2 and 50% for class 3, the overall load ρ must be less than 0.4.

Let us now discuss the consequences of the above values of burst blocking probability on end-to-end performance. Figure 4 shows the throughput of TCP Reno given by (2) as a function of RTT for $T_0 = 1.0$ s, $b = 2$, $MSS = 1500$

bytes and $p = 1\%$. As mentioned in Section 3, for slow TCP sources p can be approximated with the burst blocking probability. Also, the RTT values of interest fall reasonably in the $[60 - 200]$ ms range, depending on the type of access network. This figure says that even if a very high speed core network is employed the best we can get is a throughput of 1.3 Mbps and it remarkably decreases at 750 kbps as soon as the RTT doubles, or at 500 kbps when the RTT becomes three times, i.e. 180 ms.

Figure 5 shows the improvement in burst blocking probability when end-to-end re-route is utilized with only one preplanned alternative path for each source-destination pair. It is important to notice that the proposed end-to-end re-route scheme is able to strongly decrease the burst blocking probability guaranteed from deflection routing. Therefore, as shown in figure 6, edge-to-edge TCP throughput can significantly increase even though the RTT increases because of the utilization of a longer path.

5. CONCLUSIONS

This work investigates burst blocking probability and overall performance of an OBS network, with buffer-less nodes, adopting JET reservation mechanism.

The authors numerically demonstrated that the optical node architecture under examination can provide acceptable values of burst blocking probability and end-to-end TCP throughput, once the offered load is properly limited. Moreover the proposed burst routing policy avoids that network congestion deteriorate the end-to-end TCP throughput.

ACKNOWLEDGMENTS

This work has been partially supported by MIUR within the framework of the national project "INTREPIDO: Traffic Engineering and Protection for IP over DWDM".

REFERENCES

[1] B. Bostica, F. Callegati, M. Casoni, C. Raffaelli, "Packet Optical Networks for High Speed TCP-IP Backbones" -*IEEE Communications Magazine*, January 1999, pp.124-129.

[2] C. Qiao, M. Yoo, "Optical Burst Switching (OBS) - a New Paradigm for an Optical Internet," *Journal of High Speed Networks*, No.8, pp.69-84, 1999.

[3] M. Yoo, C. Qiao, S. Dixit, "QoS Performance of Optical Burst Switching in IP-Over-WDM Networks," *IEEE Journal on Selected Areas in Communications*, Vol.18, No.10, pp.2062-2071, October 2000.

[4] M. Casoni, C. Raffaelli, "Tandem Architecture for Photonic Packet Switches" Ü Journal of Communications and Networks, Vol.1, No. 3, September 1999, pp.145-152.

[5] M. Casoni, M.L. Merani, ŞResource Management in Optical Burst Switched Networks: Performance Evaluation of a European NetworkŤ, Proc. of 1st International Workshop on

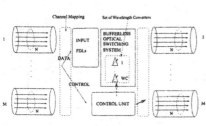

Figure 1. General core router buffer-
less architecture.

Figure 2. Reference network.

Optical Burst Switching, October 16 2003, Dallas, (Texas, USA).

[6] Y. Xiong, M. Vandenhoute, H.C. Cankaya, "Control Architecture in Optical Burst-Switched WDM Networks", *IEEE Journal on Selected Areas in Communications*, Vol.18, No.10, October 2000, pp.1838-1851.

[7] A. Fumagalli and L. Valcarenghi, "The Preplanned Weighted Restoration Scheme", Proc. of IEEE Workshop on High Performance Switching and Routing, 2001.

[8] J. Padhye, V. Firoiu, D. F. Towsley, J. F. Kurose, "Modeling TCP Reno Performance: A Simple Model and its Empirical Validation", *IEEE/ACM Trans. on Networking*, vol.8, no.2, pp.133-145, April 2000.

[9] A. Detti, M. Listanti, "Impact of Segment Aggregation on TCP Reno Flows in Optical Burst Switching Networks", Proc. of IEEE Infocom 2002, 23-27 June 2002, New York (U.S.A.).

[10] S. Gowda, R. K.Shenai, K. M.Sivalingam, H.C. Cankaya, "Perfomance Evaluation of TCP over Optical Burst-Switched (OBS) WDM Networks", Proc. of ICC 2003, 11-15 May 2003, Anchorage (U.S.A.).

54

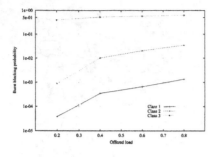

Figure 3. Total end-to-end burst blocking probability as a function of the offered load for the three burst classes, relative to (Oslo, Rome) pair.

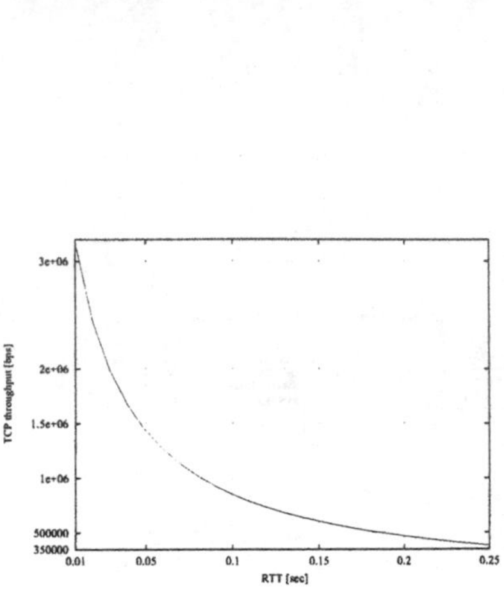

Figure 4. Throughput of TCP as a function of RTT for $MSS = 1500$ bytes and $p = 1\%$ achieved with shortest path routing.

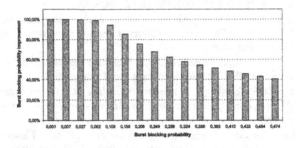

Figure 5. Burst blocking probability percentage improvement achievable utilizing one alternative path in function of bursts blocking probability p obtained with shortest path routing and deflection routing. If $p < 10\%$ the re-route scheme completely cancels block events.

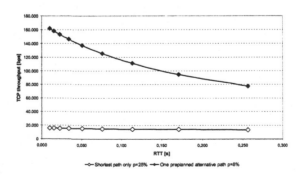

Figure 6. Throughput of TCP as a function of RTT for $MSS = 1500$ bytes and $p = 25\%$ achieved with shortest path routing, $p = 8\%$ achievable providing one alternative path.

A FRAMEWORK FOR THE ANALYSIS OF DELAY JITTER IN OPTICAL PACKET SWITCHED NETWORKS

F. Callegati,[1] W. Cerroni,[2] G. Muretto,[1] C. Raffaelli,[1] P. Zaffoni[1]

[1] *DEIS - University of Bologna*
Viale Risorgimento, 2 - 40136 Bologna - ITALY
{fcallegati,gmuretto,craffaelli,pzaffoni}@deis.unibo.it

[2] *CNIT - Research Unit at the University of Bologna*
Viale Risorgimento, 2 - 40136 Bologna - ITALY
walter.cerroni@cnit.it

Abstract: Out of order delivery and/or delay jitter are typical phenomena occurring in networks adopting a packet transfer mode and may have a relevant impact on the behavior of higher layer protocols. In this paper an original framework is proposed to analyze these phenomena in optical packet-switched networks that employ the wavelength and the time dimensions to solve congestion. This framework is the first step toward a deeper understanding of the interaction between transport networks and higher layers behaviors.

Key words: Optical networks, Optical packet switching, Delay jitter, TCP

1. INTRODUCTION

Optical Packet Switching (OPS) is a networking technology promising a huge breakthrough in terms of available capacity [1] and is considered the best candidate to provide the long-term solution for future, very high-speed networks.

Many fundamental networking problems related to OPS are still open: the interworking with higher layer functions, such as end-to-end flow control mechanisms, is one of the most significant. Some works on this issue recently appeared mainly focusing on the effects of aggregation of IP packets into longer optical bursts or packets [2, 3]. In general not many results are available regarding the effects that congestion, and the consequent congestion resolution mechanisms, in the OPS network may have on the behavior of transport protocols such as TCP.

Congestion resolution in OPS can be implemented in the time domain, using some form of delay lines, in the wavelength domain, sending contending packets to different wavelengths on the same fiber, and in the space domain, by means of deflection and alternate routing. Deflection routing is the only congestion resolution scheme considered in [3], while here we focus on the combined use of time and wavelength, that have been shown to realize a good trade-off between network control complexity and performance [4, 5].

The congestion resolution schemes affect the packet stream by modifying the time framework of the packet stream (delay jitter), changing the order of the packet transmission (out-of-order) and dropping some packets because of lack of resources (packet loss). In this paper we focus on delay jitter and out-of-order delivery, aiming at understanding how to evaluate the effects of the congestion resolution mechanisms on these phenomena.

We assume a network architecture consisting of OPS facilities exploiting a DWDM transmission infrastructure and capable to transport IP traffic by means of integration with a GMPLS control plane [6]. The network operation is therefore connection-oriented and the switching granularity of the OPS nodes is at the Label Switched Path (LSP) level. Each LSP represents a top-level, explicitly routed path formed by an aggregation of lower-level connections including several traffic flows (an Internet Draft proposes how to implement LSP hierarchies in GMPLS [7]).

We assume the availability of an all-optical switching matrix able to switch variable-length packets. Implementation issues are beyond the scopes of this work and therefore a general OPS node with full connectivity and wavelength conversion capabilities is considered. The node may also delay packets by means of buffers realized with Fiber Delay Lines (FDLs) [8].

The content of the paper is organized as follows. Section 2 provides a brief review of the congestion resolution issue in OPS. In section 3 the problem of out-of-sequence packet delivery and its influence on end-to-end protocol performance is discussed. In section 4 the proposed methodology to measure delay jitter and evaluate the degree of out-of-sequence events is presented. Conclusions are drawn in section 5.

2.   CONGESTION RESOLUTION IN THE OPS NODE

The basic assumption that makes possible the use of the wavelength domain for congestion resolution is that network paths are associated with fibers and not with wavelengths. This is motivated by the fact that all the wavelengths of the same fiber can be seen as a set of parallel links toward the same destination. Within a node, the forwarding component decides to which path (i.e. fiber) a packet must be sent and then the Switch Control Logic (SCL) decides the detailed scheduling and sets up the switching devices. The objective of the SCL is to exploit at its best the resource usage by performing two major tasks:

- choose which wavelength of the output fiber will be used to transmit the packet;

- decide whether the packet has to be delayed by using the FDL buffer or it has to be dropped.

This is called the Wavelength and Delay Selection (WDS) problem because the choices of wavelength and delay are actually correlated, being the need to delay a packet related to the availability of the wavelength selected. The WDS algorithm to solve this problem can be implemented by following different policies:

- **Static** - the LSP is assigned to a wavelength at LSP set-up and this assignment is kept constant all over the LSP lifetime. This approach requires little complexity due to processing at LSP set-up only.

- **Connectionless-like** - the wavelength is selected on a per-packet basis. This approach provides maximum flexibility in resource allocation but also requires per-packet processing, therefore it is fairly demanding in terms of load on the SCL.

- **Dynamic** - the LSP-to-wavelength assignment is executed only when congestion arises, i.e. when the time domain is not enough to solve contention due to the lack of buffering space.

Several heuristic connectionless-like WDS algorithms have been studied in the past, showing that they may significantly change the performance [4, 5]. The price to pay for this performance improvement is a non-negligible processing effort.

On the other hand, it has been observed that with a static wavelength allocation the switch performance strongly depends on the configuration of the LSP forwarding table [9] and is very sensible to the distribution of the LSP destinations, providing performance that is not easy to control.

The dynamic case is somewhat in between, simpler than the connectionless alternative but also capable of providing a more uniform and effective resource utilization than the static case. Therefore this approach is preferable in a connection-oriented network.

The major drawback of the dynamic WDS algorithms is that they do not preserve, a priori, the sequence of the packet flow neither at a global switch level nor at the LSP level. This is mainly caused by the possibility of multiplexing packets following the same LSP on different wavelengths, allowing parallel transmission. Furthermore, the adoption of dynamic routing algorithms [10] may be an additional cause of unordered deliveries.

3.   THE PROBLEM OF PACKET REORDERING

As already outlined packet loss as well as out-of-order packet delivery and delay variations affect end-to-end protocols behavior and may cause throughput impairments [11, 12].

When considering TCP-based traffic it is well known that these phenomena influence the typical congestion control mechanisms adopted by the protocol [13] and may result in a reduction of the transmission window size and consequently in bandwidth under-utilization. In particular the TCP congestion control is highly affected by loss or out-of-order delivery of bursts of segments. This is exactly what may happen in the OPS network where traffic is typically groomed and several IP datagrams (and therefore TCP segments) are multiplexed in a single optical packet, because optical packets must satisfy a minimum length requirement to guarantee a reasonable switching efficiency. Therefore out-of-order or delayed delivery of just one optical packet may result in out-of-order or delayed delivery of several TCP segments, causing multiple duplicate ACKs and/or expired timeouts and triggering congestion control mechanisms which cause unnecessary reduction in the window size.

Another example of how out-of-sequence packets may affect application performance is the case of delay-sensitive UDP-based traffic, such as real-time traffic. In fact unordered packets may arrive too late and/or the delay required to reorder several out-of-sequence packets may be too high with respect to the timing requirements of the application.

These brief and simple examples make evident the need to limit the number of unordered packets. In general out-of-order delivery is caused by the fact that packets belonging to the same flow of information can take different paths through the network and then can experience different delays. In traditional connection-oriented networks, packet reordering is not an issues since packets belonging to the same connection are supposed to follow the same virtual network path and therefore are delivered in the correct sequence, unless packet loss occurs. In an OPS network using the wavelength domain for congestion resolution, this may not be the case. Packets traveling along the same network path may use different wavelengths in order to exploit wavelength multiplexing for congestion resolution purposes. Therefore it may happen that packets of the same flow are delivered out of sequence, even though still following the same network path.

A possible solution could be to assume that this problem is solved at the egress edge-nodes that should take care of re-sequencing the various packet flows. This assumption in our view is not very realistic. It can be feasible for some flow of high-value traffic, but it is unlikely that this will happen for all the flows of best effort traffic, because of the amount of memory and processing effort that would be required. Therefore we argue that it is important and necessary to control delay jitters and out-of-order delivery of packets directly in the OPS network nodes. This is what we will discuss in the following section, where we realize that, first of all, a clear definition of this term is necessary because of the difference between the OPS network and conventional networks.

4. OUT-OF-ORDER AND DELAY VARIATION IN OPS NETWORK WITH WDM

Some authors have been dealing with measuring the degree of packet reordering [14, 15] and the impact on higher layers performance [12] in a traditional Internet scenario. However in the scenario of an OPS network using WDM the problem is different. As already explained, packets belonging to the same LSP may be transmitted on different wavelengths according to the principles built in the WDS algorithm. This may cause overtaking and/or partial overlapping of packets belonging to the same connection in a number of different ways.

A full understanding of such phenomena requires:

- to understand which cases of out-of-order packets may happen and how this phenomena can be measured in some quantitative way;
- to evaluate which are the WDS algorithms that preserve at best the time framework of the packet flows;
- to understand the effects that a specific case of out-of-order delivery may have on higher layer protocols.

This paper addresses the first issue by defining, in this section, a framework to measure, at the single node level, the modification in the data packet flows caused by the WDS algorithm. Then the paper provides an example of application of this framework to two typical WDS algorithms, to show how this tool can be used.

4.1 A framework to evaluate the delay jitter

For a generic packet P_i crossing a given OPS node, let t_i be the arrival time at the node input, s_i the departure time from the node output and $d_i = d_p + k_i D$ the delay introduced by the node itself, due to the packet header processing time (d_p, fixed) and the possible delay inside the FDL buffer ($k_i D$, where $k_i = 0, 1, \ldots, B$). Obviously $d_i = s_i - t_i$.

Let assume that two generic subsequent packets belonging to the same traffic flow P_n and P_{n+1} arrive in order, i.e. $t_{n+1} > t_n$. Let $\Delta t_n = t_{n+1} - t_n$ and $\Delta s_n = s_{n+1} - s_n$ be the relative packet offsets at the node input and output respectively. The *jitter* between packets P_n and P_{n+1}, representing the packet offset variation due to the node crossing, may be defined as

$$J_n = \Delta t_n - \Delta s_n \tag{1}$$

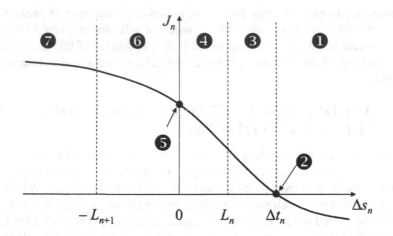

Figure 1. Behavior of the jitter depending on relative packet offset at node output

Equation (1) may also be written as

$$J_n = d_n - d_{n+1} = (k_n - k_{n+1})D = h_n D \qquad (2)$$

where $-B \le h_n \le B$. The behavior of J_n for two particular packets P_n and P_{n+1}, with length L_n and L_{n+1} respectively, is shown in figure 1, where the x axis has been divided in seven different regions:

1. $\Delta s_n > \Delta t_n$ when the packet sequence is always guaranteed since P_{n+1} experiences more delay than P_n ($J_n < 0$);
2. $\Delta s_n = \Delta t_n$ when the node is transparent and P_n and P_{n+1} have the same offset at the input and output ($J_n = 0$);
3. $L_n \le \Delta s_n < \Delta t_n$ when P_{n+1} experiences less delay than P_n ($J_n > 0$) but at the output it is still behind the tail of P_n (i.e. $s_{n+1} \ge s_n + L_n$);
4. $0 < \Delta s_n < L_n$ when the head of P_{n+1} partially overlaps the tail of P_n;
5. $\Delta s_n = 0$ when P_{n+1} completely overlaps P_n ($J_n = \Delta t_n$);
6. $-L_{n+1} < \Delta s_n < 0$ when P_{n+1} has overtaken P_n but they are partially overlapping (i.e. $|\Delta s_n| < L_{n+1}$);
7. $\Delta s_n \le -L_{n+1}$ when P_{n+1} has completely overtaken P_n (i.e. $s_n \ge s_{n+1} + L_{n+1}$).

4.2   A numerical example

The previous formalization allows to evaluate the delay jitter distribution as well as the amount of out-of-order packets, that depends on the specific definition of packet sequence. For instance, in case overlapping packets are not considered in sequence, then the out-of-sequence regions will be 1, 2, and 3. If some overlapping is allowed, then out-of-sequence is guaranteed also in region 4. The same for region 5, in case packets arriving at the same time are not considered out of order.

As an example, figure 2 shows the jitter distribution over the different regions for a static and a connectionless-like WDS policy. The results for dynamic WDS are very similar to the connectionless-like case. As expected static WDS succeeds in maintaining the packet sequence, although it gives worst performance in terms of packet loss probability. On the other hand more dynamic policies cause some packets to get out of the node unordered, but the most frequent behavior is the one related to region 2, which means that congestion happens rarely and the packets are often transmitted transparently across the node.

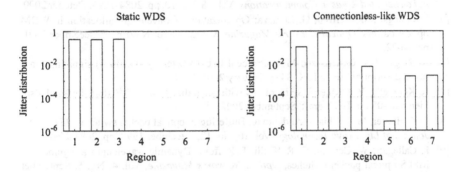

Figure 2.   Jitter distributions for static and connectionless-like WDS policies

5.   CONCLUSIONS

In this paper we have discussed the effect of scheduling algorithms on the packet sequence and time framework, in the scenario of an OPS network exploiting the time and wavelength domains for congestion resolution. We have proposed an original framework to quantitatively analyze delay jitter and out-of-order packets. This analysis can be used as a basis to compare different scheduling algorithms as shown in the numerical example provided. Moreover it represents the starting point to establish a link between the OPS network performance and the performance of higher layer protocols. This last issue is currently under investigation.

ACKNOWLEDGMENTS

This work is partially funded by the Italian Ministry of Education and University (MIUR) under the projects "INTREPIDO - End-to-end Traffic Engi-

62

neering and Protection for IP over DWDM Optical Networks" and "GRID.IT - Enabling platforms for high-performance computational grids oriented to scalable virtual organizations".

REFERENCES

[1] M.J. O'Mahony, D. Simeonidou, D.K. Hunter, A. Tzanakaki: The application of optical packet switching in future communication networks, *IEEE Communications Magazine*, Vol. 39 , No. 3, March 2001, pp.128-135.

[2] A. Detti, M. Listanti: Impact of segments aggregation on TCP Reno flows in optical burst switching networks, *Proc. of IEEE/ACM INFOCOM 2002*, Vol. 3, June 2002, pp. 1803-1812.

[3] Jingyi He, H.G. Chan: TCP and UDP performance for Internet over optical packet-switched networks, *Proc. of IEEE ICC '03*, Vol. 2,Ê May 2003, pp. 1350-1354.

[4] L. Tancevski et al.: Optical routing of asynchronous, variable length packets, *IEEE Journal on Selected Areas in Communications*, Vol. 18, No. 10, pp. 2084-2093, October 2000.

[5] F. Callegati, W. Cerroni, G. Corazza: Optimization of wavelength allocation in WDM optical buffers, *Optical Networks Magazine*, Vol. 2, No. 6, November/December 2001, pp. 66-72.

[6] L. Berger, Ed.: Generalized Multi-Protocol Label Switching (GMPLS) signaling functional description, IETF RFC 3471, January 2003.

[7] K. Kompella, Y. Rekhter: LSP hierarchy with generalized MPLS TE, draft-ietf-mpls-lsp-hierarchy-08.txt, IETF draft, September 2002.

[8] D.K. Hunter, M.C. Chia, I. Andonovic: Buffering in optical packet switches, *IEEE/OSA Journal of Lightwave Technology*, Vol. 16, No. 12, December 1998, pp. 2081-2094.

[9] F. Callegati, W. Cerroni, C. Raffaelli, P. Zaffoni: Dynamic wavelength assignment in MPLS optical packet switches, *Optical Networks Magazine*, Vol. 4, No. 5, September 2003, pp. 41-51.

[10] F Callegati, W Cerroni, G. Muretto, C. Raffaelli, P. Zaffoni: Adaptive routing in DWDM optical packet-switched networks, *Proc. of 8th IFIP ONDM 2004*, Gent, Belgium, February 2004, pp. 71-86.

[11] J. C. R. Bennett, C. Patridge: Packet reordering is not a pathological network behavior, *IEEE/ACM Transactions on Networking*, vol. 7, no. 6, December 1999, pp. 789-798.

[12] M. Laor, L. Gendel: The effect of packet reordering in a backbone link on application throughput, *IEEE Network*, vol. 16, no. 5, September/October 2002, pp. 28-36.

[13] M. Allman, V. Paxson, W. Stevens: TCP congestion control, IETF RFC 2581, April 1999.

[14] T. Banka, A. A. Bare, A. P. Jayasumana: Metrics for degree of reordering in packet sequences, *Proc. of IEEE LCN 2002*, Tampa, FL, November 2002, pp. 333-342.

[15] S. Jaiswal, G. Iannacone, C. Diot, J. Kurose, D. Towsley: Measurement and classification of out-of-sequence packets in a tier-1 IP backbone, *Proc. of INFOCOM 2003*, San Francisco, CA, March 2003, vol. 2, pp. 1199-1209.

DEMONSTRATION OF PREAMBLE LESS OPTICAL PACKET CLOCK AND DATA RECOVERY WITH OPTICAL PACKET SWITCHING

Naoya Wada[1], Hatsushi Iiduka[2], and Fumito Kubota[1]
[1]National Institute of Information and Communications Technology (NICT), 4-2-1, Nukui-Kita, Koganei, Tokyo 184-8795, Japan. wada@nict.go.jp
[2]NTT Electronics Co., 1841-1, Tsuruma, Machida, Tokyo 194-0004, Japan. ohtsuka@atsu1.nel.co.jp

Abstract: A novel preamble free optical packet 3R receiver is proposed. Optical packets stream composed of 10Gbit/s preamble-less payload data and arbitrary intervals is generated and received with instantaneous clock and payload data recovery in less than a bit period (<100ps). Preamble-less optical packet clock and data recovery with two hop 40Gbit/s optical packet switching based on all-optical code label processing is experimentally demonstrated.

1. INTRODUCTION

The next generation of optical network will need high scalability and fine granularity as well as large network capacity. Wavelength division multiplexing (WDM) technology has a huge capacity for data transmissions. However, the granularity of a WDM light-path network is coarse. Networks based on optical packet switching (OPS) can provide high scalability, fine granularity and ultrahigh-speed hopping. Despite their lack of maturity, many OPS systems have been developed because of their obvious merits [1-5].

Optical packet 3R receiver will be a very important component in a future OPS networks. Some optical burst-mode 3R receivers have been reported [6-8].

However, these burst-mode receivers require preamble pattern and/or training bits for stable clock and payload data recovery. Therefore, time for clock and data recovery of these burst-mode receivers is nanoseconds order. Such overhead time due to preamble deteriorates the network performance.

In this report, a novel preamble free optical packet 3R receiver is proposed. 10Gbit/s optical packets streams with 1000 to 4000 bits-long, preamble-less, payload data and arbitrary intervals from nanosecond to microseconds are tested. The packets stream is asynchronously received by optical packet 3R-receiver. Instantaneous clock and payload data recovery in less than a bit period (<100ps) is experimentally demonstrated. In addition, preamble-less optical packet clock and data recovery with two hop 40Gbit/s optical packet switching based on all-optical code label processing and optical time-domain multiplexing and de-multiplexing technologies is experimentally demonstrated.

2.   OPTICAL PACKET 3R RECEIVER

The block diagram of proposed optical packet 3R receiver is shown in Fig.1. Optical packet 3R-receiver consists of O/E converter and clock and data recovery (CDR) parts. Specialy tuned uni-traveling-carrier photo-detector (UTC-PD) is introduced as an O/E converter. CDR consists of phase shifter (T/2), X-OR, D-FF, and low jitter gated VCO. The block diagram of proposed low jitter gated VCO is shown in Fig.2.

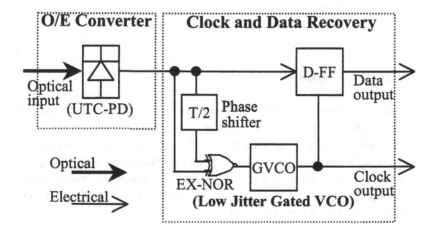

Figure 1. Block diagram of optical packet 3R receiver

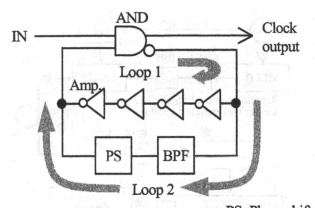

Figure 2. Block diagram of low jitter gated VCO

The low jitter gated VCO has the double phase lock loop structure. Loop 1 is for start up operation at around 10 GHz. Loop 2 is for very low jitter operation at 9.95328 GHz after loop 1.

3. EXPERIMENTAL DEMONSTRATION I: PREAMBLE LESS OPTICAL PACKET CLOCK AND DATA RECOVERY

Figure 3 represents an experimental set-up of an instantaneous clock and payload data recovery and bit error ratio (BER) measurement. The set-up consists of a 10GHz mode locked laser diode (MLLD) as a light source, LiNbO$_3$ intensity modulator (LN-IM), pulse pattern generator (PPG), optical attenuator, proposed optical packet 3R receiver, sampling oscilloscope and error detector. Optical amplifiers (EDFA) were employed to compensate insertion losses of the components in combination with optical band-pass filters (BPF) to attenuate the amplified spontaneous emission. The MLLD generates 1.5ps pulses with 9.95328GHz repetition rate at central wavelength of 1550nm. This pulse train is modulated by a LN-IM to generate optical packets streams with 1000 to 4000 bits-long, preamble-less, return-to-zero (RZ) payload data and arbitrary intervals from nanosecond to microseconds. The generated packets streams are input to the optical packet 3R receiver via optical fiber and attenuator. In the 3R receiver, Instantaneously, clock and payload data are recovered from asynchronously received preamble-less optical packet data stream. The recovery time is less than a bit period (<100ps). The waveform and BER of recovered payload data are measured by sampling oscilloscope and error detector, respectively.

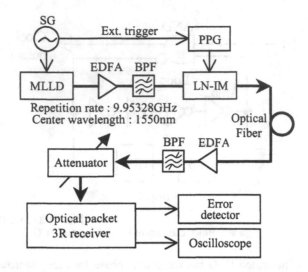

Figure 3. Experimental set-up of preamble less optical packet clock and data recovery

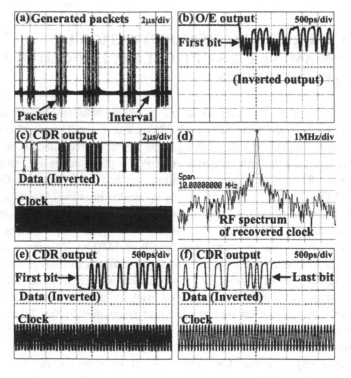

Figure 4. Measured waveforms of experimental demonstration I

Figure 4 (a) is the waveform of a generated random optical packet stream. In the O/E part of 3R receiver, input optical packets are converted to electronic signal via UTC-PD with 40GHz bandwidth. Inverted waveform of the O/E output is shown in Fig. 4(b). Figure 4 (c) represents Inverted waveforms of CDR outputs. Recovered data and clock are represented in upper and lower parts, respectively. Figure 4 (d) is RF spectrum of recovered clock by CDR. This RF spectrum shows the stable clock recovery of the optical packet 3R receiver. Figures 4 (e) and (f) are magnification of Fig. 4 (c). Figure 4 (e) is the beginning of the recovered data and clock. Figure 4 (f) shows the ending of the recovered data and clock. These results represent the instantaneous clock and payload data recovery in less than a bit period.

Figure 5 represents measured BERs in the optical packet receiving experiment. Figure 5 (a) shows measured BER and eye diagram of the recovered payload data from a random packet stream represented in Fig. 4(c). This stream consists of 4000 bits-long, preamble-less, payload data and arbitrary intervals from nanosecond to microseconds. Figure 5 (b) shows measured BERs of three different regular packet streams consist of 1000, 2000, and 4000 bits-long, preamble-less, payloads, respectively. In each case, the stream has about 100ns intervals between packets. These results guarantee instantaneous and stable clock and payload recovery for packets data.

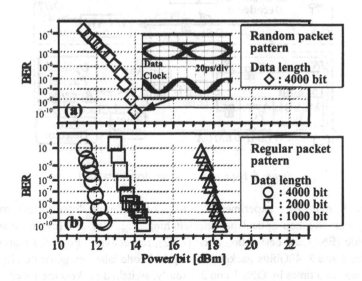

Figure 5. Measured BERs and eye diagram of experimental demonstration I

4. EXPERIMENTAL DEMONSTRATION II: PREAMBLE LESS OPTICAL PACKET RECEIVING WITH TWO HOP OPTICAL PACKET SWITCHING

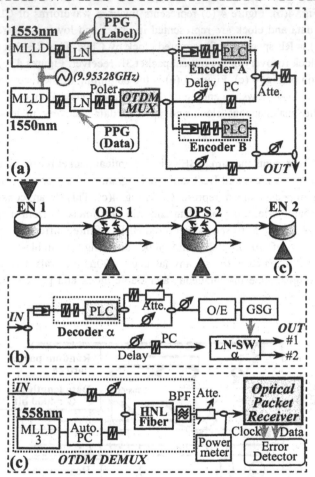

Figure 6. Experimental set-up demonstration II

Figure 6 represents experimental set-up of preamble less optical packet receiving with two hop optical packet switching. Fig. 6(a) and (c) are setup of source node (EN1) and destination node (EN2), respectively. Fig. 6(b) represents OPS nodes 1 and 2. 40Gbit/s packets with optical code label are generated in EN1 and switched two times by OPS 1 and 2. Finally, switched packets are received by EN2. In the experiment, we use packets with label *A* and *B*. EN1 consists of 10GHz mode locked laser diode (MLLD), LiNbO$_3$ intensity modulators (IM),

Planer lightwave circuit (PLC) based encoders to generate 8-chip, 200Gchip/s optical bipolar label, optical delay, and OTDM multiplexer (MUX). Second IM generates 10Gbit/s, 4000bits random packet data. This 10Gbit/s data converted to 40Gbit/s, 16000bits data by an OTDM-MUX. Generated optical label and 40Gbit/s packet data are combined to form an optical packet. OPS 1 and 2 consist of label processor and optical switch, which correspond upper and lower arm of Fig. 6(b), respectively. Label processor consists of PLC decoders as optical correlator, photo detectors (PD), gate signal generator (GSG), BPF, and optical delay. Table lookup of the packet label is performed in a parallel manner without O/E and E/O conversions [2]. Label processor control optical switches via GSG to rout packets to their designated output port. A 1x 2 LiNbO$_3$ optical switch is used in the experiment. Finally, switched packets input EN 3. In EN3, 40Gbit/s data is de-multiplexed to 10Gbit/s data by using local MLLD and high nonlinear fiber (HNL Fiber) and received by the optical packet receiver. Optical amplifiers (EDFA) were employed to compensate insertion losses of the components in combination with 5nm bandwidth optical band-pass filters (BPF) to attenuate the amplified spontaneous emission.

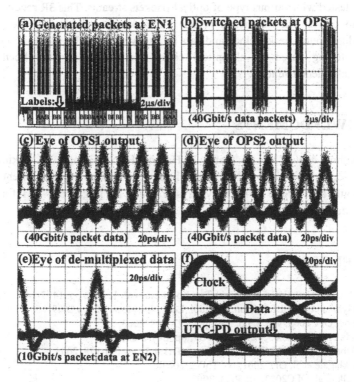

Figure 7. Measured waveforms of experimental demonstration II

Figure 7 represents measured waveforms. Figures 7(a) to (e) are generated 40Gbit/s packets with two different labels at EN1, switched 40Gbit/s packets at OPS 1, eye diagram of OPS 1 output, eye of switched 40Gbit/s packets at OPS 2, and eye of de-multiplexed 10Gbit/s packet data at EN2, respectively. Figure 7(f) shows UTC-PD output, recovered clock, and recovered data. These results show good performance of optical code label processing based optical packet switching and our proposed OPS receiver.

5.   CONCLUSIONS

A novel preamble free optical packet receiver has been proposed. 10Gbit/s optical packets streams with preamble-less payload data from 1000 to 4000 bits-long and arbitrary intervals from nanosecond to microseconds have been generated. The packets stream has been asynchronously received by optical packet 3R receiver. Instantaneous clock and payload data recovery in less than a bit period (<100ps) have been experimentally demonstrated. Clock and payload data recovery have been tested with various type of optical packets streams. This 3R receiver can be used to optical burst switched (OBS) networks as well as OPS networks. Preamble-less optical packet clock and data recovery with two hop 40Gbit/s optical packet switching based on all-optical code label processing has been also experimentally demonstrated.

ACKNOWLEDGMENTS

The authors would like to thank T. Hanyu, H. Sumimoto, T. Makino, Y. Tomiyama, and Y. Awaji of NICT for their collaboration in experiment. The authors would also like to thank Y. Matsyshima and M. Nagao of NICT for their encouragement.

REFERENCES

[1]   D.J. Blumenthal et al, IEEE Photon. Technol. Lett., vol. 11, no., pp. 1497-1499, 1999.
[2]   K. Kitayama et al, IEEE Photon. Technol. Lett., vol.11, no.12, pp.1689-1691, 1999.
[3]   S.B. Yoo, OFC2003, vol. 2, no. FS5, pp. 797-798, 2003.
[4]   N. Wada et al, OFC2003, vol. 2, no. FS7, pp. 801-802, 2003.
[5]   N. Calabretta et al, ECOC2003, vol. 2, no. Tu1.4.5, pp. 178-179, 2003.
[6]   K. Shrikhande et al, OFC2001, no. ThG2, 2001.
[7]   M. Duelk et al, OFC2003, no. PD8, 2003.
[8]   S. Kimura et al, ECOC2003, vol. 4, no. Th3.3.3, pp. 1042-1043, 2003.

COST EFFICIENT UPGRADING OF OPS NODES

J. Cheyns[1], C. Develder[2], D. Colle[1], E. Van Breusegem[1], P. Demeester[1]

[1] *Dep. of Information Technology, Ghent University- IMEC, St.-Pietersnieuwstraat 41,9000 Gent, Belgium. Corresponding author:jan.cheyns@intec.UGent.be*
[2] *OPNET Technologies ,F. Rooseveltlaan 348W, 9000 Gent, Belgium.*

Abstract: We come back on a technique to build modular switch nodes. This approach allows for a more cost effective expansion of OPS nodes. We give two example designs, showing that the method is useful only for Broadcast & Select OPS nodes when taking price decrease in function of time into account.

Key words: Optical Packet Switching, upgrade

1.   INTRODUCTION

At the end of the 20th century (D)WDM and optical amplifiers unlocked vast bandwidth, making static optical networks the carrier of growing bandwidth demands. Next step is an Automatic Switched Optical Network (ASON), dynamically allocating capacity between different nodes by wavelength paths forming logical links [1]. Still, ASONs are unable to cope with the bursty traffic of the current Internet, due to their coarse granularity, leading to inefficient bandwidth usage. Therefore, Optical Packet Switching (OPS) [2] and Optical Burst Switching (OBS) [3], where data is switched per packet/burst, receive much interest. They allow finer granularity and statistical multiplexing gains, leading to efficient bandwidth usage. Both technologies need fast optical switching matrices, of which 2 major families are Semiconductor Optical Amplifier (SOA) based Broadcast & Select (B&S) architectures and Arrayed Waveguide Grating (AWG) based designs [4]. We discussed Clos architectures for OPS nodes in [5], however, upgradeability of OPS nodes is also of key importance. Starting from SKOL, a more upgradeable modification of the Clos design [6], we evaluate applicability of the existing modular designs for the 2 OPS families.

2. REVISITING SKOL

2.1 The Clos architecture

Figure 1 shows a 3-stage Clos architecture for a NxN switch (in this paper we only consider symmetrical switches). The N in- and output ports are grouped per n, and for each such group there is a switching fabric in both the 1st and 3rd stage. The 2nd stage comprises k switches of dimension N/nxN/n, connected to each of the 1st and 3rd stage switches (size nxk, resp. kxn). The required number of 2nd stage switches (k) depends on the blocking requirements [7], e.g. a strictly non-blocking switch needs k≥2n−1 [8].

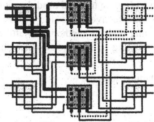

Figure 1. An example Clos architecture with N=6, n=2 and k=3.

The center stage switch size (N/nxN/n), is completely determined by the grouping factor n and the size N of the overall switching structure. Looking at the Clos structure as a modular switch, it has the disadvantage that all of the second stage fabrics have to be installed from the beginning: initial cost savings can only be made by suppressing blocks in the first and third stage. This seems to be not the most cost efficient upgrade facility.

2.2 From Clos to SKOL

To allow a more effective upgrade strategy, McDonald proposed to distribute the central fabrics' functionality over in- and output blocks [6]. For a crossbar switch, this means that an N/nxN/n switch in the second stage is split into two halves, indicated by the bold, resp. dotted lines in *Figure 1*.

This approach boils down to the following: an N/nxN/n switch in the central stage of *Figure 1* is split into 2 halves (each of size N/nxN/n), and each of those is split into N/n parts (of size 1xN/n or N/nx1). Of these halves, k are taken together (corresponding to the k original second stage fabrics), and integrated with a 1st stage switch to form a so-called SKOL building block of size N/n x (k.N/n). The other halves are joined with the 3rd stage switches to form output so-called SKOL blocks of size (k.N/n) x N/n. Note that these blocks can be exactly the same, given that the switch is reciprocal (i.e.the switch can be used in both directions, the in-and output functionality is interchangeable). This is a considerable advantage over the original Clos structure as of *Figure 1*, where at least 2 types of building blocks are required (cf. central stage fabrics have different sizes).

Yet, from an upgrade perspective, still not all of the provided building block's ports are used from the beginning. An expansion scenario is outlined in Figure 2. In Figure 2a, only 1 input and 1 output SKOL block are used, with a considerable amount of unused ports. But they are necessary to allow the switch to be expanded until its final size of NxN, as shown in Figure 2b and c. Note that the maximum possible dimension N is still limited from the beginning, as with a the original Clos switch.

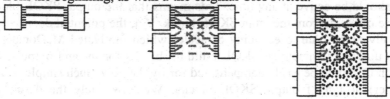

Figure 2. Upgrading using a SKOL architecture. First the full connections are present, then the grey dashed ones are added, and finally the dotted connections fully build the final node.

2.3   On the output block

As explained in 2.1, a SKOL output block has k switches of size k.N/nx1, integrated with the kxn 3rd stage switches of the original Clos design. To implement a strictly non-blocking switch, the condition $k \geq 2n-1$ needs to be fulfilled. The proof is analogue to that of the requirement of the number of 2nd stage switches in the original Clos design. Suppose that in *Figure 3* a connection from input port x in an input SKOL block A to output port y of an output SKOL block B needs to be made, while (i) all of the other n−1 inputs of block A are in already in use, and (ii) the n−1 remaining outputs of B are in use. (i) implies that n−1 of the output ports of the nxk sub-block in A are in use (and thus n−1 of the sub-blocks a). (ii) means that also n−1 inputs of the kxn sub-block in B, thus n−1 sub-blocks b. By design, each sub-block a is connected to only a single sub-block. The worst case occurs when the n−1 active blocks a in A are the ones connected to the n−1 non-active blocks b in B. To be able to connect x to y, it is required that $k \geq$ (# active a blocks) + (# active b blocks) + 1 = 2n−1.

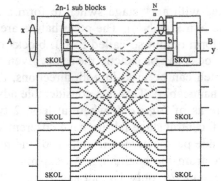

Figure 3. A SKOL node in a schematical representation

It is important to note that the sub-blocks b can be replaced by passive combiners, since they have only a single output and thus only 1 of their inputs should be active at any time. This means that the output SKOL block will be different from the input SKOL block. Yet, the potential decrease in cost advantage (due to economy of scale, which motivated McDonald to come up with symmetrical SKOL building blocks for in- and outputs, cf. section 0) should be easily compensated for by the now much simpler (thus cheaper) design of output SKOL blocks. We now apply the described SKOL design to 2 OPS node architectures, presented in more detail in [5].

3. SKOL AND AWG BASED OPS NODE

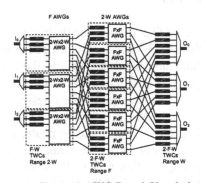

Figure 4. AWG Based Clos design

An AWG-based Clos building block (dashed box on *Figure 4*) has an AWG where input ports have Tuneable Wavelength Converters (TWC) installed, whose output wavelength determines the output port. The design isn't reciprocal: inputs and outputs can't be interchanged, so input and output SKOL blocks differ.

In *Figure 4* we chose the number of input blocks N/n=F (the number of fibres) and thus the number of ports per block n=W (the number of wavelengths on a fibre). This made the AWG at the 3rd stage unnecessary, since that AWG only switched between wavelengths on the same fibre. However, the design is dedicated (fixed) to a certain value of W. Note that we chose to have 2W inner stages instead of the minimum required 2W-1.

A possible implementation of a SKOL input block for this design is shown in *Figure 5*a. When all input blocks (F of them) would be designed like this, we need 2FW 1xF AWGs and 2FW TWCs of range F for the 'distributed' inner stage. The original Clos (*Figure 4*) has the same TWC count, but needs only 2W AWGs of size FxF. Component count for the first stage (of the SKOL block) does not change compared to *Figure 4*.

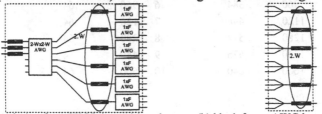

Figure 5. A design for a SKOL input (a) and output (b) block for an AWG based node

*Figure 5*b shows an output SKOL block in AWG based technology, including the considerations in section 2.3 on the passive first block in the output block, so that we can realise it by a passive combiner. We have 2n (=2W) outputs here, and not n as expected. This has the same cause as in the Clos design of *Figure 4*: we remove the AWG from the 3rd stage, as it would only switch packets from one wavelength to another within the same fibre. This way, only having the converters suffices in order to have the (at most) W packets on W different wavelengths. Using these building blocks we create an upgradeable node with respect to adding a fibre. Per fibre we need one of the above blocks, where a maximum number of fibres is set in advance to F_{max}. The switch can then grow (cfr. *Figure 2*) until F_{max} is reached. *Table 1* shows the evolution of a node for $F_{max}=10$, W=32.

Unfortunately, the SKOL method is not beneficial in the AWG-based case, as we look at the number of central (1xF in the SKOL case, FxF case without) AWGs. In the Clos case we immediately install the full middle stage, in this case all 64 10x10 central AWGs. Then, as fibre count increases, outer blocks are added as needed. Also the TWCs of the centre stage can be gradually added. So the only difference is the AWG count and their nature, i.e. 1xF vs. FxF. Roughly speaking an FxF AWG will be twice the cost of an 1xF AWG [10], so the SKOL approach is not beneficial. The reason is that the switching elements in the AWG-based technology are governed by linear cost and not quadratic as with crosspoint switches. The Clos design of *Figure 4* can thus be quite good for upgrading with extra fibres, although some provisions must be made in the beginning, which also limit possible growth. A wavelength upgrade would be a lot more complex.

Table 1. Component count evolution for an AWG based node with $F_{max}=10$ and W=32.

	With SKOL			Without SKOL		
F	*TWC*	*1xF AWG*	*2Wx2W AWG*	*TWC*	*FxF AWG*	*2Wx2W AWG*
2	320	128	2	320	64	2
3	480	192	3	480	64	3
4	640	256	4	640	64	4
5	800	320	5	800	64	5
6	960	384	6	960	64	6
7	1120	448	7	1120	64	7
8	1280	512	8	1280	64	8
9	1440	576	9	1440	64	9
10	1600	640	10	1600	64	10

4.   SKOL AND SOA BASED B&S OPS NODE

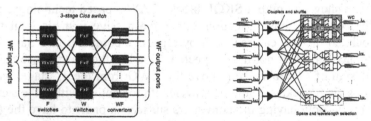

Figure 6.a) B&S SOA based Clos node design; b) Clos building block

In the B&S node architecture of *Figure 6* [9], all inputs are broadcast to all possible outputs, where a choice is made using SOA based space and wavelength selection. [9] shows that an optimised (in number of SOAs) building block with N ports has $2N^{3/2}$ SOAs, in the case of a 2-stage architecture of *Figure 6*a. We consider a slotted approach and thus W FxF stages suffices, as we can suffice with a rearrangeable node.

Again, we can use the SKOL mechanism to distribute the middle stage into the 1st stage, meaning we only have SOAs at the input stage. This means each input SKOL block would carry $2W^{3/2}+F_{max}W$ SOAs. In *Figure* 7a, the full lines show the evolution of cumulative cost as a SKOL SOA-based switch would grow, for W=32 and $F_{max}=10$. We compare this with a

Clos solution, where we immediately overbuild the central stage with $F_{max} \times F_{max}$ nodes. The initial number of SOAs of the node is lower, but as the node grows, the number of SOAs rises and becomes higher than the eventual total cost for the Clos design. Analytically:

$$\frac{\text{SKOL(final)}}{\text{Clos(final)}} = \frac{2\sqrt{W} + F_{max}}{2(\sqrt{F_{max}} + \sqrt{W})},$$

The larger W, the closer this value is to 1. However increasing F_{max} this value grows larger. Again the origin of this discrepancy with [6] is due to the fact that a quadratic law (number of crosspoints) governs the node cost.

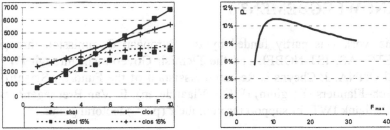

Figure 7. a) SOA count evolution for F_{max}=10 and W=32;b)Needed cost decrease

Still SKOL may be useful for this node type if the cost of the building blocks (number of SOAs) shows a steep enough decrease in time. In the test case of a switch with F_{max}= 10, where every upgrade a fibre is added and the cost is 15% lower than the previous upgrade. In *Figure 7*a the dashed curves show the cumulative cost and *Table 2* formulates the cumulative cost for the final design, so as the node has reached F_{max}, p denotes the constant relative cost decrease at every upgrade (p=15% in *Figure 7*).

Table 2. Number of SOAs for both the Clos and SKOL final design

Clos	SKOL
$2.W.(F_{max} \cdot \sqrt{F_{max}} + \dfrac{\sqrt{W}.(1-(1-p)^{F_{max}})}{p})$	$\dfrac{W.(2.\sqrt{W} + F_{max}).(1-(1-p)^{F_{max}})}{p})$

The final cost of SKOL and Clos is equal if

$$\frac{1-(1-p)^{F_{max}}}{p} = 2.\sqrt{F_{max}}$$

The condition is independent of the number of wavelengths per fibre, W. The equation's result is shown in *Figure 7*b. We see an initial increase in the needed value of cost reduction, with a maximum of 10.8% at F_{max}=10. For higher F_{max} the necessary reduction drops slowly. More importantly the needed value is not extremely high, so quite realistic, certainly for components like SOA's which still have a large margin to mature. A needed value of 10% reduction at every upgrade is a good rule of thumb.

CONCLUSION

We extended the SKOL mechanism to OPS switching nodes. A crucial difference is the non–reciprocal character of an AWG based switching node. For AWG based OPS nodes, the SKOL method doesn't result in any improvement. SOA based B&S architectures can reach a cost benefit if the price of building blocks drops sufficiently over time: 10% between every upgrade is a good rule of thumb.

ACKNOWLEGDEMENTS

This work was partly funded by the European Commission through IST- STOLAS, IST-NOBEL and the Flemish Government through IWT-GBOU ONNA. J. Cheyns is Research Assistant of the Fund for Scientific Research–Flanders (Belgium) (FWO–Vlaanderen). E. Van Breusegem and D. Colle thank IWT for support through their (post-)doctoral grants.

REFERENCES

[1] S. De Maesschalck, et al., "Intelligent Optical Networking for Multilayer Survivability", IEEE Comm. Mag., Vol. 40, 1, pp 42-49, Jan. 2002.

[2] P. Gambini, et al., "Transparent optical packet switching: network architecture and demonstrators in the KEOPS project," IEEE J. on Sel. Areas in Comm., Vol. 16, 7, pp. 1245–1259, Sept. 1998.

[3] C. Qiao, et al., "Optical burst switching (OBS) - A new paradigm for an optical internet", J. High Speed Netw., Vol. 8, 1, pp. 69–84, Jan. 1999.

[4] C. Develder, et al., "Architectures for optical packet and burst switches ", ECOC 2003, Rimini (Italy), Sept. 2003.

[5] J. Cheyns, et al., "Clos lives on in optical packet switching", IEEE Comm. Mag., Vol. 42, 2, pp. 114-121, Feb. 2004.

[6] R. I. MacDonald, "Large Modular Expandable Switching Matrices", Phot. Techn. Lett., Vol. 11, 6, pp. 668-670, 1999.

[7] A. Jajszczyk, "50 Years of Clos Networks a Survey of Research Issues", HPSR, www.tlc-networks.polito.it/HPSR2003/talks/Clos_HPSR_AJ.pdf, June 2003.

[8] C. Clos, "A study of non-blocking switching networks", Bell System Technical Journal, Vol. 32, pp. 406-424, 1953.

[9] C. Develder, et al., "Multistage architectures for optical packet switching using SOA-based broadcast-and-select switches", OFC, Atlanta (GA, USA), Mar. 2003.

[10] J. Cheyns et al. , "Evaluating Cost functions for OPS node architectures", ONDM, Ghent (Belgium), pp. 37-56, February 2004.

A SCHEDULING ALGORITHM FOR REDUCING UNUSED TIMESLOTS BY CONSIDERING HEAD GAP AND TAIL GAP IN TIME SLICED OPTICAL BURST SWITCHED NETWORKS

Takanori Ito, Daisuke Ishii, Kohei Okazaki, Naoaki Yamanaka, and Iwao Sasase
Dept. of Information and Computer Science, Keio University
Yokohama, Kanagawa, Japan 223–8522

Abstract: We propose a scheduling algorithm for reducing unused timeslots by considering head gap and tail gap newly generated by assigning a data burst in order to improve the burst loss probability and the throughput performances in Time Sliced Optical Burst Switched (TSOBS) networks. The proposed scheduling algorithm selects the timeslot in which either head gap or tail gap newly generated becomes the minimum. We show that the proposed scheduling algorithm can improve the burst loss probability and the throughput performances as compared with the conventional one.

1.   INTRODUCTION

Optical Burst Switching (OBS) has been proposed as a new scheme to realize IP over WDM networks [1]- [4]. In OBS networks, at a core router, when two or more data bursts arrive at the same output port simultaneously, the burst contention occurs. When the burst contention occurs, either burst is discarded. In order to avoid the burst loss due to the burst contention, several scheduling algorithms using a wavelength converter and an optical buffer implemented through a bundle of fiber delay lines (FDLs) have been proposed [1], [2]. The wavelength converters using O/E and E/O conversions increase the electric processing at core routers. Although the all-optical wavelength converters without O/E and E/O conversions are desirable and have been developed over a number of years, all-optical wavelength converters are not still in practical use because of the issues of the performance and cost implications. On the other hand, although FDLs can be implemented, using a large number of FDLs causes an increase of optical hardware volume and noise level due to the transmission of optical signals in FDLs. In OBS networks, since a burst length is long, the buffer size of FDLs needed in core routers becomes large. Therefore, there is the problem that optical hardware volume and noise level increase in OBS networks.

Time Sliced Optical Burst Switching (TSOBS) [3] has been proposed as an OBS system which does not use wavelength converters and can reduce the buffer size of FDLs needed in core routers. In TSOBS networks, the wavelength used for transmission of data bursts is divided into frames, and each frame is subdivided into several timeslots. In edge routers, a generated data burst is divided into the several data of one timeslot length and the divided data burst is transmitted in the same timeslot every frame. TSOBS can avoid the burst contention by using short FDLs in core routers since the data transmitted in the network are short. Therefore, TSOBS can reduce the buffer size of FDLs needed in core routers as compared with OBS. In TSOBS networks, when a certain timeslot in each frame is focused on, the empty timeslots may exist between two different data bursts already assigned to the timeslot. The group of the empty timeslots is called gap in this paper. The value of gap is defined as the number of the empty timeslots. In a core router, the data burst can be transmitted by using the gap if there is the gap which is longer than the length of the data burst. And, we call the timeslot, in which the data burst can be transmitted, the available timeslot. However, in TSOBS networks, a data burst is assigned to the available timeslot which is found first without considering gaps newly generated by assigning the data burst, and hence, a large number of small gaps where only a short burst can be assigned are generated. Therefore, the conventional scheduling algorithm increases the possibility that a long burst will be discarded. As a result, it causes the degradation of the burst loss probability and the throughput performances.

In this paper, in order to improve the burst loss probability and the throughput performances, we propose a scheduling algorithm for reducing unused timeslots by considering head gap and tail gap newly generated by assigning a data burst. The proposed scheduling algorithm selects the timeslot in which either head gap or tail gap newly generated becomes the minimum. The proposed scheduling algorithm can increase the possibility that a long burst will be assigned, since longer gaps can be generated in other timeslots by assigning the data burst to the timeslot in which the gap newly generated becomes the minimum. Therefore, the proposed scheduling algorithm can improve the burst loss probability and reduce unused timeslots on data wavelengths. We compare the performances of the proposed scheduling algorithm with that of the conventional one with respect to the burst loss probability and the throughput by computer simulations. As a result, we show that the proposed scheduling algorithm can improve the burst loss probability and the throughput performances as compared with the conventional one.

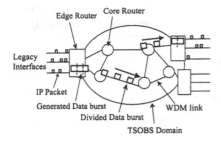

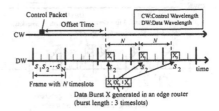

Figure 1. A TSOBS network.

Figure 2. An example of burst transmission at an edge router.

2. TSOBS NETWORKS

Fig. 1 shows a TSOBS network. Edge routers assemble several IP packets with the same destination into a burst. And edge routers divide a generated data burst into the several data of one timeslot length and transmit the divided data burst every fixed interval. Core routers carry out the scheduling and switching of data bursts. In core routers, the divided data burst is transmitted to the next router with the divided form held. Fig. 2 shows an example of burst transmission at an edge router. In TSOBS networks, the wavelength used for transmission of data bursts is divided into frames, and each frame is subdivided into N timeslots. Edge routers first transmit a control packet to reserve an output timeslot in a core router before a data burst is transmitted. Subsequently, edge routers transmit a data burst on a separate wavelength after some offset time. The offset time should also allow core routers at each hop along the path to have enough time to process the control packet before its corresponding data burst arrives. In this paper, the offset time would be proportional to the number of hops which the burst will traverse. Therefore, the offset time differs in each burst. Each control packet includes the address information, the wavelength and timeslot in which the data burst is transmitted, the offset time which identifies the frame in which the first data of the divided data burst is transmitted, and the burst length which identifies the number of timeslots used to transmit the data burst. As shown in Fig. 2, in an edge router, a generated data burst is divided into the several data of one timeslot length and is transmitted in the same timeslot every frame. In core routers, the switching of data bursts is done entirely in the optical domain. Core routers dynamically switch a divided data burst from an incoming timeslot to an output timeslot on the appropriate outgoing link. In core routers, when the burst contention occurs, the output timeslot of the data burst is changed by using FDLs in order to avoid the burst contention. The unit delay time of FDLs is one timeslot and the maximum delay which can be given to the data bursts by FDLs is N timeslots.

In the conventional scheduling algorithm, core routers search the statuses

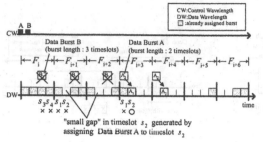

Figure 3. An example of the conventional scheduling algorithm ($N = 4$).

of the timeslots in sequence from the timeslot in which a data burst will arrive, and the data burst is assigned to the available timeslot which is found first [3]. Core routers can search the statuses of N timeslots at the maximum in sequence from the timeslot in which a data burst will arrive. If all the timeslots are unavailable, the data burst is discarded. Fig. 3 shows an example of the conventional scheduling algorithm ($N = 4$). The control packets of data burst A and B arrive at a core router in order, as shown in Fig. 3. First, the core router analyzes the control packet of data burst A, and obtains the information that the burst length of data burst A is 2 timeslots and data burst A arrives in timeslot s_1 of frame F_{i+3}. Next, the core router searches the statuses of the timeslots in sequence from timeslot s_1 in which data burst A will arrive. Since other data burst is already assigned to timeslot s_1 in F_{i+3}, the core router searches the status of timeslot s_2. Timeslot s_2 is empty in two consecutive frames of F_{i+3} and F_{i+4}. Therefore, data burst A is assigned to timeslot s_2. Then, the core router analyzes the control packet of data burst B, and obtains the information that the burst length of data burst B is 3 timeslots and data burst B arrives in timeslot s_3 of frame F_i. Timeslot s_3, s_4, s_1, and s_2 are not empty in three consecutive frames. Therefore, data burst B is discarded. In the conventional scheduling algorithm, a data burst is assigned to the available timeslot which is found first without considering gaps newly generated by assigning the data burst, and hence, a large number of small gaps where only a short burst can be assigned are generated. Therefore, the conventional scheduling algorithm increases the possibility that a long burst will be discarded. As a result, it causes the degradation of the burst loss probability and the throughput performances.

3. PROPOSED SCHEDULING ALGORITHM

In order to improve the burst loss probability and the throughput performances, we propose a scheduling algorithm for reducing unused timeslots by considering head gap and tail gap newly generated by assigning a data burst. The proposed scheduling algorithm selects the timeslot in which either head gap or tail gap newly generated becomes the minimum. The proposed scheduling algorithm can increase the possibility that a long burst will be assigned, since longer gaps can be generated in other timeslots by assigning the data

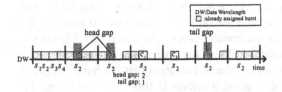

Figure 4. An example of head gap and tail gap ($N = 4$)

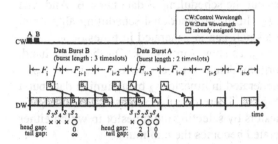

Figure 5. An example of the proposed scheduling algorithm ($N = 4$).

burst to the timeslot in which the gap newly generated becomes the minimum. Therefore, the proposed scheduling algorithm can improve the burst loss probability and reduce unused timeslots on data wavelengths. Fig. 4 shows an example of head gap and tail gap ($N = 4$). When data burst C is assigned to timeslot s_2, the head gap is 2 since timeslot s_2 is empty in two consecutive frames which exist ahead of data burst C, and the tail gap is 1 since timeslot s_2 is empty in one frame which exists behind data burst C.

Here, we explain the flow of the proposed scheduling algorithm. First, core routers search the statuses of N timeslots in sequence from the timeslot in which a data burst will arrive. Simultaneously, if the searched timeslot is available, core routers compute head gap and tail gap newly generated by assigning the data burst to the timeslot. After core routers search the statuses of all the timeslots, core routers compare head gap and tail gap in all the available timeslots and select the timeslot in which either head gap or tail gap becomes the minimum. When there are two or more timeslots in which a gap becomes the minimum, the proposed scheduling algorithm selects the timeslot in which tail gap becomes the minimum in order to use gaps as early as possible. Furthermore, when there are two or more candidate timeslots, the proposed scheduling algorithm selects the timeslot in which the delay given to the data burst by FDLs is the minimum in order not to increase the delay of data bursts. Fig. 5 shows an example of the proposed scheduling algorithm ($N = 4$). In Fig. 5, the control packets of data burst A and B arrive at a core router in order, like Fig. 3. First, the core router analyzes the control packet of data burst A, and obtains the information that the burst length of data burst A is 2 timeslots and data burst A arrives in timeslot s_1 of frame F_{i+3}. Next, the core router searches the statuses of all the timeslots in sequence from timeslot s_1 in which data burst A will arrive. Simultaneously, the core router computes head gap and tail gap in timeslot s_2, s_3, and s_4, since timeslot s_2, s_3, and s_4 are the available timeslots. In Fig. 5, when data burst A is assigned to timeslot s_4, both head gap and tail gap newly generated become the minimum. Therefore, data burst A is assigned to timeslot s_4. Then, the core router analyzes the control packet of data burst B, and obtains the information that the burst length of data burst B is 3 timeslots and data burst B arrives in timeslot s_3 of frame F_i. Similarly, the core router carries out the scheduling of data burst B. And data burst B is assigned to timeslot s_2. In the conventional scheduling algorithm, data burst B is discarded since a small gap is generated in timeslot s_2 by assigning data burst A to timeslot s_2, as shown in Fig. 3. On the other hand, in the proposed scheduling algorithm, data burst B can be assigned to timeslot s_2 since the longer gap can be generated in timeslot s_2 by assigning data burst A to timeslot s_4, as shown in Fig. 5. Therefore, the proposed scheduling algorithm can reduce unused timeslots by selecting the timeslot in which either head gap or tail gap newly generated becomes the minimum.

4.   PERFORMANCE EVALUATION

In this section, we compare the burst loss probability and the throughput of the proposed scheduling algorithm with those of the conventional one by computer simulations. The throughput is defined as the ratio of the amount of data transmitted per unit time to the transmission rate. We assume that the switch size of a core router is 16, the transmission rate on each wavelength is 40Gbps, frames are subdivided into N timeslots and the length of timeslot is $1\mu s$. The input has uniform traffic with rate ρ. And, all bursts arriving at each input port of a core router have variable length bursts. The burst length has the uniform distribution of 1 to 5 timeslots length. The number of hops to a destination edge router has the uniform distribution of 1 to 16 hops. The processing time of a control packet in each core router is $10\mu s$ [4].

Fig. 6 shows the burst loss probability versus input load ρ. From Fig. 6, we show that the proposed scheduling algorithm can reduce the burst loss probability as compared with the conventional one. The reason is as follows. In the conventional scheduling algorithm, a data burst is assigned to the available timeslot which is found first without considering gaps newly generated by assigning the data burst, and hence, a large number of small gaps where only a short burst can be assigned are generated. Therefore, the conventional scheduling algorithm increases the possibility that a long burst will be discarded and the number of data bursts which can be assigned decreases. On the other hand, in the proposed scheduling algorithm, a data burst is assigned to the timeslot in which either head gap or tail gap newly generated by assigning the data burst becomes the minimum. The proposed scheduling algorithm can increase the possibility that a long burst will be assigned since longer gaps can be generated

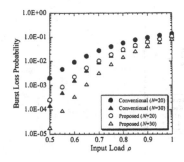

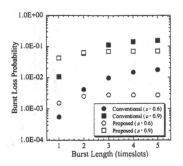

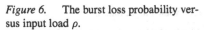

Figure 6. The burst loss probability versus input load ρ.

Figure 7. The burst loss probability by the burst length $(N = 20)$.

in other timeslots by assigning the data burst to the timeslot in which the gap newly generated becomes the minimum. Therefore, the number of data bursts which can be assigned increases. Then, in order to prove the reason mentioned above, we show the burst loss probability by the burst length in Fig. 7. From Fig. 7, we find that the conventional scheduling algorithm degrades the loss probability of a long burst as compared with that of a short burst. This is because, the conventional scheduling algorithm increases the possibility that a short burst will be assigned as compared with the possibility that a long burst will be assigned since a large number of small gaps where only a short burst can be assigned are generated. On the other hand, we find that the proposed scheduling algorithm reduces the loss probability of a long burst. This is because, the proposed scheduling algorithm can assign a short burst and a long burst with the almost same possibility since the number of small gaps where only a short burst can be assigned decreases and longer gaps can be generated. From Figs. 6 and 7, we show that the proposed scheduling algorithm can increase the number of data bursts which can be assigned since the proposed scheduling algorithm can reduce the loss probability of a long burst. And as shown in Fig. 7, in the conventional scheduling algorithm, the burst loss probability is affected by the burst length and a long burst is more likely to be discarded than a short burst. However, in the proposed scheduling algorithm, the burst loss probability does not depend on so much the burst length and the difference of the burst loss probabilities by the burst length can be small.

Fig. 8 shows the throughput performance versus input load ρ. From Fig. 8, we show that the proposed scheduling algorithm can improve the throughput performance as compared with the conventional one. The reason is as follows. In the conventional scheduling algorithm, since the scheduling is carried out without considering gaps newly generated, a large number of small gaps where only a short burst can be assigned are generated and the number of data bursts which can be assigned decreases. Therefore, the number of unused timeslots increases and the amount of data which can be transmitted per unit time de-

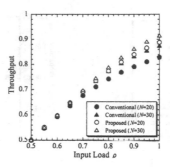

Figure 8. The throughput performance versus input load ρ.

creases. On the other hand, in the proposed scheduling algorithm, since the scheduling is carried out with considering gaps newly generated, longer gaps can be generated and the number of data bursts which can be assigned increases. Therefore, the number of unused timeslots can be reduced and the amount of data which can be transmitted per unit time increases.

5. CONCLUSION

We have proposed a scheduling algorithm for reducing unused timeslots by considering head gap and tail gap newly generated by assigning a data burst in order to improve the burst loss probability and the throughput performances in TSOBS networks. The proposed scheduling algorithm selects the timeslot in which either head gap or tail gap newly generated becomes the minimum. By computer simulations, we find that the proposed scheduling algorithm can improve the burst loss probability and the throughput performances as compared with the conventional one.

ACKNOWLEDGMENT

This work is partly supported by Keio University The 21st Century COE Program on "Optical and Electronic Device Technology for Access Network" and Japan Society for the Promotion of Science(JSPS).

REFERENCES

[1] C. Qiao and M. Yoo, "Optical burst switching(OBS)-a new paradigm for an optical internet," *J.High Speed Networks*, vol.8, no.1, pp.69-84, Jan. 1999.

[2] Y. Xiong, M. Vandehoute, and H. C. Cankaya, "Control architecture in optical burst-switched WDM networks," *IEEE J. Select. Areas Commun.*, vol.18, no.10, pp.1838-1851, Oct. 2000.

[3] J. Ramamirtham and J. Turner, "Time sliced optical burst switching," *Proc. INFOCOM 2003*, vol.3, pp.2030-2038, San Francisco, U.S.A., Apr. 2003.

[4] B. C. Kim, Y. Z. Cho, J. H. Lee, Y. S. Choi, and D. Montgomery, "Performance of optical burst switching techniques in multi-hop networks," *Proc. GLOBECOM 2002*, vol.3, pp.2772-2776, Taipei, Taiwan, Nov. 2002.

WONDER: OVERVIEW OF A PACKET-SWITCHED MAN ARCHITECTURE

A. Bianciotto and R. Gaudino
OptCom Group, PhotonLab
Dipartimento di Elettronica, Politecnico di Torino
C.so Duca degli Abruzzi 24, 10129, Torino, Italy
E-mail: alessandro.bianciotto@polito.it - Tel. +39.011.2276301 - Fax +39.011.2276299
E-mail: roberto.gaudino@polito.it - Tel. +39.011.5644172 - Fax +39.011.5644099

Abstract: This paper presents the architecture of WONDER, an advanced ring-based WDM optical packet network designed for high capacity metro environments. The network prototype is currently being developed at PhotonLab in Torino, Italy, by several Italian research groups, thanks to a financing by the Italian Ministry of University and Research (MIUR). It represents an evolution over a similar ring based prototype, named "RingO" which was previously realized in 2001-2002. The WONDER network architecture elaborates on the effectiveness of optics with respect to electronics, trying to identify an optimal mix of the two technologies. We present network architecture, physical topology and node structure of the WONDER prototype, as well as its MAC protocol. The main contribution of this article is the identification of an innovative optical network architecture, which is feasible and cost effective with technologies available today, and can be a valid alternative to more consolidated solutions in metro applications.

Key words: Optical Network Testbeds; Packet Switched Networks; Advanced Optical Network Architectures.

1. INTRODUCTION

Metropolitan area networks are one of the best arenas for an early penetration of advanced optical technologies. Indeed, their large traffic dynamism requires packet switching to efficiently use the available resources; their high capacity requirements justifies WDM use; and their limited geographical extensions lowers the impact of fiber transmission impairments such as fiber dispersion and nonlinearities. From a research view point, designing innovative architectures for metro networks often means finding cost-effective combinations of optical and electronic technologies, and new networking paradigms that better suit the constraints dictated by available photonic components and subsystems.

The OptCom and Network groups at Politecnico di Torino, Italy, have designed and prototyped network architectures for metro applications, taking an approach based upon optical packets, but limiting optical complexity to a minimum, and trying to use only commercially available components. This approach was already at the base of our previous "RingO" [2] network architecture, also developed in Turin at the end of 2002. The WONDER architecture was designed in order to find the best compromise between optical and electrical technologies. Similar solutions are currently under investigation in several other laboratories, see for instance [3]. To this end, the bulk of raw data transmission is kept in the optical domain, while network control functions are mostly implemented in the electronic domain.

In the following sections we introduce the rationale and design of the WONDER network prototype describing the physical topology, the node structure, the MAC protocol and the fault recovery mechanism, and the plans for the testbed currently under implementation in PhotonLab [1].

2.   WONDER ARCHITECTURE

The WONDER network topology, see Fig. 1, is based on two WDM fiber rings connecting N nodes. One of the rings, called the "transmission" ring, is devoted to the transmission of data, while the other, called the "reception" ring, is devoted to the reception of data. The two rings are physically interconnected by a proper fiber shortcut that can be implemented by any node in the network. The shortcut is realized by closing a loop-back fiber between the transmission and reception rings at the output of a node, which is conventionally called the "folding" node (node i in Fig. 1). The node located at the other side of the shortcut turns out to be the first on the transmission ring and the last on the reception ring. This node is conventionally called the "master" node since, by preceding all the other nodes, it is devoted to the transmission of a suitable synchronization signal to the whole network. At any time, the network must have one and only one pair of active master and "folding" nodes in order to work properly. The resulting topology can also be viewed as a folded bus on which each node has two connections in two distinct points.

The transmission of packets is time-slotted and synchronized on all wavelengths, and each packet has a fixed duration corresponding to one time slot. In our current implementation, each slot is 1 μs long, and carries a Gigabit Ethernet bit stream at 1.25 Gbit/s. Future evolutions at higher bit rates are already planned. The WONDER protocol works as follows: each node i is

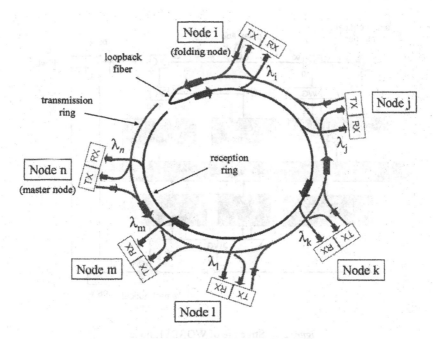

Figure 1.   The two fiber rings topology at the basis of the WONDER architecture

equipped with an optical receiver tuned on a certain wavelength λ_i on the ITU-T grid. When a node has a packet to send to another node tuned on λ_i, it must set its transmitter on λ_i and wait for the first available "empty" slot on that wavelength to transmit the packet (see following Sect. 2.2). The WONDER architecture eases some problems of the previous Ringo architecture [2], and in particular:

- Two or more nodes can share the same wavelength in reception, thus allowing a fine granularity in the allocation of the available bandwidth. Thus, a label must be associated to the optical packet to indicate its destination among the various nodes sharing the same wavelength. There are many ways to implement this kind of labeling, both in the electrical and in the optical domain. The pros and cons of each method have been evaluated, but in the end the simpler option has been chosen, consisting in the addition of an apposite field in the packet's header and a totally electronic recognition.

- Global synchronization is eased by the presence of the head-of-line master node, which sends to all other nodes a "timing" signal containing both bit and slot timing on a dedicated wavelength λ_t.

- The network is able to recover from any single fiber failure between two nodes by simply rearranging them to be "master" and "folding", i.e., by

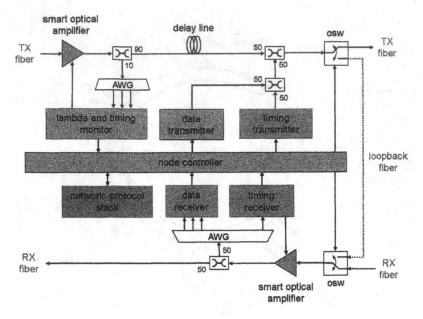

Figure 2.   Structure of WONDER nodes.

dynamically re-arranging the fiber shortcut shown in Fig.1. Each node can sense a failure by monitoring the timing signal. If this signal is not sensed on the preceding fiber on the transmission ring, an "up-hill" failure has occurred, and the node must negotiate to become the network master. If the timing signal is not sensed on the preceding fiber on the reception ring, a "down-hill" failure occurred on this fiber, and the node must become the "folding" node. In general, a node can be either in a "master", "folding" or generic "through" state: the node controller's logic can use the information about failures to switch among these three states.

2.1   Node structure

The general structure of a WONDER node is shown in Fig. 2. It is logically subdivided into a number of subsystems, each one implementing a subset of the node's functionalities. In the next subsections we'll describe the functions, interfaces and the proposed experimental setup for each subsystem.

2.1.1   Smart optical amplifiers.

In a metropolitan environment, the separation between two nodes can vary from a few kilometers to some tents kilometers, so the amount of amplification required is unknown, and has to be adapted case by case in order to have fixed optical power per channel at the output of the amplifiers, thus requiring an automatic "static" gain setting technique. Another kind of problem arises because of the bursty nature of the traffic in the network [5]. In general, when a packet enters an EDFA it sees a dynamic gain that depends on the busy/free state in the preceding slots [4]. The gain transients dynamics can seriously impact the BER performances of the optical receivers, so adequate gain transients suppression techniques are mandatory to guarantee the proper operation of the network. The optical amplifiers in the WONDER network are thus termed "smart" because they must provide both "static" nominal gain setting and fast gain transients suppression features. We envision the use of the the timing signal, which is continuous and always on, as a monitoring signal for gain settings. From an implementation point of view, two alternatives are available:

1. Use of EDFAs with a suitable control electronics which varies the pump power injected in the doped fiber as a function of the power sensed on the timing signal [6].
2. Use of Linear Optical Amplifiers (LOAs) [7], which are gain-clamped semiconductor optical amplifiers whose nominal gain can be set by varying the current injected into the active region.

The EDFAs have a very low noise figure and high output power, while the LOAs are noisier and less powerful, but more compact and potentially less expensive. In the WONDER scenario, LOAs are the preferred option for economic reasons, and their performance is currently under investigation.

2.1.2   Lambda and timing monitor.   In order to avoid collisions in the access to the transmission medium, each node must monitor the busy/free state on all wavelengths in each time slot. The monitoring extends also to the timing channel to extract slot timing, information about failures on the rings, and to provide optical power monitoring to the "smart" optical amplifiers. In our prototype, the "lambda and timing monitor" optically demultiplexes the WDM channels by an AWG, and then extract digital information on the busy/free state of channels and on eventual failures by means of an array of DC coupled photodiodes followed by threshold comparators.

2.1.3   Node controller.   The node controller is at the hearth of the node's operation and implements all the intelligence of the network. It is realized on an FPGA board exchanging data with all the other subsystems, taking care of the implementation of the MAC protocol, packets queueing, fault recovery mechanism, burst mode reception etc. We decided to base all our (electronic) architecture on an FPGA board due to the great increase in capabilities and reduction in costs that FPGA devices have shown in the last 2/3 years. The FPGA we will use for the testbed is able to handle several input/output datastream running at up to 3.2 Gbit/s, with advanced Clock and Data Recovery

(CDR) capabilities, very high processing power, and ease of integration with other external devices. The FPGA interfaces with all the upper layers of the WONDER network protocol stack to enable data exchange between nodes by means of standard protocols like FTP, HTTP and SMTP and suitable middle-ware software residing on a workstation.

As an important feature, our architecture requires an electrical data path bandwidth, on the transmitter, receiver and node controller that is equal to a single channel data rate only. In fact, the high-speed electrical interface of the transmitter and receiver need only to handle data traffic carried by a single wavelength, and not the aggregate bit rate of all wavelengths passing through the node. This is one of the advantages of our architectures with respect to current SONET/SDH circuit-switched solutions.

2.1.4   Data transmitter.   A WONDER transmitter must be able to tune on any of the available wavelengths on a slot by slot basis. Conventional trans-mitters are very different because they work with a continuous flow of data and on a fixed wavelength, so a radically different design is required. We envisioned two options:

1. Use of an array of fixed wavelength DFB lasers on the ITU-T grid, on/off switched by a digital signal coming from the node controller on a slot by slot basis. The outputs of the lasers are coupled together, then externally modulated by an external modulator, which is driven by the data bits.

2. Use of a fast tunable [8] laser which can be tuned on any desired C-band wavelength in a few tents of nanosecond by injecting a suitable current, followed by an external modulator.

In our prototype, the first option was chosen because of its component simplicity. The second option is today less feasible, but should be preferred as soon as fast tunable lasers become reliable and commercially available [8].

2.1.5   Timing transmitter.   When a node becomes the network "mas-ter", it has to transmit a timing signal containing both bit and slot time references. These signals must be combined on a single wavelength to minimize costs, so we decided to try an approach where a low extinction ratio square wave with period equal to one slot (1 μs in our implementation) is generated by directly modulating the laser with a small-signal current. The bit timing, set to 1.25 Gb/s, is then superimposed by means of an external modulator.

2.1.6   Timing receiver.   The timing receiver must extract the bit clock frequency from the timing signal, as well as providing a digital signal indicat-ing failures on the reception fiber. The first function is obtained by passing the electrical timing signal (recovered by a wide-band photodiode) into an high-pass filter with a cutoff frequency of approx. 1 GHz. The resulting signal is then directly fed to the node controller FPGA CDR circuitry, where a clean bit clock is made available. A failure signal can be issued whenever the power on the received timing signal falls below a predefined threshold.

2.1.7   Data receiver.   A WONDER data receiver must work in burst mode on a packet by packet basis, so it is quite different from conventional continuous mode receivers. The main issues in the realization of an high speed burst-mode receiver are the following:

1. The bursty nature of the received signal complicates the recovery of the phase of the clock signal needed to sample data bits at the optimal instant. The clock phase recovery must take at most some tents of nanoseconds otherwise a significant part of the packet's duration would be wasted for synchronization. The CDRs that are present on-board on the selected FPGA can automatically perform very fast phase recovery whenever the clock nominal frequency is known. This is the main reason why we send a "master" bit clock signal through the network on the timing signal.

2. Conventional high speed optical receivers are usually AC coupled. Their cutoff frequency is typically set quite low (about 30 kHz) to avoid baseline wander. Anyway, in a burst mode receiver, the cutoff frequency must be set to higher values, to have a very fast AC transient at the beginning of a packet. This in turns requires a suitable line coding (we use Ethernet 8B/10B coding).

3. A suitable "training" sequence must precede the packet useful bits (header and payload) to make sure the CDR can perform phase recovery. This is achieved by using a proper sequence, like a Barker sequence, at the beginning of the bit stream. The selected FPGA has built-in functions to implement these synchronization features.

2.2   MAC protocol

Our architecture requires a suitable MAC protocol to allocate time slots to transmitters. From the MAC protocol design perspective, WONDER is a multi-channel network, in which packet collisions must be avoided, and some level of fairness in resource sharing must be guaranteed together with acceptable levels of network throughput. A collision may arise when a node inserts a packet on a time slot and wavelength which have already been used. This is avoided by giving priority to upstream nodes, i.e., to in-transit traffic, via the λ-monitoring capability. Fairness is obtained by a Virtual Output Queuing (VOQ) structure. While standard single-channel protocols use a single FIFO (First In First Out) electrical queue, in multi-channel scenarios FIFO queuing performs poorly due to the Head-Of-Line (HOL) problem, which has been carefully studied in the literature, and can be solved using one of the VOQ [9] structures. The basic VOQ idea, applicable also to the WONDER architecture, consists in storing packets waiting for ring access in separated queues, each corresponding to a different destination, and to appropriately select the queue that gains access to the channels for each time slot.

3. CONCLUSION

Our work was motivated by the trust that optical packet transmission, though not yet standardized and commercially available, may become in the medium term a promising alternative to the current approach of building WDM networks with (at most) some degree of circuit-switching reconfigurability, but where packet switching is still completely handled at the electronic level. At the same time, we do not believe that all packet switching functions can be *completely* moved from the electrical to the photonic domain in a reliable way without fundamental improvements in optical components technology. A good compromise between the two domains (optical and electrical) is the major goal of the WONDER project presented in this paper.

REFERENCES

[1] http:\\www.photonlab.org
[2] A. Carena, V. Ferrrero, R. Gaudino, V. De Feo, F. Neri, and P. Poggiolini, "RingO: a Demonstrator of WDM Optical Packet Network on a Ring Topology", *IFIP Optical Network Design and Modeling Conference ONDM 2002*, Turin, Italy, Feb. 2002.
[3] K. V. Shrikhande, I. M. White, D. Wonglumsom, S. M. Gemelos, M. S. Rogge, Y. Fukashiro, M. Avenarius, and L.G. Kazovsky, "HORNET: a packet-over-WDM multiple access metropolitan area ring network", *IEEE Journal on Selected Areas in Communications*, Vol. 18, No. 10, pp. 2004-2016, Oct. 2000.
[4] A. Bononi, L. A. Rusch, "Doped-fiber amplifier dynamics: a system perspective", *IEEE Journal of Lightwave Technology*, Vol. 16 , no. 5, pp. 945-956, May 1998.
[5] L. Tancevski, A. Bononi and L. A. Rusch "Output Power and SNR Swings in Cascades of EDFA's for Circuit- and Packet-Switched Optical Networks", *IEEE Journal of Lightwave Technology*, Vol. 17, No. 5, pp. 733-742, May. 1999.
[6] L. Tancevski, L. A. Rusch and A. Bononi "Gain Control in EDFA's by Pump Compensation", *IEEE Photonics Technology Letters*, Vol. 10, No. 9, pp. 1313-1315, Sep. 1998.
[7] D. A. Francis, S. P. DiJaili and J. D. Walker "A single chip Linear Optical Amplifier", *OFC 2001, Anheim, CA*, paper PD13, 2001.
[8] Y. Fukashiro, K. Shrikhande et al. "Fast and fine wavelength tuning of a GCSR laser using a digitally controlled driver", *Optical Fiber Communication Technical Digest*, paper WM43, pp. 338-340, Baltimore, MD, March 2000.
[9] N. McKeown, A. Mekkittikul, V. Anantharam, and J. Walrand, "Achieving 100% throughput in an input-queued switch", *IEEE Transactions on Communications*, Vol. 47, No. 8, pp. 1260-1267, Aug. 1999.

PERFORMANCE OF OPTICAL BURST SWITCHED WDM RING NETWORK WITH TTFR SYSTEM

Yutaka Arakawa, Naoaki Yamanaka, and Iwao Sasase
Department of Information and Computer Science, Keio University, Japan

Abstract:

In this paper, we propose an architecture of Optical Burst Switched WDM ring network. In our proposed OBS ring network, every node is equipped with one tunable transmitter and one fixed-tuned receiver (TTFR) oparating on a given wavelength that identifies the node. TTFR type ring network has an advantage that no receiver collisions occur and each node can detect channel collision by transmitter side. By computer simulations, we evaluate the performance of throughput, goodput, and queueing delay. As a result, we show the performance effectiveness of our proposed architecture and access protocol.

1. INTRODUCTION

Optical burst switching (OBS) [1]- [2] is a switching technique that occupies the middle of the spectrum between the well-known circuit switching and packet switching paradigms. In OBS network, several IP packets with the same destinations are assembled into a burst, and forwarded through the network in optical domain. The transmission of each burst is proceeded by the transmission of a burst header packet, which usually takes place on a separate single channel. Recently, as an alternative of SONET/SDH based metro ring network, WDM metro ring network is focused on. SONET/SDH network represents a significant investment on the part of carries, and are currently being upgraded to support WDM. A research on the WDM metro ring network with optical packet switching, such as HORNET (Hybrid Optoelectronic Ring NETwork) [3], attracts much attention. On the other hand, the research on the WDM metro ring network with optical burst switching is also done [4]. In [4], OBS metro ring network consists of N nodes, and each node owns a home wavelength on which it transmits its bursts. The ring operates under the fixed tuned transmitter and tunable receiver (FTTR) system. In this system, every node can transmit without worrying about channel collisions since no other node can transmit with the same wavelength. However, it is possible that several packets arrive at the same node simultaneously. At that time all the bursts except one, that is randomly selected, are discarded. In this way, FTTR type ring network waste a bandwidth by the reason that bursts that might be discarded are transmitted to the destination.

In this paper, we propose an architecture of Optical Burst Switched WDM ring network. In our proposed OBS ring network, every node is equipped with one tunable transmitter and one fixed-tuned receiver (TTFR), and transmits bursts on the wavelength assigned for each destination. Since all nodes use the same wavelength for transmitting a packet to the same destination, channel collisions occur in the case that multiple nodes transmits bursts to a same destination node. However, TTFR has an advantage that causes no receiver collisions. In addition, since a burst header packet is transmitted ahead of a data burst under JET (Just Enough Time) signaling, channel collisions are detected in advance. A node that detects a channel collisions stops transmission of the burst immediately, and retransmits the burst after going through a burst of upstream node. By such upstream prioritized switching, bursts transmitted successfully will be received certainly at the destination node without being discarded during a transfer. By computer simulations, we evaluate the performance of throughput, goodput, and queueing delay. As a result, we show the performance effectiveness of our proposed architecture and access protocol.

The rest of the paper is organized as follows. We present our proposed architecture and access protocol in Section 2. The performance evaluations are shown in Section 3. Finally, we discuss the obtained results.

2. PROPOSED OBS RING NETWORK

2.1 Ring and node architecture

We consider N nodes organized in a unidirectional ring, as shown in Fig. 1. The ring can be a metropolitan area network (MAN) serving as the backbone that interconnects a number of access networks, and transporting multiple types of traffic from users, such as IP traffic, ATM traffic, frame relay traffic. Each fiber link between two consecutive OBS nodes in the ring can support $N + 1$ wavelengths. Of these, N wavelengths are used to transmit bursts, and $(N + 1)$th wavelength is used as the control channel.

Each OBS node is attached to one or more access networks. In the direction from the access networks to the ring, the OBS node acts as concentrator. It collects and buffers electronically data, transmitted by users over the access networks, which need to be transported over the ring. Buffered data are subsequently grouped together and transmitted in a burst to the destination OBS node. A burst can be of any size between a minimum and maximum value.

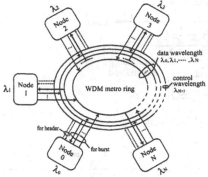

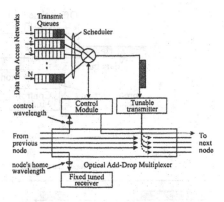

Figure 1. An architecture of Optical Burst Switched WDM ring network

Figure 2. An architecture and operation of every node

Bursts travel as optical signals along the ring, without undergoing any electro-optic conversion at intermediate nodes. In the other direction from the ring to the access networks, an OBS node terminates optical bursts destined to it, electronically processes the data contained there in, and delivers them to users in its attached access networks.

The architecture of an OBS node is shown in Fig. 2. Each node is equipped with one optical add-drop multiplexer (OADM), and two pairs of optical trans ceivers. The first pair consists of a receiver and transmitter fixed tuned to the control wavelength, and are part of the control module in Fig. 2. The control wavelength is dropped by the OADM at each node, and added back after the control module has read the control information and has inserted new infor-mation. The second pair of transceivers consists of a tunable transmitter and a receiver that is tuned to the node's home wavelength. The OADM drops the optical signal on the node's home wavelength. In OBS network, several IP packets with the same destinations are assembled into a burst, and forwarded through the network in optical domain. The transmission of each burst is pro-ceeded by the transmission of a burst header packet, which usually takes place on a separate single channel. We use Just Enough Time (JET) proposed in [1] as the signaling protocol. It starts transmitting the data burst soon after the transmission of the burst header packet. We will refer to the interval of time between the transmission of the first bit of the burst header packet and the transmission of the first bit of the data burst as the offset. The burst header packet carries information about the burst, including the offset value, the length of the burst, its priority, etc.. The purpose of burst header packets is to inform each intermediate node of the upcoming data burst, so that it can configure its switch fabric appropriately to switch the burst to the appropriate output port.

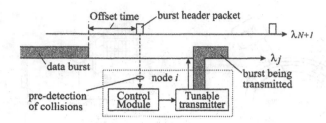

Figure 3a. Incoming burst from upstream node is detected

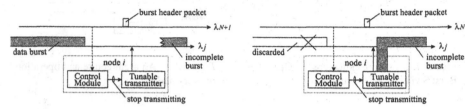

Figure 3b. Example of upstream prioritized switching

Figure 3c. Example of earliest arrival prioritized switching

2.2 Access protocol

In the proposed OBS ring network, channel collisions occur in the case of transmitting bursts to a given destination node from multiple nodes, since all nodes will transmit on the same wavelength of the destination. Therefore, an access protocol in consideration of channel collisions is required. A basic access protocol is similar to CSMA/CD (Carrier Sense Multiple Access with Collision Detection). Since a burst header packet that includes the information of the burst, such as length and destination, is transmitted ahead of a data burst, nodes can detect channel collisions in advance. Fig. 3a shows a basic access protocol with Collision Detection. When a node i transmits a packet towards node j, it confirms the availability of wavelength λ_j for a term of transmitting. If a node can transmit, a control packet is transmitted immediately on control wavelength, and after a delay of offset time, a payload is transmitted. We can classify access protocols into *Upstream Prioritized Switching* or *Earliest Arrival Prioritized Switching* according to the operation at channel collisions. Fig. 3b shows the example of upstream prioritized switching. The node which has detected a channel collision by the control packet stops transmission of the burst immediately. By upstream prioritized switching, bursts transmitted successfully at the source node will be received certainly at the destination node without being discarded during a transfer. Fig. 3c shows the example of earliest arrival prioritized switching. In the case that node i has already started to transmit a given burst, the node which has detected a channel collision will drop the burst from upstream node. By earliest arrival prioritized switching, although bursts may be discarded by the relay node, it can prevent imcomplete bursts due to the transmit interruption being transmitted downstream.

2.3 Burst Assembly

Data from access network is organized into transmit queues according to their destination. The data buffer at each OBS node is shared by $N-1$ transmit queues, each corresponding to one of the $N-1$ destination nodes. The order in which transmit queues are served is determined by the scheduler in Fig. 2. A transmit queue is available for service if its size is larger than *Min Burst Size*, or the first data of the transmit queue has waited for more than *timeout* time. If the size of available transmit queue is less than *Max Burst Size*, then a burst that includes all data in the transmit queue is constructed. Otherwise, a burst of at most size *Max Burst Size* is constructed, and the data remaining in the transmit queue is served at a later time. If two or more transmit queues are available, we choose one queue in *round-robin* or *random* manner. Although round-robin can keep the fairness for every destinaion, the synchronization of collision may occur between adjacent nodes. When channel collision occurs by the reason that a node i and a node $i+1$ transmit bursts to the same destination j, to the following timing, it's likely that they subsequently transmit a burst to the same destination, such as node $j+1$. Therefore, in order to improve the synchronization problem, we propose a random queue selection.

3. PERFORMANCE EVALUATION

3.1 Simulation model

In our simulation study we consider a ring network with 6 nodes, each with an electronic buffer of 10 MB. The distance between two successive nodes in the ring is taken to be 4 km. We assume that the control wavelength each burst wavelength runs at 2.5 Gbps. We assume that data arrives in packets, and the packet arrival process to each node is described by a modified interrupted poisson process (IPP) [5]. This modified IPP is an ON/OFF process, where both the ON and the OFF periods are exponentially distributed. Packets arrive back to back during the ON period at the rate of A_{on}=2.5 Gbps. No packets arrive during the OFF period. The packet size is assumed to follow a truncated exponential distribution with an average size of 500 bytes and a maximum size of 5000 bytes. We use the squared coefficient of variation, c^2 and of the packet inter arrival time to measure the burstiness of the arrival process. c^2 is defined as the ratio of the variance of the packet inter arrival time divided by the squared mean of the packet inter arrival time. We use the expression for the c^2 of an IPP and where the packet size is not truncated. We have $c_{IPP}^2 = 1 + \frac{2\lambda\mu_1}{(\mu_1+\mu_2)^2}$ where $1/\lambda$ = (500 bytes)/(2.5 Gbps) = 1.6 μs, and $1/\mu_1$ and $1/\mu_2$

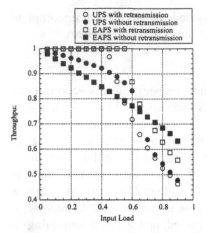

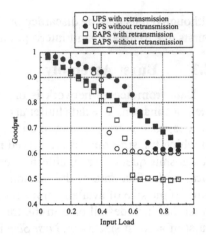

Figure 4.   Comparison of UPS and EAPS about throughput performance

Figure 5.   Comparison of UPS and EAPS about goodput performance

are the mean times of the ON and OFF periods, respectively. To completely characterize the arrival process, we use the above expression for c^2 and another equation that involves the mean times of the ON and OFF periods. We define the quantity $A_r = A_{on}\frac{\mu_2}{\mu_1+\mu_2}$ Given the c^2 and the average packet arrival rate A_r, we can calculate the quantities μ_1 and μ_2, and therefore the arrival process is completely characterized.

In this section, we evaluate throughput, goodput, delay by a computer simulation. These performance measures are estimated by varying an input load from 0.1 to 1.0 with an increment of 0.05. In this paper, an input load shows the average packet arriving rate into a single node. The destination of packets follows the uniform distribution. c^2 of packet inter-arrival time at each node is set to 20. We also set $Max\ Burst\ Size$ to 50 KB, $TimeOut$ to 4 ms and offset value to 10 μs. In each figure, UPS and EAPS represent Upstream Prioritized Switching and Earliest Arrival Prioritized Switching.

3.2   Basic performance

Fig. 4 shows throughput performance. The throughput is defined as the amount of receiving data to that of arrival data. We evaluate the performance of UPS and EAPS respectively about the case both with and without retransmission. It is shown that throughput of all deteriorates from the middle. This is because OBS has many channel collision due to the one-way signaling.

Fig. 5 shows goodput performance. The goodput is defined as the amount of receiving data to that of transmitting data. In the case of UPS, goodput performance deteriorates due to the transmit interruption. On the other hand,

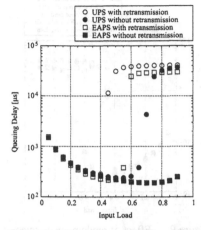

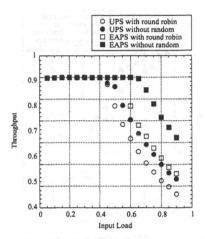

Figure 6. Comparison of UPS and EAPS about average queueing delay

Figure 7. Effect of random queue selection to throuput performance

EAPS decrease the goodput by discarding burst at intermediate nodes. It is shown that UPS maintains higher goodput than EAPS.

Fig. 6 shows average queueing delay. Queueing delay is defined as the time interval from the instance that the packet arrives at a node to the instance that the packet leaves the node. We observe that, as the average arrival rate increases, the Queueing delay first decreases, and then it increases. This is due to the fact that when the traffic intensity is low, the time for a transmit queue to reach the *Min Burst Size*. Since many collisions occur under a high load, queueing delay becomes very long (but the maximum delay is restricted by buffer size).

3.3 Effect of random queue selection

We evaluate the performance of UPS and EAPS with retransmission about the case both round-robin queue selection and random queue selection. From Fig. 7, we notice that throughput performance of both UPS and EAPS are improved by appling proposed random queue selection. Especially, it is effictive to EAPS. Moreover, the tendency is the same in Fig. 8. On the other hand, we notice that an effect to delay performance is slightly, as shown in Fig. 9.

4. CONCLUSION

In this paper, we have proposed an architecture of Optical Burst Switched WDM ring network. In our proposed OBS ring network, every node is equipped with one tunable transmitter and one fixed-tuned receiver (TTFR) oparating on a given wavelength that identifies the node. TTFR type ring network has an

102

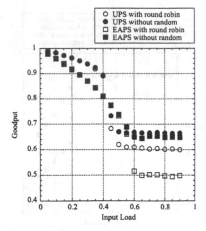
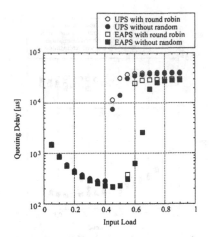

Figure 8.   Effect of random queue selection to goodput performance

Figure 9.   Effect of random queue selection to average queueing delay

advantage that no receiver collisions occur. By computer simulations, we evaluate the performance of throughput, goodput, and queueing delay. As a result,
we show the OBS ring network achieves good performance under a low load.
It is shown that our proposed random queue selection can improve the performance at a high load.

ACKNOWLEDGMENTS

This work is partly supported by Keio University COE Program in "Optical
and Electronic Device on Access Network" by Ministry of Education, Cul-
ture, Sports, Science and Technology, Japan and Support Center for Advanced
Telecommunications Technology Research, Foundation.

REFERENCES

[1] C.Qiao and M.Yoo, "Optical burst switching(OBS) - A new paradigm for an optical internet,"*J. High Speed Networks*, vol.8, pp.69-84 ,1999.

[2] Yijun Xiong, Marc Vandenhoute and Hakki C. Cankaya, "Control Archtecutre in Optical
Burst-Swtiched WDM Networks,"*IEEE Journal on Selected Areas in Communications*,
vol.18, no.10, pp.1838-1851, Oct. 2000.

[3] K. Shrikhande, I. M. White, D. Wonglumsom, S. M. Gemelos, M. S. Rogge, Y. Fukashiro,
M. Avenarius and L. G. Kazovsky, "HORNET: A Packet-Over-WDM Multiple Access
Metropolitan Area Ring Network,"*IEEE Journal on Selected Areas in Communications*,
vol.18, no.10, pp. 2004-2016, Oct. 2000.

[4] Lisong Xu, Harry G. Perros, George N. Rouskas, "A simulation study of optical burst
switching and access protocols for WDM ring networks,"*Computer Networks*, 41, pp.143-
160, 2003.

[5] W. Fischer, K. Meier-Hellstern, "The Markov-modulated Poisson process (MMPP) cook-
book,"*Performance Evaluation*, 18, pp.149-171, 1992.

PART A2:

ROUTING

ASSESSING THE BENEFITS OF WAVELENGTH SELECTION VS. WAVELENGTH CONVERSION IN WDM NETWORKS*

Nicola Andriolli, Luca Valcarenghi, and Piero Castoldi
Scuola Superiore Sant'Anna di Studi Universitari e di Perfezionamento, Center of Excellence for Communication Networks Engineering, Pisa, Italy
{nick, valcarenghi, castoldi}@sssup.it

Abstract: In this paper we consider how the performance of wavelength routed networks depends on node capabilities, such as the wavelength selection and the wavelength conversion. For this purpose we propose a modified version of the shortest path algorithm for the wavelength graph (SPAWG), where "virtual links" are added for modeling the utilization and the cost of these node capabilities.

Numerical results show the performance advantages in utilizing wavelength selection instead of wavelength conversion, when only one of the two capabilities can be implemented in each network node. When wavelength conversion and wavelength selection utilization is limited to a subset of network nodes to reduce the overall network cost, results indicate that sparse wavelength selection is not as beneficial as sparse wavelength conversion.

1. INTRODUCTION

In the latest years, the growing demand for bandwidth in core networks has been cost effectively satisfied with the deployment of wavelength division multiplexing (WDM) systems. Currently WDM is the basis of wavelength routed optical networks (WRONs), where lightpaths (i.e. all-optical end-to-end switched connections between node pairs at the wavelength granularity) are set up to generate a virtual topology, often with a higher connectivity than the physical fiber network. The lightpath establishment in WRONs involves the problem of selecting a suitable path and allocating an available wavelength

*This work has been partially supported by Italian Ministry of Education and University (MIUR) under FIRB project "Enabling platforms for high-performance computational grids oriented to scalable virtual organization (GRID.IT)" and partially by Marconi through an annual grant to Scuola Superiore Sant'Anna.

for the connection: the resulting problem is referred to as the routing and wavelength assignment (RWA) problem [1]. Network performance depends on the RWA problem solution, that in turn is affected by the network node capabilities.

In this paper two node capabilities have been investigated: (i) the possibility for a connection request of choosing any admission wavelength in source and destination nodes (*wavelength selection*, whose benefits have not been extensively studied in literature [2, 3]); (ii) the possibility of converting the wavelength of the lightpath at any intermediate node (*wavelength conversion*, which has been thoroughly investigated [4–8]). To take into account these additional node capabilities the Shortest Path Algorithm for the Wavelength Graph (SPAWG for short [9, 10]) has been extended: "virtual links" are added for modeling the utilization and the cost of any processing incurred in lightpath routing besides the hop cost. Therefore lightpaths are routed along the least cost path, that jointly minimizes the utilization of node capabilities and link capacity.

Numerical results show that, if implemented at all network nodes (full-mode), wavelength selection is more beneficial than wavelength conversion. On the other hand, if selection capabilities or conversion capabilities are deployed just in a limited subset of nodes (sparse-mode), sparse wavelength selection is not as beneficial as sparse wavelength conversion.

2.   NODE ARCHITECTURE AND RWA ALGORITHM

Node architectures incorporating various node capabilities have been investigated. As shown in Fig. 1, *wavelength selection*, which allows an outgoing (incoming) request to choose the wavelength on the first (last) hop of the lightpath, is implemented connecting with a space matrix (called wavelength selection switching fabric) the tributaries to the bank of fixed-wavelength transponders [11]. This space matrix switches the traffic from the tributaries (at the wavelength λ_t, usually at 1300 nm) to the transponder at the desired wavelength, e.g., λ_1 (in the 1550 nm range, according to ITU-T grids). On the contrary, if wavelength selection is not present, each tributary is statically linked to a fixed-wavelength transponder and it must use only a specific wavelength.

Wavelength conversion permits to change the wavelength on which a lightpath [1] is carried (relaxing the wavelength continuity constraint): in Fig. 1 this capability is hidden in the optical cross-connect. To preserve signal transparency and contain complexity and power consumption [5], an all-optical wavelength conversion scheme has been adopted, assuming that only half of

---

[1] We assume that wavelength conversion in a node cannot be performed for lightpaths added or dropped in the same node because this function is provided through wavelength selection.

the wavelengths nearest to the working one can be reached, due to technological constraints (previous studies have shown that this limitation only slightly affects network performance [6]).

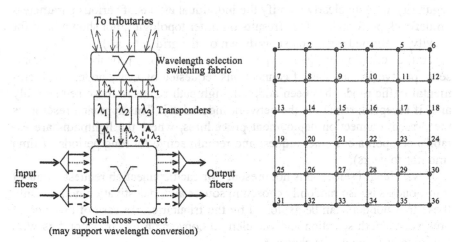

Figure 1. Architecture of an optical node with $W = 3$ wavelengths per link and nodal degree $D = 2$.

Figure 2. 6 × 6 grid topology.

To consider both wavelength selection and wavelength conversion costs, an auxiliary graph, called *wavelength graph* (WG) [9], is utilized by the SPAWG routing and wavelength assignment algorithm to find a path for the incoming connection requests. The WG is built repeating the physical topology of the network on a number of superimposed planes equal to the number of available wavelengths. On each layer the links connecting the nodes are assigned a weight that is a measure of the hop cost.

To consider the wavelength conversion, "virtual links" are added among copies of the same node on different layers. Due to the high cost and the physical constraints of the wavelength converters, the weight of these links is an order of magnitude higher than the hop cost, and proportional to the distance between wavelengths. To take into account the wavelength selection, a modified version of the WG is adopted, with an additional layer λ_0: this is a logical level that contains only the network nodes without links among them. Instead, there are "virtual links" that connect source and destination nodes of a connection request with their respective copies on one or more layers, accordingly to the absence or presence of wavelength selection. Since wavelength selection is less complex and expensive than wavelength conversion, the weight of these links has the same order of magnitude of the hop cost.

3.   NETWORK AND TRAFFIC MODEL

A 6×6 grid topology with bidirectional links and W wavelengths per link, depicted in Fig. 2, is utilized as test network. The grid topology is deployed in grid computing scenarios, and it has been chosen because its simplicity and regularity allow to clearly identify the individual effects of various parameters on network performance [12] (results on other topologies, not shown here for brevity, confirmed the conclusions drawn on the grid network).

Traffic between node pairs is generated with uniform probability: T represents the average number of connections requested by every node. An incremental traffic model has been adopted: lightpath requests arrive sequentially and, if accepted, remain in the network indefinitely. This approach resembles the present connection deployment procedures, where the lightpaths are established upon a customer request and remain active for long periods of time (months to years).

If wavelength selection is not present, for each connection request a pair of transponders is also randomly chosen in source and in destination nodes. Only these transponders can be exploited by the tributaries. In case of absence of both wavelength selection and wavelength conversion two transponders with the same wavelength are chosen [2].

The lightpath requests are processed upon arrival by the SPAWG algorithm. If a minimum cost route that connects source and destination nodes is found a bidirectional lightpath is established, and the network state is updated; otherwise the connection is blocked and rejected.

4.   NUMERICAL RESULTS

The metric considered to evaluate network performance is the blocking probability, computed as the ratio between the number of blocked lightpath requests and the number of generated requests.

In the first part of the section we compare the four architectures obtained combining the absence or the presence of both wavelength selection (WS) and wavelength conversion (WC) in each network node: NOWS-NOWC, NOWS-WC, WS-NOWC, and WS-WC.

In the second part of the section sparse wavelength selection and conversion are considered: in this case, the placement of these node capabilities must be optimized. The optimal placement problem has been proven to be NP-complete [7]. For this reason the following heuristic [8] has been utilized:

---

[2]Differently from [2], where networks with fixed transmitters and receivers and absence of wavelength conversion are not studied because full connectivity is not guaranteed, we consider this case (adding the condition that the two transponders work on the same wavelength), because this scenario is typical of many present optical networks.

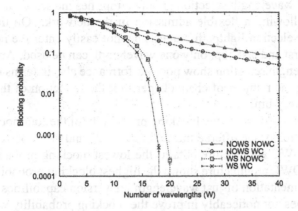

Figure 3. Blocking probability versus wavelengths per link, with $T = 10$ calls per node.

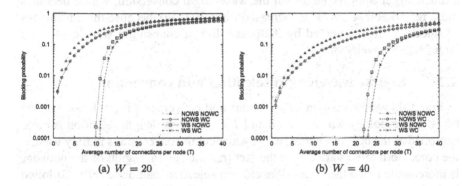

Figure 4. Blocking probability versus offered traffic.

we start from a scenario where the capability is present in all the nodes and we rank them according to its utilization; then simulations are run placing the capability only where it is significantly used.

In all the simulations at least 5000 repetitions have been run for each point; the results are plotted with the confidence interval at 95% confidence level.

4.1 Presence vs. absence of wavelength selection and conversion

Fig. 3 compares the performance of the four basic node architectures, NOWS-NOWC, NOWS-WC, WS-NOWC, and WS-WC, in terms of blocking probability as a function of the number of wavelengths W; the average number of connections T requested by every node is set to 10. This plot shows the fundamental role of wavelength selection to improve network performance, while the introduction of wavelength conversion offers only a marginal blocking decrease. Indeed wavelength selection greatly simplifies the search for a suitable lightpath by allowing a flexible admission in the network. On the contrary in absence of selection lightpath requests cannot easily enter the network because, in the first and last hop, only one wavelength can be used. Architectures without wavelength selection show poor performance also in terms of scalability: increasing the number of channels per link the reduction of the blocking probability is very limited.

Fig. 4-(a) and 4-(b) show the blocking probability of the four node architectures as a function of the offered traffic for $W = 20$ and $W = 40$ respectively. Again the WS-WC architecture obtains the lowest blocking probability while the NOWS-NOWC architecture shows the highest blocking probability. Moreover the implementation of just wavelength conversion capabilities in the network nodes does not noticeably improve the blocking probability with respect to the NOWS-NOWC architecture. Instead wavelength selection determines a much larger improvement than the wavelength conversion, which becomes more remarkable when W is increased from 20 to 40, because the advantages of selection are augmented by the possibility of choosing a wavelength in a larger set of wavelengths.

4.2   Sparse wavelength selection and conversion

Fig. 5 shows the number of selections and conversions in each node of the 6×6 grid topology with $W = 40$ and $T = 40$. Wavelength selection is more exploited at the network edges, while wavelength conversion is mainly used in the core. Both effects are due to the fact that the central region of the network is more loaded than the edges. Wavelength selection cannot be fully exploited in the inner nodes because link congestion causes a higher blocking of connections leaving or entering those nodes. However this capability is widely used in the entire network: more than 16 selections take place in every node. On the other hand wavelength conversion helps to find a route for requests that would otherwise be blocked because of the wavelength continuity constraint: hence this capability is much more exploited in the central region of the network where the majority of the traffic is routed (more than 10 conversions occur), while it is only marginally used at the edges (from 3 to 5).

In Fig. 6 the blocking probability versus offered traffic is shown with sparse wavelength selection and conversion. We assign WS and WC capabilities only to the first q_{WS} and q_{WC} fraction of the total number of nodes, in the ranking of WS and WC utilization. In particular two wavelength conversion scenarios ($q_{WC} \in \{0.25, 1\}$) are combined with three wavelength selection scenarios ($q_{WS} \in \{0, 0.5, 1\}$). From the figure we can notice that sparse wavelength selection ($q_{WS} = 0.5$) only slightly reduces the blocking probability with respect to the case without wavelength selection ($q_{WS} = 0$), but significantly increases

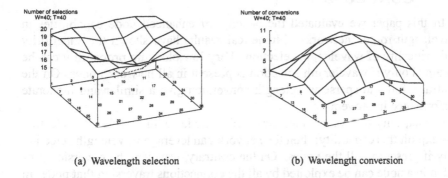

(a) Wavelength selection   (b) Wavelength conversion

Figure 5.   Number of selections and conversion in each node of the 6 × 6 grid topology.

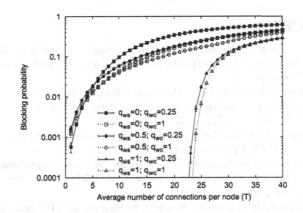

Figure 6.   Blocking probability versus offered traffic for sparse WS and WC ($W = 40$).

it with respect to the case in which wavelength selection is available in all network nodes ($q_{WS} = 1$). Indeed wavelength selection simplifies very much the search of an available lightpath, but it is a property of each specific optical node, which can be exploited only by the tributaries connected to that node. Thus it does not guarantee a fair improvement of the network performance.

On the contrary, the assignment of wavelength conversion capability to just one-fourth of the nodes ($q_{WC} = 0.25$) closely approximates the blocking probability obtained when wavelength conversion capability is present in all network nodes ($q_{WC} = 1$). This happens because even sparse wavelength conversion can be effectively utilized by all the connections traversing a node equipped with this capability, increasing in a fair manner the network performance. Thus, unlike wavelength selection, wavelength conversion provides a reduction in blocking probability approximately proportional to its exploitation.

5. CONCLUSION

In this paper we evaluated the benefits of enhanced node capabilities in wavelength routed networks. Numerical results have shown the superiority and criticality of wavelength selection. Very low blocking probability can be obtained only if wavelength selection is present in all network nodes. On the contrary, full and sparse wavelength conversion give a similar and moderate performance improvement.

In fact, only the connections admitted in nodes with wavelength selection can exploit this capability: then the network can leverage wavelength selection only if present in all the nodes. On the contrary, wavelength conversion provided in a node can be exploited by all the connections traversing that node. In this sense wavelength conversion, though placed in nodes, is a shared facility while wavelength selection is a node-peculiar facility.

REFERENCES

[1] H. Zang, J. P. Jue, and B. Mukherjee, "A review of routing and wavelength assignment approaches for wavelength-routed optical WDM networks," *Optical Networks Magazine*, vol. 1, pp. 47–60, Jan. 2000.

[2] J. Yates, J. Lacey, M. Rumsewicz, and M. Summerfield, "Performance of networks using wavelength conversion based on four-wave mixing in semiconductor optical amplifiers," *J. Lightwave Technol.*, vol. 17, pp. 782–791, May 1999.

[3] I. Widjaja and A. I. Elwalid, "Study of GMPLS lightpath setup over lambda-router networks," in *Proc. IEEE ICC'02*, vol. 5, pp. 2707–2711, 2002.

[4] J. Yates, J. Lacey, and M. Rumsewicz, "Wavelength converters in dynamically reconfigurable WDM networks," *IEEE Commun. Surveys*, vol. 2, 2nd quarter 1999. Available at http://www.comsoc.org/pubs/surveys/.

[5] B. Ramamurthy and B. Mukherjee, "Wavelength conversion in WDM networking," *IEEE J. Select. Areas Commun.*, vol. 16, pp. 1061–1073, Sep. 1998.

[6] J. Yates, J. Lacey, D. Everitt, and M. Summerfield, "Limited-range wavelength translation in all-optical networks," in *Proc. IEEE INFOCOM'96*, vol. 3, pp. 954–961, 1996.

[7] G. Wilfong and P. Winkler, "Ring routing and wavelength translation," in *Proc. ACM-SIAM Symposium on Discrete Algorithms (SODA)*, 1998.

[8] K. Venugopal, M. Shivakumar, and P. Kumar, "A heuristic for placement of limited range wavelength converters in all-optical networks," in *Proc. IEEE INFOCOM'99*, vol. 2, pp. 908–915, 1999.

[9] I. Chlamtac, A. Farago, and T. Zhang, "Lightpath (wavelength) routing in large WDM networks," *IEEE J. Select. Areas Commun.*, vol. 14, pp. 909–913, Jun. 1996.

[10] D. Saha, "Lightpath versus semi-lightpath: some studies on optimal routing in WDM optical networks," *Photonic Network Commun., Kluwer*, vol. 2, pp. 155–161, May 2000.

[11] R. Ramaswami and K. N. Sivarajan, *Optical Networks: a practical perspective*. Morgan Kaufmann, second ed., 2002.

[12] D. K. Hunter, E. D. Lowe, and I. Andonovic, "Modelling of routing in multi-fibre WDM grid networks," in *Proc. IEE Colloquium on WDM Technol. and Applic.*, 1997.

ILP BASED EVALUATION OF SEPARATE WAVELENGTH POOL (SWAP) STRATEGY

Zsolt Lakatos Ph.D. Student
Budapest University of Technology and Economics, Department of Telecommunications
Magyar Tudósok krt 2, Budapest, Hungary H-1117, e-mail:lakatos@hit.bme.hu

Abstract: The paper proposes a general approach to support the analysis and evaluation of dynamic routing and wavelength assignment strategies applicable in advanced optical networks. Some ILP based models are introduced to apply the proposed analysis support for the evaluation of Separate Wavelength Pool strategy. Results from a small network example are described and analyzed to validate the proposed ILP models in one hand and to highlight the advantages of the shared capacity related resilience and the pre-emption techniques applicable in advanced optical networks on the other hand.

1. INTRODUCTION

Recent developments in optical technology enable the implementation of complex network and management functionalities. The simple dynamic routing and wavelength assignment (RWA) strategies can be replaced by more complex and effective ones based on the advanced signaling.

Due to the various client requirements the optical network layer should support differentiated transport services. In case of dynamic optical networks this differentiation can be made according to the blocking probabilities and resilience options of the arriving optical channel requests[1].

Taking into account the typical IP client requirements two basic service classes can be distinguished in the optical layer: the premium leased optical channel service for high quality traffic (e.g. IP VPNs, QoS oriented applications, etc.); the low priority leased optical channel service for IP links carrying best-effort traffic.

The premium class is with low blocking probability and guaranteed protection in case of network element failures, however, there is no bounds for the blocking of low priority class. Even the optical channels carrying the traffic of the low priority class are pre-emptable in case of network element failures in order to support the guaranteed protection of the premium class. In case of the application of different QoP classes - mainly if a pre-emptable class is present - the availability analysis of different service classes provide further important information on the provisioning strategy[2].

In case of dynamic optical channel provisioning the permanent optical channel requests are assumed to arrive spread in time and space, and an appropriate strategy is applied to meet them under the efficient utilization of available network resources. Having these two service classes the main goal of a provisioning strategy is to restrict the uncontrolled competition between the premium and low priority traffic, since the implementation of the distinction in blocking probabilities requires efficient control during the allocation of the optical channels for requests from different classes.

2.   EVALUATION OF RESILIENT OPTICAL CHANNEL PROVISIONING STRATEGIES

To evaluate resilient optical channel provisioning strategies three aspects should be taken into account: penalty on dynamic behavior, complexity of required network consolidations, availability of low priority services.

Generally, solving a provisioning problem a random order of optical channel requests is to serve. It results in a sub-optimal decision for resource allocation, since even in case of pre-planned fixed routing the wavelength assignment is performed channel by channel subsequently according to a simple strategy (e.g. first fit). Having known the same set of optical channels in advance and routed them via the same routes an optimal solution for the wavelength assignment under fixed routes can be achieved, and the difference in the required resources can be interpreted as a penalty for the dynamic behavior. With other words this difference describes how far a random sequence based solution (or the average of numerous solutions) from the theoretically achievable global optimum.

To illustrate the above described penalty on dynamic behavior results from a small (9 nodes, 16 edges, nodal degree 3.5) network example is presented in Figure 1[3]. The left bar gives the average resource needs in case of sequential allocation, the right one represents the global optimum. The demand patterns and the routing of the optical channels are the same in both cases, the only difference is in the wavelength allocation. In the sequential case it is a simple first fit strategy [4] (sequential allocation of the idle - not yet allocated - optical channels via the fixed route), in the reference case it is an optimal - requiring minimum number of wavelengths - solution based on the ILP implementation of the graph-coloring model. The lower parts of the bars represent the amount of optical channels in use; the upper part is the amount of extra wavelengths to be installed in case of no wavelength conversion. The figures represent the total number of wavelength on links. It can be depicted on the figure that the sequential first fit allocation results in additional 30% wavelength need. It is a theoretical lower bound, since there may not be a proper sequence for a given dynamic allocation strategy to achieve this lower bound. However, the distance between the theoretical bound and the resource need for a given strategy is a good measure for the evaluation.

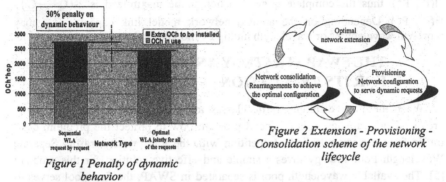

Figure 1 Penalty of dynamic behavior

Figure 2 Extension - Provisioning - Consolidation scheme of the network lifecycle

If we change our focus from the simple provisioning problem to the network development, an interesting problem can be identified. Since the traffic to be served supposed to be incremental (permanent channel requests with practically infinite holding time) the saturation of network resources are foreseen. In this case based on the experienced structure and amount of channel requests the network resources should be extended to prevent the significant increase of blocking. On the other hand, as we have seen from the previous illustration the network configuration is sub/optimal due to the channel by channel sequential solution of RWA problem. Thus, it is a quite obvious idea to consolidate the network in order to improve the resource utilization before the decisions on the capacity extension.

Figure 2 illustrates a simple Extension - Provisioning - Consolidation scheme of the network lifecycle. The network consolidation means a network reconfiguration (rearrangement) which from technological aspects can be based on the same automatic configuration capabilities as the provisioning.

The general green-field network optimization problem can be formalized as follows: the network topology is given as a $G(V,E)$ with sets of nodes V and edges E without effective capacity constraints. Edge n,m is assigned to cost function $\Phi_{m,n}(C_{m,n})$, where $C_{m,n}$ is the required capacity on the given edge. The network is have to meet demands $D=\{d_{i,j}\}$ under minimum cost $\phi(\underline{C}) = \sum \phi_{m,n}(C_{m,n})$. The edge capacities can be derived from the demands according to the applicable routing and the given topology.

The network consolidation problem can be formalized differently: in this case a complete network configuration is given initially, with its topology G(V,E), demands D=$\{d_{i,j}\}$, and routing of demands P0=$\{p0_{i,j}\}$, link capacities C(m,n) for each link m,n. The aim of the consolidation to optimize the utilization of the existing resources (link capacities) with limited network rearrangements. Therefore, the network is have to meet demands D=$\{d_{i,j}\}$ under maximum saving $\phi(\underline{C}^0 - \underline{C}) = \sum \phi_{m,n}(C^0_{m,n} - C^*_{m,n})$. To improve the basic model some penalties on the network rearrangements decreasing the savings can be taken into account. The penalty may depend on the extend of changes of the applied routes $\Pi(P^0, P*)$, thus the complete target function to be maximized is $\phi(\underline{C}^0 - \underline{C})$ - $\Pi(P^0, P*)$. Depending on the applied network model link capacity $C_{m,n}$ may represent wavelengths or wavelength multiplex modules in both cases.

3. THE SWAP STRATEGY AND SOME ILP MODEL FOR ITS EVALUATION

When there are different traffic classes to be supported in optical channel provisioning strategies, it is a general problem how to protect the premium class traffic from the uncontrolled competition with the low priority traffic. Separate Wavelength Pool strategy gives a simple and effective solution for this problem [5]. The available wavelength pool is separated in SWAP, the first pool serves to allocate working connections for premium traffic only, and the second pool is for SRLG-based shared spare capacity oriented restoration routes for premium class traffic and for working routes of low priority pre-emptable traffic. Assuming that the premium traffic can be forecasted with acceptable level of confidence, and based on this forecast the available wavelength pool can be shared. This resource sharing guarantees the high quality service (guaranteed protection, low blocking) of premium traffic independently from the amount and pattern of the low priority traffic (which assumed to be hardly forecastable).

ILP based modelling of the SWAP strategy helps to study the main features of the solution independently from the impact of random optical channel request arrival sequences. Having an assumed demand pattern the amount and configuration of required resources can be calculated to serve the given pattern or having both the network and the demands the feasibility of the specified problem can be checked.

The general ILP approach to study SWAP strategy is based on the minimum cost flow model. To different network models can be supported applying different target functions. When the target is to minimize the total amount of used optical channels it is the single fiber model (no modularity of wavelength multiplex systems is concerned). Specifying the total amount of wavelength multiplex systems to be minimized the modular network model is implemented. (In the latest case the capacity of the wavelength multiplex system in wavelengths should be specified.) The applied model is a real flow based one, where Kirchoff's laws are applied to control the flows in the source, sink and intermediate nodes. The first law assures that the given flow leaves the source, the second that it arrives to the sink, and the third is the flow conservation in the intermediate nodes.

The resilience for premium class connection is provided via a route disjoint from the working one. This approach simplifies the provisioning processes since the same restoration route can be applied in case of any single link failure. This constraint is implemented by (1)

$$\sum_{\forall h} {}^h W_{m,n} + \sum_{\forall h^*} {}^{h^*} P_{m,n} \leq 1 \quad W, P \in \{0,1\} \quad (1)$$ where $W_{m,n}$ and $P_{m,n}$

indicate the usage of h working and h* protection wavelengths on link n,m.

Applying the modular network model, the number of modules should be calculated according to (2a), (2b) and (3), as follows:

$$ {}^h L_{mn} = \sum_{\forall <s,d> \ pair} {}^h W^{sd}_{mn} \quad (2a) \qquad\qquad {}^{h^*} L_{mn} = \sum_{\forall <s,d> \ pair} {}^{h^*} P^{sd}_{mn} \quad (2b)$$

$$ M_{mn} = \max_{\forall (h \cup h^*)} {}^h L_{mn} \quad (3)$$ where ${}^h W^{s,d}_{m,n}$ indicates that the working route of

optical channel demand between s,d routed via link n,m on wavelength h, ${}^{h^*} P^{s,d}_{m,n}$ is the same for the protection route on wavelength h*. Accumulating these information ${}^h L_{m,n}$ and ${}^{h^*} L_{m,n}$ give the multiplicity of wavelengths h and h* for link m,n, respectively, and larger multiplicity sets the number of required modules on the link n,m.

With help of these basic formulas four variations are specified to model different network and technology scenarios.

The first scenario (referred later as scenario "*1+1*") is specified for reference purposes only. The premium traffic is routed according to the traditional 1+1 dedicated protection. The protection routes are calculated under fixed working routes and the wavelengths are assigned to them from the protection pool. Comparing the results from this case with the results with shared capacity oriented resilience, the efficiency of resilience oriented capacity sharing can be studied (referred later as 1+1). (No low priority traffic is routed in this case.)

The second scenario (referred later as scenario "*shared*") is dedicated to the shared capacity based resilience. Minimum disjoint paths are calculated to the fixed working ones for the premium traffic, and the spare capacities allocation is based on to the Shared Risk Link Groups (SRLG) set up according to the working routes. The calculation of spare capacities are performed according to formulas (4) and (5)

$$ {}^{h^*} F^{ij}_{mn} = \sum_{\forall <s,d>} {}^{h^*} P^{sd}_{mn} \quad (4) \qquad\qquad {}^{h^*} L_{mn} = \max_{<i,j>} {}^{h^*} F^{ij}_{mn} \quad (5)$$

where, due to the capacity sharing the failure case should be identified - i,j denote the failed link, and ${}^{h^*} F^{i,j}_{m,n}$ represents the multiplicity of wavelength h* in use on link m,n in case of the failure of link i,j. The maximum of ${}^{h^*} F^{i,j}_{m,n}$ taking into account each failure case gives the required multiplicity of the wavelength h* on link m,n, and is represented by ${}^{h^*} L_{m,n}$. (No low priority traffic routed in this case.)

The third scenario (referred later as scenario "*pre-empt*") is the extension of the previous one to include low priority traffic, as well. The premium traffic is routed according to the disjoint path SRLG based shared capacity resilience option described above, and the low priority traffic is via minimum paths and the wavelengths are assigned to the paths from the second pool. Low priority traffic is pre-emptable; thus it may use the spare resources of the high priority traffic. Therefore, the Kirchoff's laws should be formulated for low priority traffic as well, and the calculation of required wavelengths should be modified according to (6), where the larger of the premium shared protection of low priority working requirements will set up the multiplicity of a wavelength h* on a given link m,n.

$$^{h*}L_{mn} = \max\{\max_{<i,j>}{}^{h*}F_{mn}^{ij}, {}^{h*}L'_{mn}\} \quad (6)$$

The fourth scenario (referred later as scenario "*add*") is differs from the third one in the resource allocation for low priority traffic. In this case the low priority traffic is not pre-emptable, therefore its resource needs are additional in the second pool. Formula (7) gives the related calculation of the required multiplicity of wavelengths in the second pool on a given link.

$$^{h*}L_{mn} = \max_{<i,j>}{}^{h*}F_{mn}^{ij} + {}^{h*}L'_{mn} \quad (7)$$

The differences between formulas (6) and (7) express the impact of pre-emptable traffic, since in (6) the maximum in (7) the sum set up the required multiplicity of wavelengths.

4.   ILLUSTRATIVE EXAMPLES

In the chapter some illustrative examples are presented to validate the proposed study approaches and the elaborated ILP models. The network example applied for the illustrations is a 3x3 mesh-torus, which is a regular topology with uniform nodal degree and routing capabilities. The applied traffic pattern is a uniform full mesh one, both for premium and low priority traffics. The illustrations are based on a modular network model, the capacity of a wavelength multiplex system is four wavelengths, equally shared between the first and the second pool. There are no wavelength conversion functions assumed in the nodes, however, flexible wavelength selection function are assumed on the ingress sides, thus the assignment of the available wavelengths to the connections are without any restrictions. The small illustrative studies are focused on the calculation of resource needs in different scenarios. The general approach that the optical channel requests are supposed to be known in advance and the minimum amount of resources required serving them can be calculated.

Figures 3a and 3b depict the resource needs in terms of total amount of required wavelengths and wavelength multiplex modules for different scenarios optimized for the number of wavelengths and number of wavelength multiplex modules required. Comparing the corresponding results established for 1+1 dedicated and shared protection 17% gain in number of required modules, and 40% gain in number of total required wavelengths could be identified due to the spare capacity sharing.

Performing the similar comparisons for scenarios with additional (add.) and preemptable (pre-empt.) handling of low priority traffic, 15% gain in number of required modules, and 17% gain in number of total required wavelengths can be identified due to the pre-emption.

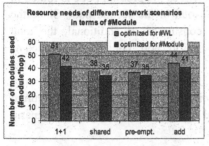
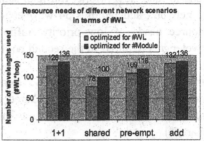

Figure 3a Resource needs of different network scenarios in terms of #Module

Figure 3b Resource needs of different network scenarios in terms of #WL

The tendencies are met the initial expectations in both cases, and the differences in number of wavelengths and in number of modules in case of the two optimization approaches - minimum total used wavelengths, minimum total used modules - are according to the general expectations, as well.

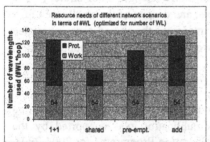
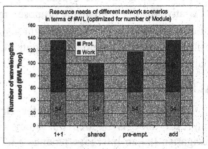

Figure 4a Resource needs of different network scenarios optimized for #WL

Figure 4b Resource needs of different network scenarios optimized for #Module

Figures 4a and 4b give the breakdown of the resource needs depicted in Figure 3b. Based on the results of Figure 4a the resource savings both in terms of wavelengths and modules can be analyzed in details. Due to the same working routes of premium traffic, the wavelengths required in the first pool are the same in all cases (lower parts of the bars). The extra resource needs dedicated to the resilience of premium traffic can be depicted, as well (upper parts of the bars). Analyzing saving of the resilience related resources due to the spare capacity sharing and pre-emption - 66% and 30%, respectively- the main positive features of the SWAP strategy can be identified more clearly.

Concerning the validation of the applied ILP models besides the confirmation of the expected tendencies, and additional cross-checking can be made. Based on the results on Figure 4a the working routes of the premium traffic require 54 wavelengths in total (lower part of the bar for scenario "shared"). The shared capacity oriented restoration of these working routes requires 24 additional wavelengths (upper part of the bar for scenario "shared") taking into account single link failures and guaranteed restoration. Note that in scenario "shared" there is no low priority traffic routed in the network. Besides the protected premium traffic in scenario "add" low priority traffic of same pattern as the premium one is routed in the network, and requires 54 wavelengths in total in the second pool. Since the resource needs of low priority traffic are considered as additional, the total wavelengths required in the second pool in scenario "add" could be interpreted as follows: 54 wavelengths for working routes of premium traffic (lower part of the bar for scenario "shared"), 54 wavelengths for working routes of low priority traffic plus 24 wavelengths for shared capacity oriented restoration of premium traffic, which makes 78 wavelengths in total (upper part of the bar for scenario "shared").

5. SUMMARY AND CONCLUSIONS

A general approach is described to study and evaluate dynamic routing and wavelength assignment strategies applicable in advanced optical networks. Based on the proposed approach some ILP based models are proposed to provide reference results for the evaluation of Separate Wavelength Pool strategy. The illustrative results validate the proposed models, and despite of the small size of the studied example the results highlight the advantages of the shared capacity related resilience and the pre-emption in optical networks providing differentiated optical channel services.

ACKNOWLEDGEMENT

The author wish to thank **Peter Braun** his valuable support implementing some ILP models applied this study. This publication was supported by the Hungarian Italian Intergovernmental S&T Cooperation Programme for 2004-2007, project reference number I-17/03.

REFERENCES

[1] S. Sengupta, R. Ramamurthy, "From Network Design to Dynamic Provisioning and Restoration in Optical Cross-Connect Mesh Networks: An Architectural and Algorithmic Overview", IEEE Network Magazine, July/August 2001

[2] L.Jereb, F.Unghváry, T.Jakab, ``A Methodology for Reliability Analysis of Multi-Layer Communication Networks", *Optical Networks Magazine,* 2 (2001), pp. 42-51.

[3] T. Jakab, H. Nakajima, H-M. Foisel, P. Szegedi, T. Zombori: Routing in translucent networks – Motivations and objectives of EURESCOM P1202 Project, NOC2002, Darmstadt, June 2002.

[4] H. Harai, M. Murata, H. Miyahara: Performance of Alternate Routing Methods in All–Optical Switching Networks, Proc. of the INFOCOM 97

[5] N. Andriolli, T. Jakab, L. Valcarenghi, P. Castoldi: Separate wavelength pools for multiple-class optical channel provisioning , Networks2004, Vienna

DISTRIBUTED WAVELENGTH RESERVATION METHOD FOR FAST LIGHTPATH SETUP IN WDM NETWORKS

Yosuke Kanitani,[1] Shin'ichi Arakawa,[2] Masayuki Murata,[3] and Ken-ichi Kitayama[1]

[1] *Graduate School of Engineering, Osaka University,*
2-1 Yamadaoka, Suita, Osaka 565-0871, Japan
[2] *Graduate School of Economics, Osaka University,*
1-7 Machikaneyama, Toyonaka, Osaka, 560-0043, Japan
[3] *Graduate School of Information Science and Technology, Osaka University,*
1-5 Yamadaoka, Suita, Osaka 565-0871, Japan

Abstract: A promising approach to the effective utilization of wavelength division multi-plexed networks is to transfer data on an on-demand basis using fast wavelength reservation. Data can then be transferred using the assigned wavelength channel. However, if wavelength reservation fails, the lightpath setup delay, which is defined as the time from when the data–transfer request arises at the source node to when the lightpath between the source–destination pair is successfully established, is seriously affected since retrials of wavelength reservation are in turn delayed by propagation delays. In this paper, we propose a new wavelength reservation method to reduce lightpath setup delay. Whereas conventional meth-ods reserve a wavelength in either the forward or backward direction, we propose to reserve it in both directions. We used computer simulations to compare our proposed method with existing methods. The results showed that our method was more efficient except under high traffic loads.

1.   INTRODUCTION

A promising approach to the effective utilization of wavelength division multiplexed (WDM) networks is to transfer the data on an on-demand basis. That is, when a data request arises at a source node, a wavelength is dynam-ically reserved between the source and destination nodes, and a wavelength channel (called a *lightpath* [1]) is configured. After the data is transferred us-ing the lightpath, the wavelength is immediately released.

Two methods have previously been presented to set up lightpaths in a distributed manner [2]. In both methods, the lightpaths are established by exchanging control packets between the source and destination nodes. The actual reservation of the link resources is performed while the control packet is traveling from either the source node to the destination node (i.e., forward direction), or from the destination node to the source node (i.e., backward direction). There have been several studies on reservation schemes aimed at reducing the blocking probability for lightpath requests [2–6]. However, a more important measure for these reservation models is *lightpath setup delay*, which is defined as the time from when the lightpath request arrives at the source node to when a lightpath is successfully configured between the source and destination nodes. Only in [6], lightpath setup delay with the retrial is evaluated but this retrial just employs existing method repeatedly and an improvement method of lightpath setup delay including the retrial is not considered. However, in order to transfer the data, the source node must keep trying to setup a lightpath until the lightpath is successfully configured. Consequently, lightpath setup delay is increased by such retrials due to the link propagation delay along the path. Thus, it is important to improve lightpath setup delay with consideration of retrials.

In this paper, we propose a new wavelength reservation method aimed at reducing lightpath setup delay by increasing the trials of wavelength reservation. More specifically, by integrating two existing reservation method, our method reserve a wavelength in both forward and backward direction, while existing reservation methods reserve a wavelength in *either* forward or backward direction.

The rest of the paper is organized as follows. Section 2 outlines wavelength–routed networks and related work, Section 3 presents our proposed method, Section 4 presents the simulation results, and Section 5 includes a brief summary.

2.   RELATED WORKS

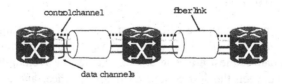

Figure 1.   Wavelength routed network

First, we will describe the structure of our wavelength–routed network. A model of the network is shown in Figure 1. It consists of *optical cross–connects* (OXCs) and optical fibers. Each fiber carries a certain set of wavelengths. Within these sets, one wavelength carries control packets and the other wavelengths are used for data transfer. The control packet controls the setup and/or tear down of lightpaths. Conventional lightpath setup methods for wavelength–routed networks are mainly based on two reservation schemes: *forward reservation* and *backward reservation*. In forward reservation, the source node sends a reservation packet (RESV) immediately when a lightpath request arises. The reservation packet reserves a wavelength from the source node to the destination node. Since the source node has no information on wavelength availability, there is no guarantee that a wavelength will be available in each link along the path. In backward reservation, the source node sends a probe packet (PROBE) toward the destination node. Information on usage of wavelengths along the forward path is collected, but no wavelengths are reserved at this time. Each intermediate node on the forward path only removes wavelengths from the list if those wavelengths are currently in use. Based on the information from the probe packet, the destination node determines a wavelength for reservation, and then sends a RESV packet toward the source node. These forward and backward reservation schemes are illustrated in Figures 2 and 3. Note that, in this paper, we do not consider wavelength conversion facilities. That is, a lightpath uses the same wavelength along the entire path, which is known as *the wavelength continuity constraint* [7].

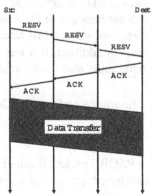
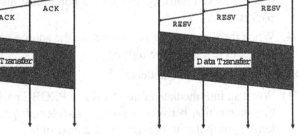

Figure 2. Forward Reservation   *Figure 3.* Backward Reservation

3. OUR PROPOSAL

In both these existing reservation methods, there is only one trial for light-path establishment during the round–trip propagation time. We therefore propose a new method for lightpath setup, based on integrating the forward and backward reservation schemes, which tries to establish a lightpath twice during the round–trip propagation time. Figures 4 and 5 illustrate our proposed scheme. In this scheme, when a lightpath setup request arises at a source node, the source node sends a PROBE packet toward the destination node, as in backward reservation. However, in contrast to backward reservation, when the destination node receives a PROBE packet, it sends not only a RESV packet (or NACK packet) but also a PROBE packet toward the source node. The PROBE packet collects information on wavelength usage from the destination node to the source node, and the source node selects a wavelength based on this information. This retrial scenario is illustrated in Figure 5. The main feature of the proposed scheme is that the edge nodes exchange PROBE packets. Below we explain the details of our proposed reservation scheme.

1. Behavior of the source node

   (S1) When a data transfer request arrives from a terminal, the source node creates a PROBE packet and sends it toward the destination node.

   (S2) When a RESV (or ACK) packet arrives, the source node informs the terminal that a lightpath has been established.

   (S3) When a NACK packet arrives, the source node sends a RESV packet and a PROBE packet. (A NACK packet is always accompanied by a PROBE packet.) This is what happens in the case of reservation failure in the backward direction; the original features of our proposal relate to this behavior. In addition, if a reservation is blocked halfway, the source node must also send a release packet (RLS).

   (S4) When the data transfer is completed, the source node sends a RLS packet to tear down the lightpath.

2. Behavior of intermediate node(s)

   (I1) When an intermediate node receives a PROBE packet, it calculates the intersection between a probed wavelength group and a wavelength group that is available in the next link.

   (I2) When a RESV or RLS packet arrives, an intermediate node reserves or releases the wavelength, respectively.

   (I3) An ACK and a NACK packet are forwarded to the next node without any processing.

3. Behavior of the destination node

    (D1) Basically, the behavior of the destination node is similar to that of the source node. When a PROBE packet arrives, the destination node sends RESV and PROBE packets simultaneously.

    (D2) When a NACK packet arrives, the destination node sends RESV, PROBE, and RLS packets simultaneously. This is similar to (S3).

    (D3) When a RSV packet arrives, the destination node sends an ACK packet toward the source node to notify that a lightpath has been established.

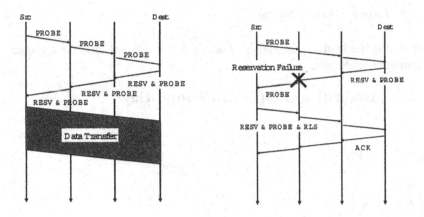

Figure 4.   Proposed scheme (successful case)   *Figure 5.*   Proposed scheme (retrial case)

4. SIMULATION RESULTS

4.1 Simulation model

To evaluate the performance of our proposed scheme, we used computer simulation to compare it with a backward reservation scheme. We used a random network as the simulation topology. Figure 6 shows the topology, which has 15 nodes. Other simulation parameters are briefly described below.

- The number of wavelengths on each link is set to 32.

- Each link has a random propagation delay with mean 1.77 $[ms]$.

- Data-transfer requests arrive according to a Poisson process, and the lightpath is held for a connection-holding period that is assumed to be exponentially distributed with mean $1/\mu$ $[ms]$.

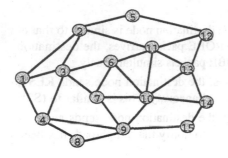

Figure 6. Random Network

We define the load ρ_{np} in Figures 7 and 8 as the offered load for a source–destination node pair.

4.2 Evaluation of Lightpath Setup Delay

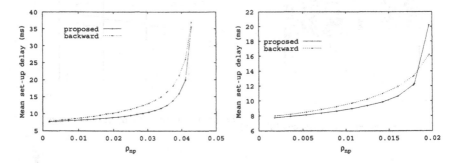

Figure 7. Lightpath setup delay ($1/\mu$ = 100ms) *Figure 8.* Lightpath setup delay ($1/\mu$ = 10ms)

In Figures 7 and Figure 8, we present the mean setup delay dependent on the load ρ_{np}. Figure 7 and Figure 8 show the results when the average of holding time $1/\mu$ is set to 100 [ms] and 10 [ms], respectively. The results showed that our proposed scheme performed better than backward reservation at almost every range, except for $\rho_{np} > 0.018$ in Figure 8. The proposed scheme was inferior to backward reservation in this situation because of the inaccuracy of the probed information. When the information collected by a PROBE packet is not accurate due to link propagation delay, the information becomes out-of-date. If the information is too old, the edge node may select a wavelength that has already been reserved by another node-pair. In the proposed scheme, the edge nodes can make several attempts to send PROBE packets compared with backward reservation, so the accuracy of the information is more important. When the connection-holding period $1/\mu$ is relatively short compared to the link propagation delay (as in Figure 8), the information tends to be less accurate.

4.3   Variation in Lightpath Setup Delay

In a wavelength–routed network, it is preferable to establish a lightpath with little variation in setup delay. If a wavelength reservation method has large variation, it is difficult to achieve stable data transmission. Therefore, it is important to consider variations in lightpath setup delay. In this section, we use the standard deviation (STD) as an indication of variation in delay.

We describe the statistical properties of $1/\mu = 100$ [ms] as an example. These include mean setup delay (Mean) and STD. Table 1 and Table 2 summarize these characteristics dependent on the number of hop–counts "H" of each source–destination node pair.

These tables indicate that the STD resulting from our proposed method is about half that resulting from backward reservation because our method can setup a lightpath in both forward and backward directions. In relation to the discussion in section 4.2, our scheme performs better, especially under a low traffic load.

Table 1.   Variations in lightpath setup delay in proposed scheme

ρ_{np}	H=1		H=2		H=3		H=4	
	Mean	STD	Mean	STD	Mean	STD	Mean	STD
0.0018	3.60	3.72	10.43	4.11	20.35	11.45	43.70	14.66
0.0125	3.82	6.57	11.17	8.02	21.91	11.22	47.31	14.93
0.0232	4.11	9.57	12.12	10.80	24.30	15.59	53.48	21.82
0.0339	4.62	11.99	13.89	14.19	29.65	22.71	68.79	36.07

Table 2.   Variations in lightpath setup delay in backward reservation

ρ_{np}	H=1		H=2		H=3		H=4	
	Mean	STD	Mean	STD	Mean	STD	Mean	STD
0.0018	3.57	7.02	10.53	17.94	20.79	25.86	45.52	27.39
0.0125	4.08	13.18	11.92	15.39	23.73	22.33	51.59	27.83
0.0232	4.64	19.74	13.79	21.72	28.26	32.36	63.19	41.91
0.0339	6.30	23.16	17.02	28.88	37.91	44.92	89.51	65.99

128

5. SUMMARY

In this paper, we presented a new lightpath setup method that reserves wavelengths in both forward and backward directions. The main objective of our method is to reduce lightpath setup delay. Our proposed method, which integrates features of two existing methods, performs lightpath establishment twice within a round–trip, while the previous methods perform it only once. The simulation results indicate that the proposed method performs better except under high traffic loads. We also evaluated other statistical properties of the methods. The results showed that the standard deviation of our method was much smaller than that of the conventional method. In future work, we plan to develop a numerical analysis of lightpath setup delay.

REFERENCES

[1] I. Chlamtac, A. Ganz, and G. Karmi, "Lightpath communications: An approach to high bandwidth optical WAN's," *IEEE Transactions on Communications.*, vol. 40, pp. 1171–1182, July 1992.

[2] X. Yuan, R. Gupta, R. Melhem, R. Gupta, Y.Mei, and C. Qiao, "Distributed control protocols for wavelength reservation and their performance evaluation," *Photonic Network Communications*, vol. 1, pp. 207–218, November 1999.

[3] L. Pezoulas, M. J. Francisco, I. Lambadaris, and C. Huang, "Performance analysis of a backward reservation protocol in networks with sparse wavelength conversion," in *Proceedings of IEEE International Conference on Communications*, vol. 2, pp. 1468–1473, May 2003.

[4] F. Feng, X. Zheng, H. Zhang, and Y. Guo, "An efficient distributed control scheme for lightpath establishment in dynamic wdm networks," *Photonic Network Communications*, vol. 7, pp. 5–15, January 2004.

[5] A. V. Shichani and H. T. Mouftah, "A novel distributed progressive reservation protocol for WDM all-optical networks," in *Proceedings of IEEE International Conference on Communications*, vol. 2, pp. 1463–1467, May 2003.

[6] D. Saha, "An efficient wavelength reservation protocol for establishment in all-optical networks (AONs)," in *Proceedings of IEEE Global Telecommunications Conference*, vol. 2, pp. 1264–1268, November 2000.

[7] Z. Zhang, J. Fu, and D. G. L. Zhang, "Lightpath routing for intelligent optical networks," *IEEE Network*, vol. 15, pp. 28–35, July 2001.

ON-ARRIVAL PLANNING FOR SUB-GRAPH ROUTING PROTECTION IN WDM NETWORKS

Darli A. A. Mello, Marcio S. Savasini,
Jefferson U. Pelegrini and Helio Waldman[1]

[1]*Optical Networking Laboratory, State University of Campinas - UNICAMP*
Av. Albert Einstein, 400 CP 6101, CEP 13083-970, Campinas - SP - Brazil
e-mail {darli}{savasini}{jpelegri}{waldman}@decom.fee.unicamp.br

Abstract:
    In this paper we incorporate the rerouting of backup paths to the Call Admission Control algorithm of Sub-Graph Routing Protection (SGRP). Sub-Graph Routing Protection with On-Arrival Planning (SGRP-OAP) has the two-fold benefit of strongly reducing the blocking probability of the protected system while eliminating the physical rerouting of established connections, which is a major drawback of the original proposal of SGRP. The new scheme is so capacity-efficient that, for all investigated topologies, at low traffic intensity the blocking probability of the system protected against single link failures is the same as the blocking probability of the unprotected system. This is possible because the new protection scheme is extremely effective in using the idle network capacity to provide backup paths.

1. INTRODUCTION

Protection and Restoration are critical issues in the design of wavelength routed WDM networks [1]. Schemes which are efficient in terms of scalability, dynamicity, class of service, restoration speed and network utilization have been extensively researched. In wavelength routed networks traffic can be either static or dynamic. With static traffic connection requests are available all at once, and the path protection problem can be solved by Integer Linear Programming [2]. Conversely, with dynamic traffic connection requests arrive sequentially and exist for a finite duration, which demands heuristic methods for solving the path protection problem [3]. In this paper we consider only dynamic traffic.

Frederick and Somani [4] have recently introduced Sub-Graph Routing Protection (SGRP), a path protection scheme suitable for dynamic traffic that exhibits improved capacity-efficiency, when compared to the already consolidated Backup Multiplexing [3]. However, the original publication on SGRP also highlighted one of its major drawbacks: upon occurrence of a link failure, even connections that do not traverse the faulty link may have to change their path or wavelength to accommodate others, causing inconvenient service interruptions. We call this *altruistic reassignment*, to distinguish it from the regular reassignment that a connection has to undergo when a link that it traverses fails. Altruistic reassignment is a cumbersome maneuver that requires care in order not to disturb costumers that are not directly harmed by the failure.

Initially conceived to support single link failures, SGRP has been also extended to support multiple, node and Shared-Risk Link Group failures [5]. In the same paper the problem involving altruistic reassignment was readdressed by imposing constraints on the RWA algorithm that suppress altruistic path or wavelength reassignment. It has been observed that if altruistic path reassignment is suppressed, but altruistic wavelength reassignment allowed, the blocking probability is slightly reduced for some network topologies, but the overall reassignment probability stays above 90%. Since the reduction in blocking probability is very slight, we will compare our results with the blocking of the original proposal, which is about the same as for the path constrained case. If both path and wavelength altruistic reassignments are suppressed, the blocking probability increases significantly.

We have recently proposed an inter-arrival planning of backup paths [6] that reduces altruistic reassignment while preserving the blocking probability. We exploited the fact that it is cheaper to recalculate backup paths than to perform altruistic reassignment, since the former does not deal with reconfiguration of connections physically established in the network, but only with logical connections stored in the system. Sub-Graph Routing Protection with Inter-Arrival Planning does not modify the Call Admission Control algorithm, exhibiting therefore blocking probabilities similar to the original proposal, and reducing the altruistic reassignment probability, but not completely eliminating it.

In this paper we present the Sub-Graph Routing Protection with On-Arrival Planning scheme (SGRP-OAP), an evolution that exhibits extremely high capacity efficiency at the expense of a moderate computational cost. Contrasting with Sub-Graph Routing Protection with Inter-Arrival Planning, SGRP-OAP modifies the Call Admission Control algorithm of the original proposal, exhibiting blocking probabilities considerably lower than the original proposal while completely eliminating altruistic reassignment, being therefore recommendable when provisioning time is compatible with the processing time of the Call Admission Control algorithm. In this paper we simulate only single link failures, but future work could extrapolate the developed concepts to other

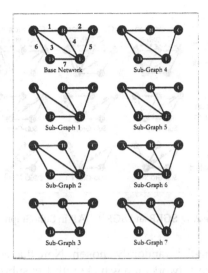

Figure 1. Sub-Graph Routing Protection

scenarios. The remainder of this paper is divided into five sections. Section 2 explains the original proposal of Sub-Graph Routing Protection. In Section 3 the Sub-Graph Routing Protection with On-Arrival Planning scheme is developed. Section 4 presents the methods used to evaluate the performance of the new scheme. Section 5 shows the simulation results, and Section 6 concludes the paper.

2. SUB-GRAPH ROUTING PROTECTION

The main idea of Sub-Graph Routing Protection is rather clever. A network topology can be represented by an undirected graph $G(V, E)$ with a vertex set V and an edge set E. The set V represents nodes and the set E represents bidirectional links. $G(V, E)$ is called the base network. A single failure of edge e_i can be represented by sub-graph $G_i = G - e_i$: the original graph G without edge e_i. In this way all possible single link failures in the network can be represented by L sub-graphs, where L is the cardinality of the edge set E, as depicted in Figure 1. In Sub-Graph Routing Protection, a connection request is only accepted if it can be successfully routed in each of the L sub-graphs and in the base network. In case of a link failure, the network immediately incorporates the state represented by the corresponding sub-graph.

In the strategy proposed by the first paper on Sub-Graph Routing Protection [4] the RWA of connections in the sub-graphs is independent of the base network configuration. In the example of Figure 2, the connection path in the base network and sub-graphs is the shortest path between source and destina-

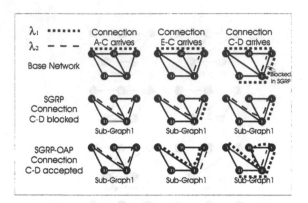

Figure 2. SGRP and SGRP-OAP in Sub-Graph 1

tion nodes, and wavelength is randomly chosen. Note that connection E-C uses wavelength 2 in the base network, and wavelength 1 in sub-graph 1. If a failure occurs on link 1, connection E-C undergoes an altruistic reassignment.

3. SUB-GRAPH ROUTING PROTECTION WITH ON-ARRIVAL PLANNING

In this section we introduce the On-Arrival Planning Strategy for Sub-Graph Routing Protection (SGRP-OAP). Here we modify the Call Admission Control algorithm of the original proposal to reroute backup routes stored in sub-graphs to better adjust the new incoming request. The algorithm is applied after the arrival of each connection request:

In the Base Network:

1. The connection is routed in the physically shortest path;

2. If this is not possible, the connection is blocked.

In each Sub-Graph G_i:

1. Initial state of G_i is saved;

2. All connections are cleared from G_i;

3. Two connection lists are generated, **including the new connection request**: C_F – connections that traverse link i in the base network, C_N – connections that do not traverse link i in the base network;

4. Connections of list C_N are routed in G_i as in the base network;

5. Connections of list C_F are routed using the shortest path among all wavelength planes, in order of arrival;

6. If it is not possible to route all connections, the connection request is blocked and the initial states of the sub-graphs are restored.

Upon the arrival of a connection request, SGRP-OAP "copies" in all sub-graphs the base network configuration of connections which do not traverse the missing link of each sub-graph in the base network. The remaining connections, which would anyway be rerouted, are then routed within the residual capacity using the shortest path among all wavelength planes. In this way altruistic reassignment, an important concern in previous works on SGRP, is eliminated. In the example of Figure 2, when connection E-C arrives, it is included in list C_N of sub-graph 1, whereas connection A-C is included in list C_F, since A-C traverses link 1 in the base network. Connection E-C is routed first, succeeded by connection A-C, according to steps 4 and 5 of the SGRP-OAP algorithm. When connection C-D arrives, it is also included in list C_N of sub-graph 1. So, after connections of list C_N (E-C and C-D) are routed in sub-graph 1, connection A-C has to be again rerouted in the shortest path among the two wavelength planes, which is 3 hops long. Note that the final configuration is free of altruistic reassignment.

4.   PERFORMANCE EVALUATION

We evaluate the performance of SGRP-OAP through simulations performed by the optical networks simulator developed at the OptiNet, Optical Networking Laboratory at the State University of Campinas - UNICAMP. The simulator was written in the Java programming language. Uniform traffic is assumed. The arrival of connection requests follows a Poisson distribution, and the holding time is exponentially distributed. We simulated three network topologies [5]: the 14-node, 23-link NSFNet; 9-node, 18-link 3x3 Mesh-Torus; and the 11-node, 22-link NJLATA. All links are bidirectional with 16 wavelengths. The curves are calculated by averaging the results of the 10 last rounds of a series of 11, each with 1000 connection requests, to simulate a steady state network occupancy.

In previous papers on SGRP mainly two performance metrics have been investigated: blocking probability and reassignment probability. The reassignment probability is the probability that a connection has to be rerouted upon the occurrence of a single link failure anywhere in the network. The reassignment probability consists of two components. The first component is the regular reassignment that a connection has to undergo when a link that it traverses fails, which depends solely on the average connection length and is close to the reassignment probability observed in Backup Multiplexing. The second component is the altruistic reassignment, which is absent when Backup Multiplexing is used. Since in SGRP-OAP no altruistic reassignment is allowed, the reassignment probability strongly decreases when compared to the original proposal, and only the regular reassignment remains. Therefore when SGRP-OAP is used the assessment of the reassignment probability metric is no longer important, and only the blocking probability is investigated. The performance of SGRP-OAP is compared with the Unconstrained RWA case because the latter possesses equal or lower blocking performance when compared to previously published RWA policies.

RWA in the Base Network

Connections are routed in the base network using Dijkstra's shortest path algorithm, in terms of hop count, applied to the physical network topology. If there are more than one shortest path, one is chosen randomly. Wavelength selection follows the Random Fit scheme.

RWA in Sub-Graphs

Unconstrained RWA: refers to the seminal paper on sub-graph routing [4], which in sub-graphs used the same RWA scheme as in the base network.

Sub-Graph Routing Protection with On-Arrival Planning: connection requests are accepted or blocked according to the steps described in Section 3. The RWA of connections in list C_F, step "5" of the SGRP-OAP algorithm, chooses the wavelength with the shortest path, which is determined by applying Dijkstra's shortest path algorithm to all wavelength planes.

We have investigated two metrics to assess the performance of SGRP-OAP: the total blocking probability, and the base network blocking probability. For a connection to be accepted, it must be successfully routed first in the base network, and then in the L sub-graphs. The total blocking probability is the probability of a connection request being blocked because it could not be routed in the base network or subgraphs. The base network blocking probability is the probability of a connection request being blocked because it could not be routed in the base network. It reflects the occupancy of the network without protection, since the allocation of protection resources is related to the routing of connections in sub-graphs.

5. SIMULATION RESULTS AND DISCUSSION

The three simulated network topologies exhibit analogous behaviors, but some effects are more pronounced in certain topologies. The results for the 3x3 Mesh Torus can be found in Figure 3. Here the benefits of SGRP-OAP in comparison with the original proposal are explicit. A remarkable result is that the total blocking probability and the base network blocking probability curves only diverge considerably after a load of 10 Erlangs per node, at approximately 1% blocking probability. This means that for blocking probabilities lower than 1%, protection can be implemented at almost zero cost in terms of capacity. The gains in using SGRP-OAP for the NJLATA topology are still considerable, but less pronounced than for the 3x3 Mesh Torus Topology, as depicted in Figure 4. Again the blocking probability is considerably less than for the Unconstrained RWA case. The total blocking probability and the base network blocking probability curves diverge after 5 Erlangs per node, at approximately 2% blocking probability. The NSFNet topology exhibits the lowest gains when using SGRP-OAP, as shown in Figure 5. The total blocking probability and the base network blocking probability curves diverge after a load of approximately 3 Erlangs per node at near-zero blocking probability.

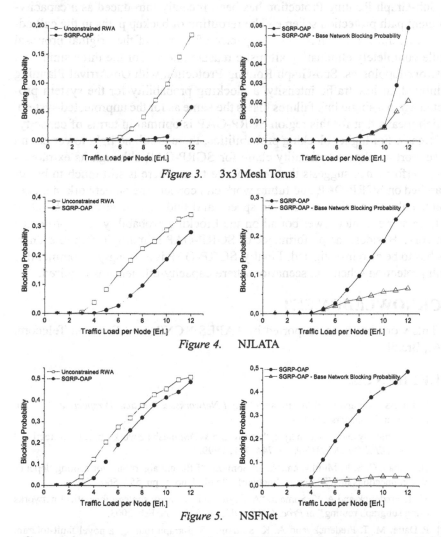

Figure 3. 3x3 Mesh Torus

Figure 4. NJLATA

Figure 5. NSFNet

Concerning computational complexity, SGRP-OAP is fairly implementable if we consider current network dinamicity. Intensive computational time is only needed for provisioning a new connection, when Dijkstra's shortest path algorithm, with worst case complexity $O(N^2)$, is applied to W wavelength planes for source-destination pairs of connections which are subject to rerouting in case of a link failure. Therefore the provisioning time scales with W, L (number of sub-graphs), the network occupancy and the complexity of the shortest path algorithm.

6. CONCLUSION

Sub-Graph Routing Protection has been recently introduced as a capacity-efficient path protection scheme. The rerouting of backup paths in the call admission control algorithm boosts capacity-efficiency of the original proposal while completely eliminating altruistic reassignment. For the three simulated network topologies, Sub-Graph Routing Protection with On-Arrival Planning exhibited, at low traffic intensity, a blocking probability for the system protected against single link failures that is the same as for the unprotected system, which means that for this region SGRP-OAP is optimal in terms of capacity-efficiency. For higher blocking probabilities, there is still no tight lower bound to support an almost-optimality claim for SGRP-OAP, although its extraordinary performance suggests that this can be true. There is still much to be researched on SGRP-OAP, and future work can concentrate on network management, failure isolation, restoration speed, and finding how close SGRP-OAP is from some tighter lower bound on the blocking probability for a protected network. Besides, the performance of SGRP-OAP in multiple-failure scenarios has to be also investigated. Finally, SGRP-OAP is a strongly recommended path protection scheme in scenarios where capacity-efficiency is required.

ACKNOWLEDGMENT

This work has been supported by FAPESP, CNPq and Ericsson Telecom. S.A., Brazil.

REFERENCES

[1] R. Ramaswami and K. N. Sivarajan, *Optical Networks: a Practical Perspective*. Morgan Kaufmann Publishers, 2002.

[2] S. Ramamurthy and B. Mukherjee, "Survivable WDM mesh networks, part I - protection," in *Proc. IEEE INFOCOM'99*, pp. 744–751, 1999.

[3] G. Mohan, C. S. R. Murthy, and A. K. Somani, "Efficient algorithms for routing dependable connections in WDM optical networks," vol. 9, no. 5, pp. 553–566, 2001.

[4] M. T. Frederick and A. K. Somani, "A single-fault recovery strategy for optical networks using subgraph routing," in *Proc. ONDM'03*, pp. 549–568, Feb. 2003.

[5] P. Datta, M. T. Frederick, and A. K. Somani, "Sub-graph routing: a novel fault-tolerant architecture for shared-risk link group failures in WDM optical networks," in *Proc. DRCN'03*, Oct. 2003.

[6] D. A. A. Mello, J. U. Pelegrini, M. S. Savasini, G. S. Pavani, and H. Waldman, "Inter-arrival planning for sub-graph routing protection in WDM networks," in *Proc. ICT'04*, Aug. 2004.

ROUTING AND WAVELENGTH ASSIGNMENT FOR SCHEDULED AND RANDOM LIGHTPATH DEMANDS: BIFURCATED ROUTING VERSUS NON-BIFURCATED ROUTING

Mohamed KOUBAA, Nicolas PUECH, and Maurice GAGNAIRE
Telecom Paris - LTCI - UMR 5141 CNRS, Paris, France

Abstract: We consider the routing and the wavelength assignment (RWA) of scheduled and random lightpath demands in a wavelength switching mesh network without wavelength conversion functionality. Scheduled lightpath demands (SLDs) are connection demands for which the set-up and tear-down times are known in advance as opposed to random lightpath demands (RLDs) which are dynamically established and released according to a random pattern of requests. Two routing strategies are proposed which process the SLDs and the RLDs separately. The first routing strategy allows to bifurcate the traffic on several routes connecting the source to the destination of a demand whereas the second strategy forces atomic routing. The routing strategies are compared through rejection ratio.

1. INTRODUCTION

Routing and wavelength assignment in optical transport networks has been extensively investigated for planning and traffic engineering purposes. RWA problems are often classified according to the nature of the considered traffic demands, namely, whether they are static or dynamic [1]. In the static RWA problem where the requests are known in advance, the problem is to set-up lightpaths while minimizing network resource usage such as the number of requested WDM channels or the number of requested wavelengths [2, 3]. The set of established lightpaths remain in the network for a long period of time. Dynamic RWA deals with connections that arrive dynamically (randomly) [4]. The performance of dynamic RWA is often measured through some network performance metric (e.g., blocking probability also called rejection ratio) [5,6].

In this paper, we deal with the routing and wavelength assignment (RWA) problem in WDM all-optical transport networks. We consider two types of traffic demands: scheduled demands and random demands. Static demands are not considered here since, once established, these demands remain in the network for a long time. This can be seen as a reduction in the number of available wavelengths on some network links. A *scheduled lightpath demand* (SLD) [7] is a connection demand represented by a tuple $(s, d, n, \alpha, \omega)$, where s and d are the source and destination nodes of the demand, n is the number of requested lightpaths, and α, ω are respectively the set-up and tear-down dates of the demand. The SLD model is *deterministic* because the demands are known in advance and is *dynamic* because it takes into account the evolution of the traffic load in the network over time. A *random lightpath demand* (RLD) corresponds to a connection request that arrives randomly and is dealt with on the fly. We use the same tuple notation to describe an RLD. To the best of our knowledge, this is the first time that both deterministic and dynamic traffic demands are considered simultaneously to address the routing and wavelength assignment problem in WDM all-optical transport networks.

The routing strategies studied in this paper aim at establishing random light-path demands on the fly, provided that the RWA for the scheduled lightpath demands has already been calculated. The first routing strategy allows *bifurcated* routing: the requested lightpaths of any demand may follow distinct paths from the source to the destination of the demand. Conversely, the second routing strategy imposes *atomic* routing also called *non-bifurcated* routing: all the requested lightpaths of a demand have to be routed on the same path. We compute the rejection ratio and discuss the advantages for each strategy by measuring the improvement of the rejection ratio via bifurcated routing.

We call a *span* the physical pipe connecting two adjacent nodes u and v in the network. Fibers laid down in a span may have opposite directions. We assume here that a span (u, v) is made of two opposite unidirectional fibers. As opposed to a span, a *link* or a *fiber-link* refers to a single unidirectional fiber connecting node u to node v. The bandwidth of each optical fiber is wavelength-division demultiplexed into a set of χ wavelengths, $\Lambda = \{\lambda_1, \lambda_2, \ldots, \lambda_\chi\}$. When a wavelength is used by a lightpath, it is *busy* (in one direction). A *lightpath* connecting a node s to a node d is defined by a physical route in the network (a *path*) connecting s to d and a wavelength λ such that λ is available on every fiber-link of this route. A *path-free* wavelength is a wavelength which is not used by any lightpath on any fiber-link of the considered path. A lightpath demand is *rejected* (*blocked*) when there are not enough available network resources to satisfy it. The *rejection ratio* (Rr) is the ratio of the number of rejected demands to the total number of lightpath demands arrived at the network.

The remainder of the paper is organized as follows. In Section 2, we describe the mathematical model and the algorithms of the studied RWA strategies. We then (Section 3) propose some simulation results obtained with them and compare our strategies in terms of rejection ratio. Finally, in Section 4, we draw some conclusions and set directions for future work.

2. THE STUDIED RWA ALGORITHMS

In this section, we describe the algorithms used for the routing and wavelength assignment of SLDs and RLDs. Both algorithms deal with the SLDs and the RLDs in two separate phases. The first phase computes the RWA for the SLDs and aims at minimizing the number of blocked SLDs. Taking the assignment of the SLDs into account, the second phase computes the RWA for the RLDs.

2.1 Mathematical model

Here are the notations used to describe a lightpath demand (LD), be it scheduled or random.

- $G = (V, E, \vartheta)$ is an arc-weighted symmetrical directed graph with vertex set $V = \{v_1, v_2, \ldots, v_N\}$, arc set $E = \{e_1, e_2, \ldots, e_L\}$ and weight function $\vartheta : E \to \mathbb{R}_+$ mapping the physical length (or any other cost of the links set by the network operator for example).

- $N = |V|, L = |E|$ are respectively the number of nodes and links in the network.

- D denotes the total number of SLDs and RLDs arrived at the network during the observation period.

- The LD number i, $1 \leq i \leq D$, to be established is defined by a tuple $(s_i, d_i, n_i, \alpha_i, \omega_i)$. $s_i \in V$, $d_i \in V$ are the source and the destination nodes of the demand, n_i is the number of requested lightpaths, and α_i and ω_i are respectively the set-up and tear-down dates of the demand.

- $P_{k,i}, 1 \leq k \leq K, 1 \leq i \leq D$, represents the kth alternate shortest path in G connecting node s_i to node d_i (source and destination of the ith demand). We compute K alternate shortest paths for each source-destination pair according to the algorithm described in [9] (if so many paths exist, otherwise we only consider the available ones).

- $\kappa_{k,i,t} = (\gamma_{1,k}^{i,t}, \gamma_{2,k}^{i,t}, \ldots, \gamma_{\chi,k}^{i,t})$ is a χ-dimensional binary vector. $\gamma_{j,k}^{i,t} = 1$, $1 \leq j \leq \chi$, if λ_j is a path-free wavelength along the path $P_{k,i}$, from s_i to d_i at time t. Otherwise $\gamma_{j,k}^{i,t} = 0$.

- $\sigma_{k,i,t} = \sum_{j=1}^{\chi} \gamma_{j,k}^{i,t}$ is the number of path-free wavelengths along $P_{k,i}$ at time t.

δ will denote an SLD whereas τ will denote an RLD. We also use n_i^δ, $P_{k,i}^\delta$ and $\kappa_{k,i,t}^\delta$ (respectively n_i^τ, $P_{k,i}^\tau$ and $\kappa_{k,i,t}^\tau$) for the parameters representing an SLD (respectively an RLD) when it is necessary to make a clear distinction between scheduled and random demands.

2.2 Bifurcated routing

Our first routing algorithm we called RWABIF [6] allows traffic bifurcation among several paths between the source and the destination of each lightpath demand. We computed K alternate shortest paths connecting the source to the destination of each demand.

2.2.1 Routing and wavelength assignment of the SLDs.

Given a set of SLDs, we want to determine a routing and a wavelength assignment that minimizes the rejection ratio. We define the following additional notations:

- $\Delta = \{\delta_1, \delta_2, \ldots, \delta_M\}$ is the set of SLDs to be established.

- (G, Δ) is a pair representing an instance of the SLD routing problem.

- a vector $(\rho_{1,1}, \rho_{2,1}, \ldots, \rho_{K,1})$ is associated to the demand δ_i. The element $\rho_{k,i}$ indicates the number of lightpaths to be routed along $P_{k,i}^\delta$, the kth alternate shortest route for SLD δ_i.

- $\rho_\Delta = ((\rho_{1,1}, \rho_{2,1}, \ldots, \rho_{K,1}), (\rho_{1,2}, \rho_{2,2}, \ldots, \rho_{K,2}), \ldots, (\rho_{1,M}, \rho_{2,M}, \ldots \rho_{K,M}))$ is called an admissible routing solution for Δ if $\sum_{k=1}^K \rho_{k,i} = n_i^\delta$, $1 \le i \le M$.

- π_Δ is the set of all admissible routing solutions for Δ.

- $\mathcal{C} : \pi_\Delta \to \mathbb{N}$ is the function that counts the number of blocked SLDs for an admissible solution. The combinatorial optimization problem to solve is:

$$\text{Minimize} \quad \mathcal{C}(\rho_\Delta)$$

$$\text{subject to} \quad \rho_\Delta \in \pi_\Delta$$

More details and examples explaining the mathematical model and the way the RWA for the SLDs is computed can be found in [6].

We used a Random Search (RS) algorithm to find an approximate minimum of the function $\mathcal{C}$. The wavelengths are assigned according to a First-Fit scheme.

2.2.2 Routing and wavelength assignment of the RLDs.

Once the RWA for the SLDs has been established, we deal with the RLDs sequentially, that is demand by demand at arrival dates. When a new RLD arrives, one tries to route all the requested lightpaths on the shortest path, if it is possible (i.e. if there are as many available path-free wavelengths along the shortest path as the requested number of lightpaths), otherwise, several paths between the source and the destination of the demand are used whenever the cumulated number of path-free wavelengths along the considered shortest paths is at least equal to the number of the requested lightpaths, otherwise, the demand is rejected. As they require fewer WDM channels, the path-free wavelengths with shorter paths are preferred to those with longer ones. More details can be found in [6].

2.3   Non-bifurcated routing

Our second routing algorithm called RWANBIF imposes non-bifurcated routing: all the requested lightpaths by a LD have to follow the same path between the source node and the destination node of the demand. Traffic splitting is thus prohibited. Again, we computed K alternate shortest paths between the source and the destination of each demand. For each lightpath demand, we try to route the requested number of lightpaths along one of the K shortest paths if this is possible otherwise the LD is rejected.

2.3.1   Non-bifurcated routing and wavelength assignment of the SLDs.

Given a set of SLDs, we want to determine a routing and a wavelength assignment that minimize the rejection ratio taking into account the fact that for each SLD, all requested lightpaths have to be routed on a same path between the source and the destination of the demand. We use the same notations as in subsection 2.2.1. Note that this time $\rho_\Delta = ((\rho_{1,1}, \rho_{2,1}, \ldots, \rho_{K,1}), (\rho_{1,2}, \rho_{2,2}, \ldots, \rho_{K,2}), \ldots, (\rho_{1,M}, \rho_{2,M}, \ldots, \rho_{K,M}))$ is called an admissible routing solution for Δ if for each SLD i $(1 \leq i \leq M)$, there exists a unique j, $1 \leq j \leq K$ such that $\rho_{j,i} = n_i^\delta$ and $\rho_{k,i} = 0$ for each path k, $(1 \leq k \leq K)$, k different from j.

Again, we use a Random Search (RS) algorithm to find an approximate minimum of the function $\mathcal{C}$.

2.3.2   Non-bifurcated routing and wavelength assignment of the RLDs.

Once the RWA for the SLDs have been calculated, we establish the RLDs sequentially. All the lightpaths of an RLD are routed through the same path as opposed to the algorithm of subsection 2.2.2. Whenever there are enough available wavelengths on two distinct paths, the shortest one is preferred as it will use less WDM channels.

3.   EXPERIMENTAL RESULTS

In this section we experimentally evaluate the algorithms proposed in the previous sections. We used two network topologies: the former is the 14-node NSFNET network, the latter is the hypothetical US backbone network of

29 nodes and 44 spans. Due to space limitation we present only the results obtained with the 29-node network. For the sets Δ, the source/destination nodes are drawn according to a random uniform distribution in the interval $[1, 14]$ (respectively $[1, 29]$) for the 14-node network (respectively the 29-node network). We also used uniform random distributions over the intervals $[1, 5]$ and $[1, 1440]$ for the number of lightpaths and the set-up/tear-down dates of the SLDs respectively. We assume observation periods of about a day (1440 is the number of minutes in a day). Random connection requests (RLDs) arrive according to a Poisson process with an arrival rate $\nu = 1$ and if accepted, will hold the circuit for exponentially distributed times with mean $\mu = 250$ much larger than the cumulated round-trip time and the connection set-up delay. The number of lightpaths required by an RLD is drawn from a random uniform distribution in the interval $[1, 5]$. We call a scenario the set of demands be they scheduled or random that occur from start to finish of a day. We assume that we compute $K = 5$ alternate shortest paths between each source/destination pair and that there are $\chi = 24$ available wavelengths on each fiber-link in the network. We want to assess the gain obtained using the RWABIF algorithm compared to the RWANBIF algorithm.

We generated 25 test scenarios, run the two algorithms on them and computed rejection ratio averages for each algorithm. Figure 1 shows the average number of rejected SLDs and RLDs when D, the total number of lightpath requests arrived at the network, varies from 50 to 500. Each couple of bars shows the average number of blocked demands computed using the RWABIF (left bar) and the RWANBIF (right bar) algorithms respectively. Each bar is divided into two segments. The height of the black segment indicates the average number of rejected SLDs whereas the height of the white one shows the number of rejected RLDs.

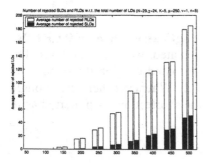

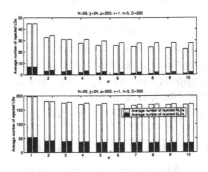

Figure 1. average number of rejected SLDs and RLDs w.r.t. the total number of LDs arrived at the network

Figure 2. average number of rejected SLDs and RLDs w.r.t. K

We notice that the global rejection ratio (total number of blocked SLDs and RLDs) increases when D increases. The figure also shows that the RWABIF algorithm computes a lower global rejection ratio than the global rejection ratio computed by the RWANBIF algorithm. A global rejection gain of 5% is observed. Multipath routing provides better usage of network resources. Note that the number of rejected SLDs computed by the RS algorithm is higher in the case when the traffic is not bifurcated.

In the following figures, we present simulation results obtained with two different values of D (250 and 500). Each figure shows two subfigures. The upper (respectively lower) subfigure shows results obtained when $D = 250$ (respectively $D = 500$).

Figure 2 shows the average number of rejected SLDs and RLDs when K, the number of alternate shortest paths, varies from 1 to 10. We notice that the average number of blocked demands falls when the value of K increases and becomes roughly constant when $K \geq 5$. Note that the global rejection ratio is always higher in the case when the RWANBIF algorithm is used.

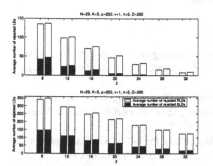

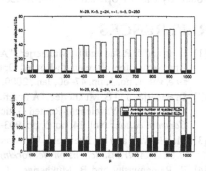

Figure 3. average number of rejected SLDs and RLDs w.r.t. χ

Figure 4. average number of rejected SLDs and RLDs w.r.t. μ

Figure 3 shows the average number of rejected LDs when χ, the number of available wavelengths on each fiber-link of the network, varies from 8 to 32. We notice that the average number of rejected LDs falls when the the number of available wavelengths goes from 8 to 32. The RWABIF algorithm still services more lightpath demands than the RWANBIF algorithm.

Figure 4 shows the number of rejected LDs w.r.t. μ, the mean duration of RLDs. As μ increases, more RLDs are rejected and hence the global rejection ratio increases. This is due to the fact that when the mean duration of RLDs increases, the network resources, when affected, are occupied by the RLDs for longer periods and no sufficient resources are available to service an arriving RLD. The average number of rejected SLDs remains roughly constant because SLDs are routed independently from RLDs.

4.   CONCLUSIONS

In this paper, we proposed two new routing algorithms to deal with the routing and wavelength assignmemnt of both scheduled and random lightpath demands in an all-optical transport network. Both algorithms are presented and compared through rejection ratio. The algorithms process in two separate steps and first compute the RWA for the SLDs and then consider the RLDs on the fly. The RWABIF algorithm allows bifurcated routing as opposed to the RWANBIF algorithm which imposes that all the requested lightpaths of a lightpath demand have to be routed on the same path. We computed the average number of rejected SLDs and RLDs for different simulation scenarios. The results show that a global rejection gain of 5% is obtained with the RWABIF algorithm at the cost of a significant complexity increase resulting from signalisation messages necessary to the establishment of the various paths in the case when bifurcated routing is allowed.

Future work will focus on how decreasing the rejection ratio. Wavelength rerouting techniques may alleviate the wavelength continuity constraint imposed by the all-optical cross-connect switches.

REFERENCES

[1] Zang and J. P. Jue and B. Mukherjee, "A Review of Routing and Wavelength Assignment Approaches for Wavelength-Routed Optical WDM Networks", *Optical Networks Magazine*, vol. 1, no. 1, pp. 47–60, Jan., 2000.

[2] D. Banerjee and B. Mukherjee, "A Practical Aprroach for Routing and Wavelength Assignment in Large Wavelength-Routed Optical Networks", *IEEE Journal on Selected Areas in Communications*, vol. 14, no. 5, pp. 903–908, Oct., 1996.

[3] M. Alanyali and E. Ayanoglu, "Provisioning Algorithms for WDM Optical Networks", *IEEE/ACM Transanctions on Networking*, vol. 7, no. 5, pp. 767–778, Oct., 1999.

[4] R. Ramamurthy and B. Mukherjee, "Fixed-Alternate Routing and Wavelength Conversion in Wavelength-Routed Optical Networks", *IEEE/ACM Transactions on Networking*, vol. 10, no. 3, pp. 351–367, 2002.

[5] R. Ramaswami and K.N. Sivarajan, "Routing and Wavelength Assignment in All-Optical Networks", *IEEE/ACM Transactions on Networking*, vol. 3, no. 5, pp. 489–500, Oct., 1995.

[6] M. Koubaa and N. Puech and M. Gagnaire, "Routing and Wavelength Assignment of Scheduled and Random Lightpath Demands", *Wireless and Optical Communications Networks*, Jun. 7–9, 2004.

[7] J. Kuri and N. Puech and M. Gagnaire and E. Dotaro and R. Douville, "Routing and Wavelength Assignment of Scheduled Lightpath Demands", *IEEE JSAC Optical Communications and Networking Series*, vol. 21, no. 8, pp. 1231–1240, Oct. 2003.

[8] I. Chlamtac and A. Ganz and G. Karmi, "Purely Optical Networks for Terabit Communication", *Proc. IEEE INFOCOM*, pp. 887–896, 1989.

[9] D. Eppstein, "Finding the k Shortest Paths", *SIAM Journal of Computing*, vol. 28, no. 2, pp. 652–673, 1998.

SEMI-LIGHTPATH APPROACH FOR BANDWIDTH GUARANTEED PROTECTION IN IP- OVER-WDM NETWORKS

R.Gangopadhyay[1] , G.Prati[2], N.Rao[3]

[1,3] *Department of Electronics and Electrical Communication Engineering*
Indian Institute of Technology, Kharagpur, India.
Tel:+913222-78028,Fax:+91 3222-55303
e-mail: `ranjan@ece.iitkgp.ernet.in`

[2] *i*
Scuola Superiore Sant'Anna, Piazza Martiri della Libertà 33, 56127 Pisa, Italy
Tel:+39-050-970719, Fax:+39-050-9711208
e-mail: giancarlo.prati@cnit.it

Abstract: The paper presents efficient algorithms for routing and wavelength assignment for dynamic lightpath establishment as well as bandwidth guaranteed path protection against single link failure in a WDM mesh network. Several protection strategies for multi-wavelength optical network have been studied to indicate their relative efficiency in terms of network blocking probability performance and resource requirement. Protection results in integrated routing scenario are also presented.

1 INTRODUCTION

The current trend in research on optical networking is focused on various issues related to IP (Internet Protocol) over WDM (Wavelength Division Multiplexing) technology. These may include transmission aspects in the physical fiber layer, and switching, routing, control and management issues in the data link layer. One of the important functionalities of the control layer is to route IP traffic from a source node (s) to the destination node (d) utilizing a lightpath (same wavelength in all the links determining the path from s to d) or a semilightpath (a chain of wavelengths assigned link by link from s to d). In general, the semilightpath algorithm (SLP) achieves simultaneous routing and wavelength assignment (RWA) for dynamic lightpath provisioning in a WDM network with or without λ-conversion facility at nodes. In the past, this algorithm [1] has been used extensively for WDM network design but without adequate consideration for network survivability.

In the present paper we have exploited the SLP algorithm for solving the RWA problem in a WDM mesh network for a single link failure case and simultaneously guaranteeing cent percent protection to the disrupted traffic. Various strategies of protection: dedicated, partial shared and complete information have been implemented. In addition to the wavelength routing scheme, various strategies of protection such as dedicated path protection and shared path protection in the integrated routing scenario have also been investigated.

2 PATH PROTECTION STRATEGIES

We have considered four path protection schemes in our work, namely, dedicated path protection, shared path protection, partial shared path protection with exact reservation and complete shared path protection We consider a network of N nodes and m links. The call setup request k is considered as a duplex (s_k, d_k). For the request k, s_k specifies the ingress node and d_k specifies the egress node. For each request, both an active path and a corresponding backup path have to be set up. Since all calls are to be protected, both active and backup paths need to use link disjoint paths.

We consider the following definitions [2] that will be required for various protection strategies we have adopted in our present studies. Let

A_{ij} denotes set of active paths that use link (i, j)

B_{ij} denotes set of back-up paths that use link (i, j)

F_{ij} denotes total bandwidth reserved by active paths that use link (i, j)

G_{ij} denotes total bandwidth reserved by backup paths that use link (i, j)

and the residual bandwidth $R_{ij} = C_{ij} - F_{ij} - G_{ij}$, where C_{ij} is the capacity of link (i,j).

2.1 Dedicated protection

In this case the only information known at the time of routing the current request is the residual bandwidth R_{ij} for each link (i, j) in the network . Since no information is known other than R_{ij} we have no means of finding out which active paths can share the backup links. Hence, the complete resource required by the requested call has to be reserved on both active and backup paths.

The algorithm works as follows:
- STEP 1: Generate the logical topology for the network.
- STEP 2: Determine the active path (shortest path) for the requested call in the logical topology using the SLP algorithm
- STEP 3: Determine the shortest link-disjoint backup path for the requested call in the network using the SLP algorithm
- STEP 4: If both the active and the backup paths are available, the call is routed through the computed path, otherwise the requested call is blocked.

2.2 Complete shared protection

In this case the information available is the set of active paths A_{ij} that use link (i, j) and the set of backup paths B_{ij} that use link (i, j). The algorithm is same as the dedicated protection in section 2.1, except STEP 3 as follows:

- STEP 3: Compute the new set of costs for each link on the network. Let δ_{ij}^{uv} represent the total number of calls that use link (i, j) for active path and link (u, v) for backup path. A failure of any link used by the active path whose backup path is to be computed would lead to a capacity demand of not more than $(\delta_{ij}^{uv} +1)$ on link (u, v) that has to be used for the backup path of the active path under consideration. So, the cost for every link may be defined as [3]

$$
\begin{aligned}
\phi_{ij}^{uv} &= 0, && \text{if } \delta_{ij}^{uv} +1 \le G_{uv} \text{ and } (i, j) \ne (u, v) \\
&= (\delta_{ij}^{uv} +1 - G_{uv}) && \text{if } \delta_{ij}^{uv} +1 > G_{uv} \text{ and} \\
& && R_{uv} \ge (\delta_{ij}^{uv} +1 - G_{uv}) \text{ and } (i, j) \ne (u, v) \\
&= \infty, && \text{otherwise.}
\end{aligned}
$$

The task is to determine the shortest link disjoint backup path for the requested call in the network with new set of weights assigned to each link.

2.3 Partial shared protection

In this case for every link (i,j) the aggregate bandwidth information F_{ij}, G_{ij} and R_{ij}. are available. The information is independent of the number of calls that are currently present in the network. The algorithm for partial shared protection case is same as dedicated protection with only the suggested change in STEP 3 as follows:

- STEP 3: Compute the cost for each link on the network as

$$
\begin{aligned}
C_{uv} &= 0, && \text{if } (M+1) \le G_{uv} \\
&= M +1 - G_{uv} && \text{if } (M+1) > G_{uv} \text{ and } R_{uv} > (M +1 - G_{uv}) \\
&= \infty && \text{otherwise}
\end{aligned}
$$

where, C_{uv} is the cost associated with link (u, v) in the backup path and M representing the largest value of F_{ij} for some link (i, j) in the active path. Determine the shortest link-disjoint backup path for the requested call in the network with new set of weights assigned to each link.

2.4 Partial shared protection with exact reservation

In the partial shared protection the link cost is based on a conservative value of M. A better performance of the algorithm can be achieved by exact reservation [3]. Let M_1 represent the active bandwidth in F_{ij} in the active part for direct (s,d) pair calls only and M_2 represent the largest value of F_{ij} excluding M_1 value in the active path whose backup path is going to be shared by current backup bandwidth and $M = M_1 + M_2$. The algorithm for partial shared protection with exact reservation case is the same as the partial shared case except STEP 3 is changed as follows:

- STEP 3: Compute the cost for each link in the network as

$$C_{uv} = 0, \qquad \text{if } M+1 \leq G_{uv}$$

$$= M + BW - G_{uv} \quad \text{if } M+1 > G_{uv} \text{ and}$$

$$R_{uv} \geq (M +1- G_{uv})$$

$$= \infty, \qquad \text{otherwise.}$$

Determine the shortest link disjoint backup path for requested call in the network with new set of weights assigned to each link.

3 PROTECTION TECHNIQUES IN INTEGRATED ROUTING

In this section we consider the problem of integrated routing for restorable connections with and without backup sharing. The routing protocol makes use of the wavelength usage on physical links and bandwidth usage on IP logical links. We consider two protection strategies for integrated routing in IP over WDM networks viz. 1) dedicated path protection and 2) shared path protection.

4.1 Dedicated path protection

We consider the following definitions that are required for dedicated path protection strategy in integrated routing. Let

$R_{uv, \lambda} \rightarrow$ Residual capacity on λ^{th} wavelength on link (u,v).
$BW \rightarrow$ Call request bandwidth.
$C_{uv, \lambda} \rightarrow$ Weight of λth wavelength on link (u,v).

In this strategy, the information available to the path selection algorithm is only the residual capacities on each wavelength link. Hence, the complete resource required by the requested call has to be reserved on both active and backup paths. The algorithm works as follows:

- STEP 1: Construct the wavelength graph.
- STEP 2: Assign the weight to each wavelength link as given below.

$$C_{uv, \lambda} = 1 \quad \text{if} \quad R_{uv, \lambda} \geq BW$$
$$= \infty \quad \text{otherwise.}$$

- STEP 3: Compute the active path.
- STEP 4: Remove the fiber links through which the active path traverses.
- STEP 5: Assign the weight to each wavelength link as given below.

$$C_{uv, \lambda} = 1 \quad \text{if} \quad R_{uv, \lambda} \geq BW$$
$$= \infty \quad \text{otherwise.}$$

- STEP 6: Compute the backup path.
- STEP 7: If both active and backup paths are found, admit the call into the network. Otherwise block the call.

4.2 Shared path protection

We consider the following definitions that are required for shared path protection strategy in integrated routing. Let

$R_{ij, \lambda} \rightarrow$ Residual capacity on λ^{th} wavelength on link (i ,j).

$F_{ij, \lambda} \rightarrow$ Active capacity on λ^{th} wavelength on link (i ,j).

$G_{ij, \lambda} \rightarrow$ Backup capacity on λ^{th} wavelength on link (i ,j).

BW $\rightarrow$ Call request bandwidth.

$C_{ij, \lambda} \rightarrow$ Weight of λth wavelength on link (i , j).

$\phi_{ij}^{uv, \lambda} \rightarrow F_{ij} \cap G_{uv, \lambda}=$ set of demands that use link (i,j) on active path and wavelength link (λ, u, v) on the back up path.

$\delta_{ij}^{uv, \lambda} \rightarrow$ Sum of all demand values of set $\phi_{ij}^{uv, \lambda}$.

M $\rightarrow$ Maximum value of $\delta_{ij}^{uv, \lambda}$ for all (i,j) belonging to active path.

The algorithm given in section 4.1 is valid here also except STEP 5.

- STEP 5: Assign the weight to each wavelength link as given below.

$$C_{uv, \lambda} = 0 \qquad \text{if } (M + BW) \leq G_{uv, \lambda}$$
$$= (M+BW- G_{uv, \lambda}) \quad \text{if } (M + BW) > G_{uv, \lambda} \text{ and}$$
$$(M + BW- G_{uv, \lambda} \leq R_{uv, \lambda})$$
$$= \infty \qquad \text{otherwise}$$

Applying the SLP algorithm to the weighted network graph, the edge disjoint backup path can be found out.

5. Simulation Results

We simulate a dynamic network environment with the assumptions that the call requests process is Poisson and the call holding time folllows a negative exponential distribution. Figure 1 shows a comparison of the blocking probability P_b versus the total network load for a 14- node NSFNET utilizing 8 wavelengths for dedicated and partial shared protection cases. It may be noted that partial shared protection with exact reservation ensures the best blocking performance compared to the other schemes considered especially at lower loads.

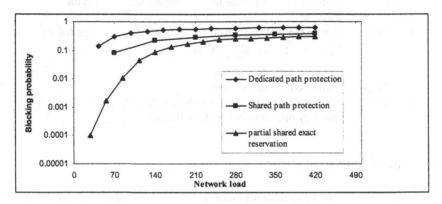

Figure. 1. P_b vs. the network load for a 14 –node NSFNET, W = 8

Figure 2 shows the resource usage in terms of required number of wavelengths versus network load satisfying $P_b \leq 0.05$ for several protection strategies. The shared protection with exact reservation demands the minimum network resource.

In the case of integrated routing using the same NSFNET topology with 16 wavelengths, the results for the blocking probability and the network resource usage against network load for the dedicated protection as a function of R and BW are shown in Figs. 3 and 4 respectively. The performance is seen to improve significantly with increased number of router nodes and sub-lambda traffic grooming. Similar trend is also noticed in the results for the integrated shared path protection case with much improved performance compared to those in dedicared protection case as depicted in Fig.5

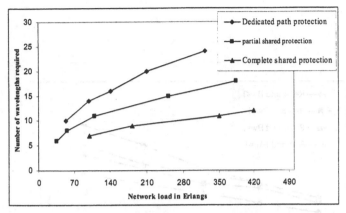

Figure 2. Resource usage versus network load, for $P_b \leq 0.05$ for
several protection strategies

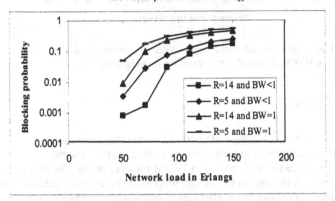

Figure 3. P_b versus network load for a 14-node NSFNET, dedicated path
protection, R ≡ number of router nodes, BW ≡ call bandwidth, W=16

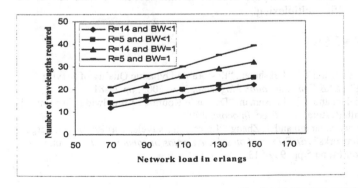

Figure 4. Number of wavelengths required versus network load for a 14-
node NSFNET dedicated path protection in integrated routing.

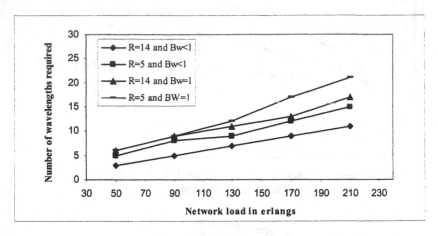

Figure 5 Number of wavelengths required versus network load for a 14-node NSFNET shared path protection in integrated routing.

5 CONCLUSION

Based on SLP algorithm in finding both active and backup paths, several bandwidth guaranteed protection strategies depending on the link state information available have been applied in WDM mesh network.The simulation results indicate that partial shared protection with exact reservation achieves the best network blocking probability performance and resource requirement. In the integrated routing scenario in IP-over-WDM network shared path protection achieves significant performance gain if the network contains more number of router nodes than pure OXC nodes and facilitates more sub-lamda granurality traffic multiplexing.

REFERENCES

[1]. M. Kodialam and T.V. Lakshman, "Restorable Dynamic Quality of Service Routing", *IEEE Communication Magazine*, June 2002, pp 72-81.
[2]. M. Kodialam and T.V. Lakshman, "Dynamic Routing of Bandwidth Guaranteed paths with Restoration", *Proc. Infocom*, 2002.
[3]. I.Chlamtac, A.Farago and T. Zhang "Lightpath (wavelength) Routing in Large WDM Networks" *IEEE Journal on selected areas in communications*, June 1996, vol. 14, no.5 pp: 909-913.

COMPARISON OF *p*-CYCLE CONFIGURATION METHODS FOR DYNAMIC NETWORKS

Dominic A. Schupke
Dominic A. Schupke, Siemens AG, Corporate Technology, Information and Communications, Otto-Hahn-Ring 6, 81730 Munich, Germany
dominic.schupke@siemens.com

Abstract: Dynamic optical networks which are protected by *p*-cycles can be operated by different *p*-cycle configuration methods. We compare the blocking probability of two dynamic *p*-cycle configuration approaches and one static *p*-cycle configuration approach (protected working capacity envelope).

1.   INTRODUCTION

In this paper we evaluate *p*-cycles [1,2,3] in optical networks with dynamic connections. A *p*-cycle is a cycle in a network and can protect working capacity of links which have both end points on the *p*-cycle. These links include the links covered by the *p*-cycle as well as those non-covered links which join nodes of the *p*-cycle.

Protected connections are an important transport network service not only for existing, but also for emerging networks [4]. Even connections with short holding times can require protection, since a failure of an established connection may slow down or disrupt the user's application. Thus, besides it is desirable to have low connection blocking, an established connection should also have high availability, which we ensure in our consideration by *p*-cycles.

This paper is organized as follows. In Section 2 we briefly recapitulate the *p*-cycles protection concept and discuss several connection management approaches for dynamic networks in Section 3. In Section 4 we describe the assumed dynamic demand model and in Section 5 we evaluate the connection management approaches using simulation. In Section 6 we draw several conclusions.

2.   *p*-CYCLES

Figure 1 depicts the protection principle of *p*-cycles for link protection. The *p*-cycle in Figure 1(a) is preconfigured as a closed connection on the cycle B-C-D-F-E-B. Preconfiguration means that the configuration is done before a failure occurs. The *p*-cycle is able to protect working capacity on its own links, called on-cycle links, as shown in Figure 1(b). Upon failure of on-cycle link B-C, the *p*-cycle offers protection by the route on the remainder of the cycle (C-D-F-E-B). The protectable capacity on on-cycle links is thus one capacity unit. The protection of on-cycle links is logically equal to multiplex-section shared protection rings (MS-SPRings) in SDH and bidirectional line-switched rings (BLSRs) in SONET.

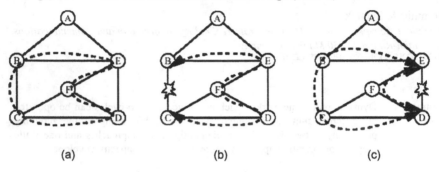

(a)   (b)   (c)

Figure 1: Protection principle of *p*-cycles for link protection.

Unlike rings, however, *p*-cycles also protect links outside the *p*-cycle path: Each link which has both its end points on the *p*-cycle can also be protected. These links are called straddling links. Figure 1(c) shows the protection of such a link (E-D). We can provide two protection routes for straddling links, in the example, routes E-B-C-D and E-F-D. In effect, we can protect two working capacity units of straddling links.

3.   CONNECTION MANAGEMENT APPROACHES

We investigate three approaches to management of dynamic connections. The first two approaches can change *p*-cycles per new connection (dynamic *p*-cycles). In the third approach we assume an unchangeable *p*-cycle configuration (static *p*-cycles). In any case, we do not reconfigure existing working connections to accommodate newly arriving connections. All operations on the *p*-cycles are hitless for the working paths, since they affect the protection configuration only.

We can describe the link resources of the network, at a decision point in time i, by their capacity vectors for working (w_i) and protection (p_i). In addition, we have a vector for the installed capacity (c). For a new demand arrival at time instance i, we have to decide whether a connection (requiring working capacity Δw_i) can be admitted to the network or not.

We propose three methods in the next subsections.

3.1 Routing in Spare Plus Protection Capacity (RSPC)

In this method, we accept a new connection demand if it can be protected by p-cycles after it is virtually routed in the non-working capacity $c - w_{i-1}$, i.e., if a feasible p_i exists for $w_{i-1} + \Delta w_i$. If it is possible, we release the p-cycle configuration, set-up the new connection ($w_i = w_{i-1} + \Delta w_i$), and configure the new p cycles (p_i). Otherwise we block the request and leave the configuration ($w_i = w_{i-1}$, $p_i = p_i-1$).

3.2 Routing in Spare Capacity (RSC)

Now, we accept a new connection if it is routable in the spare capacity $c - w_{i-1} - p_{i-1}$ and can be protected by p-cycles, i.e., if a working path (Δw_i) exists in the spare capacity and we find a feasible p_i for $w_{i-1} + \Delta w_i$. If it is possible, we set-up the new connection and reconfigure the p-cycles. Otherwise we block the request and leave the configuration. This approach may achieve shorter times, during which the p-cycles are in the process of reconfiguration, than RSPC.

3.3 Routing in Protected Working Capacity Envelope (RPWCE)

In this principle (see [2,5]), we assume a static protected working capacity envelope (PWCE) in which connections are set up and torn down. While the PWCE can also be dynamic, with less frequent change actions than for connection requests, a static PWCE is clearly the easiest solution and can model a dynamic PWCE in quasi-static operation points.

In comparison with the above two approaches, the static PWCE is even simpler, since connection management deals only with configuration of paths and not with configuration of both paths and p-cycles.

For all time instances i, we have a fixed amount of protection capacity $p_i = p$ and a fixed protectable working capacity w', i.e., the PWCE. We accept a new connection if it is routable in the PWCE, i.e., if a working path (Δw_i) exists in the PWCE which fulfils $w_{i-1} + \Delta w_i \leq w'$. Otherwise we block the request and leave the configuration. An accepted connection is automatically protected by the p-cycles.

4. DYNAMIC DEMAND MODEL

We derive a dynamic demand model from a static traffic matrix (which may be prescaled). For every demand unit, we implement one client dynamically requesting services, e.g., a general client interfacing by a user-to-network interface (UNI) or a requesting multilayer control instance setting up a connection for an IP router port pair. A single client follows an on-off-state model with a time distribution for the on-state and a time distribution for the off-state, regardless whether the network can provide a connection or not [4]. In other words, a client tries to set up a connection at off-to-on transition and releases an established connection for it at on-to-off transition.

Clients request connections independently from each other. If not enough available resources are found in the network, the request will be rejected. The estimation for the network-wide blocking probability using a sequence of connections requests is the ratio of rejected requests over generated requests.

5. EVALUATION

We evaluate the blocking performance of the hypothetical Germany network (17 nodes, 26 links) in [6]. We assume full wavelength conversion and length-based cost. We dimension it for the unscaled demand in [6] using firstly shortest path routing, yielding working capacity w_{dim}, and secondly p-cycle optimization [1,3], yielding protection capacity p_{dim}.

Dynamic and static p-cycles use this dimensioned capacity differently during network operation. For dynamic p-cycles (RSPC and RSC), we find connections and p-cycles in the aggregate capacity, i.e., $c = w_{dim} + p_{dim}$. Also for dynamic requests, we firstly route an admitted connection and secondly we configure the p-cycles. For static p-cycles, we find connections in the RPWCE, i.e., $w' = w_{dim}$. The overall capacity in the network is the same for all cases.

As in [4], the time distribution for the on-state is deterministic and the time for the off-state is exponentially distributed. The ratio of the on-state time to the mean time for the off-state is 1/12, i.e., if we take 12 dynamic traffic sources between a given pair of nodes, we will have the same total traffic demand in the dynamic network on average, as in the static traffic matrix (e.g., clients use their connections continuously for 1 hour out of 12 hours on average).

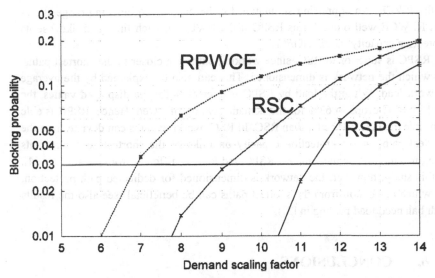

Figure 2: Blocking of dynamic connections over demand scaling factor.

Since an analytical treatment of dynamic configuration (paths and p-cycles) appears to be too complex, we evaluate the blocking performance by discrete event simulation. Although the network is dimensioned with the shortest paths, we use shortest available path routing with physical length as metric, since, except for high blocking levels, it yields lower blocking probabilities than shortest path routing. For RSPC and RSC, the computation of the p-cycles (p_i) has to finish within 10 s, which can be regarded as adequate time for an on-demand request [4]. We also allow an optimality gap of 1%.

Figure 2 depicts the simulation results for RSPC, RSC, and RPWCE. We show the blocking probability with confidence intervals (at 95% confidence level) over the demand scaling factor, i.e., the demand in [5] is prescaled by the value at the x-axis.

For an upper limit of 3% for the blocking probability, we can scale the demand matrix to integers of 6 for RPWCE, 9 for RSC, and 11 for RSPC. On the one hand, the RPWCE concept does not achieve as low blocking probabilities as dynamic p-cycles, since it is less flexible in finding capacity for a demand request. On the other hand, the higher complexity of RSPC and RSC (routing and p-cycle configuration) is opposed to the simplicity of RPWCE (routing only), which can justify to install more capacity for the RPWCE to reduce the blocking probability. For example, assume RPWCE follows the economy of scale rules of loss systems for which the Erlang-B formula apply; to shift the blocking performance curve of RPWCE in Figure 2 to the region of RSPC, i.e., to roughly half the blocking probability, by the economy of scale effect we need less than double the capacity of the PWCE. An appraising simulation for the considered blocking range shows that RPWCE well outperforms RSPC if the PWCE of each link (and likewise its protection capacity) is scaled by 1.75.

RSPC is superior to RSC, since working paths are closer to the shortest paths, for which the network is dimensioned. This can also be explained by the average network load. While the load by RSPC is over 80% for the displayed values, the load by RSC drops to 63% for the demand scaling of seven. Hence, RSPC is able to utilize the network better than RSC. In RSC, working paths can deviate from the shortest paths, since protection capacity on links of the shortest paths prohibits using it. The relative outcome for RSPC and RSC is different from the study in [4]. In this study, however, the network is dimensioned for dedicated path protection, for which dispersion from the shortest paths can be beneficial (see also the results with balanced load routing in [3]).

6.   CONCLUSIONS

We evaluated p-cycles in dynamic optical networks by simulation. Among the configuration approaches routing in spare plus protection capacity, routing in spare capacity, and routing in protected working capacity envelope, a case study has shown that the blocking probability increases in the order mentioned, given that the overall network capacity is the same for the three approaches. This is because the former two approaches use dynamically configured p-cycles which can adapt flexibly to the traffic situation. The latter approach, however, is significantly less complex in operation than the former two approaches.

Further research could investigate in detail how much capacity is needed for the protected working capacity envelope concept in excess to the one needed for an optimal assignment of dynamic p-cycles, given that the blocking probability is equal.

ACKNOWLEDGMENTS

This work is part of the work when the author was with the Institute of Communication Networks at Technische Universität München, Munich, Germany.

REFERENCES

[1] W. D. Grover and D. Stamatelakis. Bridging the ring-mesh dichotomy with p-cycles. In Proceedings of the International Workshop on Design of Reliable Communication Networks (DRCN), (Munich, Germany), April 2000. Invited Talk.

[2] W. D. Grover. Mesh-based Survivable Networks: Options and Strategies for Optical, MPLS, SONET and ATM Networking. Prentice Hall, Upper Saddle River, 2003.

[3] D. A. Schupke, C. G. Gruber, and A. Autenrieth. Optimal Configuration of p-Cycles in WDM Networks. In Proceedings of the IEEE International Conference on Communications (ICC), (New York City, NY, USA), April–May 2002.

[4] D. A. Schupke, M. Jaeger, and R. Huelsermann. Comparison of Resilience Mechanisms for Dynamic Services in Intelligent Optical Networks. In Proceedings of the International Workshop on Design of Reliable Communication Networks (DRCN), (Banff, AB, Canada), October 2003.

[5] W. D. Grover. The Protected Working Capacity Envelope Concept: An Alternate Paradigm for Automated Service Provisioning. IEEE Communications Magazine, 42(1):62–69, January 2004.

[6] R. Hülsermann, S. Bodamer, M. Barry, A. Betker, C. Gauger, M. Jäger, M. Köhn, and J. Späth. A Set of Typical Transport Network Scenarios for Network Modelling. In Proceedings of ITG-Fachtagung Photonische Netze, (Leipzig, Germany), May 2004.

PART A3:

GMPLS AND NETWORK CONTROL

AN EXPERIMENTAL GMPLS-BASED WAVELENGTH RESERVATION PROTOCOL FOR FLOODING GLOBAL WAVELENGTH INFORMATION IN UNI-RING-BASED MAN

Raül Muñoz1, Ricardo Martínez1, Jordi Sorribes2, Gabriel Junyent1,2
1: Centre Tecnològic de Telecomunicacions de Catalunya (CTTC)
C/ Gran Capità 2-4 08034 Barcelona, Spain
2: Universitat Politècnica de Catalunya (UPC)
C/ Jordi Girona 1-3 08034 Barcelona, Spain
{raul.munoz, ricardo.martinez, jordi.sorribes, gabriel.junyent}@cttc.es

Abstract: The accelerating growth of Internet traffic, together with its bursty traffic pattern is specially motivating the research on not only high-bandwidth but also dynamic metropolitan networks based upon recent advances in optical networking technologies such as R-OADM and OXC. The dynamism of the wavelength-routed networks can be achieved by means of a distributed control plane (i.e signalling for wavelength reservation and routing for dissemination of topology and optical resource state), which can be based in GMPLS. The objective of this paper is to propose a GMPLS-based signalling protocol which allows to have a global wavelength resource information without any routing protocol when provisioning bidirectional connections in uni-ring-based MAN. Performance evaluation has been carried in a GMPLS test-bed composed of Linux routers named ADRENALINE.

1.   INTRODUCTION

The existing metropolitan area networks (MANs), primarily made up of TDM technology, are not optimized for tomorrow's demands. Built to reliably and efficiently transport voice traffic, the TDM metro network is quickly reaching its capability to grow with the fast-changing data-centric world. The accelerating growth of Internet traffic, together with its bursty traffic pattern is motivating the research on not only high-bandwidth metro networks but also dynamic metro networks based upon recent advances in optical networking technologies such as *Wavelength Division Multiplexing* (WDM), reconfigurable *Optical Add Drop Multiplexers* (OADMs) and *Optical Cross Connects* (OXCs), capable of providing reconfigurable high-bandwidth end-to-end optical connections. The automation of the future optical MAN is achieved by means of a distributed optical-control plane (i.e, routing and signalling), which can be based in the Generalized Multiprotocol Label Switching (GMPLS), an extension to MPLS for fiber, wavelength, waveband and TDM switching..

Under distributed control each node makes its decisions based on the network state information (topology and wavelength resources) it maintains, which can be either local o global. In GMPLS-based networks, enhancement to IP interior gateway protocols (e.g. extended OSPF-TE or IS-IS) can be used to flood (periodically or threshold-based) network state information so that each node in the network can have a global knowledge of the network state, using link-state advertisement (LSA) update messages. The need to broadcast update messages may result in significant control overhead, and furthermore, it is possible for a node to have outdated information, and make incorrect routing decision based on this information. For the case in which a node only knows the status of its immediate links (local information), collisions are likely to occur if attempts to establish lightpaths for two contemporary connection requests are initiated over a particular link from both directions simultaneously, specially when no wavelength converters are available and a lightpath must be establish using the same wavelength on all the links along the path (wavelength-continuity constraint). So efficient distributed wavelength reservation protocols are needed for dynamic WDM networks with rapidly changing wavelength availability.

The objective of this paper is to propose an enhancement to GMPLS which allows to have a global wavelength resource information using a fixed routing scheme without LSA update messages, based on a GMPLS-based distributed control scheme for bidirectional lightpath establishment in unidirectional-ring-based MAN named Salmon Reservation Protocol (SRP), proposed by the authors in [1]. Performance evaluation, in terms of blocking probability and average lightpath set-up time, has been carried in ADRENALNE test-bed composed of Linux-based routers acting as GMPLS optical connection controllers (OCCs).

The remainder of this paper is organized as follows. In section 2 we describe GMPLS-based distributed Lightpath establishment schemes. In section 3 we present the SRP proposal. Section 4 describes the enhancement to the GMPLS signalling protocol in order to flood the wavelength state information throughout the whole network. An overview ADRENALINE testbed is described in section 5. Finally section 6 presents the experimental performance and section 7 concludes the paper.

2. GMPLS-BASED DISTRIBUTED LIGHTPATH ESTABLISHMENT SCHEMES

In order to set up a lightpath, a signalling protocol is required to exchange control information among nodes and to reserve resources along the path, such as GMPLS extensions to RSVP-TE and CR-LDP. In this paper we will only consider RSVP-TE. Likewise we concentrate on wavelength reservation schemes with wavelength continuity-constraint. GMPLS-based reservation protocols are categorized based on whether the resources are on a hop-by-hop basis along the forward path (*Forward Reservation Protocol*, FRP), or reserved on a hop-by-hop basis along the reverse path (*Backward Reservation Protocol,* BRP) [2,3].

In GMPLS, the signalling phase consists of a generalized label request, sent in a RSVP Path message, traversing hop-by-hop from the source node to the destination, followed by a generalized label assignment, sent in a RSVP Resv message, traversing in the opposite direction back to the source. When a new connection request arrives to the source node, it initiates a RSVP Path Message containing a Generalized Label Request Object. Then a strict path to the destination is determined, recorded on an Explicit Route Object (ERO). In this work we assume fixed routing scheme, that is, a fix route is specified between each source-destination pair. In order to guarantee the wavelength-continuity constraint when no global information about optical resources is available, a Label Set object is included in the Path message at the source node. The Label Set Object allows an upstream node to restrict the set of labels that a downstream node can choose, ensuring that a downstream node will assign a label that is acceptable to an upstream node. The source node includes a Label Set Object specifying the available wavelengths on its outgoing fiber link. If a Forward Reservation scheme is used, all the wavelengths specified in the Label Set are locked at each hop in the path, over-reserving temporally the resources. In contrast a backward reservation scheme does not reserve the wavelengths of the Label Set, it just collects the usage information of wavelengths in the path. Each intermediate node updates the received Label Set removing currently unavailable wavelengths. If no wavelength of the Label set is available, the request is blocked and the reserved wavelengths (only for FRP) on the partially established path are immediately released. If the Path message reaches the destination node, one label is selected based on a wavelength assignment algorithm such as the First-Fit or Random, and initiates a Resv message including the Generalized Label Object with the selected wavelength that will be configured at each hop toward the source node. For FRP, in addition, the temporally reserved wavelengths will be set free at each hop.

Using the GMPLS signalling extensions it could be possible to integrate forward reservation and backward reservation into one process. This is accomplished through the Suggested Label Object. This object is used to provide a downstream node with the upstream node's label preference. This permits the upstream node to start reserving and configuring its hardware with the proposed label. Therefore combining a Label Set Object based on a Backward Reservation Scheme, and a Suggested Label Object (based on a Forward Reservation Scheme), it could be possible to do a Forward Reservation based on a single label (conservative), and if it fails, the Backward Reservation would continue the connection request. In [5] it is shown that the combination of BRP & FRP excels both FRP and BRP working separately, but this study is not GMPLS-based. Therefore we will work with this reservation scheme assuming that is the best one according this study.

3. SALMON RESERVATION PROTOCOL (SRP)

Salmon Reservation Protocol (SRP) is a GMPLS-based distributed control scheme for bidirectional lightpath establishment in unidirectional-ring-based metropolitan networks supporting both Soft Permanent Connections (SPC) and Switched Connections (SC) in multi-domain environments, proposed by the authors in [1]. In the basic GMPLS architecture [4], bidirectional optical connections are established using a single set of Path and Resv messages, but this mechanism does not work in unidirectional rings, since one fiber is dedicated as working fiber and the other is dedicated as protection fiber. Working and protection fiber operate is opposite directions: the working ring operates on the clockwise direction on the protection ring and the protection rig on the counter clockwise direction.

When a new end-to-end SC or SPC request crossing multiple domains arrives to the unidirectional ring (figure1), the source node separates the request connection into two RSVP connection segments between the input/source node and the output/destination node located in the same ring. Both connection segments in the ring are associated using the "Session Name" attribute of the Session Attribute object. The input/source node inserts in the Session Name filed a unique value to allow unique identification of both RSVP connection segments at the output/destination node. Therefore the input/source node initiates two RSVP Path messages identified by its Session Object, containing a Generalized Label Object, a Label Set Object (based on a Backward reservation scheme), a Suggested Label Object (based on a Forward Reservation scheme), an Explicit Route Object (ERO) to determine a strict path to the destination node, and other relevant objects. The first RSVP Path message requests the downstream wavelength, therefore the ERO specifies an strict path to the destination in the same direction of the ring transmission, whereas the second Path message request the upstream wavelength. In this case the ERO specifies a strict path to the output/destination in the opposite direction of the ring transmission. Moreover the second Path message includes an Upstream Label Object (UL) in order to indicate that the wavelength resources are requested in the opposite direction of the optical transmission. Then Both RSVP Path message are sent along their respective explicit routes to the destination. When one of the two Path messages reaches the destination/output node, it waits until the associated Path message also reaches the destination, checking the Session Attribute. Once both Path messages have reached the destination/output node, it is checked whether the destination of the requested connection is out of the Ring. In the first case, the node will generate a single Path message to the next node and the request connection will continue its journey. In the second case, two RSVP Resv messages, one for each connection segment will be generated.

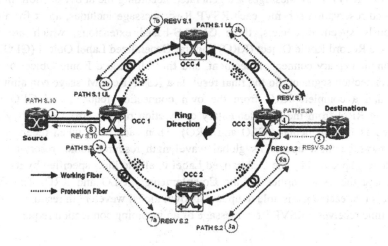

Figure 1. SRP. Bidirectional Lightpath establishment in uni-ring-based MAN

4. AN EFFICIENT GMPLS-BASED WAVELENGTH RESERVATION PROTOCOL FOR MAN

The main drawback of combining FRP and BRP as explained above is that the suggested label is chosen by the source node that only have local information. In this case the wavelength of the Suggested Label is chosen randomly from the label set. So there is no guarantee that the label suggested by the source node be available along every link in the path.

In order to solve this drawback we propose a simple enhancement to the GMPLS signalling protocol that can be used to flood global wavelength state information when bidirectional connections are requested over unidirectional-based WDM rings. In our proposal each node in the ring has a global wavelength resources table, indicating which wavelengths are available in all the links of the ring. The first time the algorithm is used, all the wavelengths of the table are set as free. Then the source node that initiates the connection request inserts in both RSVP connection segments a Record Route Object in order to record all the hops in both routes. The Suggested Label is chosen by the first node according its global wavelength resource table. It must accomplish the wavelength continuity constraint, so only those wavelengths that are available at each link form the source node to the destination node are suitable. From the valid wavelengths, one of them is chosen randomly and is used by the Suggested Label Object. Once both RSVP connection segments (upstream and downstream) arrive to the destination node, two RSVP Resv messages are generated according the above section. In the proposed reservation scheme, each RSVP Resv message includes, apart from the previously specified objects, three GMPLS-based extensions, which are the complete Record Route Object (RRO) and the Generalized Label Object (GLO) of the complementary connection segment, and the full Record Route Object of the own connection segment. So the final result is a RSVP Resv Message containing a GLO and a complete RRO from the own connection request, and GLO and complete RRO from the complementary connection request (from now they are referred as Complementary GLO and RRO). Then each node when receives the Resv message must update its global wavelength resource table, reserving the wavelength specified by the Generalized Label in all the links specified by its own RRO, and the same applies for the Complementary GLO and RRO. The final Result is that each node is able to update its own global wavelength resources table every time receives a RSVP Resv message of an incoming connection request.

5. ADRENALINE TESTBED

ADRENALINE testbed is based on an experimental ASON/GMPLS control plane composed by 9 distributed GMPLS-based Optical Connection Controllers (OCC), allowing the establishment of real-time, dynamic, end-to-end optical connections. Each OCC has been implemented on a Linux-based router with a Pentium IV 2,6 GHz processor. OCCs are interconnected by fast Ethernet point-to-point links, using both simulated and real links. The real links are bidirectional optical fiber links with a distance of 35Km each (control channels are carried on 1310nm). The simulated links are based on an additional PC with a network emulation package that allow to emulate the link delay between two OCCs. The network topology is based on a unidirectional-based ring, using 3 real optical fiber link and 6 simulated links that emulates a delay of also 35Km. Note that the circumference of the ring is about 300km, suitable for a metro core network.

6. EXPERIMENTAL PERFORMANCE EVALUATION

In this section, we investigate the performance of the proposed GMPLS-based reservation protocol for unidirectional-based metro networks over the control plane of the experimental testbed, comparing when the Suggested Label is chosen from local (named SRP Local) or global (named SRP Global) wavelength resource information. Firstly, we describe the main assumptions adopted and then we present the results and discussions. All the lightpath requests have been assumed as bidirectional connections. Lightpath requests arrive according to a Poisson process, and the lightpath holding time is exponentially modeled with a mean of 100 ms. To avoid having almost zero size lightpath holding time (holding time inferior to setup delay), we have added a small fixed time of 10ms to the lightpath holding time (*offset time*). The traffic is uniformly distributed among all node pairs (typical behaviour in metro core rings). Each data point is obtained over a simulation of 100.000 connection requests. Each link supports 8 wavelengths. The time to configure an OADM node is 10ms.

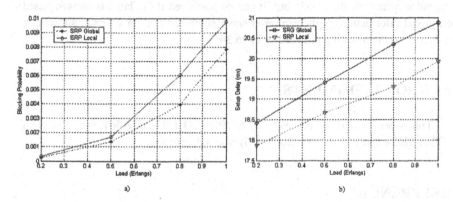

a)   b)

Figure 2. Figure 2. a) Blocking Probability. b) Set-up delay

Load is measured in Erlangs, which can be calculated by multiplying the connection arrival rate with the average connection holding time. To study the network's behaviour under different loads, the arrival rate of connection requests is varied as a parameter. Figure 2.a) plots the obtained blocking probability vs. load for global and local SRP. Global SRP always exhibits an upgraded behaviour respect to the local WRP approach, for high loads. For example the blocking reduction of the global SRP is about 34 percent compared to the local SRP when the total offered load is fixed at 0,8Er. For low loads there are very few differences, since almost all the resources are free and therefore the probability of failure when selecting a wavelength based on local knowledge is low. Figure 2.b) shows the setup delay vs. load. Obviously, Local SRP presents lower delays than SRP, reducing the setup delay up to 8%. As shown, the connection setup delay increases as load increases due to the fact that each OCC has to support more RSVP sessions, causing the increase of the queuing delay at each OCC.

7. CONCLUSIONS

This paper presents a GMPLS-based wavelength reservation scheme that allow to have a global wavelength resource information using a fixed routing scheme without LSA update messages, based on a GMPLS-based distributed control scheme for bidirectional lightpath establishment in unidirectional-ring-based MAN named SRP. Using the GMPLS signalling extensions it could be possible to integrate FRP and BRP into one process, through the Suggested Label (based on FRP) and the Label Set (based on BRP) objects. The Suggested Label is chosen by the first node randomly based on local information. This integration reports an upgraded behaviour than BRP, but it can be improved if the label is chosen based on global information by the proposed wavelength reservation scheme, showing a reduction of the blocking probability up to 34%.

ACKNOWLEDGMENTS

This work is part of the NetCat, EMPIRICO (PU-2002-56) and TBONES (ITEA 02024 and FIT-070000-2003-936) research projects.

REFERENCES

[1] R. Muñoz, R.Martínez, G. Junyent, C. Pinart, A. Amrani, "Performance Evaluation of two new GMPLS Lightpath Setup Proposals over an Unidirectional OADM Ring Implemented on a Testbed", Proceedings of 29th ECOC, Rimini, September 21-25 2003.

[2] [D. Saha, "A comparative study of distributed protocols for wavelength reservation in WDM optical networks", SPIE Optical Networks Magazine, Vol. 3, No. 1, 2002, pp.45-52

[3] [A. V. Sichani, H. T. Mouftah "A Novel Distributed Progressive Reservation Protocol for WDM All-optical Networks", Proceeding of ICC 2003, Anchorage, Alaska, USA.

[4] E. Mannie et al, Generalized Multi-Protocol Label Switching architecture, IETF draft-ietf-ccamp-gmpls-architecture-05.txt (to proposed standard).

[5] F. Feng, x. Zheng, H. Zhang, Y. Guo "An Efficient Distributed Control Scheme for Lightpath Establishmment in Dynamic WDM networks", Photonic Network communications, 7:1, 5-15, 2004.

GMPLS WITH INTERLAYER CONTROL FOR SESSION-UNINTERRUPTED DISASTER RECOVERY ACROSS DISTRIBUTED DATA CENTERS

Tetsuo IMAI[1], Soichiro ARAKI[2], Tomoyoshi SUGAWARA[3], Norihito FUJITA[4], and Yoshihiko SUEMURA[5]

[1-5] System Platforms Research Laboratories, NEC Corporation,
1753, Shimonumabe, Nakahara-ku, Kawasaki, Kanagawa 211-8666, Japan,
{t-imai@ct[1], s-araki@cj[2], tom-sugawara@ap[3], n-fujita@bk[4], y-suemura@bp[5]}.jp.nec.com

Abstract: We propose a session-uninterrupted disaster recovery system using a novel session migration technique as a GMPLS application. Existing disaster recovery systems have a problem of a service interruption. The session migration based on an interlayer control of GMPLS, VLAN change-over, and process migration, maintains continuous TCP service between a user and a virtualized server, even when the service migrates from a primary data center to a backup one. We developed a prototype system and showed that BoD (bandwidth on demand) by GMPLS improved the recovery time from 80.10 sec to 9.85 sec, during transmitting a process data of 40MByte.

1.   INTRODUCTION

It is important for mission-critical businesses to continue even in the event of a disaster that may bring a whole data center to a halt. In this paper, we propose a novel session-migration technique for the disaster recovery system in data centers using the BoD (bandwidth on demand) service provided by GMPLS (generalized multi-protocol label switching) [1].

Existing disaster recovery systems are classified into following 3 types.

- Remote data backup: Make a copy of data at backup site by users(low-class)
- Data replication: Make a copy of data automatically(middle-class)

- Clustering: The same system is set up at backup site, and make a copy of data automatically(high-class)

Our session-migration technique is classified as the best-class. It can rapidly change a service-providing server to an alternative backup one and change a user's network to another one at the same time, enabling users to keep their own TCP sessions. It enables data center managers to provide users with continuous service. We developed a session migration controller to control change-overs in the service-providing server and users' networks, and dynamic reservation of a bandwidth by GMPLS. And we demonstrated the effectiveness of our session-migration system.

2. SESSION MIGRATION

2.1 Problems with existing method of disaster recovery

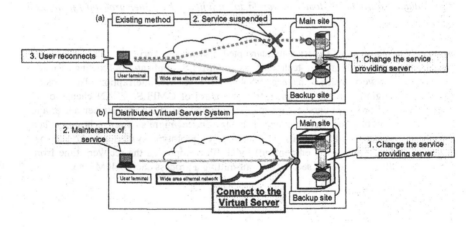

Figure 1. (a) existing disaster recovery system, (b) our distributed virtual server system.

It is essential that mission-critical businesses have continuous access to services. However, because existing recovery methods mainly use systems within the same network (subnet), they can only change the server running a service from the main site to a backup site. Therefore, there is a following problem. When a disaster occurs, the server running the process to be recovered is moved from the server at the main site (primary server) to one at a backup site (backup server) which is widely separated. To continue using the service, users have to carry out troublesome task for recovery, including closing the TCP (transmission control protocol) session to the primary server, and re-connecting to the backup server(Fig. 1 (a)). The problem is that service to users is suspended until the recovery procedures are completed. Our proposed disaster recovery system intends not to affect users' access to services. Our session-migration technique enables TCP services between users and a virtualized server to continue, even when the service migrates from a primary server to a backup one(Fig. 1 (b)).

2.2 Session migration and current performance problems

To provide continuous service even in the event of a disaster, we migrate service-providing processes from the primary server to a backup one, which is widely separated, with maintaining the TCP session between the user and the process. A "process migration"[3] is used to swap the active server to another server by migrating a memory image of the process from one server to the other.

There are several problems in maintaining users' TCP sessions during process migration. First, we have to move the server's IP address between the servers at the same time as the process migration occurs, but sharing the same IP address in the same wide-area network creates difficulties. Secondly, when we carry out process migration between widely separated servers, we have to minimize transmitting time. Session migration is triggered by the occurrence of some sort of disasters, for example, a major electrical power failure or a fire alarm at a data center. Therefore, session migration has to begin with these warnings, and be completed before the disaster eventuates. In addition, long downtime of the process may cause a performance degradation or a session disconnection.

To solve the first problem, we distribute the same IP address to different VLANs, and we change-over the "user VLAN" (the VLAN that the user belongs to) in synchronization with process migration in a session-migration sequence. This enables users to connect quickly with the new network without affecting their use of the service. It can maintain the TCP session without interruption.

To solve the second problem, we rapidly reserve a wide bandwidth for process migration in advance by GMPLS technique.

There has been considerable work recently on new services such as BoD for optical networks using GMPLS techniques. GMPLS will enable service providers to quickly deliver various types of paths to customers[2]. In our session-migration technique, we use GMPLS to reserve a bandwidth for the process migration, which requires rapid transmission of a huge amount of data to minimize the length of time that services provided by a data center are suspended.

And it is important to coordinate these three techniques of different layers, the change-over of user VLAN, the process migration, and the GMPLS. In our session migration, an MCS (migration-control server) controls each of them.

A detailed description of our session-migration technique follows.

174

2.3   Overall system architecture

Figure 2 shows the architecture of the session-migration system. "Server 1" is at the main site, and "Server 2" is at the backup site; a PMC (process migration controller) is implemented in both of them. There are routers at the edge of each site. Users and sites are connected by a wide-area Ethernet network consisting of L2SWs (layer 2 switches that support VLANs), and an underlying L1NW (layer 1 network) consisting of L1SWs (layer 1 switches) controlled by respective CP (control plane)-devices. In a wide-area Ethernet network, user sites and the main site are connected by a VLAN 1, and user sites and the backup site are connected by a VLAN 2. In the dedicated control network, an MCS is connected to a PMC, which controls process migration, a NAGW (network access gateway), which controls changes in the user VLAN, a CP-device, which directs the GMPLS to reserve a bandwidth in the L1NW, and the L2SW, which supports the VLANs.

Ordinarily, a user connects to a virtual IP address, 192 168.2.1, at Server 1 in Main site via a NAGW to utilize a service. In the following description, we use "192.168.2.1" as the virtual IP address, but it's tentative.

The components in Fig. 2 are described below.

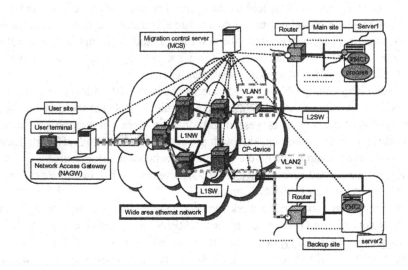

Figure 2. System Architecture of the session migration.

The MCS manages the session-migration sequence, and directs the behavior of other elements of the session-migration process through remote commands using a Telnet or other methods. The PMC 1 and PMC 2 control process migration when a trigger is given by the MCS. The process migration technique transports a running task (process) from one server to another, including the state of communication and virtual IP address, and enables the process to keep running on the other server. The NAGW manages the rapid switch-over of server required to be connected by changing the user VLAN; i.e., the NAGW changes the user VLAN from VLAN 1 to VLAN 2, as shown in Fig. 2, when a trigger is given by the MCS. The CP-devices control the L1SWs via the GMPLS. They carry out reservation/deletion of the bandwidth between Server 1 and Server 2 when a trigger is given by the MCS, and can dynamically establish a L1-VPN (layer 1-virtual private network). Both of the L2SWs support VLANs and can dynamically establish a L2-VPN between Server 1 and Server 2 when a trigger is given by the MCS. The routers at the main site and the one at the backup site are in subnets of the same address architecture. Therefore, there is no need to change the routing table for the routers when a virtual IP address is moved from Server 1 to Server 2.

2.4   Detailed sequence of session migration

The flow of the session-migration sequence is shown in Table 1. Ordinarily, a user connects to the virtual IP address 192.168.2.1 of Server 1, and utilizes a service by communicating with a process on Server 1.

When the alarm sounds for a disaster, the data center manager, or automatic equipment, commands the MCS to begin session migration (STEP 1). The MCS commands the CP-devices and L2SWs to configure a L1/L2-VPN (STEP 2, 3). It set up the dedicated and wideband VPN for process migration.

When the VPNs are set up, the MCS starts the sequence for process migration. The MCS commands the PMC 1 to freeze the process (STEP 4), and communication between the user and Server 1 is suspended. Next, the MCS commands PMC1 and PMC 2 to migrate the process (STEP 5), and to move the virtual IP address 192.168.2.1 from Server 1 to Server 2 (STEP 6). This causes temporary disappearance of the IP address 192.168.2.1 to the user because the user is connected to VLAN 1 at this point. At STEP 7, the MCS commands the NAGW to change the user VLAN from VLAN 1 to VLAN 2, and the user rediscovers the IP address 192.168.2.1. Throughout this sequence, the user simply loses the IP address and rediscovers it, and is unaware of the change in the VLAN to which the user is connected. Then the MCS commands the PMCs to resume the process(STEP 8); the user can then resume communicating with the server (Server 2) and using the service.

Finally, the MCS commands the CP-devices and L2SWs to delete L1/L2-VPN (STEP 9, 10), and reports the completion to the data center manager (STEP 11).

In the aspect of effect of session migration on users, the following issues can potentially affect users during service migration.

Processing period from STEP 4 to STEP 8: User communication to the server is suspended during this time. However, depending on the type of service, if this period is very brief, it does not affect service utilization in general.

Processing time required for STEP 7: The user loses the virtual IP address 192.168.2.1 of the servers at this point. Again, depending on the type of service, if this time is very brief, the user's application will not detect packet convergence, and will not stop the use of the service.

If both these processing periods are sufficiently brief, the user will be unaware of the session migration and will experience continuous service.

Table 1.Sequence for session migration.

STEP 1:	The manager commands the MCS to start session migration.	Pre process
STEP 2:	The CP-devices reserve a bandwidth for process migration.	
STEP 3:	The L2SWs configure the VLAN for process migration.	
STEP 4:	The PMC freezes the process to be migrated.	Main process
STEP 5:	The PMC carries out process migration.	
STEP 6:	The PMC changes over the virtual IP address.	
STEP 7:	The NAGW changes the user VLAN from VLAN 1 to VLAN 2.	
STEP 8:	The PMC resumes the process.	
STEP 9:	The CP-devices deletes the bandwidth.	Post process
STEP 10:	The L2SW delete the VLAN.	
STEP 11:	The MCS reports the completion of session migration.	

3.   PROTOTYPE SESSION MIGRATION SYSTEM AND ITS PERFORMANCE EVALUATION

We built a prototype system (Fig. 3) to evaluate the session migration technique described above. The MCS, L2/L3SW, and CP devices were constructed on a rack on the right of the photograph; on the left, there is a rack of SDH nodes (SpectralWave U-node, NEC) as the L1SW controlled by the CP devices. The NAGW is installed in the PC at the right. The features of these components are listed in Tables 3 and 4. Server 1 and Server 2 have a Linux OS, each PMC is built as a daemon on Linux. The process migrated in the demonstration was a video data transmission process, which uses 40 MB of memory; the user receives streaming data via a TCP at a bit rate of 2 Mbps. Server 1 and Server 2 are connected by a gigabit Ethernet to the individual SDH nodes, which are connected to each other by a 2.4-Gbps link. During the session migration sequence, the GMPLS commands the SDH nodes to reserve a bandwidth of 1 Gbps. It then establishes a bandwidth of 1 Gbps between servers.

We carried out session migration using this architecture and found that the procedure did not have any effects on users browsing a streaming video.

Below, we discuss the effects of session migration on users' service utilization based on the discussion in section 2.4.

For comparison, we carried out a demonstration of session migration without GMPLS. In the case without GMPLS for reservation of bandwidth, servers carry out the process migration through a public network. We emulate that by connecting the L2/L3SWs with a 10-Mbps link directly; the bandwidth between the servers was then 10-Mbps. Table 2 shows the processing time required for each step of the session-migration process, with and without GMPLS. Note that STEP 2, 3, 9, 10 aren't necessary in the case without GMPLS.

As shown in Table 2, with GMPLS, the processing time from STEP 4 to STEP 8 (the period for which user communication with the server is suspended) was about 2.7 sec. We used Real One Player (Vers. 2.0) as the video browser, and it has a 30-sec buffer by default, which is sufficient to cover the suspension period.

Without GMPLS, the period was about 79.6 sec. long, which was too long to be covered by the buffer. The video stream was therefore suspended briefly.

The processing time required for STEP 7, when the user loses the virtual IP address of the server, is about 0.1 sec. in each demonstration. This demonstration showed that the video browser did not suspend service and the TCP congestion controller did not decrease its throughput.

Table 2. Processing time required for each step.

Table 3. Features of components

Table 4. Features of switches

. . . .	. . . .
. . . .	. . . .

Figure 3. Demonstration system.

4. CONCLUSION

We proposed a session-uninterrupted disaster recovery system as a GMPLS application. Using the interlayer controller, our session migration technique coordinates three methods: BoD by GMPLS, process migration between servers, and a change in the user VLAN, and avoids interruption of users' accesses to the services provided by the servers. In a demonstration of the system, we showed that the technique worked well and that GMPLS improved the recovery time from 80.10sec to 9.85 sec during transmitting a process data of 40MByte.

ACKNOWLEDGMENTS

This work includes a part of the result of Key Technology Reseach Promotion Project (the research project on large-scale high reliability server) supported by NEDO.

REFERENCES

[1] E. Mannie, "Generalized Multi-Protocol Label Switching Architecture," IETF Internet Draft, draft-ietf-ccamp-gmpls-architecture-07.txt.
[2] H. Ishimatsu, S. Tanaka, M. Akashi, T. Hashimoto, E. Yamaguchi, Y. Oyama, H. Nakano, A. Inomata, M. Murakami, Y. Ashikaya, S. Ryu, "Prototype Demonstration of On-Demand/Scheduled Wavelength Path Service," OFC2003, ThR2, 2003.
[3] Dejan S. Milojicic, F. Douglis, Y. Paindaveine, R. Wheeler, S. Zhou, "Process Migration," In ACM Computing Surveys, Volume 32, Issue 3, September 2000.

MONITORING SERVICE "HEALTH" IN INTELLIGENT, TRANSPARENT OPTICAL NETWORKS

Carolina Pinart[1], Abdelhafid Amrani[1], and Gabriel Junyent[1,2]
[1]Centre Tecnològic de Telecomunicacions de Catalunya, c/ Gran Capità 2-4 08034 Barcelona, e-mail: {carolina.pinart, abdelhafid.amrani}@cttc.es
[2]Universitat Politècnica de Catalunya, c/ Jordi Girona 1-3 D4 08034 Barcelona, e-mail: junyent@tsc.upc.es

Abstract: This paper focuses on the design and implementation of a low-complexity dialogue mechanism between the management and transport planes of an all-optical network. This dialogue aims at exchanging relevant information from monitoring the performance and degradation of optical signals with minimal disturbance to the optical services and minimum knowledge of the transport history of data, with a view to ensure service quality. Complexity of the dialogue is measured in terms of response delays in the event of failures.

1. INTRODUCTION

The accelerating growth of data traffic is motivating the research for more efficient, flexible, intelligent optical network architectures. Optical networks promise to be the underlying next generation technology for the future Internet and broadband networks, being Internet Protocol (IP) over Wavelength Division Multiplexing (WDM), IP/WDM, one of the most promising candidates.

Performance monitoring plays a fundamental role in the deployment and future evolution of the optical network industry, because assuring service quality (QoS) will be key for the success of next-generation networks, which are to enable dynamic, differentiated optical services. Therefore, efforts are underway to investigate new ways to monitor the performance of data on an optical network with minimal conversion to electronics and minimal disturbance to the signal on the fiber, as well as to integrate fiber-optic networks, communications and signal processing, and optical network protocols.

As for protocols in IP/WDM, Generalized Multi-Protocol Label Switching (GMPLS) is thought to be an integral part of next-generation optical networks, especially as control plane of the Automatic Switched Optical Network (ASON) [1], because it renders optical networks intelligent. Concerning the management information protocol, the Simple Network Management Protocol (SNMP) [7] has been the industry reference for network management since the last 1980s; SNMP agents are installed in almost any system to enable remote access to its components, making SNMP a de facto standard for networked hardware management [6].

This paper focuses on the mechanisms for Optical Performance Monitoring (OPM) in an IP/WDM network, especially the dialogue of the transport and management planes to exchange relevant information obtained from monitoring the performance and degradation of signals with minimal disturbance to the signals and minimum knowledge of the transport history of the data to assure QoS.

The remainder of this paper is organized as follows. Section 2 is devoted to the reference model and parameters for monitoring the health of optical services, as well as the different inputs for such monitoring. Section 3 describes the experimental ASON/GMPLS testbed, focusing on the data model and the SNMP agent implementation for service monitoring. Section 4 deals with preliminary performance evaluation results of the dialogue. In Section 5 we draw conclusions.

2. REFERENCE MODEL AND MONITORING PARAMETERS

The reference model of OPM has three layers. Starting from the WDM input, we encounter channel management, then channel quality and protocol performance [6]. These layers are known as transport, signal quality and protocol monitoring, respectively. In this work, we only deal with transport monitoring due to the fact that the method chosen for implementing OPM is a non-disruptive all-optical dedicated monitor that taps the optical signal on a Dense WDM (DWDM) fiber. In this case, the monitor is shared among the wavelengths carried in the fiber. At this layer, the most commonly suggested monitoring parameters are [6]: aggregate power, channel power, spectral optical signal-to-noise ratio (SNR) and channel wavelength. In contrast, time domain parameters such as Bit Error Rate (BER) and Q-factor cannot be obtained by the considered OPM owing to its transparent nature.

In this work, an optical service, understood as the provisioning of an optical channel (wavelength), encompasses the following QoS parameters: establishment and release delays [7], which may be monitored directly in the manager, channel "health" (obtained through OPM) and status of optical components (component failures) that might lead to service degradation, such as lasers, receivers and active switching and control components. To retrieve these QoS parameters, several monitoring points must be set in the optical network. Major component failures may take place in the control plane (faults in the GMPLS-based Optical Connection Controllers, OCC) and in the transport plane. Therefore, monitoring points for component failures are split in the optical node controller (*NC* in Figure 1) and management agents (OCC and all-optical add-drop multiplexer, OADM, illustrated as *Agent* in Figure 1) at each node. Moreover, the impairments that may affect the optical signal quality, causing undesirable variations of the channel power, frequency or OSNR, are monitored by a transport monitor (Figure 1). Note that degraded BER and restoration delay are usually considered in the event of failures [7], but since we use all-optical monitoring at WDM layer and fast physical protection at Optical Multiplex Section level (OMS), these parameters are not considered in this work.

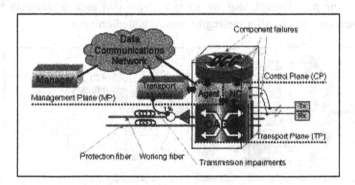

Figure 1. Monitoring in an IP/WDM node

3.   EXPERIMENTAL ASON/GMPLS TESTBED

The experimental testbed[1] used in this paper is a DWDM transport network enabled with optical intelligence through a GMPLS control plane to allow real-time, dynamic configuration of optical services between multiple clients (ASON). Due to economic reasons, 8 wavelengths per fiber, spaced 100 GHz (ITU channels from 30 to 37) at speeds up to 2.5 Gbps are available. Due to its relevance in this

---

[1]   ADRENALINE testbed: All-optical Dynamic REliable Network hAndLINg IP/Ethernet Gigabit traffic with QoS. http://www.cttc.es/adrenaline/

work, the transport plane's architecture is summarized in Section 3.1, whereas the overall architecture of the testbed is further detailed in [5].

3.1 Transport plane

The testbed's transport plane is an Optical Transport Network (OTN) [2] that provides uni and bidirectional optical channels transparent to the format and payload of client signals. Moreover, it is the source of information about the state of connections and their performance, which is central to OPM. All laser sources are fully tunable, and no wavelength converters are contained in the OADMs [5] (architecture depicted in Figure 2). In this work, the configuration of the OTN is a unidirectional ring with OADMs. So, each link has two unidirectional fibers, one of which is used for OMS protection [2]. In the architecture of an OADM, an incoming signal from the working fiber is demultiplexed into the 8 wavelengths before entering the 2x2 all-optical switches, each of which may either drop a wavelength (maximum 4 wavelengths dropped at each OADM) and optionally add a new one, or pass through that wavelength. Note that each wavelength has a switch associated, so that all wavelengths may be added/dropped.

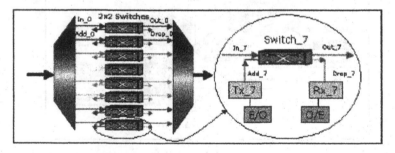

Figure 2. Architecture of an OADM

Once all output wavelengths are ready, they are multiplexed and the DWDM signal is inserted in the outgoing fiber (Figure 2). At this stage, 1% of optical power of the DWDM signal is tapped and brought to an all-optical signal performance monitor. This equipment allows a first estimation of the optical signal quality by measuring the channel power, frequency, OSNR, and their respective drifts in a totally transparent manner, not only at each optical node output but at the input as well. Consequently, node or link failures can be detected. Moreover, degradation thresholds allow proactive fault management by detecting and avoiding or correcting a failure before it occurs. In contrast, conventional digital methods of optical signal monitoring such as BER test, Bit Interleaved Parity (BIP) check or Cyclic Redundancy Check error (CRC) enable the detection of a failure in an optical network without identifying its origin, or its localization. In addition, the failure detection is only at the electrical edge nodes resulting in a high granularity of the detected fault, higher network resource utilization for restoration and slower recovery. Moreover, digital methods of fault detection are specific to the digital characteristics of the optical signals and generally require optical demultiplexing, optical to electrical conversion and synchronization to the bit rate.

A description of the processing mechanisms for OPM data, as well as of the most relevant network elements for optical service monitoring follows.

3.2 Optical node controller

Active components of the transport plane are the 2x2 optical switches, the transmitters (lasers), the transceivers (E/O and O/E) and the photonic receivers (Figure 2). Moreover, if the OADM is reconfigurable, an additional (active) matrix switch is needed for the distribution stage [5], not considered in this work. Last but not least, optical amplifiers (OA in Figure 1), located in the optical links, are also active. Then, the optical NC is responsible for monitoring the state of these active components, as well as for changing their state according to the needs of the control and management planes, which come mainly from service provisioning [3]. In this work, the testbed's NCs contain a 32-bit micro-processor with RS232 and RJ45 ports. The serial port communicates with the active optical components at a speed of 115,2 Kpbs. The NC interacts with these components through a Field Programmable Gate Array (FPGA) circuit that ensures electrical connections both for information exchange and electrical supply to the optical components. The micro-processor is a Linux platform that performs the following:

Proxying. Execution of requests from the OADM agent (Section 3.3) and the manager (through the agent), such as full status information of receiver Rx_7. Retrieval of transmission information from the agent and/or the optical monitor.

Connection Controller Interface (CCI). Execution of commands from the OCC of the optical node (control plane) related to a provisioning process. For instance, add a channel with wavelength 1553.33 nm to output Out_7, which means changing status of laser Tx_7 from OFF to ON, setting the laser channel to 30, setting the laser variable optical attenuator (VOA), changing the status of the optical 2x2 switch from bar to cross, and setting the matrix switch input/output.

Monitoring. Periodic retrieval of status information from the optical components. Alarms and notifications in case of failures to the OADM agent. Status information involves the characteristics of the following components:

- Transceivers: input/output power.
- Lasers (Transmitters): laser temperature, optical power, wavelength stability, laser current, laser temperature, aging failure and port failure, in addition to output power from external modulators.

- Avalanche photodiodes optical receivers (APD): loss of power alarm detection.

3.3 OADM agent

ITU-T Recommendations of X.700 Series or M.3010 are examples of conceptual (information) modeling of the management plane, including service provisioning. Such modeling is independent of specific implementation, and defines relationships between managed objects. Since control plane and transport plane elements (OCCs and OADMs, respectively) contain management agents (Figure 1), as well as Management Information Bases (MIB) and a message communication function, encompassing a management information protocol, the manager is able to access the information models of both the control and transport planes. Therefore, we focus on a data model for service monitoring. As for the OADM agent's MIB, it contains the IETF module OPT-IF-MIB [4] and the module OHW-CTTC-MIB (Figure 3, right), developed in this work. OPT-IF-MIB defines objects for managing optical interfaces associated with WDM systems or characterized by the OTN architecture. The main object of OHW-CTTC-MIB is opticalHwTable, complementary to OPF-IF-MIB because it contains information of transmitters, receivers and switches, indexed by opticalSwitchNumber. The example on Figure 2 (right) would have all parameters of opticalHwTable indexed by 7, since the 2x2 switch is Switch_7.

The OADM management agent (Figure 3b) is based on SNMP (manager-agent paradigm) [8], which has been chosen as management information protocol because it is the industry reference. This agent has four main functions: responding to monitoring information queries, sending alarms/notifications of resource status, proxying with the NC and the monitor (retrieval of frequency assignment and stability, power level and OSNR for each optical channel, at the input and at the output of each node) and retrieval of power level of optical amplifiers (status information), via SNMP. The transport monitors also respond to SNMP queries made by the manager and send alarms caused by transmission impairments. Instead of overloading the network with status information, messages are only sent when a failure is likely to occur (warning), so that the control and management planes may both react proactively without service disruption, and reactively after a failure (fault). The monitor used in this work is Digital Lightwave's Optical Wavelength Monitor, which is all-optical and monitors power, wavelength (every 10 ms) and optical SNR (every 100 ms) for multiple channels on DWDM networks. Alarms are sent via SNMP according to the monitor's DIGL-OWM-MIB module.

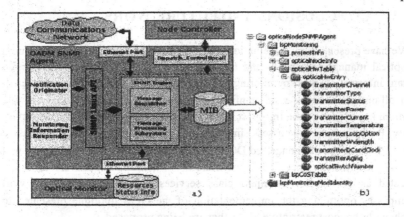

Figure 3. a) OADM architecture and b) OHW-CTTC-MIB module (transmitter objects)

4. PRELIMINARY PERFORMANCE EVALUATION

We consider the worst-case failure scenario depicted in Figure 5a. Laser 1, which transmits data of a Premium service [7] in channel 30, experiences a power loss that degrades the Service Level Agreement (SLA). As soon as the NC detects the fault, it informs the OADM agent, which forwards the event to the manager (SNMP Trap). Due to the severity of the fault, the manager requests (SNMP Set) that data be transmitted in the same channel by idle laser 2. The Data Communications Network (DCN) is congested and the distance between the manager and agent is 400 km. Figure 5b plots OPM performance in terms of delay.

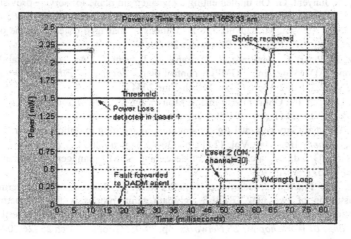

Figure 5. a) Test setup and b) Channel power vs. time (ITU channel 30, congested DCN)

5.　　CONCLUSIONS AND FUTURE WORK

We have presented implementation issues to achieve a simple dialogue among the optical management plane and the transport and control planes to forward relevant information from monitoring the performance and degradation of signals in an all-optical network with minimal disturbance to the optical services and minimum knowledge of the transport history of data, with a view to ensure service quality. Preliminary results assess simplicity in terms of low delays (worst case is around 50 ms in heavily congested DCN). As far as we know, recovery times with degraded SLA are not specified in the literature, whereas full recovery times are suggested as 50 ms for Premium class services [7]. Therefore, future work encompasses optimizing the implementation of the OPM dialogue presented to comply with the most restrictive SLAs that are being proposed.

ACKNOWLEDGMENTS

This work is part of the NetCat, EMPIRICO (PU-2002-56) and TBONES (ITEA 02024 and FIT-070000-2003-936) research projects.

REFERENCES

[1] *ITU-T G.8080*, Architecture for the Automatically Switched Optical Network (ASON), November 2001.
[2] *ITU-T G.872*, Architecture of optical transport networks, November 2001.
[3] Pinart, C.; Junyent, G.; On implementing a management plane for service provisioning in IP over reconfigurable WDM networks, *in Proceedings ONDM 2004*, pp. 465-480.
[4] Lam, H.-K.; Stewart, M.; Huynh, A.; Definitions of managed objects for the optical interface type, *IETF RFC 3591*, September 2003.
[5] Muñoz, R.; Pinart, C.; Martínez, R.; Amrani, A.; Junyent, G.; An experimental ASON based on OADM rings and a GMPLS control plane, *Journal of Fiber and Integrated Optics*, Vol. 23, Nb. 2-3, pp. 67-84, March-June 2004.
[6] Kilper, D. C., et al. ; Optical Performance Monitoring, *IEEE/OSA Journal of Lightwave Technology*, Vol. 22 (1), pp. 294-304, January 2004.
[7] Fawaz, W., et al; Service Level Agreement and provisioning in optical networks, *IEEE Communications Magazine*, Vol. 42 (1), pp. 36-43, January 2004.
[8] Harrington, D. Ed.; An architecture for describing SNMP management frameworks, *IETF RFC 2571*, April 1999.
[9] Muñoz, R.; Pinart, C.; Martínez, R.; Sorribes, J.; Junyent, G.; Experimental demonstration of two new GMPLS lightpath setup proposals for soft-permanent connections over a unidirectional OADM ring implemented on EMPIRICO testbed, *in Proceedings III Workshop MPLS Networks*, pp. 79-95. Girona, March 25 -26 2004.

A CENTRALIZED PATH COMPUTATION SYSTEM FOR GMPLS TRANSPORT NETWORKS: DESIGN ISSUES AND PERFORMANCE STUDIES

Gino Carrozzo[1], Stefano Giordano[2] and Giodi Giorgi[1,2]

[1] *Consorzio Pisa Ricerche, Computer Science and Telecommunication Dept.,*
  c.so Italia 116, 56125 Pisa, ITALY, Telephone: (+39) 050 915 811, Fax: (+39) 050 915 823
[2] *Information Engineering Dept., University of Pisa,*
  via Caruso, 56122 Pisa, ITALY, Telephone: (+39) 050 2217 511, Fax: (+39) 050 2217 522

Abstract: The GMPLS standardization is paving the way for new configurable Traffic Engineering (TE) policies and new survivability schemes for transport networks. In this context, a centralized Path Computation System (PCS) has been implemented, suited for transport networks with a GMPLS control plane. After a brief description of the requirements for a PCS in a GMPLS network, some design issues for the proposed implementation are drawn, with particular emphasis on the centralized approach and on the strategies for achieving the connection survivability. Some results of an intensive testing campaign are shown for the validation of the design choices.

1. INTRODUCTION

The international standardization committees (e.g. ITU-T, OIF and IETF) are all converging in the design of an integrated network with a common Generalized Multi-Protocol Label Switching (GMPLS) control plane. GMPLS will manage all the network data planes [1, 2], providing the required automation in the computation, the setup and the recovery of circuits for next-generation Automatically Switched Optical Network (ASON).

GMPLS is an extension to devices capable of performing switching in time, wavelength and space domains of the MPLS control plane architecture. The core GMPLS architecture is based on a set of extensions to protocols for routing (e.g. OSPF and IS-IS) and signalling (e.g. RSVP), just available in IP networks. Moreover, other signalling protocols have been proposed (e.g. LDP, CR-LDP) and a new link management protocol (i.e. LMP) has been designed from scratch in order to relate properly the de-coupling between Data Plane and Control Plane. From a routing perspective, the GMPLS extensions provide new information for circuit computation and they enable configurable Traffic Engineering (TE) policies and new recovery strategies. In such a context, this paper takes aim at describing the implementation of a centralized Path Computation System (PCS) suited for transport networks with a GMPLS control plane. In details, Sec. 2 is focused on the requirements for the PCS in a centralized GMPLS network scenario. In Sec. 3 the design issues for the proposed implementation are drawn, with particular emphasis on the centralized approach and on the strategies for achieving the connection survivability. Some results of an intensive testing campaign are shown in Sec. 4 while conclusions and future directions are given in Sec. 5.

2. REQUIREMENTS FOR A GMPLS PCS

Within GMPLS standardization, the main focus is on protocols objects and mechanisms, while only high level requirements are proposed for traffic engineering (TE) and survivability. In the GMPLS context, a set of properties is assigned to each link for routing purposes (e.g. TE metric, available/used bandwidths, resource colours, SRLG list, inherent protections, etc.), transforming a traditional links in a Traffic Engineering link (TE-link). A GMPLS path calculator is expected to return Label Switched Paths (LSPs), i.e. sequences of nodes, TE-links and – in case – labels, which try to match some constraints derived from the TE information above. Once an LSP is computed, it describes univocally a unidirectional or bi-directional connection between a source and a destination node.

The standard Shortest Path First – SPF– algorithms [3] are not suited for such a computation, since they cumulate only the standard link metric along the path. A modified SPF algorithm is needed, called Constraint-based Shortest Path First (CSPF) and routes should be the shortest among those which satisfy the required set of constraints [4]. Another important issue, raised by the great traffic amount carried on a LSP, is the connection survivability after a fault. The solution for this problem depends on the recovery strategy implemented in the network. New recovery strategies are enabled by the GMPLS control plane according to the overall taxonomy sketched in Figure 1 (ref. [5–8] for details). Span level strategies are prone to waste resources in the network, because of the sub-optimality of the resulting backup paths; whereas end-to-end recovery strategies are more efficient. Moreover, restoration fits better the dynamical assignment/release of the network resources with respect to protection; but, in case of a fault, a higher blocking probability for the restoring traffic might be experimented, due to the failure handling by control plane mechanisms instead of hardware ones (e.g. detection, notification and miti-

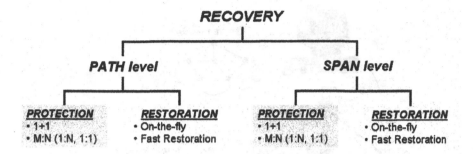

Figure 1. Recovery taxonomy.

gation). A common requirement for all the recovery strategies shown above is the disjointness between the resources (links or nodes) used by the primary route and by its backup. This is needed to minimize the blocking probability of the dynamical recovery action in case of fault. In the GMPLS architecture different levels of disjointness for LSPs are defined [4, 8]:

- *node*, in which different nodes (and different links) are crossed by the primary-backup pair of LSPs;

- *link*, in which only different links are crossed by the two LSPs;

- *SRLG*, in which the Shared Risk Link Group lists of the two LSPs have no intersection.

3. DESIGN ISSUES FOR A GMPLS PCS

In the GMPLS architecture the specification for PCS is considered implementation dependent and no preference can be derived by the standards on the choice of a centralised or a distributed implementation, besides of the intrinsic distributed approach of the GMPLS control plane. The architectural choice we made in the context of the TANGO project is for the implementation of the PCS module inside a centralized network manager (NM). This solution promises to be the most effective for a full and flexible handling of traffic engineering and survivability into the network, particularly when these requirements need to be extended to a multi-area (or multi-domain) scenario. In our implementation the PCS acts as a path computation server for the GMPLS network (ref. Figure 2), receiving from the GMPLS Network Elements (NE) the topology information and the computation requests across a single- or multi-area/AS. The LSPs computed by PCS (if any) are communicated to the ingress GMPLS NE, triggering a standard GMPLS signalling session (e.g. via G.RSVP-TE). The communications between the NM and the NEs are carried out by means of COPS protocol with proper extensions (ref. [9] for details).

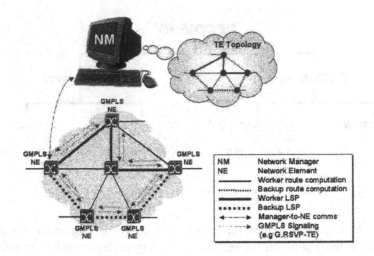

Figure 2. GMPLS network model with a centralized PCS.

We based our routing engine on an implementation of the Dijkstra SPF algorithm, properly modified with a TE-constraints validation step for handling Constrained SPF computations. One of these constraint validations is the check on the bandwidth availability of the candidate link. Moreover, in order to let the algorithm converge towards an optimal SPF solution (e.g. between those which satisfies the required TE-constraints), the metric we chose to minimize during computation is bandwidth-dependent, according to the equation:

$$link\_weight = std\_metric + TE\_metric + F(avail\_bw) \qquad (1)$$

where *std_metric* is the standard OSPF link metric, *TE_metric* is the GMPLS metric for TE purposes, *avail_bw* is the available bandwidth on the TE-link and *F(x)* is a proper Traffic Engineering function designed for balancing bandwidth consumption on the links.

The main feature of our PCS is in the processing of LSP requests with survivability requirements. Based on the recovery taxonomy at path-level and on the requirements shown in Sec. 2, we identified three Classes of Recovery (CoR) for the LSP requests, distinguishing the survivability of the connections according to an Olympic model:

- *Gold*, when the lowest blocking probability for the recovery and the fastest reaction times (e.g. around 50ms) are required. This CoR applies to the protection strategies; it implies the computation of a pair of maximally disjoint paths in order to guarantee the total resource redundancy and uncorrelation.

- *Silver*, when fast reaction times (e.g. less than 1s) are required. This CoR applies to Fast Restoration strategies; it implies the computation of a pair of disjoint paths in order to guarantee a level of resource uncorrelation.

- *Bronze*, when reaction times around 1s are acceptable. This CoR applies to On-the-fly restoration strategies; it implies the computation of an optimal worker path when the request is received and the computation of an optimal backup one at failure occurrence.

In this model are not included *Unprotected* LSP, with no request for survivability of the carried traffic (e.g. preemptable or best-effort traffic). The bronze service is the slowest approach because of the time spent for failure detection, localisation, notification and mitigation. The silver service is sub-optimal as well, because the backup paths do not take into account any modification of the network load/topology occurred in the meanwhile. In this case, we chose to compute the backup LSP trying to satisfy the required link/node disjointness, but with no guarantee for success and for optimality (i.e. the resulting backup might not be maximally disjoint w.r.t its primary and the pair might not have the cumulative shortest cost in the network). For this kind of computation we used the Two Step Approach (TSA) algorithm, based on two Dijkstra SPF runs [3] and on a simple temporary network transformation for avoiding links/nodes of the worker path. The gold service is the most exacting in terms of optimality of the computation, because a least cost pair of disjoint paths is required, providing that it is also the maximally disjoint pair in the network. Many algorithms have been proposed in literature for such a computation [5, 10, 11], most of them in the general context of the K-shortest path theme. We focused on the work by R. Bhandari because it promised lower theoretical complexity w.r.t. other algorithms (e.g. the famous Suurballe's one), when only $K = 2$ shortest paths are searched. The Bhandari's algorithm implemented in this work is the optimal counterpart to the sub-optimality of TSA one. However this optimality is paid for a higher complexity introduced by the required network transformation, by a modified SPF running also on negative graphs and by an interlacing/re-ordering procedure for the final paths.

4.   PERFORMANCE STUDIES

This section is aimed at highlighting the performance of the routing scheme adopted when computing a pair of maximally disjoint shortest paths with respect to links or to nodes. Measures have been collected on different topologies with increasing meshing degrees (ref. Figure 3), ranging from an interconnected rings topology derived from the Interoute I-21 network (meshing degree 3.14) to a number of Manhattan topologies variously connected (meshing degrees from 3.43 to 6.84.

192

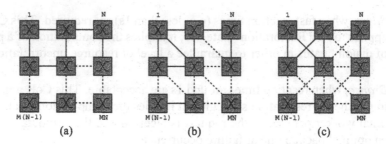

Figure 3. Generic lattice topologies: (a) simple Manhattan; (b) half-meshed Manhattan; (c) meshed Manhattan.

All these topologies have been modelled with generic nodes, configurable as fully connectable SDH 4/4 Cross-Connects. Adjacent nodes have been connected by a bi-directional TE-link with random values for its TE-information (e.g. TE metric, available/used bandwidths, resource colours, SRLG list, etc.). All the TE-links have been configured with 4 STM-64 ports VC4-multiplexed; so, a maximum of 256 allocable labels has been obtained in each adjacency. The computational environment for all the tests has been based on a Pentium III 800MHz PC with Linux RedHat 7.1 OS. A large number of bi-directional LSP computations (*req_path*) have been requested on each topology (e.g. up to a connection request from each node towards all the others), trying to test the algorithms in an overloading condition and observing: (a) the number of computed LSPs (*comp_path*); (b) the number of totally link-/node-disjoint pairs of LSPs (*disjoint*); (c) the mean computation time for each LSP request (*time*).

The overall performance of the two types of algorithms has been measured by evaluating a Global Performance Factor (GPF), defined as:

$$GPF = \frac{disjoint}{req\_path} \cdot \frac{comp\_path}{req\_path} \qquad (2)$$

The first term is related to the algorithm's effectiveness in creating maximally disjoint paths while the latter represents the algorithm's effectiveness in computing valid paths, according to the resource availability on TE-links. This latter term has no effect on GPF in case of a theoretical infinite resource availability. In Figure 3-a and Figure 3-b the GPF is drawn at the different meshing degrees, showing a mean higher performance for the Bhandari's algorithm w.r.t. the TSA one (+0.57% for link- and +2.79% for node-disjointness). The higher complexity in Bhandari's algorithm is responsible for an increase in mean computation times for each LSP request (+3.59% for link and +4.14% for node-disjointness) as shown in Figure 5 only for the link because the two trends are similar. This is due to the higher complexity introduced by Bhandari's algorithm (e.g. the graph transformation, the modified SPF for negative graphs and, last but not least, the interlacing/re-ordering procedure for the fi-

INTELLIGENT OTN IN THE TLC OPERATOR INFRASTRUCTURES

Ovidio Michelangeli WIND Telecomunicazioni S.p.A.Via C.G.Viola 48
Roma ovidio.michelangeli@mail.wind.it,
Alberto Mittoni: WIND Telecomunicazioni S.p.A.Via C.G.Viola 48 Roma
alberto.mittoni@mail.wind.it.

Abstract: The operators are today facing the migration of their transport infrastructures toward intelligent optical layer by introducing in the core optical switches based on GMPLS/ASON protocol. The paper describes the advantages can be achieved by using the next generation Optical Platforms by comparing the ring and mesh network.
Network restorations in the distributed and centralised control plane are also compared by showing some lab test results. Finally are outlined the current limitations and the issues still open to be solved by the industry and the standard bodies.

1   INTRODUCTION

The continued growth in high-speed Internet applications will drive the backbone transport network migration for many TLC operators in the coming years. The existing infrastructures designed on ring topology are not more efficient to support the traffic demand due to the fixed and mobile broadband services. The technological platform in the core are moving toward next generation network based on packet switching rather than circuit switching and the traffic behaviour changes from the traditional voice traffic (constant) to Internet traffic (instantaneous with large peak).

2   DRIVERS TO MIGRATE THE BACKBONE TRANSPORT NETWORK AND OPERATOR REQUIREMENTS

The traffic demand for telecommunications services increases the transmission flows in long distance transport network. Fixed and mobile services based on multimedia applications will require an additional bandwidth to the transport network.

The Operators ask for new mesh layer having the following functionality:

- Real time Bandwidth allocation: Network solutions have to cope with uncertainty in bandwidth needs.

- Faster Provisioning: As new services come into effect, clients demand faster access to these services.

- Seamless migration to an Intelligent Transport Network: Network Service Provider with existing SDH or DWDM networks need to protect their investment with a smooth migration path of their networks to the optical layer.

- Service differentiation: In a highly competitive environment, prices become lower and equalized. The Network Service Provider that can offer a wider array of services becomes the winner.

- Migration to new services support: new services are in demand by most clients who want to benefit from the ITN new features (Ethernet, transparent wavelength services, etc.).

  ✓ Optical Virtual Private Networks

  ✓ Transparent Wavelengths leasing

  ✓ Ethernet Services

  ✓ Bandwidth adapting Data Service

- Operational cost reduction: More and more complex networks start to take a toll on maintenance, provisioning, performance control, etc.

TLC Operators are demanding ways to achieve that all these operational issues do not translate into higher costs.

3   GMPLS/ASON ARCHITECTURE

The GMPLS/ASON architecture [2] is based on mesh topology that is more suitable for large network with different traffic protection requirements. [2,3] It requires less capacity and then are cheaper than ring topology. Figure 2 illustrates the GMPLS/ASON network architecture that is divided in three main domains:

- Control plane: performs the call and connection control functions. Through signalling, it sets up and releases connections, and may restore a connection in case of a failure.

- The management plane: performs management functions: fault, configuration, accounting, performance and security.
- The transport plane: provides flow transfer for user information from one location to another. It can also provide flow transfer for some control and network management information. The transport plane is layered as it is equivalent to the Transport Network defined in ITU Rec. G.805.

ASON/GMPLS is a transport network that, via control plane, can provide automatic network resource and state information discovery. Furthermore it allow continue monitoring of the network and the connected links. It can set up and delete connection.

Different regulatory and organisation bodies have been involved to develop the new intelligent optical transport network. The most important are the following:

- IETF (Internet Engineering Task Force) -> GMPLS
- ITU-T (International Telecommunication Union-Transmissions)->ASON framework
- OIF (Optical Internetworking Forum) -> OIF-UNI/NNI

Other organisations (such as ANSI, MPLS Forum) still take part in the GMPLS standardisation activity, but the ones cited above are the most important.

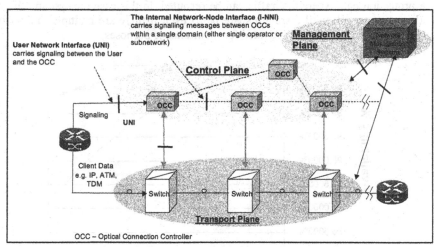

Figure 2: GMPLS/ASON network architecture

4 ADVANTAGES FOR THE TLC OPERATOR

The new architecture can offer different advantages for the network operators need. The automatic services in the GMPLS/ASON Architecture allow to reduce OPEX (manual operations are limited) in terms of provisioning and management aspects. With automatic provisioning, request might be directly sent by end user

198

and managed by ASON control plane (new services for end user, immediate implementation of customer order). The operation of the network is cheaper and have a more efficient network management (faults are directly handled by control plane and traffic is immediately restored.). below are described the further advantages for the Network operators:

- Network Availability (in case of multiple faults)
- CAPEX reduction
- Network flexibility
- New revenue generating services (Value added services improve revenues)

4.1 Network availability

Figure 3 shows the Availability versus MTTR. By increasing the MTTR (i.e increasing of multiple faults in the network) the restoration improve the availability figures.

In the multiple rings an equipment fault cause an automatic switch in the protection path. As consequence (for instance in the maintenance works) the path is not protected.

In the restoration the control plane in case of path failure can dynamically find a protection path where the traffic can be rerouted. In this case the system remain always protected (until there are spare path) even if there are multiple faults (for instance some links are de activated for maintenance purposes)

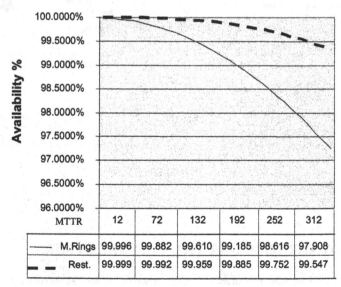

	MTTR	12	72	132	192	252	312
——	M.Rings	99.996	99.882	99.610	99.185	98.616	97.908
– –	Rest.	99.999	99.992	99.959	99.885	99.752	99.547

Figure 3: Availability comparison between mutiple ring and mesh network with restoration.

4.2 Capex reduction

TLC operators are now concentrated at the core business. They are today driven by business case to deploy new technological platform in their network to minimize CAPEX (maximize the existing infrastructure) and reduce OPEX (simplified service configuration and provisioning). A simple Capex reduction is not more sufficient. New Investment should provide not only revenue but also a reduction of operational expenses. Before to implement new infrastructure in their networks, TLC operator prefer to make a business case in order to be sure that new technology can respond their need and optimising their investment.

WIND achieved a case study to evaluate the total cost of its transport backbone network by considering a time frame of 3 years. As input the existing network topology and the traffic matrix for three years has been provided. As output the total network cost in two different scenario has been evaluated:

a) Migration of the existing network following the traditional ring topology;

b) Migration of the existing network with the introduction of a mesh layer based on GMPLS/ASON architecture.

4.3 Case study results

In the following are showed the main results.

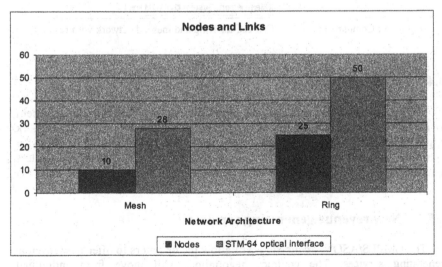

Figure 4: Number of nodes and STM 64 optical interfaces for ring and mesh network with restoration

With reference to Figure 4 the introduction of mesh network with restoration allows to save DWDM links and the number of the equipment in the nodes. The restoration technique allows to share the transmission resources dedicated to the protection and then decreasing the number of 10 Gbps transponder in the network. The mesh networks with restoration allow to save 70% of capacity for protection. In Figure 5 is illustrated the CAPEX with its main item: Electro – Optical Cross Connect, Network Management System and the DWDM links.

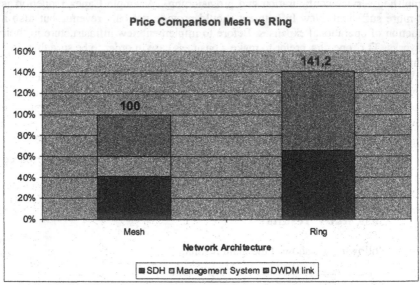

Figure 5 : Comparison between CAPEX for Ring and meshed network with restoration

4.4   Network Flexibility

The network upgrade in the mesh network is very simple because is sufficient to add only I/O interface in the equipment. In the ring topology the network upgrading require to add a new network element. Typically, in case of upgrade, the number of adding section required in the mesh network with restoration are less than 50% respect to those necessary on ring network.

4.5   New revenue generating services

The GMPLS/ASON architecture will allow the operators to offer new revenue generating services. The contract negotiations will move from individual connections to service level agreement. The end to end connectivity will be automated and will be Client-driven.

In addition services such as bandwidth on demand, optical virtual private network, differentiated services based on QoS will be easily provided to the customers.

5 WIND TRANSPORT NETWORK MIGRATION

On the basis of the business case results WIND decided to introduce a mesh network layer to gather only very big nodes and to transport high level traffic (STM-N N≥1) as show in Figure 6.

The ring network will be used for national/regional area and for transport of low level traffic (VC12 and VC3). The new layer is realised by a electrical – optical – switch at high capacity connected with other nodes (from 2 to 5) via DWDM link at 10 Gbps. The network protection will be configurable for each path with different protection level by using the resources dedicated to the shared protection. Control Plane will be centralised and it will migrate in distributed configuration in the medium term.

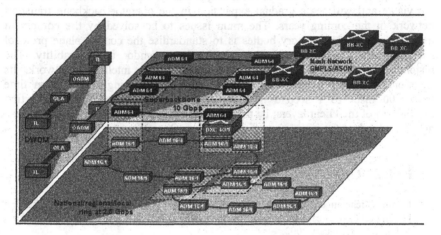

Figure 6: Wind backbone transport network evolution

5.1 Test results

Wind achieved in the vendor laboratories some functional test in order to verify the main Technical performance. Table 1 shows the restoration time for multiple faults in tree different restoration technology: one based on the distributed control plane, one based on the centralised control plane and the last one based on Pre – Planned configuration where the restoration is distributed and the provisioning is centralised.

Table 1: Restoration time versus distributed and Pre – Planned control plane.

Restoration Technology	On The flight		Pre-Planned
Control Plane	Distributed	Centralised	Restoration : distributed, Provisioning: Centralised
Restoration time for multiple paths	500-1000 msec.	300-500 msec.	500-850 msec.

6   CONCLUSION AND OPEN ISSUES

The all-optical core vision is becoming a reality. Optical Intelligent networks are viable and will see a gradual expansion in the operator backbone transport network in the coming years. The main issues to be solved by the equipment manufacturers and regulatory bodies is to standardise the control plane protocol still proprietary, in order to allow the multi - vendor interoperability. The restoration algorithm is the key to design the mesh intelligent network. Its functionality should cover two important aspects: Obtaining a restoration time from 200 msec to 1 second in order to avoid conflict with the protection mechanism of the client layers; be able to minimise the DWDM links in the mesh network to reduce the investment of the Telecom Operators.

REFERENCES

[1] Ovidio Michelangeli, "Choosing between OEO and OOO solutions in the core and making a business case for DWDM in the MAN at the edge level" Optical Switching Summit 30-31January 2002.
[2] E.Mannie et al.,"Generalized Multi-protocol label Switching (GMPLS) architecture " Internet Draft <draft-ietf-ccamp-gmpls-architecture-05.txt>, Mar.2003
[3] Ovidio Michelangeli , Evolving WIND's Transmission Core Network towards an intelligent meshed network based on GMPLS/ASON. IIR conference "Evolving SDH to support data services. London 3oth March- 1st April
[4] Ovidio Michelangeli; Alberto Mittoni; "Inserimento di reti ottiche magliate nella rete di backbone" Riva del Garda – Fotonica 2003

NOVEL ACTIVE MONITORING OF CUSTOMER PREMISES USING BLUETOOTH IN OPTICAL ACCESS NETWORK

S.B.Lee, W.Shin, and K.Oh
Dept. Of Information and Communications, Gwangju Institute of Science and Technology (GIST), 1 Oryong-dong, Buk-gu, Gwangju 500-712, SOUTH KOREA and sam@gist.ac.kr

Abstract: A new rerouting scheme and spare capacity planning for optical link failures is demonstrated employing Bluetooth monitoring. Optical networks require adequate fault monitoring in order to accurately identify and locate network failures. Detailed physical layer information as well as link surveillance is carried to OLT for in-situ monitoring of QoS.

1. INTRODUCTION

Progress in optical networking has stimulated development in optical performance monitoring (OPM), particularly regarding signal quality measures such as optical signal-to-noise ratio(SNR), Q-factor and dispersion[1-2]. The need for fault management capability in the transmission media is also driven by expansion of optical access networks. One possible fiber network architecture for FTTH(Fiber To The Home), PON(Passive Optical network) system consists of a number of cascaded passive optical power splitters or AWGs(Arrayed Waveguide Gratings), originating from a single optical fiber and it may be terminated at the customer premises with optical nodes. The maintenance of such passively split networks presents a new set of challenges to service providers, as due to the star topology, the various branches of the network are likely be equidistant from the transmitter such that they cannot be sensibly probed and addressed from the head-end via a conventional maintenance instrument such as the optical time domain reflectometer(OTDR)[3]. Additionally, present monitoring techniques have been mainly concerned on optical path itself and detailed functional information of the physical layer in the customer's premises, such as optical network unit (ONU), has not been thoroughly attempted due to complexity in optically multiplexing the monitored information over the main signals.

Bluetooth is an open wireless specification that enables short-range connections between communication devices[4] at a frequency of 2.4 GHz within the maximum link distance of 1.2km. We propose a novel monitoring technology in optical access networks by transporting the monitoring signal of ONU functional information over the bluetooth, based on which optical path is can be self-healed, for the first time.

2. EXPERIMENTS

A schematic of the proposed system is shown in *Figure 1*. We assumed bi-directional PON (Passive Optical Network) system for fiber to the home (FTTTH) environment where downward signal is carried over 1550nm and upward signal over 1310nm. Fast ethernet media converters(MC) operating at 100Mbps were designed and fabricated for fiber to UTP conversion at the ONU in the customers. Information on key operating parameters for physical medum dependents(PMD) in ONU are monitored in real time and sent over Bluetooth using a synchronized antenna pair, one at ONU and the other at Optical Line terminal (OLT). Bluetooth supports only 780kb/s, which may be used for 721kb/s unidirectional data transfer(57.6kb/s return direction) or up to 432.6kb/s symmetric data transfer. The proposed system also include self-healing network function, by employing optical MEMS switch for optical routing in the case of fault and error in either optical path or PMD in ONU. Note that the communication over Bluetooth will not interfere with optical signals in the wired network and the its communication range of 1 km do cover most of deployed FTTH systems.

In ONU, the downstream signal is monitored in both optical and electronic domain. Optically 1510nm power is monitored using a broadband tap coupler, which reflects the real time optical link power budget. In the electronics, the fault-error detection output from the MC is monitored for QoS. Furthermore inside temperature and humidity of ONU are monitored to cope with an abrupt surge. These downstream monitoring information is time division multiplexed for every 8 bit using RS232C and sent to the corresponding antenna at OLT over Bluetooth. At OLT, the optical power of the upstream signal at 1310nm is monitored, which serves a redundant check for the real time optical link power budget. Similar to ONU, temperature and humidity of OLT are monitored. These upstream monitoring signals at OLT are multiplexed in RS232C and sent to the antena in ONU.

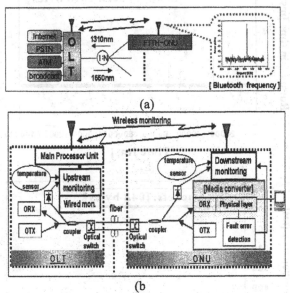

Figure.1. (a) Experimental set-up for bluetooth monitoring
(b) Physical layer status and optical power monitoring

2.1   Self healing mechanism using bluetooth

An example Note that in the proposed scheme, transmittion failure due to optical link damage can be detected in redundant manner, firstly by optical power monitoring of 1550nm at ONU, and its Bluetooth signal and sencondly by optical power monitoring of 1310nm upstream at OLT. In the case of failure of these monitoring signals, the optical path can be self-healed by switching to the spare optical route using a 2x2 MEMS switch. Local switching capability at ONU and OLT would be highly beneficial for faster response of system restoration[5-6].

In order to test the self-healing capability, we have simulated fiber cut by plugging in and out the fiber connectors and the temporal responses of the system to switch to other optical path have been measured. Ping protocol that travels to a distant IP address back and forth over the link, was used to measure the round trip time for the packets of 2048 bytes in the link. For 10 simulated occasions of fiber cut, distribution of the roundtrip time is shown *in Figure. 2*(a). Average round trip time was around 1ms and restoration time for the network was in average 380 ms, which included MEMS response time of 0.5 msec as well as bluetooth transmission and protection-information processing time.

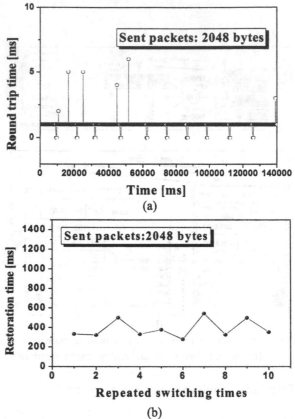

Figure.2 (a) Round trip time for fiber cut occasions
(b) System restoration time for fiber cut

The distribution of restoration time is shown *in Figure. 2*(b). The minimum was 270msec and the maximum was 550msec. The latency of the system was found to be reliable in spite of wireless transmission.

Figure.3 represents distribution of the number of lost packets during the system restoration as described earlier. Individually, 32bytes packets were sent. Maximum of 10 packets were lost and the average packet loss was approximately 18% in the case of 32 bytes packet transmission. It is observed that the loss rate significantly reduces as the transmitted packet size increases, so that about 0.3% loss rate for 2048 byte packets.

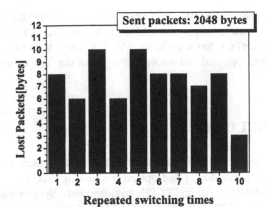

Figure.3 The number of lost packets for damage

2.2 Physical layer monitoring technique

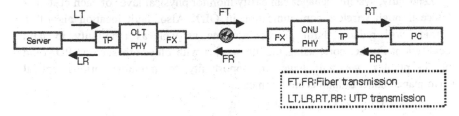

Condition (auto-negotiation	OLT Physical layer				ONU Physical layer			
	TP		FX		FX		TP	
	Link	Fault	Link	Fault	Link	Fault	Link	Fault
LT fault	off	on	on	off	on	on	off	off
LR fault	off	off	on	off	on	off	on	off
LT & LR fault	off	on	on	off	on	off	off	off
FT fault	off	off	on	on	off	on	off	off
FR fault	off	off	off	on	on	on	off	off
FT & FR fault	off	off	off	on	off	on	off	off
RT fault	on	off	on	off	off	off	off	off
RR fault	off	off	on	on	on	off	off	on
RT & RR fault	off	off	on	on	on	off	off	on

Figure.4 Fault location algorithm

Fiber-optic signal detects signal which indicates whether or not the fiber-optic receive pair is receiving valid signal levels. So far, although the network fault manager receive alarms for some network failures, it's not easy to know which line or part causes faults between OLT and customer premises. As shown in *Figure.4* , this algorithm can minimize the time to find and to correct fault location. For example, TP(link-on),TP(fault-off),FX(link-on)and FX(fault-off) on OLT physical layer and FX(link-on),FX(fault-off),TP(link-off) and TP(fault-off) on ONU physical layer let fault manager know UTP line(especially RT fault) connected to the customer device(PC) has a problem. Also, through wireless monitoring, fault location could be recognized and resolved in spite of the fiber or the other lines cut.

3.   CONCLUSIONS

In summary, the novel self-healing system was demonstrated by providing a composite Bluetooth monitoring and optical monitoring. The system can be readily applied to OLT-ONU links in FTTH environment. Symmetric upstream and downstream signals at 100Mbps were monitored and along with monitoring signals for temperature, humidity and fault error detection they are multiplexed over Bluetooth within 1km. The restoration time for optical path cut was less than about 500ms and less than 0.7% of packet was lost for 2048 byte packets.

Especially, system manager can easily monitor physical layer of each customer device through wireless communication in OLT. Also, fault location algorithm solution will reduce alarm processing time as well as ambiguity in fault localization. Furthermore, using mobile service with bluetooth such as PDA and mobile phones, we will have an opportunity to monitor various optical performance under ubiquitus environments.

ACKNOWLEDGMENTS

This work was supported in part by the KOSEF through the UFON research center, the Korean Ministry of Education through the BK21 program, and ITRC-CHOAN program.

REFERENCES

[1] D.C.Kilper, et. al., "Optical performance monitoring", *J. Lightwave Technol.*, vol.22, 294-304, 2004
[2] Giammarco Rossi, et. al., "Optical performance monitoring in reconfigurable WDM optical networks using subcarrier multiplexing", *J. Lightwave Technol.*, vol. 18, 1639-1648, 2000
[3] B Selvan, et. al., "Network monitoring for passively split optical fibre networks", *The institution of Electrical Engineers*, 1997
[4] Paulo Bartolomeu, et . al., "Distributed monitoring subsystems based on Bluetooth implementation", *Emerging Technologies and Factory Automation*, vol. 2, 16-19, 2003

[5] Sava Stanic, et. al., "On monitoring transparent optical networks", *International Conference on Parallel Processing Workshops*, 217-223, 2002

[6] Mr. Michael, et. al., "Optical switching for automated test systems", *AUTOTESTCON proceedings IEEE*, 140-151, 2002

SHARED MEMORY ACCESS METHOD FOR A λ COMPUTING ENVIRONMENT

Hirohisa Nakamoto,[1] Ken-ichi Baba,[2] and Masayuki Murata[1]

[1] *Department of Information Networking, Graduate School of Information Science and Technology, Osaka University, Suita, Osaka 565-0871, Japan*
[2] *Cybermedia Center, Osaka University, Ibaraki, Osaka 567-0047, Japan*

Abstract: Although optical transmission technology for high-speed broadband networks is being studied actively, the conventional packet-based switching technology cannot assure high-quality communication for each connection. We therefore propose a new computing environment, called λ computing environment, that provides virtual channels utilizing optical wavelength paths connecting computing nodes. These channels provide the reliable high-speed paths needed for recently deployed distributed applications including SAN and Grid computing. In this paper, we propose and evaluate a method for accessing the virtual ring network for realizing a shared memory in a distributed fashion on WDM-based photonic networks.

1.   INTRODUCTION

In recent years, as users of networks such as the internet are increasing, the amount of traffic is increasing steadily. Various applications utilizing images become to be used, and the demand on the technology which enables the high speed and large scale transmission in networks is increasing. Research on optical transmission technology has therefore been expanded in efforts to develop WDM technologies, which use light of various wavelengths. Investigators are actually exploring the feasibility of a new WDM technology that can use 1000 wavelengths [1]. IP over WDM networks are being studied in order to provide high-speed Internet transmission based on WDM technology. An Internet routing technology called Generalized Multi-Protocol Label Switching (GMPLS) is also being standardized [2], and research on the optical packet switch has begun [3].

Many of these technologies, however, presuppose the present Internet technology. That is, the granularity of information is assumed to be the IP packet.

And the assumption of an architecture based on packet switching technology makes it very hard to provide high-quality communication to each connection. New application technologies such as SAN and Grid computing need to provide the end user with a high-speed and reliable communication pipe, and that cannot be done without setting up a mass wavelength path between end users. That is, the end users can be provided an ultrahigh-speed and ultrahigh-quality communication pipe by building a photonic network that uses established fibers, or newly laid fibers if needed, and by using the wavelength multiplexed in the fiber as the minimum particle size for information exchanges.

OptIPuter is middleware proposed for the high-speed distributed computation environment provided by an optical network [4]. It was developed for the Grid environment to be established on optical networks. It provides virtual communication paths but is based on the present Internet technology and treats a packet as the particle size of information. As a result, the packet-processing problem mentioned previously arises.

We therefore propose a new architecture, called λ computing environment, that has virtual channels utilizing wavelength paths on WDM-based photonic networks for connecting computing nodes. In the conventional Grid environment, data is exchanged by TCP/IP message-passing. On the other hand, The λ computing environment provides reliable high-speed communication between nodes on the Grid by using established wavelength paths. Distributed computation on high-speed channels is thus provided by making virtual channels in the mesh of a photonic network of optical fibers connecting the network nodes and the computing nodes. Moreover, it is possible to use wavelengths as a shared memory by constituting a virtual ring in the λ computing environment. As a result, it is not necessary to distinguish the shared memory from a communication channel in a wide-area distributed system. We expect this to result in high-speed data exchange between computing nodes (see Fig. 1).

In this paper, we propose and evaluate a method for accessing the virtual ring network utilized as a shared memory. Specifically, we propose that the cache in the CPU and local memory of each computer group are used as a cache of a shared memory. When a virtual ring is used as a shared memory, it is necessary to consider restrictions on the timing and frequency of access because the shared memory is spread out on a long-distance optical fiber and also to take into consideration the coherency between the shared memory in the virtual ring and the cache of each computer group. It is not necessary, however, to distinguish the shared memory in a wide-area distributed system from a communication channel, and this seems to be what makes it possible to exchange data between computing nodes at extremely high speeds. In consideration of such features, we propose a shared memory access method for the λ computing environment and evaluate the method through simulations.

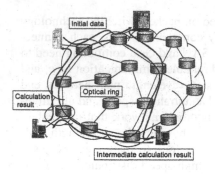

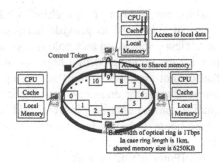

Figure 1. Virtual ring on a photonic network.

Figure 2. Network model.

2. SHARED MEMORY AND ACCESS METHOD IN A λ COMPUTING ENVIRONMENT

2.1 Network model

In the network model we use (see Fig. 2), the computing nodes constituting the λ computing environment are connected by optical fibers that make a ring network virtually. This paper presupposes that each computing node has one CPU, a level 1 cache, and a local memory. A local memory is used for storing programming codes, and a shared memory is used for storing the shared data used by all the computing nodes.

The optical ring network has a wavelength path for shared memory and a wavelength path for control signals. The bandwidth of the wavelength path for the shared memory is 10 Tbps and the propagation delay is 5 nano-second per meter. The processing delay in the intermediate nodes, such as network devices that are part of the optical ring network, is not taken into consideration here but is simply included in the propagation delay. Therefore, when this optical ring network is used as a shared memory, its capacity is 6250 KBytes per kilometer of fiber length.

2.2 Shared memory access method

In conventional shared memory systems, each processor's access to the shared memory is restricted by contention in a shared bus. Access to the shared memory is therefore usually improved by using a cache to increase the speed of access. Every processor usually has a cache system consisting of a level 1 cache, a level 2 cache, a level 3 cache, and many stages. When each processor has a cache, it is necessary to fully take into consideration the consistency between the data in those caches and the data in the shared memory [5, 6].

In the λ computing environment, the application program calculates intensively within a computing node. And it exchanges data among computing nodes after synchronous process. So it does not have to exchange data during the processing going on within a computing node. We therefore use the write-back invalidation protocol because the number of write-back accesses to the shared memory is smaller for this protocol than for the any of the snoop cache protocols. Considering the restrictions on memory access timing and cache coherency, we propose the protocol described in Sec. 2.2.1. The cache coherency problem arising when two or more computing nodes synchronously update the same local cache data is solved by preparing a token for control messages. Parallel computers also collaborate by using such synchronous operations as atomic operation, caching of synchronous variable, waiting on a shared memory, a memory lock, the barrier synchronization method, and so on [5, 6]. Application programs evaluating a proposed method first calculate locally and then perform synchronous operations. So in the work reported in this paper we used the barrier synchronization method. The barrier synchronization method on an optical ring is explained in Sec. 2.2.2.

2.2.1 Write-back invalidation protocol for an optical ring network.
As mentioned above, we use the write-back invalidation protocol to solve the cache coherency problem. In adapting this protocol to our network model, we must take a care of control messages, the read-miss process, and the write-miss process.

In the conventional method, when a read miss occurs and the other computing node has relevant data, that node will send the data to the demanding computing node. In the shared memory using an optical ring network, however, the demanding computing node can directly access the shared memory. This is because the delay for direct access to the shared memory is shorter than the waiting for transmission of relevant data from another computing node. In the write-back invalidation protocol, the data in the local cache has three possible states: Invalid (I), Clean (C), and Dirty (D). When a copy demand message arrives and the relevant data is in state C, none of the computing nodes returns a response message. When the relevant data is in state D, in contrast, the node returns a response message. In this case, it is not necessary to access a shared memory.

We next consider the processing for the write-out from a processor to a cache. When data in the cache is in state C and a processor writes out the data to local cache, the data will change from state C to state D. Then the invalid demand message for relevant data is added to the control token, and the token is sent out to the wavelength path for control. At this time, other computing nodes with the data of a corresponding address know that the relevant data has been written out, and they change the state of the data that they have in the self-cache into state I. If two or more computing nodes write out the C state data at the same address simultaneously, two or more computing nodes may hold the data in state D. This problem is solved by having a computing node that needs to update the cache data from state C to state D first catch the control token. After catching the control token, it checks that another computing node has not added the invalidation message to a relevant address. After checking the control token, it adds the invalidation demand message to the control token, and sends the control token to wavelength for control. It can then change the state of the cache of relevant data from state C to state D. Moreover, when the control token is caught and another computing node has already added the invalidation message to a relevant address, the state of the relevant caches is changed to state I state, and the cache update is yielded to another computing node.

2.2.2 **Barrier synchronization.** Barrier synchronization [5, 6] in the shared memory on the optical ring network is done by first allocating part of the shared memory to a synchronous memory area. When a synchronous memory is accessed, a Fetch&Decrement operation like that in the conventional method is performed indivisibly. That is, it ensures that access to a synchronous memory indivisibly causes subtraction processing to the relevant data. Since only one computing node at a time can access the synchronous memory, the execution of an atomic operation is easy when an optical ring network is used as a synchronous memory. When an application program is establishing synchronization among some computing nodes, each node accesses the synchronous memory. The number of processors is stored there. The value of the synchronous memory will be set to 0 if all nodes access it. If the value of a synchronous memory is set to 0, all nodes will finish their synchronous processing and begin the next processing.

3. PERFORMANCE EVALUATION

In this section, we report our evaluation of the performance of the proposed shared memory access method by using simulation. We show the results of using not only the shared memory in the λ computing environment but also results of using the conventional TCP for the distributed computation. We show them here so they can be easily compared. When coding our simulation program, we referred to the ISIS library [7] currently being developed at the Amano Laboratory at Keio University.

3.1 Simulation model

We used the following network model in our simulation. Each computing node in the λ computing environment is interconnected with optical fiber and configures the ring network virtually. Each computing node has one CPU, a level 1 cache, and a local memory. The CPU clock frequency is 3 GHz, the capacity of a level 1 cache is 512 KB, and capacity of a local memory is 2G Bytes. The computing nodes are assumed to be equally spaced around the optical ring network. An optical ring network has a wavelength path for shared memory and a wavelength path for control signals. The bandwidth of the wavelength path for a shared memory is 10 Tbps, and propagation delay time is 5 nano-second per meter. The processing delay in the interface of each computing node and the intermediate nodes is not taken into account here but is assumed to be included to the propagation delay. We also assume that the shared memory system exchanges data according to the TCP. This system has one shared memory server and each computing node is connected to that shared memory server by an Ethernet. The performance of a computing node is the same as that in a photonic shared memory model, and the distance from each of the computing nodes to a shared memory server is set to 1 kilometer. The Ethernet transmission speed is set to 1G bps.

We evaluated the performance by using some of the Splash2 benchmark programs, such as the radix sort program that sorts the sequence of an integer value by using the radix-sort algorithm, the matrix-product program that calculates the product of an $n \times n$ matrix, and the queen-problem program that solves the n-queen problem.

3.2   Result of the radix-sort program

We first show in Fig. 3 the number of CPU execution clocks when we apply a radix-sort program to the simulator of the shared memory model in the λ computing environment. The numbers of keys for sorting are 32768, 65536 and 131072. Even if it increases the number of computing nodes in the case of the problem size 32768, parallel processing is not taken advantage of because the number of synchronous operations is a large fraction of the total number of operations. When problem size becomes large, however, like 65536 or 131072, parallel processing is advantageous whenever there are fewer than eight computing nodes. The effect of parallel processing becomes weaker, however, as the number of nodes increases. We show in Fig. 4 the number of CPU execution clocks when we use radix-sort programs in the simulator of the TCP message passing model. Although the same tendency as that seen in the shared memory simulator in the λ computing environment is shown, the number of execution clocks is, on the whole, larger than that in the shared memory simulator in the λ computing environment. From these results we found that the shared memory and the access method for the λ computing environment have advantages when performing distributed processing of a problem of sufficient size.

3.3 Result of the matrix-product program

We show in Fig. 5 the number of CPU execution clocks when we apply a matrix-product program to the simulator of the shared memory model in the λ computing environment. The matrix sizes are from 32×32 to 256×256. When the problem size is small, parallel processing is not taken advantage of. But when problem size becomes large, the effect of parallel processing is evident when the number of computing nodes is less than eight. Though we cannot show the graph of a TCP model because of lack of space, even if problem size becomes large, parallel processing is not taken advantage of. This is unlike what we see in the results we obtained with the radix-sort program. From these results, we can see that when performing distributed processing of a problem of sufficient size, the shared memory method for the λ computing environment performs better than the method for TCP message-passing does.

3.4 Result of the queen-problem program

We show in Fig. 6 the number of CPU execution clocks when we apply a queen-problem program to the simulator of the shared memory model in the λ computing environment. The problem sizes are from 4×4 to 32×32. In the case of a queen problem, there is no advantage of parallel processing even when the problem size is large. This is because the number of synchronous accesses is larger than in the other application programs, though we cannot show the graph because of lack of space.

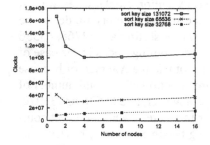 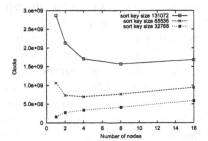

Figure 3. Processing time for a radix-sort program using photonic shared memory.

Figure 4. Processing time for a radix-sort program using TCP message passing.

4. CONCLUSION

In this paper, we proposed a method for accessing shared memory on a photonic network. We also evaluated the performance of the proposed method by

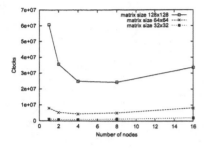

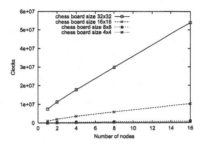

Figure 5. Processing time for a matrix-product program using photonic shared memory.

Figure 6. Processing time for a queen problem program using photonic shared memory.

using benchmark programs for parallel computing. As a result, we showed that the effectiveness of using an optical ring as a shared memory and of parallel processing implemented by increasing in the number of nodes is seen when the number of synchronous processing is small. A more efficient method for accessing a shared memory and a practical use of a local memory should be considered. Furthermore, since neither the processing delay due to the interface nor simultaneous access to a photonic ring network from two or more node computers was taken into consideration in the work reported here, these are subjects that should be investigated. We are also going to examine the shared memory system and the memory access technique at the time of using an optical ring network as a high-speed channel.

REFERENCES

[1] M. Murata and K. Kitayama, "Ultrafast photonic label switch for asynchronous packets of variable length," in *IEEE INFOCOM 2002*, June 2002.

[2] E. L. Berger, "Generalized multi-protocol label switching (GMPLS) signaling functional description," in *IETF RFC3471*, Jan. 2003.

[3] T. Yamaguchi, K. Baba, M. Murata, and K. Kitayama, "Scheduling algorithm with consideration to void space reduction in photonic packet switch," *IEICE Transactions on Communications*, vol. E86-B, pp. 2310–2318, Aug. 2003.

[4] T. DeFanti, M. Brown, J. Leigh, O. Yu, E. He, J. Mambretti, D. Lillethun, and J. Weinberger, "Optical Switching Middleware for the OptIPuter," *IEICE Transaction on Communication*, vol. E86-B, Aug. 2003.

[5] H. Amano, *Parallel Computer*. Shoukoudou, June 1996.

[6] N. Suzuki, S. Shimizu, and N. Yamanouchi, *An Implemantation of a Shared Memory Multiprocessor*. Koronasha, Mar. 1993.

[7] M. Wakabayashi and H. Amano, "Environment for multiprocessor simulator development," *I-SPAN 2000*, pp. 64–71, Dec. 2000.

PART A4:

TRAFFIC ENGINEERING

A MULTILAYER-ROUTING-STRATEGY WITH DYNAMIC LINK RESOURCE ADAPTATION

Robert Prinz[1], Dr. Andreas Iselt[2]

[1]*TU-Muenchen, Institute for Communication Networks, Arcisstr 21, 80333 Munich, prinz@lkn.tum.de*
[2] *Siemens AG, Corporate Technology, Information and Communication, Otto-Hahn-Ring 6, 81730 Munich, Germany, andreas.iselt@siemens.com*

Abstract: With the automation in optical networks based on the introduction of a control plane (GMPLS, ASON) it will be possible to adapt the transport network dynamically to the traffic requirements of packet based client layers. This allows reducing leased line costs and improving the throughput. In some cases the network throughput limitation comes from the limited switching capacity of transit routers. This is especially the case when Multiservice-network elements are used, which usually are implemented as an extension of SDH network elements with packet switching capabilities. Hereby the packet switching matrix usually only covers a fraction of the whole interface capacity of the node, since the main switching is expected to be carried out in the TDM (SDH) domain. An intelligent Multilayer-Routing-Strategy can relief the packet layer by circumventing packet switching with transport shortcuts. In this paper we present such a new Multilayer-Routing-Strategy, that allows to route demand with a distributed routing mechanism. This is complemented by a centralized optimization instance, that optimizes the overall network status. Simulative investigations allow a first quantification of the advantage of this approach.

1.   INTRODUCTION

The introduction of automated connection control in optical transport networks using ASON (Automatically Switched Optical Networks) [3] or GMPLS (Generalised Multi-Protocol Label Switching) [4] together with the standardisation of interfaces like UNI (User Network Interface) and NNI (Network Network

Interface) will allow establishing connections immediately on customer demand. If the "customer" is a network operator (e.g. IP/MPLS network operator, ISP) he can adapt the capacity of his network to the actual load pattern. It is also possible that this network operator does not even own the transport network infrastructure, but leases it dynamically based on the online offer of competing suppliers.

The challenge is to automate the bandwidth adaptation process of the client network with the target of a stable, cost-efficient and dynamic network, that optimally distributes the load and can react on resource constraints. The Multilayer-Routing-Strategy (MLRS) that is presented here is described using a MPLS network (client layer) on an optical network (server layer), but is also applicable to other network technologies.

2. ARCHITECTURE

In Figure 1 the regarded network architecture is depicted. A two-layer network with packet switched MPLS over circuit switched (optical) transport is assumed. The elements of the client and the server layer are collocated at some positions. This may be realized by integration of packet switching and circuit switching capabilities, as it is often found in Multiservice-Nodes.

We further assume that an User-Network-Interface (UNI) between the layers is available, that allows the client layer to request the setup and the release of connections from the server layer. This mode of operation is usually called Overlay Model. The principle of operation is that the links between nodes in the client layer, which are realised as connections in the underlying server layer are adapted dynamically in accordance to the requirements (i.e. traffic demand) of the client layer.

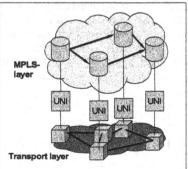

Figure 1. Network architecture

In the following sections a strategy for this adaptation is proposed and evaluated.

3. MULTILAYER ROUTING STRATEGY

We developed a multilayer routing strategy (see also [1]) which is divided in two separate processes. The first process is a distributed algorithm which is implemented in all routers of the client network. It is responsible for routing new connections in the client network. This process can add new link resources (optical channels which are dynamically requested from the transport network) to the existing client network in reaction to a capacity bottleneck if necessary. The distributed algorithm is faster and more reliable than a central management solution. The second process which has a lower priority runs on one or more central servers and tries to reoptimize the current routing and link capacity adjustment on specific events with the objective to minimize the transport cost. This process is not time critical and even in case of a server failure network operation is not impaired.

The following assumptions are considered:
- The client network uses MPLS and all MPLS paths have a reserved bit rate.
- The available capacity of links and routers is distributed via OSPF-TE in the whole network. Each node knows the currently available network resources.
- Source routing is used and could be done by RSVP-TE using the Explicit Route Object.
- Each node has a transit connection data base which matches each transit connection to its originating node. When using RSVP-TE for MPLS path setup its data base can be used.
- The central servers for the reoptimization have access to all connection data of the client network.

3.1 The Algorithm for new Connection Requests

If a new client layer connection request arrives at node S with the maximum bit rate $d_{S,T}$ between source node S and target node T, the algorithm for a new connection request is applied, which distinguishes four different cases:

Case 1: There is a path p that has enough free capacity on each involved element (source-, target- and transit-routers as well as the links of the path p). The connection will use this path p. If there is more than one possible path, the connection will use the best path (e.g. the path with the lowest hop count and with the most available capacity). This is illustrated in Figure 2(a).

Case 2: Source node S and target node T have enough free capacity but there is no path with sufficient free capacity between them.

Now the client network can add a new link to its topology or increase the bit rate on an existing link by creating a new optical channel by the UNI to resolve the capacity bottleneck in the network. The problem of the selection of the node pair to

224

be connected by a new optical channel is solved by optimization. The objective is
to find a cost minimized solution with the greatest benefit for the current network.
The condition which has to be met here is, that a path p is generated which has
enough free capacity on all involved elements for the current connection request
(see Figure 2(b)).

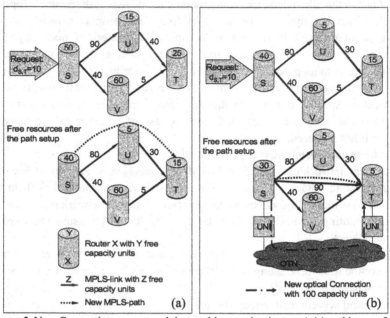

Figure 2. New Connection request and the resulting routing in case 1 (a) and in case 2 (b)

Case 3: At least one of the nodes S or T has not sufficient free capacity and it is
not possible to free capacity on this router by rerouting existing connections (e.g.
all existing connections on this highly loaded router begin or end at it). The new
connection request must be blocked (see Figure 3(a)).

Case 4: Source node S or target node T or both have not enough free capacity,
but it exists one or more transit connections on each highly loaded router with (in
sum) more than $d_{S,T}$ bit rate usage. Now this router signals the originating node of
selected transit connections (e.g. the transit connections with the greatest bit rate
reservation) to find an alternative path for these connections without using this
router. Provided that there is enough capacity available in the optical layer, it must
be possible to reroute the connection. For this rerouting also a new link can be
added if necessary. After the rerouting procedure the connection request can be
routed through the network, possibly adding a new channel to the network as
described in case 2 (see Figure 3(b)).

To case 3 and 4: If target node T has not enough resources for the new connection request source node S can not decide whether the target node can solve this problem by rerouting or not. So node S has to signal an inquiry to target node T (see in Figure 3(a) and (b)).

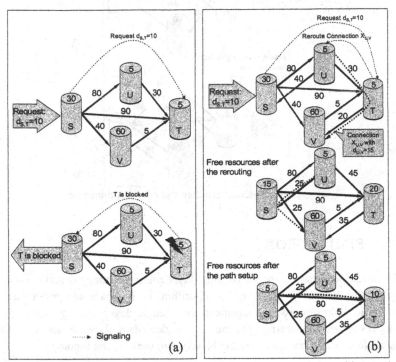

Figure 3. New Connection request and the resulting routing in case 3 (a) and in case 4 (b)

3.2 The Reoptimization Process

In addition to the algorithm for a new connection request described in chapter 3.1 a second mechanism with a lower priority is involved. It tries to reoptimize the routing with the objective to free and release optical channels. It may be executed periodically or triggered by a connection release. Besides the required constraints for this optimization other constraints like limiting the maximum number of reroutable connections per iteration are imaginable. The reoptimisation process runs on one or more synchronized central servers and has access to all necessary data for the reoptimization (e.g. connection data and leased link data). A reoptimization step is shown in Figure 4.

226

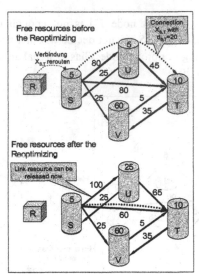

Figure 4. Network resources before and after reoptimization

4. SIMULATION

We developed a simulator for the multilayer routing strategy described above to analyze the dynamic behavior of the algorithm. To compare our strategy using the dynamic link resource management to a single layer strategy with static network resources we modified the mechanism described above in that way, that for the static case no new resources can be added to the existing topology.

4.1 Simulation Conditions

For the simulation we use a pan-European network from [2] which is displayed in Figure 5. The used bidirectional traffic intensity matrix from [2] is shown in Table 1 together with a bidirectional link capacity adjustment optimized for this traffic in the static case (each link capacity must be a multiple of 10 capacity units). Also the capacity NC of each network node is optimized and must be a multiple of 40 capacity units (see in Table 1). The capacity of the nodes is used for the dynamic case as well as the static case. In the dynamic case the network has no links at the beginning of the simulation and the algorithm can add new channels with the granularity of 10 capacity units from any node to any other node at any time always with the same cost.

The connection requests are generated randomly. The mean value $i_{S,T}$ of the neg. exponentially distributed inter arrival time of the node pair S-T is given by the equation (1). Thereby s is the mean value of the also neg. exponentially distributed service time, which has the fixed value of 30 time units in our simulations. $A_{S,T}$ is the traffic intensity value of node pair S-T out of Table 1. To simulate different network loads the traffic intensity $A_{S,T}$ is multiplied with a filling ration f, which we vary in a range of 0.1 to 0.4 in 0.05 steps. The filling ratios between 0.1 and 0.4 lead to values between 1419 and 5828 connection requests within the 500 time units simulation period.

$$i_{S,T} = \frac{s}{A_{S,T} \cdot f} \tag{1}$$

Every 10 time units the network is reoptimized by the described mechanism of section 3.2. The path for each connection is restricted to a maximum of two transit nodes and the capacity of each connection has a fixed value of one capacity unit.

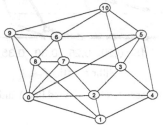

Figure 5. COST239 Network

Table 1. Bidirectional traffic matrix, node capacities NC and bidirectional link capacities

Node	NC	0	1	2	3	4	5	6	7	8	9	10	
0	320		20	40			70		10	50	60		
1	160	12.5		20		40				20			
2	200	15	15		10	40			20				
3	80	2.5	2.5	2.5		10	10		10			10	
4	200	5	7.5	7.5	2.5		80						
5	280	27.5	22.5	27.5	5	22.5		40				10	
6	160	12.5	5	7.5	2.5	2.5	20		20	30	20	10	
7	80	2.5	2.5	2.5	2.5	2.5	5	2.5		0			
8	160	17.5	5	15	2.5	2.5	15	10	2.5		20		
9	160	25	7.5	7.5	2.5	5	20	12.5	2.5	10		10	
10	80	2.5	2.5	2.5	2.5	2.5	7.5	2.5	2.5	2.5	2.5		
								Demand (Traffic Intensity)					

(Right margin, rotated: Link capacity for the static case)

4.2 Results

Figure 6 shows the results as a function of the filling ratios f. In the static case the blocking ratio b is zero up to the filling ratio of f=0.25. It can be seen that the blocking in the static case is higher than in the dynamic case where even for a filling ration of f=0.3 no connection is blocked. Regarding the resource requirements it can be seen, that in the dynamic case the network requests links in the simulation period which rises from 126 links for f=0.1 up to 377 links for f=0.4 in a nearly linear manner. It is remarkable that the mean link capacity $ØLC$ of the

228

network in the dynamic case lies between 13% for f=0.1 and 35% for f=0.4 of the link capacity used in the static case, which has a value of 1360 capacity units.

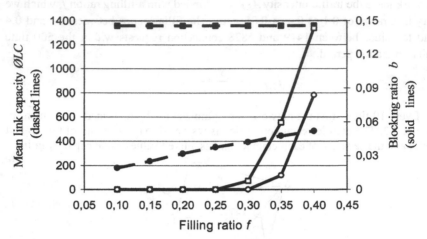

Figure 6. Simulated mean link capacity ØLC (dashed lines) and blocking ratio b (solid lines)

5.   SUMMARY

In this paper we have presented a multilayer routing strategy that enables dynamic adaptation of link resources in the client network. This dynamic adaptation to changing demands bears potential for saving cost (required link capacities and switching matrix sizes) and improving blocking. The strategy has been investigated by simulation. The simulation results show that the resource requirements (mean link capacities) can be reduced remarkably. Additionally the blocking probabilities can be improved allowing a higher network utilization without experiencing higher blocking ratios.

REFERENCES

[1] R. Prinz, A. Iselt, "Eine Multilayer-Routing Strategie für GMPLS-Netze mit dynamischer Linkressourcenanpassung," *5. ITG-Fachtagung Photonische Netze*, Leipzig, 2004, pp. 53-57.
[2] P. Batchelor et al., "Ultra High capacity optical transmission networks: Final report of action COST239," in *Faculty of Electrical Engineering and Computing*, Zagreb, 1999.
[3] *ITU-T Rec. G8080/Y.1304*, "Architecture for the Automatically switched Optical Network (ASON)," November 2001
[4] E. Mannie, D. Papadimitriou et al., "Generalized MPLS Architecture," *Information Draft, draft-ietf-ccamp-gmpls-architecture-01.txt*, November 2001

EVALUATION OF BANDWIDTH-DEPENDENT METRICS FOR TE LINKS IN A GMPLS PATH COMPUTATION SYSTEM.

Gino Carrozzo[1], Stefano Giordano[2], and Giodi Giorgi[1,2]

[1] *Consorzio Pisa Ricerche, Computer Science and Telecommunication Dept.,*
  c.so Italia 116, 56125 Pisa – ITALY, Telephone: +39 050 915 811, Fax: +39 050 915 823
[2] *Information Engineering Dept., University of Pisa,*
  via Caruso, 56122 Pisa – ITALY, Telephone: +39 050 2217 511, Fax: +39 050 2217 522

Abstract: The GMPLS standardization is paving the way for the implementation of new configurable traffic engineering (TE) policies for transport networks. This paper takes aim at evaluating the effects of using bandwidth-dependent TE metrics in a centralized Path Computation System (PCS), suited for handling the routing requests in an operational transport network with a GMPLS control plane. The results of an intensive testing campaign show an evident improvement in the utilization of network resources when such TE metrics are enabled, whatever survivability requirement is imposed on the LSP (e.g. classical 1+1 protection, pre-planned or On-the-Fly restoration, etc.). Moreover, a simple policy function is suggested as a good trade-off between the achievable performance and the computing load on CPU.

1. INTRODUCTION

The international standardization committees (e.g. ITU-T, OIF and IETF) are all converging in the design of an integrated network with a common Generalized Multi-Protocol Label Switching (GMPLS) control plane. GMPLS will manage all the network data planes [1], providing the required automation in the computation, the setup and the recovery of circuits for next-generation Automatically Switched Transport Network (ASON). The core GMPLS architecture is based on a set of protocols for routing (e.g. G.OSPF-TE and G.ISIS-TE), signalling (e.g. G.RSVP-TE) and link management (e.g. LMP) in order to manage correctly the separation between the Data Plane and the Control Plane.

From a routing perspective, the GMPLS extensions provide new information for circuit computation and they enable configurable traffic engineering (TE) policies and new recovery strategies.

In such a context, this paper takes aim at evaluating the different bandwidth-dependent TE metrics in a centralized GMPLS Path Computation System (PCS). In details, Sec. 2 is focused on the requirements for the PCS in a centralized GMPLS network scenario. In Sec. 3 the implemented bandwidth-dependent metrics are defined, while some results of an intensive testing campaign are shown in Sec. 4, deriving conclusions in Sec. 5.

2. ROUTING REQUIREMENTS FOR A GMPLS PCS

Within the GMPLS standardization, traffic engineering (TE) and survivability are still discussed in terms of high level requirements, though they are fundamental for load balancing and traffic resiliency.

Path computation systems for standard IP networks are generally based on distributed, fast and simple routing algorithms (e.g. of the Shortest Path First class -SPF- [2]), integrated into the routing protocol module. These algorithms operate on a graph derived from the real network. The graph contains only the routing-capable nodes (a.k.a. vertices) and the links between them (a.k.a. edges) with an appropriate link metric. In the GMPLS context, a link connecting two ports of neighbouring nodes may consist of more than one consecutive physical resource (e.g. fibres), possibly crossing routing incapable devices (e.g. regenerators, optical amplifiers, optical mux/demux, etc.). For this reason a set of properties is assigned to each link for routing purposes (e.g. TE metric, available/used bandwidths, resource colours, SRLG list, inherent protections, etc.), transforming the traditional links in traffic engineering links (TE-links). A GMPLS path calculator is expected to return Label Switched Paths (LSPs), i.e. sequences of nodes, TE-links and labels, which try to match some constraints derived from the TE information above. Once an LSP is computed, it describes univocally a unidirectional or bi-directional connection between a source and a destination node. The standard SPF algorithms are not suited for such a computation, as they cumulate only the link metric along the graph. A modified SPF algorithm is needed, called Constraint-based Shortest Path First (CSPF), as routes should be the shortest among those which satisfy the required set of constraints [3]. In the GMPLS architecture no specification is available for the implementation of a PCS module and no preference can be derived by the standards on the choice of a centralized or distributed implementation, in spite of the intrinsic distributed approach of the GMPLS control plane.

The architectural choice we propose in this work is for a PCS module inside a centralized network manager (NM), since this solution promises to be the most effective for the full and flexible management of traffic engineering into the network, as well as for survivability purposes. This feeling is easily sustainable above all when operating in a multi-area (or multi-domain) scenario. In our implementation the PCS acts as a path computation server for the GMPLS network, receiving from the GMPLS Network Elements (NE) the topology information and the requests for computation across a single- or multi-area/AS. The LSPs computed by PCS (if any) are communicated to the ingress NE and trigger a standard GMPLS signalling session (e.g. via G.RSVP-TE). The communications between the NM and the NEs are carried out by means of COPS protocol with proper extensions [4]. Focusing on LSP requests with survivability requirements [5], we identified three Classes of Recovery (CoR) for the LSP requests (e.g. Gold, Silver, Bronze), respectively related to the request for LSPs with path protection (e.g. SDH/SONET 1+1), Fast Restoration, or On-the-fly restoration [6]. In our PCS different algorithms are used for the different CoRs, ranging from optimal implementations (e.g. in case of Gold CoR) to the sub-optimal ones (e.g. in case of Silver and Bronze CoR). In details, for the Gold CoR we focused on the R.Bhandari's algorithms [8] for computing a pair of least cost maximally disjoint paths; these algorithms represent the optimal counterpart to the sub-optimality required for the Silver CoR, for which we chose the Two Step Approach (TSA) algorithm. TSA is based on a double Dijkstra SPF run and on a simple temporary network transformation for avoiding links/nodes of the worker path. The main advantages of such an algorithm are in the easiness of implementation and in the limited complexity both of the SPF algorithm (e.g. the Dijkstra complexity in our implementation) and of the network transformation. Further details on this issue are in [4] and [6,7].

3. BANDWIDTH-DEPENDENT TE METRICS

The routing engine inside our PCS module is based on an implementation of the Dijkstra SPF algorithm. Constrained SPF computations are obtained by adding a TE constraints validation step to the well-known Dijkstra flow [2] (e.g. check on the bandwidth availability of the candidate link). Moreover, in order to let the algorithm converge towards an optimal SPF solution (e.g. between those which satisfy the required TE-constraints), we chose to make bandwidth-dependent the link metric minimized during computation, according to the equation [9]:

$$total\_cost = metric_{std} + metric_{TE} + policy_x(bw) \qquad (1)$$

where $metric_{std}$ is the standard OSPF link metric, $metric_{TE}$ is the GMPLS metric for TE purposes, bw is the allocated bandwidth on the TE-link and $policy_x$ ($1 \leq x \leq 2$) is a proper traffic engineering function designed to balance the traffic load (e.g. bandwidth consumption) in the topology.

The policy functions we define for this work are detailed in Eq. 4 and Eq. 5, where bw_{free} is the free bandwidth, while bw_{th} is a threshold with respect to the total bandwidth (bw_{tot}) of the TE-link. The constants K and t are respectively a resource cost per unit and a smoothing factor of the overall TE function.

$$f_1(x) = K \cdot \left(1 + \left(\frac{x}{bw_{tot}}\right)^2\right) \tag{2}$$

$$f_2(x) = K \cdot \left(1 - \frac{x}{1 + bw_{tot}}\right)^{-t} \tag{3}$$

$$policy_1 = K \cdot \frac{1 + bw_{tot}}{1 + bw_{free}} \tag{4}$$

$$policy_2 = \begin{cases} 5 \cdot K & \text{if} \quad bw = 0 \\ f_1(bw) & \text{if} \quad 1 \leq bw \leq \frac{bw_{th}}{2} \\ f_2(bw) - f_2(\frac{bw_{th}}{2}) + f_1(\frac{bw_{th}}{2}) & \text{if} \quad \frac{bw_{th}}{2} \leq bw \leq bw_{th} \\ 10 \cdot K & \text{if} \quad bw \geq bw_{th} \end{cases} \tag{5}$$

In Figure 1-a the two policy functions are plotted, both with $K = 2$ and, only for $policy_2$, $t = 1.5$ and $bw_{th} = 75\%$ of bw_{tot}.

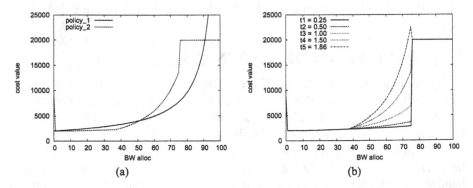

(a)   (b)

Figure 1. Policy functions (a) and effect of the smoothing factor t in $policy_2$ (b).

These policy functions increase the total cost of the link as allocated bandwidth increases, in order to avoid the overloading of the link and the resulting network congestion.

The $policy_1$ function has been designed in order to discourage the link picking as the available bandwidth decreases on it.

The $policy_2$ function has been conceived with the same aims, but with a further requirement on the tunability of its behaviour when increasing the cost of the TE-link. Indeed, $policy_2$ increases the link cost less than $policy_1$ till the reference bandwidth of $bw_{th}/2$, encouraging LSP to pick resources on it; when the $bw_{th}/2$ is reached, a "tunable" trend is configurable (ref. Figure 1-b) depending on the value of the smoothing factor, encouraging or discouraging the link picking more or less than $policy_1$. Moreover, an extra cost is assigned to totally free TE-links (i.e. $5 \cdot K$) in order to discourage their utilization in presence of just used links: this allows to fill up the most of the TE-links in a balanced way, bounding the number of used links and the need for their installation (i.e. a kind of feedback to the network planning).

4. PERFORMANCE STUDIES

In this section the performance of the different bandwidth-dependent TE metrics proposed in Sec. 3 is evaluated. The computational environment for all the tests is based on an Intel Celeron 500MHz PC with Linux Slackware 8.1 OS. Measures have been collected on different topologies with increasing meshing degrees (ref. Figure 2 and 3).

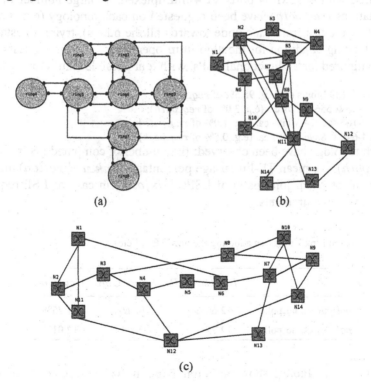

Figure 2. Test topologies: (a) interconnected rings; (b) Italian; (c) NSFNET.

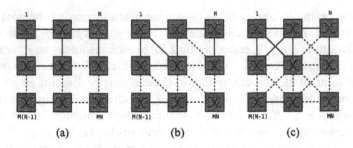

Figure 3. Manhattan topologies: (a) simple; (b) half-meshed; (c)meshed.

All these topologies have been modelled with generic nodes, configurable as SDH 4/4 Cross-Connects. Nodes have been assumed to be fully connectable, i.e. any of their ingress port may be cross-connected to an egress one. Adjacent nodes have been connected by a TE-link with random values for its TE-information (e.g. TE metric, available/used bandwidths, resource colours, SRLG list, etc.). All the TE-links have been assumed bi-directional and each configured with 4 STM-64 ports VC4-multiplexed. A large number of LSP computations (*req_path*) have been requested on each topology (e.g. up to a connection request from each node towards all the others), trying to establish an overloading condition for the algorithm operations. All the requests have been configured for bi-directional LSPs, with four possible values for the bandwidth:

- VC4 @ 139 Mbit/s ca. (e.g. 95.5% of *req_path*);
- VC4-4c @ 556 Mbit/s ca. (e.g. 3.0% of *req_path*);
- VC4-16c @ 2224 Mbit/s ca. (e.g. 1.0% of *req_path*);
- VC4-64c @ 8896 Mbit/s ca. (e.g. 0.5% of *req_path*).

For each topology have been observed: the number of computed LSPs (*comp_path*), the mean TE-link usage percentage (*link_utilization*) and the number of totally disjoint pairs of LSPs (*disjoint*) in case of LSP requests with recovery requirements.

Table 1. Gain in Link Utilization adopting the new TE policies.

	Bronze CoR	*Silver CoR*	*Gold CoR*
policy$_1$ vs. no_policy	+2.66%	+0.76%	+1.75%
policy$_2$ vs. no_policy	+2.75%	+0.80%	+2.01%

The results in Table 1 show that higher link utilization are achieved when *policy*$_1$ or *policy*$_2$ are used. This behaviour is much evident in topologies with lower meshing degrees, which are the common case for currently opera-

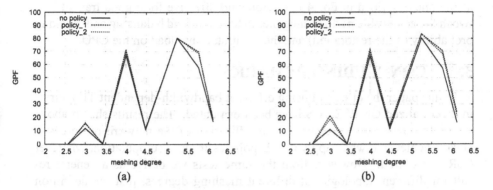

Figure 4. GPF at different TE policies (a) for Silver CoR; (b) for Gold CoR.

Table 2. GPF gain adopting the new TE policies.

	Silver CoR	*Gold CoR*
$policy_1$ vs. no_policy	+2.86%	+5.28%
$policy_2$ vs. no_policy	+2.49%	+4.01%

tive networks. However, the advanteges of using TE policies are related to the optimality of the algorithm adopted for path computation: indeed, when a sub-optimal strategy is used (e.g. TSA in case of Silver CoR), TE enhancements are lower, due to the worst resource picking; this behaviour is not observed in case of optimal algorithms (e.g. Bhandari's for Gold CoR). In order to summarize the performance for the different CoRs, a Global Performance Factor (GPF) has been defined as:

$$GPF = \frac{disjoint}{req\_path} \cdot \frac{comp\_path}{req\_path} \qquad (6)$$

in which the first term is related to the algorithm's effectiveness in creating maximally disjoint paths, while the latter term represents the algorithm's effectiveness in computing valid paths, according to the resource availability on TE links. In Figure 4 the GPF is drawn at the different meshing degrees, showing a mean higher performance in case of utilization of the TE policies, both for Gold and for Silver CoRs (ref. Table 2 for numerical details). However, the advantages of defining complex TE policies (ref. Eq. 5) do not pay for the introduced load on CPU, as demonstrated by the improvement in link utilization of less than 0.25% and by the worsening of the GPF value obtained when using $policy_2$ w.r.t. $policy_1$. Therefore, a simple policy function such as the one described in Eq. 4 is suitable and effective for a good tradeoff between the achievable performance (i.e. in terms of load balancing and blocking probability of the restoration) and the computational load on the CPU.

5. CONCLUDING REMARKS

In this paper the effects of using different bandwidth-dependent TE metrics in a centralized GMPLS PCS have been evaluated. The results shown above highlight how an improvement in the utilization of the network resources is always achievable when applying TE policies in path computation, whatever CoR is required. However, from the same tests we derive as a general result for different topologies at different meshing degrees, that the definition of fine-grain TE policies do not pay for the achieved performace enhancment, degraded also by the additional CPU load. Further investigations are ongoing to study the performance of a fine TE tuning in a fixed GMPLS topology with real expected traffic loads.

ACKNOWLEDGMENTS

This work has been supported in part by the Italian Ministry of Education, University and Research through the project TANGO (MIUR Protocol No. RBNE01BNL5).

REFERENCES

[1] E. Mannie (Editor) et al., *Generalized Multi-Protocol Label Switching (GMPLS) Architecture*, draft-ietf-ccamp-gmpls-architecture-07.txt, Internet Draft, Work in progress, May 2003.

[2] R.K. Ahuja, T.L. Magnanti, and J.B. Orlin, *Network Flows - Theory, Algorithms and Applications*, Prentice Hall, 1993.

[3] J. Strand et al., *Issues for Routing in the Optical Layer*, IEEE Communications Magazine, pages 81-87, Feb. 2001.

[4] *Traffic models and Algorithms for Next Generation IP networks Optimization (TANGO) Project*, http://tango.isti.cnr.it/.

[5] P. Lang (Editor) et al., *Generalized MPLS Recovery Functional Specification*, draft-ietf-ccamp-gmpls-recovery-functional-00.txt, Internet Draft, Work in progress, Jan. 2003.

[6] G. Carrozzo, *Algorithms and Engines for On line Routing in Generalized MPLS Networks*, PhD dissertation, Dept. Information Eng., University of Pisa, ITALY, 2004.

[7] G. Carrozzo et al., *A Pre-planned Local Repair Restoration Strategy for Failure Handling in Optical Transport Networks*", Photonic Network Communication, vol. 4 (no. 3-4), pages 345-355, Jul./Dec. 2002.

[8] R. Bhandari, *Survivable Networks - Algorithms for Diverse Routing*, Kluwer Academic Publisher, 1999

[9] B. Awerbuch et al., *Throughput-Competitive On-Line Routing*, IEEE Symposium on Foundations of Computer Science(FOCS), 1993, pp. 32-40.

A NEW TRAFFIC AGGREGATION SCHEME IN ALL-OPTICAL WAVELENGTH ROUTED NETWORKS

Nizar Bouabdallah[1,2], Emannuel Dotaro[1], and Guy Pujolle[2]
[1]Alcatel Research & Innovation, Route de Nozay, F-91460 Marcoussis, France
[2]LIP6, University of Paris 6, 8 rue du Capitaine Scott, F-75015 Paris, France
e-mail: nizar.bouabdallah@lip6.fr

Abstract: In wavelength-division multiplexing (WDM) optical networks, the bandwidth request of a traffic stream can be much lower than the capacity of a lightpath. Efficiently grooming low-speed connections onto high-capacity lightpaths will improve the network throughput and reduce the network cost. In this paper, we propose and evaluate a new concept of traffic aggregation in mesh networks that aims to eliminate both the bandwidth underutilization and scalability problems existing in all-optical wavelength routed networks. Our objective is to improve the network throughput while preserving the benefits of all-optical wavelength routed networks.

1.   INTRODUCTION

In the current nomenclature of optical networking, a network is referred to as *transparent* when its constituent nodes are all-optical cross connects (OXCs) where no conversion into the electrical domain is performed. In other words, within transparent networks, lightpaths are routed from source to destination in the optical domain, optically bypassing the intermediate nodes. Whereas, in *opaque* networks, lightwave channels are detected at each node, then electronically processed, switched and reassigned to a new outgoing wavelength when needed.

Realizing connections in an all-optical (transparent) wavelength routed network involves the establishment of point-to-point (PtoP) lightpaths between every edge node pair. These lightpaths may span multiple fiber links. Hence, some sort of virtual adjacency is created between the ingress and egress nodes via the established lightpath, even if these two nodes are geographically far apart.

In order to successfully instantiate connections, network resources (e.g., wavelengths, transceivers) have to be attributed. This issue relative to the assignment of network resources is well known as the routing and wavelength assignment (RWA) problem. A number of RWA studies have been conducted in the optical networking domain [1]-[3]. But, most previous studies assumed that a connection requests the entire bandwidth capacity of a lightpath channel. In this study, we consider the case where a connection can request either the whole or some fraction of the lightpath capacity. This makes the problem more practical and general.

The all-optical wavelength routing approach presents two obvious advantages. The first advantage stems from the fact that the optical bypass eliminates the need for Optical-Electrical-Optical (OEO) conversion at intermediate nodes. As a result, the node cost decreases significantly, since in this case the number of required expensive high-speed electronics, laser transmitters and receivers is reduced. The second advantage is due to all-optical routing which is transparent with regard to the bit rate and the format of the optical signal.

Nevertheless, wavelength routing presents two drawbacks. First, routing at a wavelength granularity puts a serious strain on the number of wavelengths required in a large network. For instance, if PtoP lightpaths needs to be established between every edge node pair in a network presenting N edge nodes, then $O(N^2)$ lightpaths are required. The second drawback behind wavelength routing is the rigid routing granularity entailed by such an approach. This granularity is large which could lead to bandwidth waste especially when only a portion of wavelength capacity is used. For operators, an efficient use of network resources is always a concern. In wavelength routed networks, this efficiency is possible only when there is enough traffic between pair nodes to fill the entire capacity of wavelengths.

To alleviate the aforementioned problems, we propose a new solution based on the distribution of the aggregation process. This solution combines the advantage of the optical bypass in transparent wavelength routed networks and statistical multiplexing gain in sub-wavelength routed networks. The lightpath, remaining entirely in the optical domain, is shared between the source node and all intermediate nodes up to the destination. So, we deal with multipoint-to-point (MptoP) lightpaths. Here, one lightpath represents routes from multiple ingress nodes to a single egress node.

A detailed description of this new approach will be outlined in the next section. In section 3, we investigate the node architecture needed to support traffic-aggregation feature within the WDM optical network. In section 4, the comparison between distributed aggregation and the classical strategies is performed based on a simple provisioning algorithm. Finally, some conclusions are drawn in section 5.

2.   DISTRIBUTED AGGREGATION IN SHEME

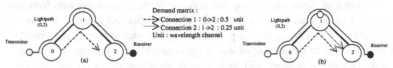

Fig. 1 A simple demonstration network. (a) All-optical Wavelength routed network. The connection request (1,2) is rejected.
(b) Distributed aggregation scheme. All connection requests are satisfied.

The key idea behind our proposed scheme of distributed aggregation is to allow lightpath channels sharing among several access nodes. Instead of limiting the access to lightpath capacity at the ingress point, each node along the path can fill on the fly the optical resource (wavelength) according to its availability. In this case, a lightpath channel can be shared by multiple connections traveling towards a common destination. This approach deals with MptoP connections.

Wavelength routing is performed in a similar way as in all-optical networks, i.e. signals remain in the optical domain from end to end and are optically switched by intermediate OXCs. Since the lightpath remains transparent at intermediate nodes, a MAC (Medium Access Control) protocol is required to avoid collision between transient optical packets and local ones injected into the lightpath. We have already proposed a simple MAC protocol based on void/null detection in [4]. This mechanism guarantees collision free packet insertion on the transient wavelength at the add port of an intermediate node.

To illustrate the distributed aggregation mechanism, we will present a simple three-node network example as depicted in figure. 1. Each fiber is supposed to have one wavelength channel and each node is considered as being equipped by a fixed transmitter and a fixed receiver. Two connection requests are to be served: (0,2) with a bandwidth requirement equal to half of the wavelength capacity; and (1,2) with a bandwidth requirement equal to quarter of the wavelength capacity. If the classical all-optical network case (i.e. networks that do not perform distributed aggregation) is taken into account (figure1.a), only one connection (0,2) would be served because of the resource limitations (the wavelength between (1,2) and the receiver at node 2 are busy). The connection requested between node pair (1,2) will be rejected in spite of the fact that the wavelength between these two nodes is not fully used. Hence, a supplementary wavelength between pair nodes (1,2) and a new receiver at node 2 would be required in order to satisfy all the connection requests.

However, using the distributed aggregation scheme (figure1.b), the traffic demand could be satisfied by establishing one lightpath from node 0 to node 2. In this case, both connections will share the same lightpath. Indeed, the second connection (1,2) would be carried by the spare capacity of the existing lightpath.

Note that the lightpath $0 \to 1 \to 2$ is still routed in the optical domain at node 1, preserving the benefit of optical bypass.

The merit of distributed aggregation is that multiple connections with fractional demands can be multiplexed into the shared lightpaths. As a result, the wasted bandwidth problem confronted in pure wavelength routed networks is eliminated. In addition, due to the sharing of wavelength channels, the number of admissible connections in the network will be increased. Furthermore, as connections from different nodes to the same destination are aggregated on the same lightpath, the destination node will receive less lightpaths. So, less physical components (wavelength, transceiver) would be used, resulting in the save of a great deal of equipment costs. Moreover, in order to provide connections between all edge node pairs using MptoP lightpaths, a total number of O(N) lightpaths is required since only one lightpath per individual egress node could be sufficient. Thus, we alleviate the scalability issue encountered in all-optical wavelength routed networks. However, still additional control mechanisms would be required to arbitrate the access to the shared lightpaths.

3.   NODE ARCHITECTURE

We consider a network of N nodes connected by unidirectional optical links constituting an arbitrary physical topology. To carry connection requests in such a WDM network, lightpath connections have to be established between pairs of nodes. In all-optical networks a connection request is carried by only one lightpath before it reaches the destination in order to avoid extra signal processing at intermediate nodes along the path. Two important functionalities must be supported by the WDM network nodes: one is wavelength routing and the other is multiplexing and demultiplexing. figure. 2 depicts a sample node architecture that can be employed in a WDM optical network.

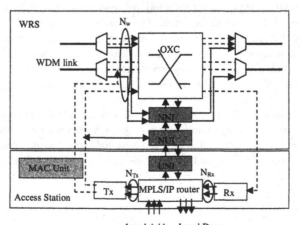

Local Add   Local Drop

Figure. 2. Node architecture.

The node architecture comprises two components: the wavelength-routing switch (WRS) and the access station. While the WRS performs wavelength routing and wavelength multiplexing/demultiplexing, the access station performs local traffic adding/dropping and low-speed traffic-aggregation functionalities. The WRS is composed of an OXC, which provides the functionality of wavelength-switching.

Each OXC is connected to an edge device, i.e. access station, which can be the source or the destination of a traffic flow. In figure. 2, each access station is equipped with a certain number of transmitters and receivers. Traffic originated at the access station is transmitted as an optical signal on one wavelength channel by virtue of a transmitter. Note that the access station can be either the origin of a lightpath or an intermediate node using an already established lightpath. In the latter case, the injected traffic by an intermediate node should have the same destination as that of the traversing lightpath. Moreover, as described in [4], a MAC unit is required to avoid collision between transient packets and local ones.

Aggregating low-speed connections to high-capacity lightpaths is done by the MPLS/IP router according to the MAC unit decision. The advantages of this model are that: 1) it provides flexible bandwidth granularity for the traffic requests and 2) this MPLS/IP-over-WDM model has much less overhead than the SONET-over-WDM model, widely deployed in optical networks. Usually, the potential disadvantage of such a model is that the processing speed of the MPLS/IP router may not be fast enough compared to the vast amount of bandwidth provided by the optical fiber link. However, our scheme alleviates this issue since each MPLS/IP router processes only its local traffic. In other words, the transit traffic travelling through a WDM node remains at the optical layer, and it is not processed by the intermediate access nodes.

4.   NETWORK DIMENSIONNING

4.1   Procedure and Algorithm for Network Dimensioning

It is well known that the RWA optimization problem is NP-complete [5]. In [6], we used a small network topology as an illustration where it was possible to obtain results using the Integer Linear Programming (ILP) methodology. Nonetheless, in order to achieve the same study for large networks, we will use heuristic approach. In this paper, the comparison between distributed aggregation and the classical strategies is tackled from a different perspective. Here, we aim at dimensioning the optical US backbone (figure. 3), under both strategies, so as to serve all traffic requests between any pair of optical nodes. The network topology consists of 29 nodes and 43 links. The network planning has been achieved following the logical process shown below. The inputs of the analysis are:

1) The physical network topology.
2) The traffic matrix.
3) The adopted routing scheme, which is the shortest path algorithm in our case.
4) The adopted wavelength assignment approach, which is the first fit (FF) strategy in our work.

The network dimensioning is achieved by evaluating the OXC and IP/MPLS router dimensions by means of the heuristic algorithm used to map the different lightpaths. When dealing with the distributed aggregation strategy we will use a novel heuristic algorithm, called *Maximizing Traffic Aggregation (MTA)*, in order to plan the MptoP lightpaths.

Let λ_{sd} denote the aggregate traffic between node pair s and d, which has not been yet carried. As explained before, λ_{sd} can be a fraction of the lightpath capacity. In our study, we suppose that $\lambda_{sd} \in [0,1]$, so at most one lightpath between every node pair (s,d) is required to carry all the traffic requests. Let $H(s,d)$ denote the hop distance on physical topology between node pair (s,d).

The MTA algorithm attempts to establish lightpaths between source-destination pairs with the highest $H(s,d)$ values. The connection request between s and d will be carried over the new established lightpath. Afterwards, the algorithm will try to carry, as long it is possible, connections originating from intermediate nodes and travelling to the same destination d, based on the currently available spare capacity of the lightpath (s,d). This heuristic tries to establish lightpaths between the farthest node pair in an attempt to allow the virtual topology collecting the maximum eventual traffic at the intermediate nodes. The pseudo-code for this heuristic is presented hereafter:

Step 1: Construct virtual topology:

1.1: Sort all the node pairs (s,d) (where $\lambda_{sd} \neq 0$) according to the hop distance $H(s,d)$ and insert them into a list L in descending order.

1.2: Setup the lightpath between the first node pair (s',d') using the first-fit wavelength assignment and the shortest-path routing, let $\lambda_{s'd'} = 0$.

1.3: Sort all the node pairs (i,d') (where $\lambda_{id'} \neq 0$ and i is an intermediate node traversed by the lightpath (s',d')) according to the hop distance $H(i,d')$ and insert them into a list L' in descending order.

1.4: Try to setup the connection between the first node pair (i',d') using the lightpath (s',d'), subject to the current available bandwidth on lightpath (s',d'). If it fails, delete (i',d') from L'; otherwise, let $\lambda_{i'd'} = 0$, update the

available bandwidth of the lightpath (s',d') and go to Step
1.3 until L' becomes empty.
1.5: Go to Step 1.1 until L becomes empty.

Step 2: Route the low-speed connections over the virtual
topology constructed in Step 1.

The dimensioning is accomplished so that all the traffic is forwarded within the network. In the classical case, the routing algorithm is simply applied to the traffic matrix to find all the PtoP lightpaths required to forward the traffic. However, when the distributed aggregation is considered, the MTA heuristic is applied to map all the MptoP lightpaths. Then the number of OXC ports, transmitters and receivers is determined.

As explained before, the optical node consists of an electrical IP router part and an OXC part. Each port of the electrical IP router is connected to the OXC port via an internal wavelength. A lightpath is established between nodes by setting up the OXCs along the route between nodes. Some lightpaths transit to the electrical IP router after traversing the OXC part and others exit the node through the OXC without electrical processing (figure. 2). Each lightpath needs a dedicated OXC port when traversing an intermediate node along the path. In addition, a transmitter is required at the ingress node and a receiver is needed at the egress node of the lightpath. Moreover, in the distributed aggregation case, each intermediate node along the path using the traversing lightpath to transmit its traffic needs also a transmitter. Recall that, the access station of a node can be either the origin of a lightpath or an intermediate node using an already established lightpath. Let N_{tx} and N_{Rx} denote the number of transmitter and receiver ports per node. Let N_w denote the number of OXC ports per node as shown in figure. 2.

4.2   Dimensioning Results and Comparison

Table 1 reports the dimensioning results of the network in the classical and the distributed aggregation cases. It is clear that each node in the network needs 28 transmitters in order to forward its traffic destined to all the other nodes in both cases. Furthermore, each node requires 28 receivers to receive all the PtoP lightpaths emanating from all the other nodes in the classical case. However, as multiple connections travelling from different nodes to the same destination could be aggregated on the same lightpath, when the distributed aggregation scheme is considered, the destination node will receive less lightpaths. Consequently the number of MptoP lightpaths and required receivers to handle all the traffic requests is reduced as depicted in table 1. We record a save of above 55% of the receivers.

Moreover, as the number of lightpaths is reduced whith the distributed aggregation strategy, the number of OXC ports is also reduced. The gain recorded is beyond 49%. This latter gain is less important than the one obtained when dealing with receivers since the number of OXC ports depends not only on the number of established lightpaths but also on the number of hops per lightpath. Indeed, the mean hops number per lightpath is 3,58 in the classical case, whereas it is 4,2 in the other case. These results show how the distributed aggregation scheme relieves the scalability issue encountered with the classical all-optical networks.

Finally, it is meaningful to compare the lightpath load in both strategies. In the classical case, the average load of a lightpath is 27,28%. This result emphasizes the resources under-utilisation problem already mentioned. Distributed aggregation scheme alleviates this issue, while the average load of a lightpath is 61,20% in this case. Figure. 4 depicts the dimensioning results corresponding to each node of the network. They represent a detailed description of the results reported in table 1.

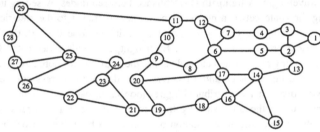

Figure. 3. The US optical backbone.

Table 1. Dimensioning results.

	# Tx	# Rx	#lightpaths	#OXC ports	Load/lightpath	Hops/lightpath
Classical approach	812	812	812	3722	27,28%	3,58
Distributed aggregation	812	362	362	1886	61,20%	4 ,20

5. CONCLUSIONS

We have proposed a distributed aggregation approach in all-optical wavelength routed networks. This approach combines the merits of both the optical bypass of transparent wavelength routing and the multiplexing gain of sub-wavelength routing. In this approach, we tend to aggregate traffic traveling from different nodes to a common destination in the same lightpath channels. As a result, the number of managed lightpaths is significantly reduced and the utilization percentage of the optical channels is improved as well. The dimensioning results of the US optical backbone revealed that the proposed approach reduces significantly the network cost. In this specific case, around 50% of the receiver and OXC ports are saved when the distributed aggregation is considered.

REFERENCES

[1] I. Chlamtac, A. Faragó, and T. Zhang, "Lightpath (wavelength) routing in large WDM networks", IEEE J. Select. Areas Commun., vol. 14, pp. 909–913, June 1996.
[2] D. Banerjee and B. Mukherjee, "Wavelength-routed optical networks: Linear formulation, resource budgeting tradeoffs, and a reconfiguration study", IEEE/ACM Trans. Networking, vol. 8, pp. 598–607, Oct. 2000.
[3] D. Banerjee and B. Mukherjee , "A practical approach for routing and wavelength assignment in large wavelength-routed optical networks", IEEE J. Select. Areas Commun., vol. 14, pp. 903–908, June 1996.
[4] N. Bouabdallah et al., "Matching fairness and performance by preventive traffic control in optical multiple access networks", Proc. of OptiComm '2003, Dallas, October 2003.
[5] B. Mukherjee, Optical Communication Networks. New York: Mc-Graw-Hill, 1997.
[6] N. Bouabdallah et al., "Distributed aggregation in all-optical wavelength routed networks", ICC '2004, Paris, France, June 2004.

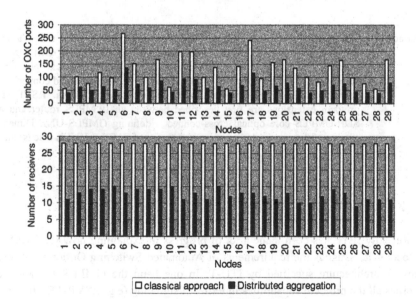

Figure.4. The OXC port and receiver needs per node for the distributed aggregation and classical cases.

MULTI-LAYER RECOVERY ENABLED WITH END-TO-END SIGNALING

D. Verchere[1], D.Leclerc, A.Noury, B.Ronot, M.Vigoureux, O.Audouin, A.Jourdan
D. Papadimitriou[2], B.Rousseau, G. Luyts
S. Brockmann[3] W. Köber, G. Eilenberger

[1]*Alcatel R&I, Route de Nozay, 91460 Marcoussis, France,Dominique.Verchere@alcatel.com*
[2]*Alcatel Bell, Alcatel Bell, Francis Wellesplein 1, B-2018 Antwerpen, Belgium,*
[3] *Alcatel SEL, Holderaeckerstrasse 35, 70499 Stuttgart, Germany*

Abstract: Within GMPLS framework, the signaling protocol Resource reSerVation Protocol with Traffic Engineering extensions (RSVP-TE) is extended to support the requirements of an Automated Switched Optical Network architecture. This paper presents the extensions of the end-to-end connection services in an overlay network built on two control planes. RSVP-TE protocol extensions are first described between an IP/MPLS router and a SDH/GMPLS core optical cross-connect, defining GMPLS-UNI. Dimensioning of three scenarios proving the benefits of GMPLS-UNI is discussed.

1.   INTRODUCTION

Recently IETF has been made a lot of progress in explaining how its GMPLS protocol suite satisfies the requirements of Automated Switching Optical (ASON) Network architecture specified by ITU-T. In one hand the GMPLS framework specifies all the protocol capabilities in terms of signaling (e.g. RSVP-TE), routing (e.g. OSPF-TE) and link management (e.g. with LMP). And in the other hand, ITU-T ASON recommendations specify the architecture and requirements for the Automatic Switched Transport Network as applicable to SDH transport networks.

GMPLS generalizes the label switching concepts introduced in MPLS to all switching technologies. GMPLS extends the label switching capabilities to network elements hosting non-packet based switching matrix, from labeled packet, frame, cell switching technologies to circuit switching technologies including SDH and Optical Transport Hierarchy (OTH).

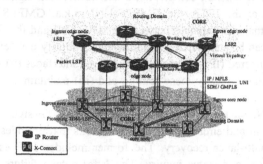

Figure 1. Multi-Layer network architecture: IP/MPLS over SONET-SDH/GMPLS

E.g. an IP/MPLS-over-SDH/GMPLS network architecture is considered within this paper, all the results presented apply to an SONET/GMPLS transport network. The core network comprises two control planes: IP packet transport layer on top of the time-division-multiplexing (TDM) transport layer. The TDM layer consists of optical cross-connects (OXCs) interconnected by one or more optical fibers (physical links). Each SDH connection corresponds to a logical link for the IP/MPLS layer. These links interconnecting Label Switching Routers (LSRs), constitute a virtual topology. The control plane interactions i.e. the functional coordination capabilities between the layers are described in the following section. The section 2 provides the new signaling extensions required at the router-OXC interface to enable end-to-end connection coordination. In section 3, three multi-layer network recovery scenarios are presented and the section 4 allows to position the new User-to-Network Interface GMPLS-UNI capabilities.

2. CONTROL PLANE INTERCONNECTION MODELS

As defined in [1], different control plane interconnection configurations can be deployed at the User-to-Network Interface (UNI) reference point. Optical Internetworking Forum did specify the OIF UNI version 1.0 implementation agreement [2]. This signaling interface is applicable when the IP/MPLS client and SDH/GMPLS server belong to separate administrative domains. End-to-end signaling functions, the exchange of reach-ability, topology or explicit route connection information between the control plane instances are not possible. The messages exchanged through the OIF UNI are limited to requests for and acknowledgements of the establishment and tear-down of transport connections.

OIF UNI introduces GENERALIZED_UNI object enabling address separation, both types and spaces increasing complexity. More GENERALIZED_UNI object functionality can be addressed with EXPLICIT_ROUTE object.

From the OIF-UNI inherent complexity, a user-to-network signaling interface is being developed at the IETF: the GMPLS-UNI (a.k.a. GMPLS for overlay networks [3]). The major differences between the OIF UNI and the GMPLS UNI signaling interface can be summarized as follows: (i) simplify end-reference point identification to numbered (IPv4/IPv6) or unnumbered interfaces (ii) allow client-driven explicit routing, typically loose routing and (iii) maintain a single end-to-end RSVP[4] signaling session (see the more in [1]).

The GMPLS UNI interface [3] is defined as an RSVP-TE signaling interface for packet-over-circuit and circuit-over-circuit networks that provides: connection provisioning and multi-layer recovery. The former includes LSP establishment, deletion, modification, and status inquiry. The latter entails failure notification, establishment of resource disjoint paths in response to a failure.

For end-to-end recovery, GMPLS-UNI enables to associates explicitly one or more working connections with one protecting connection with the ASSOCIATION object [4]. This object is carried in the Path message of the working LSP and in the protecting LSP. Each transit node along the LSP route transmits the ASSOCIATION object without any modification. GMPLS UNI allows the usage of the PROTECTION object as extended in [4]. The client LSR's explicitly specify the desired end-to-end LSP recovery level starting at the ingress and terminating at the egress router. The protection types defined are: dedicated protection, protection with extra-traffic, (pre-planned) re-routing without extra-traffic and dynamic re-routing. The Notify message provides a fast mechanism to notify non-adjacent nodes of events such as LSP failures [4]. Notify messages are only generated when explicitly configured during LSP establishment with the NOTIFY_REQUEST object. The Notify message does not have to follow the same path as the Path/Resv messages used to establish the LSP and is not processed at intermediate nodes. From GMPLS-UNI signaling function such as (i) end-to-end explicit routing and label control, (ii) end-to-end LSP association, (iii) end-to-end LSP protection, (iv) end-to-end route diversity, and (v) multi-layer fast notification, three recovery scenarios are developed.

3.   MULTI-LAYER NETWORK RECOVERY SCENARIO

To dimension the recovery in IP/MPLS-over-SDH networks, three scenarios have been developed. For each, the OIF-UNI and the GMPLS-UNI interfaces are compared. The Italian network topology was used, composed with 30 point-of-presences (PoP's) interconnected with 62 physical links. At each PoP, a single backbone router office is set-up. The line side interfaces of the router are supposed to be the same as OXC ones i.e. STM-X with X=1,16 or 64. The total number of STM-X ports is summed from add/drop ports and in/out ports. Through each scenario, the interface STM-X sends/receives VC-4-Xc virtual containers.

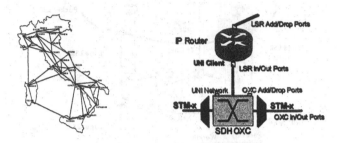

Figure 2. Network Topology (left), a single backbone router office architecture (right)

To allow packet LSP to be provisioned, TDM LSP are triggered following a bottom-up sequence.

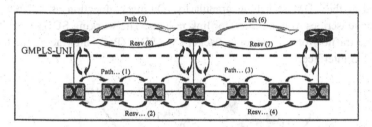

Figure 3. Packet LSP over TDM LSP multi-hop provisioning

In the scenario 1 (figure 4), a packet LSP is established between router B and F across the virtual topology. For the TE link between B and F, there exits different route from edge OXC b to f. The working SDH connection carries the traffic from B outgoing interface to F ingoing interface. The recovery connection is established between the edge OXC b to f and is route diverse from working connection. Each recovery connection can protects several connections.

The TDM recovery connections are pre-planned (see [1]) i.e. network resources are computed, selected and reserved but not cross-connected. The recovery resources can not carry any traffic because they are not provisioned at the data plane but only at the control plane level (i.e. "soft-provisioned"). Soft-provisioned connections enable the sharing of the recovery resources. In case of failure, switching from the working to the recovery sub-connection occurs at the edge i.e. OXCs b and f. The issue intrinsic to scenario 1 is at UNI reference point between the router and its adjacent OXC, it requires link protection. The protection only works between the edge cross-connects inside the transport network. The recovery resources can not transport router driven best-effort traffic because the recovery switching point is located inside SDH/GMPLS network.

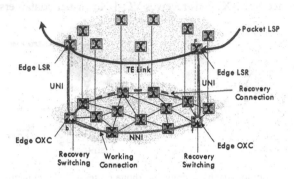

Figure 4. Scenario 1: SONET/SDH sub-connection recovery

The second recovery scenario (figure 5) has its recovery switching point at the edge router interface. The connection provisioning and recovery switching are both processed at the TDM switching granularity. During connection establishments, the router subscribes to be notified for link failures. The working SDH connection carries the traffic from B outgoing Packet-over-SDH (PoS) interface until the F ingoing PoS interface.

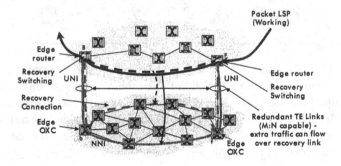

Figure 5. Scenario 2: end-to-end SONET/SDH recovery connection.

The recovery connection is established from B to F and is route diverse. The protecting TDM connections are soft-provisioned and can be shared between several source-destination pairs. For a link failure event, after reception of Notify message, the ingress router signals the corresponding soft-provisioned recovery connection. This bottom-up end-to-end connection recovery requires a coordination based on the Notify message from the OXC to the router. This scenario requires to implement GMPLS-UNI between each router-OXC pair. Finally this scenario enables dual homing protection at UNI reference point i.e. the edge devices when an edge router is connected to two edge routers. In the scenario3 represented in fig. 6, the provisioning of the packet LSP's is completely processed on the logical topology created by the IP/MPLS network[5]. Each packet LSP and its detour LSP (i.e. its protecting LSP) can use one or more TE links.

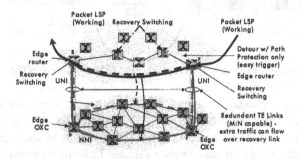

Figure 6. Scenario 3: Packet Label Switched Path Local Recovery

In the transport network, the recovery SONET/SDH connection used to protect the working SONET/SDH connection is used by the detour LSP protecting the packet LSP. This packet LSP is nested in the working TDM connection and its detour LSP is nested in the recovery TDM connection. In this scenario, the recovery SONET/SDH connection are fully reserved i.e. the network resource along the TDM recovery connection route are computed, selected and reserved and cross-connected (see [1] for details). Each detour LSP reserves the same amount of bandwidth than its attached protected packet LSP. A detour LSP can be shared between one or more packet LSP's i.e. the bandwidth reserved for the detour LSP can be shared by other detour LSP's. This technique is referred as "Facility backup" in [5]. In scenario 3, there is no dedicated TDM connection either for working connection or for recovery connection. Each TDM LSP can nest a set of packet LSP's and detour LSP's. Inside a TDM connection, the set of packet LSP's are route diverse from the other packet LSP's protected by the detour LSP's.

4.   GMPLS-UNI DIMENSIONINGS

The results prove that the GMPLS-UNI based multi-layer recovery is superior in terms of recovery speed and/or resource utilization (Figs. 7–9). Fig.7 illustrates, the total number of interfaces i.e. router+OXC for STM-1 (*left*) and STM-16 (*right*) interfaces. Fig.8 gives STM-64 filling ratio with VC-4 (*left*) and VC-4-16c (*right*) and Figure 9, the traffic recovery ratio in function of the recovery delay (ms). For the same recovery level (IP/MPLS protection), sce.2 is the optimal for resource usage and scenario 3 is the fastest.

252

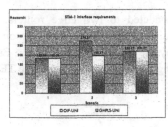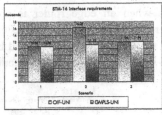

Figure 7. Number of router + OXC interfaces: STM-1 (*left*) & STM-16 (*right*).

Figure 7 is obtained with an average of TE link connectivity degree 10.77 logical link interface per router (see fig.1). The requirements are provided depending of the LSR-OXC interface:OIF-UNI vs GMPLS-UNI. The difference between sce.1 and 2 is due to the link interconnection at OXC-LSR, in sce.1 no interface protection is implemented. In sce.3, working packet LSP and its detour are routed over link diverse connections. Fig.8 compares STM-64 filling ratio that is fraction of STM-64's bandwidth reserved to carry VC-4 or VC-4-16c granularities. For connectivity degree 4.13, 7.43, 10.77 and 29.00, average number of optical hops between the edge routers for SDH connections is 3, 3.5, 3.8 and 5.1, resp. "3" corresponds to the connection explicit route: ingress router → ingress OXC → egress OXC → egress router. In fig. 9, the recovery ratio (% of traffic affected by failure & recovered) is plotted. In sce. 3, traffic recovery is processed by router. Recovery of connection translates into recovery of all working packet LSP's.

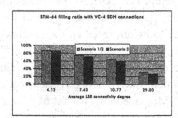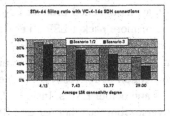

Figure 8. STM-64 filling ratio: VC-4 (left) / VC-4-16c (right) SDH connections

The recovery time is the duration elapsed from a failure detection and the time the failed connection is recovered, $T^{(i)}$. The recovery of the first connection is completed at $T_0^{(i)}$ (i=1,2,3), the recovery of the ultimate connection is at $T_{100}^{(i)}$. It appears that the recovery speed of SDH connection (scenario 1 & 2) is slower than packet LSP-based recovery, SDH/GMPLS recovery requires an additional round trip time to signal network resource compared to IP/MPLS one: $T_0^{(3)} \leq T_0^{(1)} \approx T_0^{(2)}$. Due to TDM soft-provisioning applied in sce. 1 & 2, protecting SDH connection requires to be cross-connected before traffic can be switched onto them. The difference between sce.1 and 2 (~5 ms) is due to one additional hop along the route of the Notify message. For scenario 1 & 2, IP/MPLS recovery mechanism can be a fallback mechanism of SDH/GMPLS recovery failure.

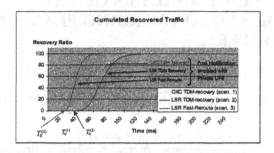

Figure 9. Traffic Recovery Ratio (%) in function of the Recovery delay (ms)

These recovery mechanisms are resource optimal, provide fast recovery and yield a deterministic network status.

5. CONCLUSIONS

GMPLS UNI benefits were proved qualitatively and quantitatively facilitating sequences of recovery actions between layers which in turn allows optimizing the resource usage and the delays inferred by the restoration of the traffic. New contributions are further extending the capabilities of the GMPLS UNI by allowing routing information to be exchanged to evolve towards an augmented model. The GMPLS UNI as defined by the IETF, is an important step in the continued evolution towards the delivery of integrated and intelligent network solutions.

REFERENCES

[1] D. Papadimitriou, W. Körber, B. Rousseau, S. Brockmann, D. Verchère, "The Private User Network Interface", OSA-Journal. Optical. Network. Vol.3, No.3, pp119-132
[2] "User Network Interface (UNI) 1.0 Signaling Specification", OIF, October 2001.
[3] "GMPLS-UNI RSVP Support for the Overlay Model", G. Swallow et al., Internet Draft (work in progress), draft-ietf-ccamp-gmpls-overlay-04.txt, April 2004.
[4] "RSVP-TE Extensions in support of End-to-End GMPLS-based Recovery", J.P. Lang et al., Internet Draft, draft-ietf-ccamp-gmpls-recovery-e2e-signaling-01.txt, May 2004.
[5] "Fast Reroute Extensions to RSVP-TE for LSP Tunnels", P. Pan (Ed.) et al., Internet Draft (work in progress), draft-ietf-mpls-rsvp-lsp-fastreroute-02.txt, February 2003.

PERFORMANCE ANALYSIS OF THE CONTROL AND FORWARDING PLANE IN AN MPLS ROUTER

D.Adami[1], N.Carlotti[2], S.Giordano[2], M.Pagano[2], M.Repeti[2]

[1]*CNIT Research Unit - Dept. of Information Engineering - University of Pisa (ITALY)*
E-mail: davide.adami@cnit.it

[2]*Dept. of Information Engineering - University of Pisa (ITALY)*
E-mail: {s,giordano, m.pagano}@iet.unipi.it, {m.repeti, n.carlotti}@netserv.iet.unipi.it}

Abstract: Multiprotocol Label Switching (MPLS) was originally designed to provide higher packet forwarding rates in network equipment. Nevertheless, it was soon realized that it could also provide other advanced features, such as Traffic Engineering and Virtual Private Networks capabilities. The key feature of MPLS is a strict separation between control and forwarding operations, which reflects on the software and hardware architecture of the routers. The paper presents the results of an experimental study aimed at evaluating the performance of a Label Switching Router (LSR). In particular, the behaviour of the LSR control and forwarding planes has been analyzed in different working conditions as it concerns both the processing and computational effort due to the control plane and the data traffic to be forwarded.

I. INTRODUCTION

Network layer routing can be partitioned in two basic components: control and forwarding. The former is responsible for the construction and maintenance of the forwarding table, the latter is concerned with the forwarding of packets from input to output interfaces, on the basis of the forwarding table maintained by the router and the information carried in the packet itself.

At present, network routers implement the control and forwarding planes in a distributed way. In particular, the control component consists of one or more routing protocols for the exchange of routing information among the routers as well as the algorithms used to convert the collected information into a forwarding table. The forwarding component consists of a set of algorithms used to take a forwarding decision on a packet. Traditional IP routers typically use destination-based forwarding to determine the next hop of a packet. The longest prefix-match, required in IP address look-up to perform destination-based forwarding, was originally implemented in software, but it is too expensive for core routers, because it requires a high computational effort. To improve the performance of a traditional IP router, a trade-off between scalability and flexibility is necessary: on the one hand, coarse forwarding granularity assures system scalability, on the other hand a system supporting only coarse forwarding granularity may be fairly inflexible, making available a number of differentiated treatments which could be insufficient.

In this context, Multiprotocol Label Switching (MPLS) was originally introduced as a fast switching technique. The key architectural principle of MPLS is a clean functional separation of network layer routing into control and forwarding components. The label switching forwarding component makes a forwarding decision on a packet using a label carried in the packet and a forwarding table maintained by the Label Switching Router (LSR). Just like a control component of any routing system, the label switching control component must provide for consistent distribution of routing information among LSRs as well as consistent procedures for constructing forwarding tables from this information. To support label switching, the control component has also other functionalities, such as creating binding between labels and Forwarding Equivalence Classes (FECs), informing other LSRs of the binding created, taking into account of bindings to construct and maintain the forwarding table. Under ideal conditions, a LSR should be able to forward data at whatever speeds the label switching forwarding component runs, regardless of the computational and processing effort required to the control component by routing and signaling protocols. Therefore, the hardware architecture of a LSR should be designed and implemented in order to satisfy also the new functional requirements of the control and forwarding planes. Concerning this issue, the system architecture of Juniper routers has been split in two portions: the Routing Engine and the Packet Forwarding Engine, which have been respectively designed to handle the general routing operations and the forwarding of packets. The experimental analysis, described in this paper, aims at verifying in practice the ideal separation, both in terms of functionalities and performance, between the control and forwarding plane of an MPLS commercial router. To this purpose, an M10 Juniper router has been configured as a LSR and tests have been performed to investigate how the processing and computational load, due to a control plane based on the RSVP-TE signaling protocol, affects the forwarding performance of the router. The paper is organized as follows: section II and III highlight the main features of MPLS and. RSVP-TE. Afterwards, section IV describes the RSVP-TE performance tests. Finally, section V contains the experimental results and section VI sums up the work.

II. MULTIPROTOCOL LABEL SWITCHING

An MPLS domain is a contiguous set of nodes, called Label Switching Routers (LSRs), which are capable of switching and routing packets on the basis of a label appended to each of them. An LSR that interfaces to a traditional router is called an Edge LSR (E-LSR). With respect to the direction of the traffic flow, it is possible to distinguish between Ingress LSR and Egress LSR. An Ingress LSR receives traffic from a non-MPLS router, while an Egress LSR sends traffic to a non-MPLS router. It is relevant to highlight that each traffic flow has its own Ingress and Egress LSR. In a network supporting MPLS, a label-switched path (LSP) is a unidirectional connection through multiple LSR. The Forwarding Equivalence Class (FEC) is defined in [1] as "a group of IP packets which are forwarded in the same manner (e.g., over the same path, with the same forwarding treatment)". In general, the granularity of FECs within a router can vary from very coarse to extremely fine, according to the level of information used in assigning an IP packet to a FEC: at one hand, an FEC can be associated with the flow generated by a particular application for a particular source and destination host pair, at the other hand, a FEC can be associated with all the flows destined to an Egress LSR. A packet is assigned to a FEC only at an Ingress LSR, where the label for the packet is also created. Although MPLS was originally designed to speed up the IP packet-forwarding process and retain the flexibility of an IP-based networking approach, the deployment of MPLS enables *Traffic engineering, Resilience and Virtual Private Networks (VPNs) support* capabilities. MPLS is frequently mentioned among major Quality of Service (QoS) technologies for packet networks. It is worthwhile pointing out that MPLS plays a key role in enabling QoS, but QoS is supported only if MPLS is combined with other technologies such as RSVP-TE and Differentiated Services.

III.   THE RSVP-TE PROTOCOL

In an MPLS network, an LSP must be set up and labels assigned at each hop before traffic can be forwarded. Two different classes of LSPs may be established:
- *Control-driven LSPs*: each LSR determines the next interface for the LSP according to its forwarding table and requests the label to the next-hop router;
- *Explicitly-routed LSPs* (or Constraint-based Routed LSPs, CR-LSPs): the set up message specifies the route for the LSP.

To establish an LSP, a signaling protocol for coordinating the label distribution, explicitly routing the LSPs, reserving bandwidth and preventing loops is necessary. Since the MPLS architecture does not suppose the mandatory use of a single signaling protocol, the following protocols have been proposed for label distribution: Label Distribution Protocol (LDP), Resource Reservation Protocol extensions for MPLS (RSVP-TE), Constrained-based Routed LDP (CR-LDP).

In more details, LDP [2] is suitable for establishing a control-driven LSP, but traffic-engineering capabilities are not supported.

RSVP-TE is an extension of RSVP [3] to establish traffic-engineered LSPs: it satisfies the requirements for traffic engineering, but, due to its soft-state nature, it suffers from inherent scalability problems when the number of LSPs per LSR increases. Moreover, the use of an unreliable transport protocol doesn't guarantee fast failure notification to the endpoints affected by the failure, even though an explicit teardown message is sent.

CR-LDP [4] [5] [6] is a hard-state protocol that extends LDP to carry the explicit route information, the traffic parameters for resource reservation and the options for CR-LSP resilience. It is relevant to highlight that CR-LDP has been designed to address and solve the main drawbacks of RSVP-TE, ensuring reliable transport of signaling messages and providing better scaling properties. Nevertheless, some networking devices manufactures (e.g. Juniper Networks), have not yet implemented CR-LDP in their operating systems, so that RSVP-TE is the only available signaling protocol to be used in MPLS networks. This is the reason why, in our work, we focused on RSVP-TE only.

IV.   RSVP-TE PERFORMANCE TESTS

The RSVP-TE performance tests have been carried out by using the Adtech AX/4000 from Spirent Communications. The AX/4000 Broadband Test System [8] has been designed for testing the performance and evaluating the QoS level provided by broadband networks. This device is a modular multi-port system capable of testing simultaneously multiple network technologies such as ATM, IP, Frame Relay and Ethernet at speed up to 10 Gbps. Moreover, the system is capable of generating data traffic with different profiles, performing full-rate analysis, stressing the network nodes in critical working conditions, emulating routing and signaling protocol. Our experimental analysis has been performed by means of the RSVP-TE Forwarding Performance During Tunnel Establishment Test [9], a performance test available within the Spirent Connect software package. This test allows measuring the capability of a Device Under Test (DUT) to correctly forward MPLS labeled packets to stable RSVP tunnels, while other tunnels are continuously set up and torn down (flapped). Figure 1 shows the scenario where the performance tests have been carried out. In particular, two Gigabit Ethernet interfaces of the test device, emulating an Ingress and an Egress LSR, have been connected to an M10 Juniper router, which acts as an LSR. The performance test has been organized as it follows:

1. Two different groups of LSPs are established: the first one consists of the LSPs which, after being created, remain stable for all the duration of the test The second one is the set of LSPs that are flapped.

2. The Ingress LER generates MPLS labeled packets and sends them to the Egress LER using the LSPs belonging to the first stable group.

3. The Egress LER analyses each packet received.

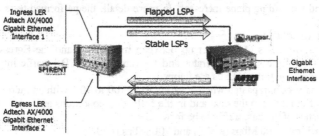

Figure 1: The testbed

In particular, the test consists of two subsequent phases (Fig. 2):

1. **Prior Flapping Phase** (duration time: 15 seconds): a number equal to M of stable LSPs are established between the Ingress LER and the Egress LER and Constant Bit Rate (CBR) data traffic is sent through the M LSPs. At the same time, N LSPs are established and prepared for flapping even if, in this test phase, remain stable.

2. **During Flapping Phase** (duration time: 12 seconds): M LSPs are stable and data traffic is sent over them, whereas N LSPs are flapped twice, with the flapping interval set at 5 seconds. During the first flapping event (duration: 5 seconds), the LSPs established for flapping are torn down and, subsequently, K ($K \leq N$) LSPs are re-established. During the second flapping event (duration: 2 seconds), the group of K LSPs is torn down and set up again, while, simultaneously, the $N-K$ LSPs, which have been torn down in the first flapping event, are re-established. At the end of the test, if all the N LSPs are active, there has not been any failure in the flapping test.

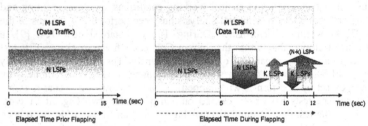

Figure 2: Prior and During Flapping run time

V.   EXPERIMENTAL RESULTS

Two different categories of performance tests have been carried out: the first one (Test 1.x.x and 3.x.x) aims at evaluating the behaviour of the router forwarding plane when the processing and computational effort required to the control plane progressively increases; the second one (Test 2.x.x and 4.x.x) focuses on analysing the router behaviour when the complexity of the operations involving both the control and forwarding plane increases. In more detail, the performed tests are the following:

1. Tests 1.x.x and 3.x.x

100 data traffic LSPs (M) are set up between the Ingress and the Egress LER. Moreover, the Ingress LER generates and forward labeled CBR traffic into these LSPs. At the same time, N LSPs are flapped.

The tests have been repeated when N varies from 100 to 900, with an increment of 100 LSPs from a test to the next and in the following conditions as it concerns the characteristics of the data traffic to be forwarded:

- Total Bit rate: 500 Mbps ($\rho=0.5$) and 900 Mbps ($\rho=0.9$).
- Packet size: 64, 512 and 1500 bytes.

It is relevant to outline that:

1. the overall data traffic is fairly subdivided among M CBR data sub-stream, so each of the *M* LSP carries the same share of traffic to be forwarded;
2. the packet size includes the Gigabit Ethernet header, payload and trailer;
3. for IP over Gigabit Ethernet, when the packet size of the data stream is equal to 64 bytes, the maximum bit rate which can be really reached is limited to 761.9 Mbps, due to the minimum inter-packet gap (96 bits/time) and the preamble (8 bytes) to add to each frame.

2. Test 2.x.x and 4.x.x

M data traffic LSPs are set up between the Ingress and the Egress LER. Moreover, the Ingress LER generates and sends labeled CBR traffic into these LSPs. At the same time, *N* LSPs are flapped. The tests have been repeated when *M* and *N* vary from 100 to 500, with an increment of 50 LSPs from a test to the next, in the following operating conditions as it concerns the characteristics of the data traffic to be forwarded:

- Bit rate: 500 Mbps and 900 Mbps.
- Packet size: 64, 512 and 1500 bytes.

Analysis of the control plane behaviour

To characterize the behaviour of the control plane, we have taken into consideration only the first flapping event (5 second).

Figure 3 and 4 respectively show, when the data rate is equal to 500 Mbps and 900 Mbps, the number (K) of LSPs that the router is able to re-establish, after tearing down N LSPs and before the second flapping event.

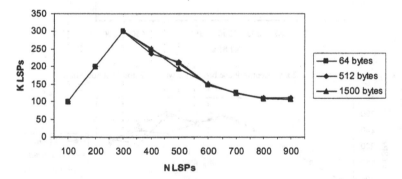

Figure 3: Test 1.x.x - Control Plane behaviour with CBR traffic at 500 Mbps

If N ≤ 300, all the LSPs are torn down and re-established before the second flapping event, regardless of the data streams bit rate and packet size. On the contrary, if N > 300, the number of LSPs re-established after the first flapping event is always less than the number of LSPs torn down, because the time interval between the two flapping event is not long enough to tear down and re-establish all the LSPs. Moreover, the number of LSPs re-established after the first flapping event slightly changes both with the packet size and data stream bit rate.

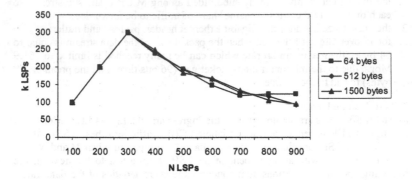

Figure 4: Test 3.x.x - Control Plane behaviour with CBR traffic at 900 Mbps

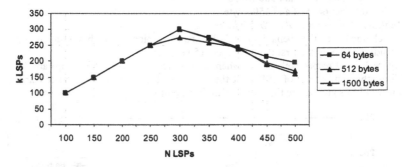

Figure 5: Test 2.x.x - Control Plane behaviour with CBR traffic at 500 Mbps

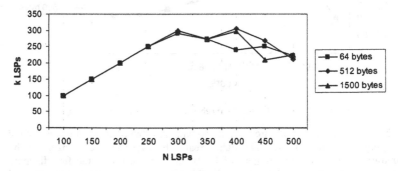

Figure 6: Test 4.x.x - Control Plane behaviour with CBR traffic at 900 Mbps

Figure 5 (data rate equal to 500 Mbps) and 6 (data rate equal to 900 Mbps) respectively show the number (K) of LSPs re-established before the beginning of the second flapping event, when the flapped and data traffic LSPs number is proportionally increased. We can observe that, if $N \leq 250$, all the LSPs are torn down and re-established, independently of the bit rate and packet size of the data stream. On the contrary, if N (and M) > 250, K is always less than N and the number of LSPs re-established after the first flapping event changes both with the packet size and bit rate of the data stream.

Analysis of the forwarding plane behaviour

To evaluate the performance of the forwarding plane, two metrics have been chosen:

- Packet loss prior and during flapping.
- One-way delay prior and during flapping, computed as the average of the one-way delay measured for the M data sub-streams.

In all the tests performed, no packet loss was ever experienced.

Table I and II show the results obtained for the one-way delay:

- When the bit rate and the packet size remain equal, the number of flapped LSPs (N) doesn't affect the performance of the forwarding plane. For this reason, each tables row reports the value of the one-way delay measured for any value of N.
- In all the tests performed, there is no degradation in the data plane performance prior and during flapping.
- The one-way delay is independent of the test category as well as the bit rate of the data stream, when the packet size is equal, except for 64 bytes packet size.
- When the packet size is equal to 64 bytes and the bit rate is 761.9 Mbps, the one-way delay prior and during flapping dramatically increases. This behaviour may be explained considering that, as previously said, 761.9 Mbps is the maximum theoretical throughput for IP over Gigabit Ethernet in these operating conditions and the packet rate assumes the maximum value (1488085 pkt/s).

First TypeTests One-Way Delay			Second TypeTests One-Way Delay		
Test	Prior	During	Test	Prior	During
1.1.x	0.007 ms	0.007 ms	2.1.x	0.007 ms	0.007 ms
1.2.x	0.016 ms	0.016 ms	2.2.x	0.016 ms	0.016 ms
1.3.x	0.03 ms	0.03 ms	2.3.x	0.03 ms	0.03 ms
3.1.x	0.156 ms	0.156 ms	4.1.x	0.157 ms	0.156 ms
3.2.x	0.016 ms	0.016 ms	4.2.x	0.016 ms	0.016 ms
3.3.x	0.031 ms	0.031 ms	4.3.x	0.031 ms	0.031 ms

Table I:
Average One-Way Delay of the M data flows

Table II:
Average One-Way Delay of the M data flows

The analysis of the control and forwarding plane behaviour clearly shows that the introduction of the MPLS architecture doesn't weigh on the router performance.

In particular, the functional separation between the data and the control plane, typical of the MPLS architecture, allows solving the performance issues of traditional IP routers, which are related to the management of signaling protocols and fast forwarding rates.

The hardware and software architecture of next generation routers allows simultaneously handling a high data traffic and a heavy control traffic, due to signaling, exchange of routing information and soft-state refresh messages generated by RSVP-TE.

Finally, we can also state that the introduction of the DiffServ architecture functional components, such as classifier, meter, marker, policer and scheduler shouldn't significantly affect the behaviour of the control and data planes.

VI. CONCLUSIONS

The paper presents the results of an experimental study aimed at evaluating the performance of the control and forwarding components of an M10 Juniper router that supports Multiprotocol Label Switching. To characterize the behaviour of the control plane, it has been taken into consideration the number of LSPs that the router tears down and re-establishes during the time interval between two subsequent flapping event: the results of the performance tests show that the behaviour of the control plane is not linear and doesn't depend on the bit rate and packet size of the data traffic to forward. As far as the forwarding plane is concerned, using the packet loss and the one-way delay as performance metrics, we found that the forwarding behaviour of the router is independent of the control plane. In fact, no packet loss was ever experienced and the one-way delay of the data streams is independent of the LSP number. Moreover, the RSVP-TE performance tests highlight that there is no degradation in the data plane performance from the prior flapping phase to the during flapping phase. The one-way delay increases with the packet size and is independent of the tests category as well as the bit rate of the data streams for each packet size taken into consideration, except for 64 bytes packets when the bit rate is equal to 761.9 Mbps. In this case, the one-way delay prior and during flapping significantly increases in comparison to the values obtained when the bit rate is equal to 500 Mbps.

REFERENCES

[1] E. Rosen, A. Viswanathan, R. Callon "Multiprotocol Label Switching Architecture", RFC 3031, January 2001

[2] L. Andersson, P. Doolan, N. Feldman "LDP Specifications", RFC 3036, January 2001

[3] R. Braden, L. Zhang, S. Berson "Resource ReserVation Protocol (RSVP)", RFC 2205, Sept.1997

[4] B. Jamoussi, L.Andersson, R. Callon "Constraint-Based LSP Setup using LDP", RFC 3212, January 2002

[5] J. Ash, M. Girish, E. Gray "Applicability Statement for CR-LDP", RFC 3213, January 2002

[6] J. Ash, Y. Lee, P. Ashwood-Smith "LSP Modification Using CR-LDP", RFC 3214, January 2002

[7] D. Awduche, L. Berger, D. Gan, T. Li, V. Srinivasan, G. Swallow "RSVP-TE: Extensions to RSVP for LSP Tunnels", RFC 3209, December 2001

[8] http://www.spirent.com about "AX/4000 Broadband Test System"

[9] Spirent Communications Test Methodologies, "RSPV-TE Forwarding Performance During Tunnel Establishment Test"

INTER-DOMAIN ROUTING IN OPTICAL NETWORKS

Américo Muchanga[1], Lena Wosinska[2], Fredrik Orava[3], Joanna Haralson[4]
Royal Institute of Technology, KTH/IMIT, Sweden,
[1]*americo@imit.kth.se,* [2]*Lena.Wosinska@imit.kth.se,* [3]*fredrik@it.kth.se,* [4] *it03_jdr@it.kth.se*

Abstract: In this paper we present a mechanism for obtaining an abstraction of network topology in optical networks in order to compute an end-to-end lightpath across multiple domains.

1.   INTRODUCTION

The main benefit of the Multiprotocol Label Switching (MPLS) is to provide traffic engineering in IP networks. The connectionless operation of IP networks becomes more like a connection-oriented network where the path between the source and the destination is pre-calculated based on user specification.

Generalized Multiprotocol Label Switching (GMPLS) extends MPLS to provide the control plane (signaling and routing) for devices that switch in any domain, i.e. packet, time, wavelength and fiber. This common control plane is expected to simplify network operation and management by automating end to end provisioning of connections, managing network resources, and providing the level of QoS that is expected in the new, sophisticated applications.

Thus future data and telecommunication networks are likely to consist of elements that will use GMPLS to dynamically provision resources and to provide network survivability using protection and restoration techniques.

There is also a great interest in extending IP-based protocols to control optical networks. To replace the existing solutions the new optical transport networks must provide similar or higher QoS, protection and restoration mechanisms that are available in current techniques. This implies that network operators have to be able to provide end-to-end QoS guarantees for its customers. Thus the network operators or carriers need to have traffic engineering capabilities to set lightpaths in networks that are beyond their administration domains, which means that they have to be able to exchange resources and topology information through an Exterior Gateway Protocol (EGP).

Domain boundaries exist for the purpose of abstraction from irrelevant details, e.g. domain related information, to the outside world. This information might reveal to the network architecture between two competing entities. However it might be vital to enable network or service providers to achieve diversity and protection that is essential for optical networks survivability

There are only a few attempts that address the issues of inter-domain routing and optical networks [1,2]. They are still in the early phases. In order to utilize the full potential of optical networks information about optical network resources has to be conveyed through inter-domain routing protocols. However, the IETF GMPLS architecture draft [3] does not cover extensions for inter-domain routing (e.g. BGP).

This paper addresses the information that has to be shared between domains through an inter-domain signaling mechanism and its level of relevance to support the development of optical networks. This information could be conveyed through traffic engineering extensions to the existing inter-domain routing protocols. Therefore we are proposing a method of network abstraction that would enable extensions to BGP to achieve traffic engineering information exchange required to facilitate the operation of networks. The ability to diversely route optical connections is very important for the reliability and resilience of the optical network.

Our approach is to use BGP for sharing not only the information about network reachability but also to share information about essential data in internal networks that is required to enable peers to setup end-to-end lightpaths with requested quality.

2. RATIONALE FOR INTER DOMAIN ROUTING IN OPTICAL NETWORKS

In this paper we concentrate on the signaling and routing issues when networks belonging to different carriers are interconnected. We consider services that make two competing companies to be involved in optical inter-domain routing relationship.

In inter-domain routing relationship today, service providers share basically only the network reachability information. Occasionally they do exchange information that allows them to reach some degree of diversity, in situations where multi-exit discrimination is possible. In these situations the inter-domain routing protocol is not used to carry information that has any significance for lower layers.

However, inter-domain relationship in optical networks can involve exchange of information that can reveal the network architecture of the competing peers, and

Figure 1. Network model and components of provider networks
interconnected through a wavelength routing network

thus affecting the competitive advantage of the participating entities. If end-to-end light path provisioning is to be achieved, service providers need to exchange some information in the interconnecting network interfaces, which is beyond the network reachability information that is carried today.

Figure 1 shows network model and components. We focus on the relationship between provider A and provider B, and in particular on the cooperation between the edge routers that interconnect the networks of the providers. Figure1 shows a situation in which the networks are connected by a point-to-point optical link. Our proposed method, however, is not limited to this simple topology and we consider also more general network topologies.

Note, that if two IP clients are connected through two optical domains (as shown in Figure 1) the conventional IP adjacency for IP layer operations as well as adjacencies for circuit switched operations have to be maintained.

The optical control plane has been proposed in the GMPLS architecture [3]. Protocols and functions of an optical control plane are illustrated in Figure 2. The optical network control plane is divided in routing protocols and signaling protocols. Routing protocols are used for topology and resource discovery and

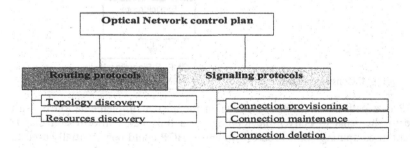

Figure 2. Protocols and functions of an optical control plane

dissemination while signaling protocols are used for connection provisioning, maintenance and termination.

Topology and resource discovery and dissemination can be done both within domain and between domains. Protocols that achieve that within a domain have been proposed [4, 5] and are now going through a maturing phase. They have been presented in a form of extensions to existing protocols such as Open shortest path first (OSPF) and Intermediate System to Intermediate System (IS-IS). These extended protocols have already been implemented in a number of routing engines. However, so far there is no protocol implementation that provides the same functionality for inter-domain operations.

3. ROLE OF INTER-DOMAIN ROUTING PROTOCOLS

To support devices that switch in time, wavelength and fiber, MPLS was extended with protocols that advertise the availability of these resources in order to allow the establishment of a label switched end-to-end path. OSPF and IS-IS have been extended to advertise, within a domain, the availability of optical resources in the network (e.g. generalized representation of various link types, bandwidth on wavelengths, link protection type, fiber identifiers). Once that information is known signaling protocols such as RSVP[6] and LDP[7] have been extended for traffic engineering purposes that allow a label-switched path (LSP) to be explicitly specified across the optical core. Figure 3 shows a constraint-based routing flow.

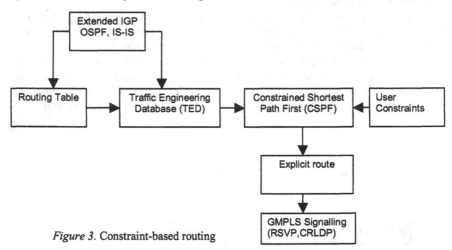

Figure 3. Constraint-based routing

BGP is a path vector protocol, which in the current implementation doesn't include mechanisms to deliver the topology and link status information that is important in computing optical routes. However, BGP could be potentially used to support some or all the services provided by intra-domain routing protocols in order to enable a proper operation of signaling protocols.

4. OPERATORS' PERSPECTIVE

In order to offer similar services as provided today by intra-domain traffic engineering protocols, BGP has to be modified to carry all the database containing link and resources information that is created by these intra-domain routing protocols. In addition it has to support the LSP setup process that is accomplished through GMPLS signaling.

BGP, after receiving information, would have to digest it in the internal network in order to allow internal routers to create a database that reflects the global network encompassing different network operators.

Exporting the link state information from one domain to another domain would consume a lot of resources during the initial flooding process and after a change of the network topology or availability resources. Announcing all this information would make the networks not to scale, because the abstraction used today between domains (i.e., only announce reachable networks and not the full topology) is what makes Internet to be able to scale to its current size. Additionally it would enable the peering entities to gain a detailed view of how the other network is implemented that would require applying expensive routers in order to handle huge amount of information. Getting access to this kind of information would enable one entity to gain competitive advantage by being able to engineer his network to offer superior quality of service guarantees.

The above reasons make it unlikely that network providers will be willing to share the entire view of their network to a competing entity, because it does not scale, it is costly and is not attractive from a business point of view. However, the traffic engineering characteristics of GMPLS and its ability to use one forwarding mechanism for multiple applications makes it an attractive technology that could enable providers that are connected at several points to achieve for instance load balancing in the inter connection links and to design services that are based on differentiated QoS parameters.

Therefore, BGP/GMPLS should be used to allow the peering entities to create an abstracted view of the network and share only a few abstract links or paths in the network that would provide to the peering entities some degree of choice to reach destinations inside or outside that network. Thus, we propose a way to achieve the abstracted topology and how BGP could carry that information.

As mentioned, there are strong reasons why all learned information should not be exported to BGP. Therefore we suggest that a carrier or network provider should create abstract aggregations of virtual domains inside his networks to be advertised to peering Exterior Gateway Protocol (EGP) speakers in order to allow them to have some traffic engineering capabilities. This concept is also raised in [8] and is illustrated in Figure 4 where two providers own a mesh network with links characterized by their resources and capacities.

There are many possible paths to go from say R1 to R6. These possible paths are discovered inside the domain of the providers through the use of any extended Interior Gateway Protocol (IGP). Thus the border routers, e.g. R7 and R6, contain a traffic engineering database describing their view of the network within provider A. Our suggestion is that BGP should, for instance, advertise only two paths to the

268

peers. For instance R7 located in provider A could advertise to R8 and R9 located in provider B that R1 can be reached through the path R7-R2-R1. In the same way R6 located in provider A would advertise to R9 located in provider B that R1 can be reached through the path R6-R3-R4-R1. Note that there are more possible paths to go from R6 to R1, for instance, R1 could be reached by the route R6-R5-R4-R1. However, the explicit paths are not exported to the peer, only the knowledge about the existence of two paths, the associated properties and also how to name them should be communicated. In this way some degree of diversity or choice can be offered to the peer provider while maintaining some abstraction within the network domain, enabling the providers to hide the details about their network that they feel would impact their competitive advantage.

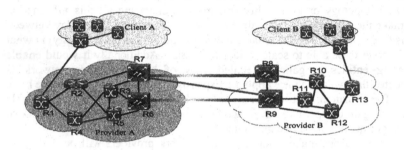

Figure 4. An example of abstraction that border routers could provide to each other

5. MECHANISM FOR CREATION OF ABSTRACTED PATHS

In order to create the abstracted network topology with only few possible paths between the edge nodes we suggest introducing the following constraints:
1. Node diversity
2. Minimum bandwidth available within the paths
3. Degree of protection
4. Shortest paths

As a result the edge nodes could be asked for instance to compute a shortest path that is node diverse and has a minimum of 10Mbps with links along the path protected by a dedicated protection.

The idea is that the resulting paths are the ones that could be advertised to other EGB speakers in the border routers. This result could be tightened to a quality of service that might be requested by a peer provider. Assume for instance that client A and B in Figure 4 is one company located in two geographically separated areas.

In location A client A is connected through Provider A and in location B client B is connected through provider B. Providers A and B have to agree on the end-to-end QoS they want to provide to their respective customers. Thus the two providers should decide to advertise to each other paths selected along the constraints that enable them to guarantee the required QoS.

The advantage of this method is that creating the abstracted network topology inside a domain is based on IGP and signaling protocols such as RSVP-TE. As a result the constructed paths are just like any other LSPs and can be treated like any LSPs that have satisfied certain computation criteria.

This procedure can be extended for interconnection of multiple domains. Assume two domains are connected through a third domain as shown in Figure 5. Provider B can advertise to Provider A and C only a limited set of paths to reach both domains. In this scenario we abstract a domain as if it was a node, one can decide to advertise either all possible paths connecting domains A and C or only the constrained paths.

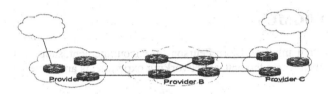

Figure 5. Interconnection between three providers

In this latter case BGP needs to be modified in order to carry only the abstracted information to the peer, which is the main advantage of our proposal.

6.   CONCLUSIONS

Routing and signaling protocols play an important role in optical networks. We have proposed that Interior gateway protocols can be extended in order to advertise not only link state information that is basically used to provide reachability information but also to convey resource information that is required for traffic engineering. In this way Routing and Signaling protocols can be extended to support packet switching devices as well as time, wavelength and spatial switching devices.

In order to allow for traffic engineering to be done on end-to-end basis across multiple domains link state information as well as traffic engineering (TE) information between domains needs to be exchanged. This should be done through an Exterior Gateway Protocol such as BGP.

Exporting the entire network map with all TE parameters is not reasonable due to business, operational, capacity and scalability purposes. Therefore we propose a method to obtain an abstraction of the network topology. In this method only a few optional paths satisfying some acceptable QoS guarantees are provided. These QoS can be negotiated between peering entities and used to create the visible LSPs. The paths can be computed using constraints such as diversity, bandwidth and protection.

A similar thinking can be applied for links inside a domain as well as for links connecting domains. In this case a domain is seen as if it was a node with several interfaces that are connected to other domains.

We also propose that once the abstracted set of paths have been obtained and announced to the peers, a streamed down mechanism can be applied in the similar way as presented in [1].

Finally, we believe that the proposed mechanism is promising and we are going to evaluate its performance by OPNET simulations as well as by experiments made on our testbed.

ACKNOWLEDGMENTS

The authors wish to thank Prof. Björn Pehrson for encouragement, support and for providing the testbed where the optical networking experiments are carried out.

REFERENCES

[1] Y.Xu, A. Basu, Y.Xue, "A BGP/GMPLS Solution for Inter-domain Optical Networking", I-D, (Expired October 2003)

[2] G. Bernstein, L.Ong, B.Rajagopalan, D.Pendarakis, A.Chiu, F.Hujber, J.Strand, V.Sharma, S.Dharanikota, D.Cheng, R.Izmailov, "Optical Inter domain routing Considerations", I-D, (Expired Augusti 2003)

[3] Eric Mannie (Editor), et.al, "Generalized Multi-protocol Label Switching (GMPLS) Architecture", I-D, Work in Progress (Expired November 2003).

[4] K. Kompella, Y. Rekhter, "OSPF Extensions in Support of Generalized MPLS", I-D, Standards Track, Work in Progress (Expired June 2003).

[5] K. Kompella, Y. Rekhter, "IS-IS Extensions in Support of Generalized MPLS", I-D, Work in Progress (Expired June 2003)

[6] L. Berger, "RFC 3473 - Generalized Multi-Protocol Label Switching (GMPLS) Signaling Resource ReserVation Protocol-Traffic Engineering (RSVP-TE) Extensions", *Standards Track, January 2003*

[7] P. Ashwood-Smith and L. Berger, "RFC 3472- Generalized Multi-Protocol Label Switching (GMPLS) Signaling Constraint-based Routed Label Distribution Protocol (CR-LDP) Extensions", *Standards Track, January 2003*

[8] G.M.Bernstein, V.Sharma, L.Ong, "Interdomain Optical Routing", *Journal of Optical Networking,* Feb 2002/Vol. 1, No.2

PART A5:

TECHNIQUES FOR OPTICAL NODE

OPTICAL NETWORK UNIT BASED ON A BIDIRECTIONAL REFLECTIVE SEMICONDUCTOR OPTICAL AMPLIFIER

Josep Prat[1,2], Cristina Arellano[1,3], Victor Polo[1,4], Carlos Bock[1,5]
[1] Dep. of Signal Theory and Communications, Universitat Politècnica de Catalunya,
C/ Jordi Girona D-4, 08034 Barcelona
[2] jprat@tsc.upc.es
[3] cristina.arellano@tsc.upc.es
[4] polo@tsc.upc.es
[5] bock@tsc.upc.es

Abstract: An optical access network transceiver based on a reflective semiconductor optical amplifier (RSOA) operating as modulator and photodetector is demonstrated. The system shows proper operation at 1.25 Gbit/s to 30 km reach.

Keywords: Access networks, Fiber-to-the-home, Optical modulation, Optical signal detection, Reflective semiconductor optical amplifier.

1.   INTRODUCTION

Fiber-to-the-Home technology is one of the main research objectives in the "Broadband for all" concept that encourages the development of optical access infrastructure [1]. In order to fulfill this concept, cost effective solutions must be developed to be able to offer broadband connections to end users at a reasonable cost. A key element in access networks is the Optical Network Unit (ONU) of the Customer Premises Equipment (CPE). This has a direct impact on the cost per customer and in general on the overall cost of the access network. Therefore, simple ONUs needs to be designed. Also, a desirable characteristic of the outside plant of an access network is the use of one single fiber for both upstream and downstream transmission to reduce network size and connection complexity [2].

Some ONU designs that avoid the local generation of light have been lately demonstrated that fulfill the above simplicity criteria. The laser and its stabilization electronics are avoided, and being all ONUs equal while maintaining the necessary transparency for the WDM operation of the network. This is accomplished by using different modulating formats for upstream and downlink transmission or some amplitude signal regeneration, and advanced devices: Phase-Shift Keying/Intensity Modulation [3], Electro-Absorption Transceiver for signal remodulation [4], polarization rotation remodulator [5] and IM remodulation using semiconductor optical amplifiers [6].

We here demonstrate the use of a Reflective Semiconductor Optical Amplifier (RSOA) as a potential candidate for an electro-optic transceiver at the ONU, operating in burst mode. The advantages of this device are inherent optical amplification of the incoming signal and the use of a single fiber in the outside plant. Incoming light can be modulated by the RSOA injection current carrying the upstream user data [7]. The RSOA can also act as a photodetector by sensing the voltage variation of the electrode [8]. In this paper we integrate both functionalities, detection and modulation, in a FTTH network with single fiber access.

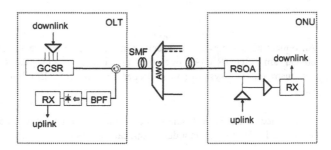

Figure 1. Schematic of the FTTH Network

2. SYSTEM SCHEME

Figure 1 depicts the network topology and the design of the proposed Optical Layer Termination (OLT) and ONU. A Grating-assisted codirectional Coupler with rear Sampled Reflector (GCSR) is used at the OLT as a light source. Downstream data is modulated up to 1.25 Gbit/s using a LiNbO$_3$ modulator, and sent through an optical circulator that will play its role when receiving upstream data. A 1x8 AWG routes optical signals depending on input wavelength to the desired ONU. The ONU is built just by one RSOA, which acts as modulator and photodetector.

Electrically, an interface circuit for the RSOA has been designed to separate the received downstream signal and the ONU upstream data, combining a low noise preamplifier and a laser driver both connected at the current electrode. Two electrical switches are used to activate receive or transmit mode, and avoid the saturation of the preamplifier.

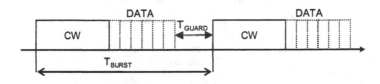

Figure 2. Downlink Burst

Downstream data and upstream carrier are sent from the OLT time multiplexed in a single burst. The first burst section is downstream data and then, after a guard band, optical carrier is sent for upstream modulation purposes. Once upstream data is modulated it is sent back to the OLT; there, the optical circulator routes the incoming signal to the detection branch and isolates the laser source. The data source and receiver are synchronized in burst mode to the corresponding packet period.

3. EXPERIMENTAL SETUP

The setup that has been implemented to demonstrate the feasibility of the transmission is very similar to the proposed network topology.

Regarding the RSOA optical specifications, the device has chip length of 500µm, reflective facet with 90% refection and front facet reflecting less than 0.1%. RSOA optical gain has been adjusted to 10 dB injecting 65mA of current. This is the current that has been used during testing because offered enough gain, modulation efficiency, detection sensitivity and enough tolerance against reflections.

Tests have been performed using two separate fibers for upstream and downstream and one single fiber for both directions. Also, detection and modulation tests have been performed.

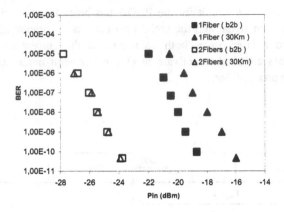

Figure 3. RSOA acting as modulator (Uplink)

Figure 3 presents the results obtained using the RSOA as a modulator. The -25 dBm sensitivity (at a BER of 10^{-9}) obtained for two different fibers are almost identical in both back-to-back and using 30 km SMF. The reason is that fiber dispersion is negligible at 1.25 Gbit/s. When the system becomes bidirectional, that is to say, in a single fiber scenario, a 5 dB degradation of sensitivity has been observed. And also there is a 2 dB of penalty when SMF fiber is introduced, due to the rayleigh backscattering effect which is one of the most important signal perturbations when transmitting using a single fiber on both directions [9]. With reference to the difference in sensitivity when using one or two fibers, it can be deduced that it is caused by reflections in connectors because of isolation between the circulator ports is not perfect despite APC connectors have been used to minimize reflections.

RSOA detection capabilities have been tested by modulating a 1.25 Gbit/s data signal using a LiNbO$_3$ modulator and a GCSR laser source located at the OLT. We observe that the RSOA needs a higher optical power (about -19 dBm) to be able to produce the detectable variation of voltage. This is because of the certain level of saturation or carrier density reduction in marks. The results that have been obtained in both modulation and detection mode are summarized in Figure 4.

It can be seen that downstream signal is not severely degraded when transmitting using a single fiber along 30 km, though a 2 dB penalty in sensitivity was measured when the RSOA was working as modulator, in the uplink. The explanation to this difference is that upstream carrier is sent from the OLT and needs to be transmitted twice along the fiber, being more affected by rayleigh scattering than the downstream signal, which just travels in one direction along the fiber; also, the upstream optical power is lower than the downstream, producing a lower interference.

We have also tested several optical carriers and have found that when using wide spectrum light sources, rayleigh scattering is less critical than with narrower sources, confirming that backscattering interference is a coherent process.

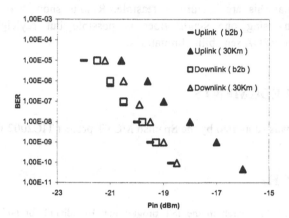

Figure 4. Downlink / Uplink BER comparison

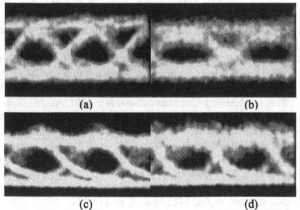

Figure 5. RSOA detection/modulation output eye diagrams: (a) detection (downlink) back-to-back, (b)detection 30 km SMF, (c) modulation (uplink) back-to-back, (d)modulation 30 km SMF.

4. CONCLUSIONS

A RSOA as Optical Network Unit has been presented as a potential cost effective solution for a FTTH passive optical platform with full-duplex operation along single fiber outside plant.

Modulation and photo detection tests have been implemented in order to demonstrate that this architecture is feasible. Results show that bidirectional communication using one single fiber is possible but rayleigh scattering interference and RSOA sensibility limitations.

ACKNOWLEDGMENTS

This work was supported by the Spanish MCYT project TIC2002-00053.

REFERENCES

[1] A. Houghton, "Research in the IST programme: Broadband for all", ECOC 2003 Rimini, Mo3.1.1.

[2] J. Prat, P. E. Balaguer, J.M. Gené, O. Díaz, S. Figuerola, *Fiber-to-the-Home Technologies*, Kluwer Ac. Publishers, ISBN1-4020-7136-1.

[3] W. Hung, C-K Chan, L-K Chen and C. Lin, "System characterization of a robust re-modulation scheme with DPSK downstream traffic in a WDM access network", Proc. ECOC 2003 Rimini, We3.4.5.

[4] M. Schneider, T. Reimann, A. Stöhr, S. Neumann, F.-J. Tegude, D. Jäger, "Monolithically Integrated Optoelectronic Circuits using HBT, EAM, and TEAT", Proc. MWP 2002, Awaji, Japan, 2002, pp. 349-352, ISBN 4-88552-187-4 C3055.

[5] T. Koonen, K. Steenbergen, F. Janssen, J. Wellen, "Flexibly Reconfigurable Fiber-Wireless Network using Wavelength Routing Techniques: The ACTS Project AC349 PRISMA", Photonic Network Communications, July 2001, pp. 297-306.

[6] H. Takesue and T. Sugie, "Data rewrite of wavelength channel using saturated SOA modulator for WDM Metro/Access networks with centralized light sources", Proc. ECOC 2002 Copenhagen, 8.5.6.

[7] N. Buldawoo et al., "A Semiconductor Laser Amplifier-Reflector for the future FTTH Applications", ECOC 1997 Edinburgh, p196 – 199.

[8] N. Buldawoo et al., "Transmission Experiment using a laser amplifier-reflector for DWDM access network, ECOC 1998 Madrid, pp.273-274.

[9] P. Gysel and R. K. Staubli, "Statistical properties of Rayleigh backscattering in single-mode fibers", J. Lightwave Technol., vol. 8, pp.561-567.

OPTICAL LABEL RECOGNITION BASED ON ADDITIONAL PRE-SPREAD CODING

Hideaki Furukawa,[1] Tsuyoshi Konishi,[1] Kazuyoshi Itoh, [1]
Naoya Wada,[2] and Fumito Kubota [2]

[1] *Graduate School of Engineering, Osaka University, 2-1, Yamadaoka, Suita, Osaka 565-0871, Japan,* [2] *National Institute of Information and Communications Technology, Tokyo, Japan*

Abstract: We propose novel optical label recognition based on additional pre-spread coding. This method can realize both high recognition power in label recognition and data transmission efficiency in network paths. We present the optical implementation of this method and show the recognition power by simulation.

1. INTRODUCTION

In the next-generation photonic networks based on a packet switch, optical label recognition is a key technology to route optical packets according to label without any electronic bottlenecks. Since an optical correlation system can recognize temporal optical codes only by passive optical devices without a complex synchronization, it is considered to be a powerful solution for optical label recognition [1-4]. In an optical label recognition system based on optical correlation, the recognition can be accomplished by comparing the maximum intensity of correlation signals between optical codes and a matched filter. To improve its recognition power, it is desirable that the maximum intensity ratio (SNR : signal to noise ratio) of correlation signals of the target optical code to non-target ones is high. Since the bit-pattern of long-bit spread codes are generally much different from each other, spread codes, such as M-sequences or Gold codes, are used as optical codes to raise the SNR [2,4]. However, since long-bit spread codes occupy large part of packets, data transmission efficiency becomes much low in network paths.

In the network, it is said that the number of labels of 10^4 is required [2]. Generally, the bit length of optical codes is 14-bits enough to represent labels of 10^4. However, 256-bits or more are necessary for spread codes to represent labels of 10^4 [4]. From the viewpoint of data transmission efficiency, it is desirable that the bit length of transmitted optical codes is as short as possible in network paths. Here, if transmitted short-bit optical codes can be expanded into long-bit spread codes just before label recognition in network nodes, both high SNR and high data transmission efficiency can be realized. In this paper, we propose optical label recognition based on additional pre-spread coding.

2. OPTICAL LABEL RECOGNITION USING ADDITIONAL PRE-SPREAD CODING SYSTEM

Figure 1 shows the conceptual diagram of the proposed optical label recognition based on additional pre-spread coding. In network paths, short-bit optical codes are used to achieve high data transmission efficiency. In network nodes, these short-bit optical codes are expanded into long-bit spread codes by an additional pre-spread coding system. After the additional pre-spread coding, since the obtained long-bit spread codes are much different from each other, high SNR can be achieved in the label recognition based on optical correlation.

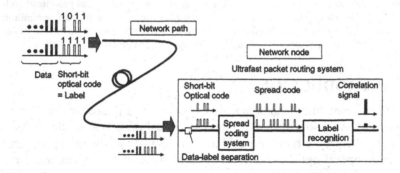

Figure 1. Optical label recognition using additional pre-spread coding.

Here, we adopt Gold codes as one of spread codes to realize the proposed additional pre-spread coding system. The generating procedure of Gold codes is shown in Fig. 2. First, M-sequence 1 is generated from an input short-bit optical code. M-sequence 2 is an arbitrary M-sequence, which is prepared in advance. Secondly, a Gold code is generated by executing EXOR operation between M-sequence 1 and 2. Here, a set of M-sequences can be generated from the initial bit-pattern of M-sequence 1 by shifting it with different amount depending on each input short-bit optical code. By executing EXOR operation between this set of M-sequences (for example M1-001, M1-010, and so on) and M-sequence 2, we can obtain a set of Gold codes from different input short-bit optical codes. Here, there is a feature that M-sequence of a short-bit optical code can be generated by executing EXOR operation between M-sequences generated from each bit of the optical code. For example, M-sequence of [011] can be obtained by executing EXOR operation between that of [001] and [010]. Thus, we have only to construct an optical setup which generates M-sequence from each bit and executes EXOR operation for spread coding.

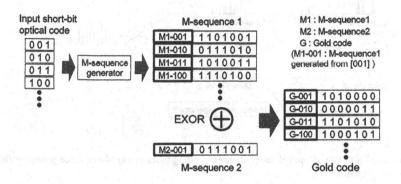

Figure 2. Generation procedure of Gold code.

The optical setup of spread coding system is shown in Figs. 3 and 4. As shown in Fig. 3a, an arbitrary M-sequence can be generated from one bit by adequately setting fiber delay lines in the M-sequence generator. In case of several bits amplitude shift keying (ASK) optical codes, M-sequences are generated from each bit by this generator. (Fig. 3b.) The summation of these M-sequences is output as the M-sequence 1 of several bits optical codes. Moreover, the M-sequence 1 is added to M-sequence 2 generated from one clock signal. Instead of directly executing EXOR operation between these M-sequences, we achieve equivalent processing to EXOR by using this summation and the intensity conversion. As shown in Fig. 4a, to realize equivalent processing to EXOR, the intensity of optical pulses is converted into 0 or 1 at the even or odd intensity level respectively. The intensity conversion is implemented using optical multilevel thresholding [5]. In the high nonlinear fiber (HNLF), self-frequency shifting generates, and the center frequency of optical pulses is shifted depending on the intensity. (Fig. 4b.) As a result, each optical pulse can be separated by arrayed waveguide grating (AWG). (Fig. 4c.) The intensity level of each optical pulse is adjusted to 0 or 1 by appropriate filters. Consequently, ASK Gold codes can be generated by this setup. This method can also obtain binary phase shift keying (BPSK) Gold codes by converting the intensity level of each pulse into 1 and modulating the phase to 0 or π with phase shifters.

282

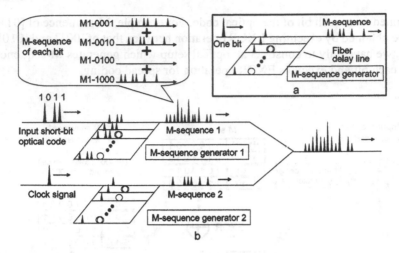

Figure 3. Front part of optical setup of spread coding system. (a) M-sequence generator.(b) Gold code generator.

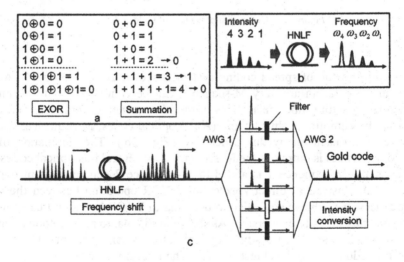

Figure 4. Rear part of optical setup of spread coding system. (a) EXOR operation and summation of light intensity. (b) Self-frequency shift in the HNLF. (c) EXOR operation using intensity conversion.

3. SIMULATION OF LABEL RECOGNITION USING SPREAD CODING

We calculate correlation signals of 16 different types of 127-bits BPSK Gold codes generated from 4-bits ASK optical codes. The Gold code generated from

[1010] is set as the target Gold code. Figure 5 shows the maximum intensity of correlation signals of Gold codes with a matched filter. We can successfully obtain considerably high minimum SNR 15.2 : 1 compared with SNR 8.0 : 1 in the previous work [4].

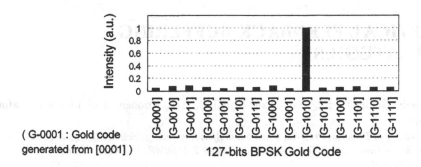

(G-0001 : Gold code generated from [0001])   127-bits BPSK Gold Code

Figure 5. The maximum intensity of correlation signals of 127-bits Gold codes.

4. CONCLUSION

We proposed optical label recognition based on additional pre-spread coding, which can realize both high SNR and data transmission efficiency. From a calculation result, it is confirmed that high SNR can be realized using Gold codes generated by additional pre-spread coding.

REFERENCES

[1] Cardakli, M.C., Lee, S., Willner, A.E., Grubsky, V., Starodubov, D., Feinberg, J.: 'Reconfigurable optical packet header recognition and routing using time-to-wavelength mapping and tunable fiber Bragg gratings for correlation decoding', IEEE Photon. Tech. Lett., 2000, **12**, (5), pp. 552-554.

[2] Kitayama, K., Wada, N., and Sotobayashi, H.: 'Architectural considerations for photonic IP router based upon optical code correlation', J.Lightwave Technol., 2000, **18**, (12), pp. 1834-1844.

[3] Konishi, T., Kotanigawa, T., Tanimura, K., Furukawa, H., Oshita, Y., and Ichioka, Y.: 'Fundamental functions for ultrafast optical routing by temporal frequency-to-space conversion', Opt. Lett., 2001, **26**, (18), pp. 1445-1447.

[4] Oshiba, S., Kutsuzawa, S., and Nishiki, A.: 'Optical label processing using a SSFBG encoder/decoder', IEICE Technol. Rep., 2003, **PN 2003-15**, (2003-09), pp. 29-34.

[5] Konishi, T., Tanimura, K., Asano, K., Oshita, Y., and Ichioka, Y.: 'All-optical analog-to-digital converter by use of self-frequency shifting in fiber and a pulse-shaping technique', J. Opt. Soc. Am. B, 2002, **19**, (11), pp. 2817-2823.

OPTICAL FEEDBACK BUFFERING STRATEGIES

Ronelle Geldenhuys[1,2], Jesús Paúl Tomillo[2], Ton Koonen[2] and Idelfonso Tafur Monroy[2]
[1]University of Pretoria, Pretoria, 0002, South Africa, ronelle.geldenhuys@eng.up.ac.za
[2]Eindhoven University of Technology, Den Dolech 2, 5600MB, The Netherlands, i.tafur@tue.nl

Abstract: This paper considers the performance of fixed and incremental feedback buffers in an optical switch, and compares this with the performance of a feedback buffer configuration implementing switchable delay lines. It is shown that for a medium sized switch the switchable delay line implementation outperforms the other configurations, although for a large switch, the incremental structure has the best performance.

1.   INTRODUCTION

Contention resolution in optical packet switches is implemented using fibre delay lines (FDLs) in either a travelling or a feedback (also called recirculating) configuration[1]. Travelling buffers are either input or output buffers, and have limited performance with respect to the packet loss ratio because of the huge amount of fibre required for sufficient buffering of bursty traffic. Feedback buffers have the advantage that the FDLs are reused thus decreasing the total amount of fibre required to achieve an acceptable packet loss ratio. There are however various disadvantages: the optical signal must be amplified on each recirculation resulting in an increase in amplified spontaneous emission (ASE) noise from the optical amplifiers; and the buffered signal has to traverse the switch fabric on each recirculation resulting in crosstalk and optical loss[2].

There are 2 factors supporting the implementation of feedback buffers in optical switches. The first is the improved performance of switch technology, for example decreased crosstalk and optical loss in all-optical switch technology[3]. The second is the self-similar nature of Internet traffic resulting in a situation where it is almost impossible to buffer traffic satisfactorily, even when using feedback buffers with a lot of fibre[4].

This paper compares feedback buffering (Figure 1) with 3 configurations: fixed, incremental, and using switchable delay lines (SDLs).

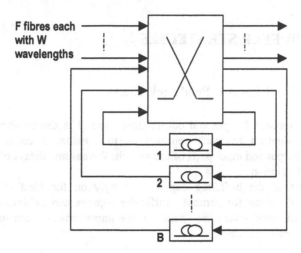

Figure 1. Schematic representation of a non-blocking switch with a feedback buffer. There are B FDLs and can be implemented in one of three configurations: 1. fixed, 2. incremental, 3. SDLs

2. TRAFFIC MODEL

Traditionally, short-range dependent models are used to model traffic, but real traffic displays burstiness on a wide range of time scales. Large-scale correlation refers to correlations that last across large time scales. Long-range dependence refers to values at any instant being positively correlated with values at all future instants. These characteristics result in Internet traffic being described as selfsimilar. Self-similarity is usually defined in statistical or qualitative terms, loosely including anything that "looks like itself" when magnified.

In this paper, three traffic models are used: Bernoulli traffic is the simplest traffic pattern; geometrically distributed ON and OFF periods are used to model bursty traffic; and a Pareto distribution is used to simulate self-similar traffic. To simulate a heavy-tailed distribution, a Pareto distribution can be used to produce "pseudo-self-similar" arrival processes[5]. This traffic has large-scale correlations but the traffic is not actually long-range dependent. When using a self-similar model to describe long-range dependence, only a single parameter is required: self-similarity is characterised by the Hurst parameter, H, which relates linearly to the shape parameter, α, of the heavy-tailed file size distribution in the application layer. 0.5<H<1.0, and as H approaches 1, both selfsimilarity and long-range dependence increases. 0<a<2, and if a<2 then the distribution has infinite variance, and if a≤1 then the distribution has infinite mean. According to [6] a typical value for a is 1.2.

3. BUFFER STRATEGIES

3.1 Fixed Length Feedback Lines

F is the number of input and output fibres and B is the number of feedback fibres. The simplest feedback approach is to consider B equal length FDLs connecting output and input ports of the switch. We assume delays of one time slot for this configuration.

The effect of the buffering depends strongly on the kind of traffic used. Simulations show that for Bernoulli traffic the improvement achieved by adding an extra feedback fibre is very important, but the improvement is not so considerable for the other traffic models.

3.2 Incremental Length Delay Lines

In this configuration the delays are distributed from 1 to B time slots. Here, a scheduling strategy is required to decide to which feedback fibre contending packets should be delivered. The three strategies simulated, based on [7], are as follows:
1. *minDelay* – the delay of the packets is minimised by sending bursts through the minimum delay fibre available.
2. *noOvr* – reservation of resources is done in advance so that when a new burst arrives at the switch its destination, route and the resources used are calculated in advance.
3. *avoidOvr* – this is similar to *noOvr* except that if no buffer is found the packet/burst is not dropped, it is sent to the minimum delay line available.

3.3 Buffering with Switchable Delay Lines

The switchable delay line structure is shown in Figure 2. This structure can switch the length of a fibre and get delays ranging from 0 to 7T. T depends on the length of the delay loops.

The following three scheduling strategies were used:
1. *useBusy* – using the minimum possible delay, first try to use a busy (already configured) SDL, and then an idle SDL. If there is no available minimum delay, the packet/burst is dropped.

2. *minBusy* – using the minimum possible delay, first try to use an idle SDL configuring it with that delay and if not possible, use a busy SDL. If this is not possible, the packet/burst is sent through a minimum delay available SDL.

3. *minIdle* – similar to *useBusy*, but if the required minimum delay is not found, then the packet/burst is sent through a minimum delay available SDL.

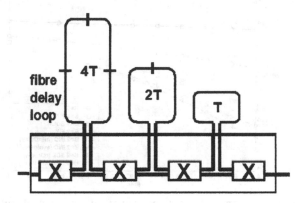

Figure 2. Schematic representation of a switchable delay line.

4.   RESULTS

Event-driven simulation based on slotted operation was used. Both Bernoulli and self-similar traffic models are used. A Bernoulli traffic model can be used to simplify configurations where relative results are required.

A burst is a sequence of one or more packets sent in consecutive time slots from the source and all with the same destination. Two different approaches were used for modelling bursts:

1. *Packet trains*: The sequence of packets cannot be segmented and all the packets that compose the burst follow the same route as the first one[8].

2. *Packet wagons*: Packets that compose the burst are considered independently[9].

The influence of the following factors are shown in Figures 3 – 8:

-   Traffic type and load
-   Number of feedback fibres
-   Fixed versus incremental FDLs
-   Buffering strategies
-   Switch size – Medium: F = 4, W = 8; Large: F = 8, W = 32

Note that for the incremental delay structures the strategy *avoidOvr* is an upgraded version of *noOvr* and always behaves better, so to improve readability of

288

figures *noOvr* is not plotted.

Figure 3 shows the performance with Bernoulli traffic. B=0 means no feedback is used and it is plotted as a reference curve. Differences with 1 feedback fibre are insignificant, but with 2 fibres the incremental strategy *avoidOvr* has the best results.

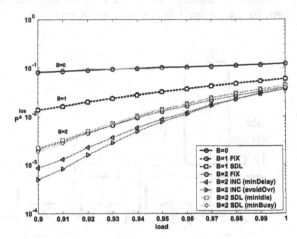

Figure 3. Performance of a medium sized switch simulated with Bernoulli traffic. F=4, W=8.

Figure 4 shows the same configurations with self-similar traffic and using packet trains. For B=1 results for fixed feedback and SDL are almost the same. For B=2 the best algorithm is *minBusy* using SDL and the difference is bigger as the traffic load is smaller.

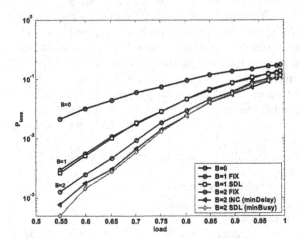

Figure 4. Performance of a medium sized switch simulated with self-similar trains. F=4, W=8.

Figure 5 compares self-similar performance assuming wagons. For B=1 results are almost identical and the best configuration for B=2 is *minBusy* using SDL.

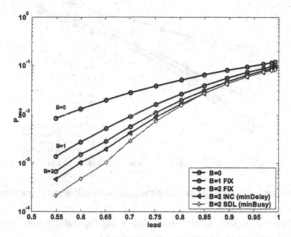

Figure 5. Performance of a medium sized switch simulated with self-similar wagons. F=4, W=8.

Figure 6 shows the performance of a large switch with F=8 and W=32 using Bernoulli traffic. For B=1 the fixed feedback structure behaves slightly better than the one based on SDL. For B=2 the difference is more notable and the best algorithm is *avoidOvr*, except for loads very near to 1, where the fixed delay structure is preferred. Figure 7 shows similar results for self-similar traffic.

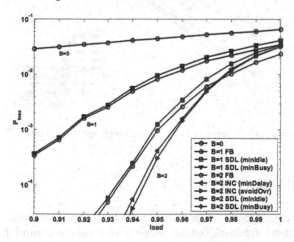

Figure 6. Performance of a large switch simulated with Bernoulli traffic. F=8, W=32.

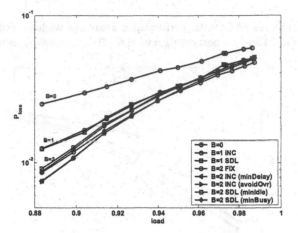

Figure 7. Performance of large switch simulated with self-similar wagons. F=8, W=32.

Figure 8 shows that the incremental delay configuration also outperforms the others when self-similar traffic trains are simulated.

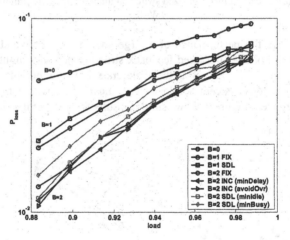

Figure 8. Performance of large switch simulated with self-similar trains. F=8, W=32.

5. CONCLUSION

Three different feedback buffer architectures have been examined:
1. Fixed delay feedback, which needs a very simple control for the switch fabric.
2. Incremental delay feedback, that by just using more fibre allows for improvement of the performance.

3. Switchable delay lines, dynamic devices that allow a finer control over the routing process.

··· Two main scenarios were simulated: a medium and a large switch. The results obtained show that for the medium sized switch and self-similar traffic, the switch with SDL feedback and using the *minBusy* strategy obtains better performance, especially for low loads. For the large switch the results show that incremental structures behave better than those based on SDL. The reason for this is that the SDL is not a very dynamic system; it switches the whole fibre and not each single wavelength and thus all packets that come into the SDL at the same time slot will be assigned the same delay. Therefore the bigger W is, the less dynamic the SDL will be and then the worse its performance will be.

ACKNOWLEDGMENTS

This work was done as part of the European IST-project STOLAS.

REFERENCES

[1] De Zhong W., Tucker R. S., "Wavelength Routing-Based Photonic Packet Buffers and Their Applications in Photonic Packet Switching Systems", *IEEE Journal of Lightwave Technology*, Vol. 16, No. 10, October 1998.

[2] R. Langenhorst et al., "Fiber Loop Optical Buffer", IEEE Journal of Lightwave Technology., Vol. 14, pp. 324-335, March 1996.

[3] M. T. Hill, A. Srivatsa, N. Calabretta, Y. Liu, H. de Waardt, G.D. Khoe, H.J.S. Dorren, "1 x 2 optical packet switch using all-optical header processing", *IEEE Electronic Letters*, vol. 37, pp. 774-775, 2001.

[4] C. Develder, M. Pickavet, P. Demeester, "Choosing an appropriate buffer strategy for an optical packet switch with a feed-back FDL buffer", in *Proc. ECOC*, 2002, Sep. 2002, Vol. 3, pp. 8.5.4.

[5] V. Paxson and S. Floyd, "Wide Area Traffic: The Failure of Poisson Modeling", *IEEE/ACM Transactions on Networking*, vol. 3, pp. 226-244, June 1995.

[6] M.E. Crovella and A. Bestavros, "Self-Similarity In World Wide Web Traffic: Evidence And Possible Causes", *IEEE/ACM Trans. Networking*, vol. 5, Dec. 1997.

[7] Chris Develder, Mario Pickavet and Piet Demeester, "Strategies for an FDL Based Feed-Back Buffer for an Optical Packet Switch with QoS Differentiation", *COIN 2002*, Cheju, Korea, July 2002.

[8] R. Jain and S. A. Routhier, "Packet Trains – Measurements and a New Model for Computer Network Traffic", *IEEE Journal on Selected Areas in Communications*, Vol. SAC-4, No. 6, pp . 986-995. September 1986.

[9] Chris Develder, Mario Pickavet, Piet Demeester, "On trains and wagons: switching variable length packets in a slotted OPS network", *COIN 2003*.

40GB/S WDM-MULTICASTING WAVELENGTH CONVERSION FROM 160GB/S OTDM SIGNAL

Yoshinari Awaji, Tetsuya Miyazaki, Fumito Kubota
National Institute of Information and Communications Technology, 4-2-1 Nukuikita, Koganei, 184-8795 Tokyo, Japan: yossy@nict.go.jp

Abstract: We demonstrated a wavelength conversion experiment from 160Gb/s:OTDM signal to 4λ x 40Gb/s:WDM signals with multicasting function by using four-wave mixing in a highly-nonlinear fiber. This technique was based on the OTDM/WDM converter using FWM and the multicasting function is easily available by aligning mutual timing of WDM clock pulses.

1. INTRODUCTION

Multi-channel OTDM de-multiplexers are indispensable for a practical implementation of OTDM networks. The multi-channel de-multiplexers can be classified two categories. One is a combination of several optical gate switches and splitters, which needs a lot of components [1-2]. Another type utilizes a parametric process between OTDM signal and multi-color clock pulses by using single nonlinear medium. This type can be also considered as OTDM/WDM converter. Several investigations about such converter have been reported [3-5]. We demonstrated 160/4x40Gb/s:OTDM/WDM converter by using four-wave mixing (FWM) in response to the increase of channel rate up to 40Gb/s, either [6].

By the way, considering a border node between OTDM network and WDM network, conventional OTDM/WDM gateway offers an aggregation function of peer to peer data links. For example, a communication channel assigned to one of tributary channel within OTDM signal is assigned to a specific wavelength path uniquely, and only one WDM port can receive a specific tributary channels. On the other hand, multicasting function of node can deliver same information to several wavelength paths, simultaneously. In this report, we demonstrated the

160/40Gb/s:OTDM/WDM converter with WDM-multicasting function, for the first time. It is able to supply higher functionality to OTDM/WDM gateway.

2. EXPERIMENT

2.1 Basic concept

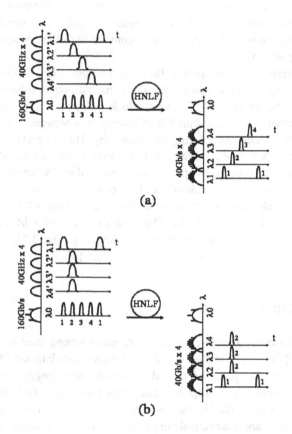

Figure 1. Proposed OTDM/WDM converter

(a) OTDM/WDM converter, full-DEMUX

(b) OTDM/WDM converter, multicasting

Figure 1 shows the principle of proposed OTDM/WDM converter. A 160Gb/s OTDM signal acts as pump light of FWM. 40GHz clock pulses on four wavelengths were wavelength-converted to four 40Gb/s data on different wavelength-channels as a result of FWM in highly-nonlinear fiber (HNLF). If four clock pulses are overlapped on each tributary channels of 160Gb/s:OTDM signal one-on-one, it can be said as full de-multiplexing or OTDM/WDM full-conversion (Fig. 1(a)). However, if some of clock pulses have same timing, the corresponding wavelength-channels share same information simultaneously (Fig. 1(b)). This multicasting configuration is easily changed simply by moving the clock pulse timing. That is to say, a wavelength channel can request to connect with a specific tributary channel ignoring the other wavelengths channels. This flexibility can be realized by using individual clock pulses in contrast with the configuration using single chirped pulse [3].

A pulse width of clock pulses have to be less than OTDM time slot (6.25ps@160Gb/s) to suppress an interchannel crosstalk come from neighbouring tributary channels, hence FWM between clock pulses on same time slot can interfere each other. Incidentally, the FWM interference between clock pulses was negligible for the full de-multiplexing condition (Fig. 1(a)) in previous experiment [6] because the each clock pulses on four wavelengths occupied time slots exclusively. Reducing the peak power of clock pulses is important issue to overcome the crosstalk. We demonstrated the worst case in following experiment. That is, all the clock pulses located at same time slot and one of tributary channel was distributed to four wavelengths. This condition can also be considered as broadcasting. Any channel number of multicasting, e.g. Fig. 1(b), can be easily obtained from following results and our previous work [6].

2.2   Setup and results

Figure 2 shows an experimental setup. A mode-locked laser diode (MLLD) oscillating at 10GHz was launched into 2km of dispersion-flattened fibre (DFF) to generate SC light. The pulse width and the center wavelength were 1.5ps and 1568nm, respectively. The average injected power was 20mW (estimated as 1.3W of peak power). Half of the SC light was quadrupled to generate 40GHz pulses using a PLC circuit and spectrum-slicing four wavelength channels at 1552.524 ~ 1557.363nm using a 200GHz spaced arrayed-waveguide-grating (AWG-1). The pass band of this AWG was 0.88nm of Gaussian (FWHM). As a result of spectrum-slicing, the pulse width in every channel was 4.3ps with Gaussian fitting. These four lights were adjusted each timing and polarization as same. Another part of the SC light was filtered at 1546.12nm, modulated by 10Gb/s PRBS: 2^7-1 data, and multiplexed up to 160Gb/s using a PLC multiplexer that had 200ps offset delays in each multiplexing segment to de-correlate the tributary channels. The bandwidth of this 160Gb/s light was adjusted by following cascaded 3nm band-pass filters (BPFs) resulting in a pulse width of 2.8ps. The residual chirp was compensated by the single-mode fibre (SMF).

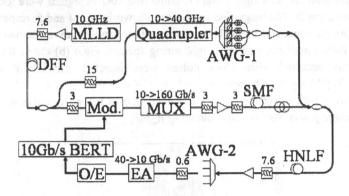

Figure 2. Experimental setup

The 160Gb/s light and four 40GHz lights were coupled and launched into 400m of HNLF to cause FWM. The nonlinear coefficient was $11.8W^{-1}km^{-1}$, the zero-dispersion wavelength was 1546.05nm, and the dispersion and dispersion slope at 1550nm were 0.209ps/nm/km and $0.038ps/nm^2/km$, respectively. The average power of 160Gb/s and 4x40GHz signals was +20 and +10dBm, respectively. The 4x40Gb/s idler signals generated at 1535.036 ~1539.766nm were filtered and spectrum-sliced using a 200GHz spaced AWG-2 (designated $\lambda1 \sim \lambda4$). To eliminate WDM-interchannel crosstalk, a 0.6nm BPF was applied. We evaluated the bit error quality at a 10Gb/s BERT set, hence we had to de-multiplex each 40Gb/s signal to 10Gb/s using an electro-absorption (EA) modulator after OTDM/WDM conversion.

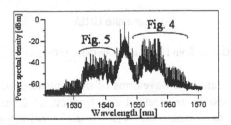

Figure 3. Spectral allocation

Center: 160Gb/s OTDM signal, Right: 4 x 40GHz clock, Left: 4λ x 40Gb/s converted data

Figure 3 shows the spectral allocation. The center peak was 160Gb/s:OTDM signal. In spite that the injected 160Gb/s signal into HNLF had restricted bandwidth due to the cascaded BPFs, the output 160Gb/s spectrum broadened because of intraband FWM. Four 40GHz clock pulses were located at longer wavelength than 160Gb/s signal, and the converted 40Gb/s signals were located at shorter wavelength. The magnified spectra are shown in Fig. 4 and 5, respectively. In Fig.4 (a), the spectra of input clock pulses on four wavelengths are shown in two different colors to distinguish the neighbouring spectra. Fig.4 (b) shows the FWM components occurred when clock pulses were injected into HNLF without 160Gb/s:OTDM signal. The spectral peak of FWM components caused by clock pulses were -30dB lower than the clocks itself due to the wider pulse width and lower average power than 160Gb/s pumping light.

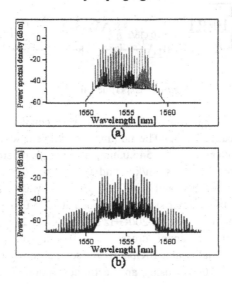

Figure 4. Spectra of clock pulses

(a)   Input into HNLF

(b) Output from HNLF clock without 160Gb/s signal

Figure 5 shows spectra and waveforms of the converted 4λ x 40Gb/s signals. The solid lines on spectra represented the location of 0.6nm BPF for every wavelength channels to eliminate the WDM interchannel crosstalk. The 40Gb/s waveforms were observed by optical sampling system. In spite of several causes of crosstalk, the eye opening was very clear. The variation of the pulse peak decreased when the timing of clock pulses were different (full de-multiplexing configuration), hence it must be caused by FWM interference between clock pulses. The clarification of this phenomenon is future issue.

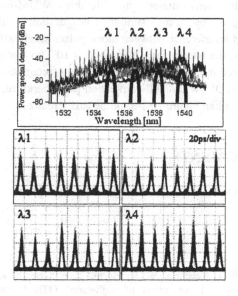

Figure 5. Spectra and waveforms of converted 4λ x 40Gb/s data

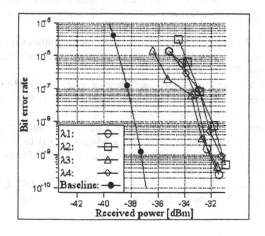

Figure 6. Measured bit error rates for the worst case

Finally, we measured bit error rates (BERs) at 10Gb/s as shown in Figure 6. We achieved the BER < 10^{-9} for every channel, and the worst cases of each wavelength channels and the 10Gb/s baseline were represented. The maximum MUX/DEMUX penalty was -5.8dB.

3.   CONCLUSIONS

We successfully demonstrated the 40Gb/s WDM-multicasting from 160Gb/s:OTDM signal by using similar configuration with OTDM/WDM converter. The FWM interference between clock pulses surely affect the converted waveforms, however, the bit error rate were below 10^{-9} for every channel. In this experiment, we showed four wavelengths multicasting (broadcasting), and we have already shown the OTDM/WDM full-conversion [6]. Therefore, any combination of channel selection as like Figure 1(b) could be easily realized.

ACKNOWLEDGMENTS

The authors wish to thank Mr. H. Sumimoto for his support.

REFERENCES

[1] M. L. Dennis, W. I. Kaechele, W. K. Burns, T. F. Carruthers, I. N. Duling III, "Photonic serial-parallel conversion of high-speed OTDM data," *IEEE Photon. Technol. Lett.*, vol. 12, pp. 1561–1563, November 2000.

[2] I. Shake, H. Takara, K. Uchiyama, I. Ogawa, T. Kitoh, T. Kitagawa, M. Okamoto, K. Magari, Y. Suzuki, T. Morioka, "160 Gbit/s full optical time-division demultiplexing using FWM of SOA-array integrated on PLC," *IEE Electron. Lett.*, vol. 38, pp. 37–38, January 2002.

[3] K. Uchiyama, T. Morioka, "All-optical time-division demultiplexing experiment with simultaneous output of all constituent channels from 100Gbit/s OTDM signal," *IEE Electron. Lett.*, vol. 37, pp. 642–643, May 2001.

[4] H. Sotobayashi, W. Chujo, K. Kitayama, "Photonic gateway: TDM-to-WDM-to-TDM conversion and reconversion at 40 Gbit/s (4 channels x 10 Gbits/s)," *JOSA B.*, vol. 19, pp. 2810–2816, November 2002.

[5] L. Rau, W. Wang, B-E. Olsson, Y. Chiu, H-F. Chou, D. J. Blumenthal, J. E. Bowers, "Simultaneous all-optical demultiplexing of a 40-Gb/s signal to 4 x 10 Gb/s WDM channels using an ultrafast fiber wavelength converter," *IEEE Photon. Technol. Lett.*, vol. 14, pp. 1725–1727, December 2002.

[6] Y. Awaji, T. Miyazaki, F. Kubota, "160/4x40Gb/s OTDM/WDM Conversion Using FWM Fibre," ECOC 2004, now submitting.

MULTIPLE WAVELENGTH CONVERSION FOR WDM MULTICASTING BY MEANS OF NON-LINEAR EFFECTS IN SOAs

G. Contestabile[1, 3], M. Presi[2, 3], and E. Ciaramella[1, 3]

[1] *Scuola Superiore Sant'Anna, Via Cisanello 145, 56124,Pisa (Italy)*
[2] *Dipartimento di Fisica, Università di Pisa (Italy),*
[3] *CNIT - Photonic Networks National Lab, Pisa (Italy), giampiero.contestabile@cnit.it*

Abstract: Using Four Wave Mixing in SOAs, we produce six simultaneous optical copies of an input NRZ data signal, all signals are compliant with a 200 GHz channel grid.

Index Terms: Multicasting, Nonlinear Optics, Semiconductor Optical Amplifier, Wavelength Conversion.

1.   INTRODUCTION

Future *transparent* all-optical networks may take serious advantage by the introduction of the optical multicast functionality [1]. Multicast is a well-known feature of IP protocol, used when one source sends the same information to several different destinations (note: multicast is similar to broadcast, the two corresponding to *one-to-many* and to *one-to-all* situations, respectively). Currently, multicast is implemented in IP digital routers, but the effectiveness of the all-optical network will further increase when actual optical multicast will be performed in the optical nodes.

At first, optical multicast will be required for circuit-switched data at fixed wavelength: in that case it could be easily implemented by means of power splitters. However the natural evolution will soon ask for the optical WDM multicast, i.e. multicast to different wavelengths.

300

WDM multicast could be achieved by means of a 1:N power splitter and N-1 wavelength converters, but clearly we would prefer a single device providing the N-1 wavelength converted copies simultaneously. A common demultiplexer can then be used for routing the channels on a wavelength basis.

Although wavelength conversion has been largely investigated in the past, there are very few results on optical multicast demonstrations. To the best of our knowledge, actual WDM multicast was demonstrated in [2], using a nonlinear SOA-based interferometer, in [3], using injection locking of a Fabry-Perot laser, in [4], using Cross Phase Modulation in a Dispersion Shifted Fiber (DSF) and, finally, in [5], with an electro-absorption modulator.

Here, we introduce a new technique for multicasting an input signal by exploiting multipump Four Wave Mixing (FWM) in a Semiconductor Optical Amplifiers (SOA).

2. OPERATING PRINCIPLE

Our scheme is a modified version of a FWM configuration that was proposed in [6] in order to obtain a flat wavelength conversion efficiency in SOAs. The scheme of this configuration is depicted in Fig. 1. A first pump (P1) has the same State Of polarization (SOP) as the incoming signal, while the other two (P2 and P3) have orthogonal SOP. Thanks to the orthogonal SOP scheme no efficient FWM contribution arises due to mixing of the pumps, as indeed modulation of gain and refraction index in the semiconductor medium is practically only due to the beating between S and P1. This modulation affects P2 and P3 and leads to the generation of two copolarized sidebands for both pumps, thus producing Ch. 1, 2, 5 and 6, respectively. Moreover, the usual FWM process produces the channel 3 signal.

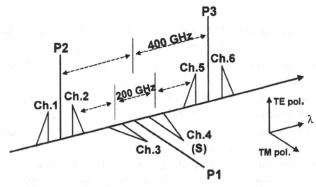

Fig.1: *Scheme of the interacting signals in the SOA and their SOPs. P1, P2 and P3 are the CW pumps. Ch.4 is the input signal S.*

3. EXPERIMENT

In this experimental demonstration, we produce six copies of an input 10 Gbit/s Non-Return-to-Zero (NRZ) signal S.

Our setup is illustrated in Fig. 2. The signal S, at λ_S=1554.3 nm, is generated by modulating a CW laser at λ_S, by means of a LiNbO$_3$ Mach-Zehnder intensity modulator driven by a 2^{31}-1 long PRBS sequence at 9.95328 Gbit/s (STM-64). The signal passes then in the multicast device where it is coupled together with the three local pumps. The orthogonal SOP between the pumps is accomplished by means of Polarization Controllers (PCs) and a Polarization Beam Splitter (PBS). The main experimental parameters are summarized in Table 1, reporting the wavelengths of the signals and pumps (2nd column), together with the input power (3rd column), output power (4th column), Optical Signal to Noise Ratio (5th column) and the single channel conversion efficiency (6th column) of all the generated signals.

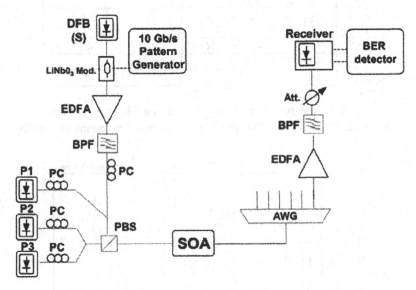

Fig.2: *Experimental setup. PC: Polarization Controller; PBS: Polarization Beam Splitter; BPF: Band Pass Filter; AWG: Array Wave-guide Grating.*

Moreover we outline that, due to the physics of the process [6], when the wavelength of P2 or P3 is changed, the corresponding couple of multicast signals is wavelength-shifted by the same amount, without any change in the conversion efficiency as long as the pumps wavelength lies within the gain curve of the SOA. This fact allows flexible selection of output wavelenths. When we set the values

302

of the pumps as in the Table I, all the six channels comply with an equally spaced 200 GHz frequency (see Fig. 3).

Properly balancing pumps and signal power levels, we optimized conversion efficiency and OSNR controlling deleterious effects that could arise either from spurious Cross Gain Modulation (XGM) or from in-band cross-talk due to high-order FWM products. Obtained optical powers and OSNRs of the various copies are greater than -22.5 dBm and 23 dB, respectively (Tab. 1).

Table 1

	λ(nm)	Input Power (dBm)	Output Power (dBm)	Estimated OSNR (dB)	Conversion Efficiency: P_{OUT}/P_S (dB)
Ch.1	1549.5		-17.3	28	-24.3
P2	1550.3	9.3			
Ch.2	1551.1		-22.5	23	-29.5
Ch.3	1552.7		-15.4	30	-22.4
P1	1553.5	9.8			
Ch.4 (S)	1554.3	7	3.7	48	-3.3
Ch.5	1555.9		-16.5	25	-23.5
P3	1556.7	8.9			
Ch.6	1557.5		-20.7	23	-27.7

To check the quality of the multicasted signals we performed BER measurements on each of them. Every channel is extracted by means of a WDM

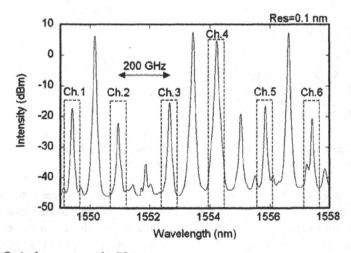

Fig.3: *Optical spectrum at the SOA output.*

demultiplexer and sent to a common STM-64 SDH receiver. The obtained Bit Error Rate (BER) curves are shown in Fig. 4. As can be seen, with respect to the input signal in back-to-back configuration, all channels exhibit a limited penalty (less than 2.5 dB for BER=10^{-9}). A slight penalty is partially due to degraded OSNR and, principally, to spurious XGM.

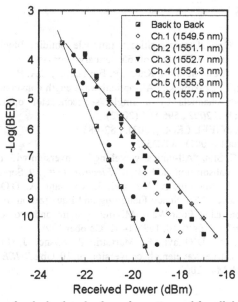

Fig.4: *BER curves for the back-to-back configuration and for all the converted signals.*

4. CONCLUSIONS

We demonstrated 1:6 all-optical WDM multicast of a 10 Gb/s NRZ signal. By exploiting a new scheme for Four Wave Mixing in Semiconductor Optical Amplifiers, we obtained six copies of an input signal, all compliant with a 200 GHz grid. Although in the present demonstration we used a polarization controller to match the SOP of the input signal and P1, our scheme can be made polarization-independent by using a more involved setup (analogous to what shown in [7] for the wavelength conversion). Finally, we outline that a higher number of multicast channels could be produced by the same scheme, using additional copolarized pumps and a SOA device with higher output saturation power.

ACKNOWLEDGMENTS

This work was partially supported by Marconi Communication SpA under a grant.

REFERENCES

[1] G. N. Rouskas, "Optical layer multicast: rationale, building blocks, and challenges," *IEEE Network*, Volume 17 Issue 1, 60-65, January-February 2003.

[2] J.L. Pleumeekers, J. Leuthold, M. Kauer, P. Bernasconi, C.A. Burrus, M. Cappuzzo, E. Chen, L. Gomez, E. Laskowski, "All-optical wavelength conversion and broadcasting to eight separate channels by a single semiconductor optical amplifier delay interferometer," *OFC 2002* , 596-597 (2002).

[3] Chow et al. Proc. of IEEE LEOS (2003), 682-683.

[4] Rau et al. OFC 2001 (2001), WDD52.

[5] K.K. Chow and C. Shu, "All-optical wavelength conversion with multicasting at 6x10 Gbit/s using electroabsorption modulator," *Electron. Lett.*, 39, September 2003

[6] G. Contestabile, F. Martelli, A. Mecozzi, L. Graziani, A. D'Ottavi, P. Spano, R. Dall'Ara and J. Ecker, "Efficiency Flattening and Equalization of Frequency Up- and Down-Conversion using Four-Wave Mixing in Semiconductor Optical Amplifiers," *IEEE Photon. Technol. Lett.*, 10, 1398-1401, October 1998.

[7] G. Contestabile, A. D'Ottavi, F. Martelli, P. Spano, J. Eckner, "Broadband, polarization-insensitive wavelength conversion at 10 Gb/s," *IEEE Photon. Technol. Lett.*, 14, 666-668, May 2002.

PART B1:

TRANSMISSION SYSTEM

OPTIMAL SPAN LENGTH DETERMINATION IN TRANSMISSION SYSTEMS WITH HYBRID AMPLIFICATION

J.D. Ania-Castañón [1], I.O. Nasieva [1], S.K. Turitsyn[1], C. Borsier[2], and E. Pincemin[2,]
[1] Aston University, School of Engineering and Applied Science, Birmingham B4 7ET, United Kingdom, e-mail: aniacajd@aston.ac.uk
[2] France Telecom R&D, 2 Aven. Pierre Main, 22307 Lannion, France, e-mail: erwan.pincemin@francetelecom.com

Abstract: The existence of an optimal span length for 40 Gbit/s WDM transmission systems with hybrid Raman/EDFA amplification is demonstrated. Optimal lengths are obtained for specific amplifier configurations and different fibre arrangements based on SSMF/DCF and SLA/IDF implementation, using a simple nonlinearity management theory.

1.   INTRODUCTION

The recent availability of reliable high power laser pumps has made possible the comeback of distributed Raman amplification (DRA) in DWDM transmission systems [1-3]. Compared to traditional lumped amplifier schemes, the DRA improves significantly the optical signal to noise ratio (OSNR). The OSNR margins which are released by the implementation of DRA can be used for extending the transmission distance and/or decreasing the signal power injected into the fibre span (thus limiting the impact of nonlinearities). Combined with dispersion management and Erbium-doped fibre amplification (EDFA), DRA can be used to better control the signal power evolution inside the amplification spans and, thus, nonlinear effects along the optical line, effectively performing "nonlinearity management" [4, 5]. In this case, the properties of the fibre used (Raman gain, attenuation, effective area, Rayleigh backscattering coefficient) and

308

the design of the dispersion map have all to be taken into account when configuring the amplification scheme.

The performance of a given system depends on the trade-off between the requirements of a high OSNR and low nonlinear impairments. Amplifier spacing is a very important design parameter as the amount of noise introduced into the system depends on the span length. Longer amplifier spans are desirable for economical system design. However, in general, they lead to faster OSNR degradation. As we will show in this paper, DRA and a proper nonlinearity management enable system design with a longer optimal amplifier span <u>without degradation of the output OSNR.</u> More specifically, by applying the approach recently developed in [4, 5] we investigate here the impact of the periodic span length on the optimal configuration of the amplification scheme in 40 Gbit/s WDM transmission systems.

2.   THEORY AND SYSTEM CONFIGURATIONS

Under the assumption that nonlinearities always contribute to the degradation of the performance of the transmission system, nonlinear phase shift (NPS) can be considered as a general measure of the nonlinear impairments [4-8]. The optimal system configuration can then be determined by performing a conditional minimization of the NPS under a fixed OSNR, or vice versa, a maximization of the OSNR under a fixed NPS. In this paper we perform numerical modelling using the average power equations for the Raman pump, signal and noise, in order to find the optimal parameters (gain split between the amplifiers and span length) that allow for the minimization of the NPS under a fixed OSNR. This approach ensures that all important effects, including double Rayleigh backscattering (DRS) noise and pump depletion, are accounted for [5].

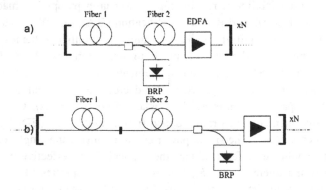

Figure 1. System configurations considered

We focus here on two basic configurations. In both cases, a two-step dispersion map with hybrid Raman/EDFA amplification is considered, but the position of the Raman amplifier is varied. In configuration a), the backward Raman pump is placed immediately after the section of positive dispersion fibre (fibre 1) and an EDFA, with a typical noise figure of 4.5 dB, is used for post-amplification at the end of the span. In configuration b), both amplifiers are placed together at the end of the periodic transmission cell. The combined gain from the two amplifiers compensates exactly for the attenuation of the periodic cell. We consider WDM transmission at 40 Gbit/s rate with channels equally spaced (by 100 GHz) and symmetrically distributed at both sides of 1550 nm. Two different dispersion maps are considered for configuration a): one based on standard single-mode fibre (SSMF) with dispersion-compensating fibre (DCF), and another one based on super large area fibre (SLA) with inverse dispersion fibre (IDF). For the configuration b), only the SLA/IDF is considered. A useful parameter to characterize hybrid Raman/EDFA amplification is the ratio η, defined as the quotient between the on-off gain provided by the DRA and the total gain, both in dB. The length of both positive and negative dispersion fibres within the span are considered as variables, with the length of the negative dispersion fibre always automatically adjusted to compensate exactly for dispersion at the end of each periodic cell. The total transmission distance is fixed to 900 km over SSMF for the first pair of fibres, or 900 km total length over SLA+IDF, for the second one. The number of spans varies with the cell length.

3.   RESULTS AND DISCUSSION

Figure 2 displays NPS in a contour plot versus the length L of the span and the gain ratio η, for a fixed output OSNR of 22 dB (the bandwidth for the noise measurement is fixed at 1 nm) in configuration a), for both fibre pairs. We can observe that there is a clear optimal length for the periodic span, for which the NPS is minimal, which in this case is about 50 km for the SSMF/DCF case, and about 90 km (60 km of SLA) for the SLA/IDF scheme. This optimal cell length can be understood as the one that allows us to find the best balance between the two basic variants:

- short cell regime, in which the signal is transmitted in quasi-lossless conditions, so high nonlinearities are produced from even relatively low input powers injected into the spans.
- long cell regime, in which the long distance between amplifiers helps reducing NPS, but leads to an increase of the amplified spontaneous emission (ASE) noise, so the input signal power has to be increased in return, leading to higher nonlinearities.

310

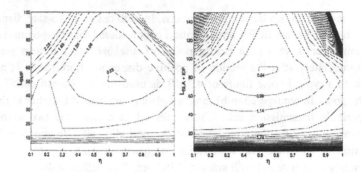

Figure 2. NPS vs. L (in km) and η for a fixed OSNR of 22 dB after 900 km total
transmission distance through SSMF/DCF (left) and SLA/IDF (right) with configuration a).

It can also be derived from figure 2 that the optimal η varies slightly with the
cell length, becoming smaller for long spans. As the span length grows, so does the
Raman pump power and gain, and with it the contribution of DRS noise (due to a
worse distribution of the gain within the span). In this situation, the gain fraction
from the Raman amplifier has to be reduced in order to recover the best possible
performance. On the other hand, the relative importance of finding the optimal η is
increased together with the length of the periodic cell, going from superfluous for
the shortest spans, to crucial for the case of 100 km SSMF section.

It is also important to study how dependent the optimal span length that
minimizes the nonlinear impairments is on the targeted output OSNR. By varying the
required OSNR, we find that the optimal cell length variation is negligible so it can
be considered independent of the output OSNR requirement. On the other hand, the
optimal η changes slightly, in particular for long span lengths. This phenomenon
can be explained by the necessity to avoid the penalties induced by DRS noise for
long cells.

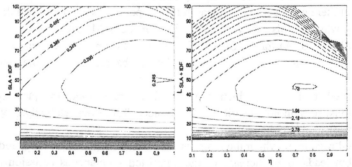

Figure 3. NPS vs. L (in km) and η for a fixed OSNR of 22 dB (left) and 30 dB (right) after
900 km total transmission distance SLA/IDF with configuration b).

Finally, figure 3 shows the results obtained for configuration b): SLA/IDF used with amplifiers located at the end of the transmission cell. The change of configuration has a clear effect on the optimal transmission length, which is reduced to about 50 km of combined SLA+IDF. For low fixed output OSNR, the optimal amplifier configuration corresponds to values of η close to 1 (i.e. full Raman amplification), but when OSNR or span length are increased, so is the Raman gain and with it the effect of DRS, so the optimal configuration requires a higher participation from the EDFA to limit the DRS impact. Finally, we can also observe that, for the same fixed output OSNR, nonlinear impairments are in general lower in scheme b) than in scheme a).

4.   CONCLUSION

We demonstrate the existence of the optimal span length for 40 Gbit/s WDM transmission systems considering different fibre mapping arrangements and hybrid Raman/EDFA amplification schemes. The results of the optimization are basically independent of the desired output OSNR, so the optimal span length determined for each system through this method is applicable to a wide range of input signal powers.

REFERENCES

[1]   Christian Rasmussen, Tina Fjelde, Jon Bennike, Fenghai Liu, Supriyo Dey, Benny Mikkelsen, Pavel Mamyshev, Peter Serbe, Paul van der Wagt, Youichi Akasaka, David Harris, Denis Gapontsev, Vladlen Ivshin, and Peter Reeves-Hall, "DWDM 40G transmission over trans-Pacific distance (10,000 km) using CSRZ-DPSK, enhanced FEC and all-Raman amplified 100 km UltraWave fiber spans", OFC 2003, PD18, 2003

[2]   T. Tsuritani, K. Ishida, A. Agata, K. Shimomura, I. Morita, T. Tokura, H. Taga, T. Mizuochi, and N. Edagawa, "70GHz-spaced 40x42.7Gbit/s transmission over 8700km using CS-RZ DPSK signal, all-Raman repeaters and symmetrically dispersion-managed fiber span", OFC 2003, PD23, 2003

[3]   M. Mehendale, M. Vasilyev, A. Kobyakov, M. Williams and S. Tsudal, "All-Raman transmission of 80×10 Gbit/s WDM signals with 50 GHz spacing over 4160 km of dispersion-managed fibre" Electron. Lett., 38, p. 648, 2002

[4]   S.K. Turitsyn, M.P. Fedoruk, V.K. Mezentsev and E.G. Turitsyna, "Theory of optimal power budget in quasi-linear dispersion-managed fibre links", Electron. Lett., 39 (2003), p. 29

[5]   J. D. Ania-Castañón, I. O. Nasieva, N. Kurukitkoson, S. K. Turitsyn, C. Borsier and E. Pincemin, "Nonlinearity management in fiber transmission systems with hybrid amplification", Opt. Comm., 233, p. 353, 2004

[6]   J.-C. Antona, S. Bigo and J.-P. Faure, "Nonlinear cunulated phase as a criterion to assess performance of terrestrial WDM systems", OFC 2002, WX5, p. 365, 2002

312

[7]   V. E. Perlin and H. G. Winful, "On trade-off between noise and nonlinearity in WDM systems with Raman amplification", Proc. OFC 2002, WB1, p. 178, 2002

[8]   A. Kobyakov, M. Vasilyev, S. Tsuda, G. Giudice and S. Ten, "Raman noise figure in dispersion-managed fibers", Proc. ECOC 2002, P1.13, 2002

SEPARATE EVALUATION OF NONLINEARITY-DUE Q PENALTIES IN LONG-HAUL VERY DENSE WDM OPTICAL SYSTEMS

Livio Paradiso[1], Pierpaolo Boffi[1], Lucia Marazzi[1], Nicola Dalla Vecchia[1], Massimo Artiglia[2] and Mario Martinelli[1,3]

[1]CoreCom, via G. Colombo 81 20133 Milan Italy, email: paradiso@corecom.it
[2]Pirelli Cavi S.p.A.,viale Sarca 222 Milan Italy, email: massimo.artiglia@pirelli.com
[3]Politecnico di Milano,piazza Leonardo Da Vinci, email: martinelli@corecom.it

Abstract: Nonlinearity-due Q penalties are experimentally evaluated on a 2000-km, 33-GHz spaced DWDM system, dependently on system length and channel spacing. SPM, FWM and XPM, which is found to be the major constrain, are separately addressed.

1.   INTRODUCTION

In long-haul DWDM systems inter-channel nonlinear effects represent a limit in system capacity increase. While NZDS fibers are designed to prevent Four Wave Mixing (FWM), Cross-Phase Modulation (XPM) still represents the major obstacle in increasing transmission systems capacity [1].

XPM was studied both theoretically [1,2,3,4] and experimentally with pump-probe [5,6,7,8] and multichannel [9] techniques. Pump-probe scheme allows evaluating separately Self-Phase Modulation (SPM) and XPM impairments. In a WDM system, multichannel measurements account for the global effect of the interfering channels, but it is difficult to quantitatively isolate penalties due to XPM only from those due to other 3rd-order nonlinearities.

In this paper nonlinearity-due Q penalties in a DWDM system are experimentally investigated using a 39 spans optical fiber line 2000-km long. For the first time, to the best of our knowledge, SPM, FWM and XPM penalties are separately addressed in a system with 33-GHz channel spacing. Results are described and discussed as a function of optical path length and per-channel input power. XPM-due impairment appears to overweight both SPM and FWM.

2. EXPERIMENTAL SETUP

The employed DWDM system is 2000 km long, with 64 power-equalized, 33-GHz spaced, polarization scrambled channels, in the 1543-1560 nm window. All channels are 10 Gbit/NRZ-IMDD. The total launch power is 12.8 dBm. The system employs 55-km fiber spans with D= -2.82 ps/(nm km) at 1550 nm and a 55-km Step index fiber for in-line dispersion compensation every 6 spans. Residual dispersion is fully compensated at the receiver end. This periodical dispersion map allows evaluating penalties at different system lengths. Nonlinear effects are excited at different levels by progressively increasing the per-channel power, by turning off some of the propagating channels (channel count ranges from 64 to 18).

SPM is analyzed first: a probe channel operating at 1551.25 nm propagates without the 32 neighbouring channels, which are turned off (see the optical spectrum in the inset of Figure 1). In this situation only SPM takes place: we experimentally verified that a given channel is not impaired by other channels farer than 133 GHz. By progressively turning off the remaining channels, per-channel power and probe Optical Signal to Noise Ratio (OSNR) is increased and the received Q factor is measured.

Pump and probe measurements are then performed by introducing a pump channel 100, 66 or 33 GHz away from the probe, thus inducing on it XPM additional penalties. Again, received Q factor is measured, while per-channel power is varied by changing the total channel number. EDFA gain can be considered flat over the band relevant for nonlinear effects, thus minimizing errors in evaluating neighbouring channels power. FWM penalties are not observed owing to the fact that only two channels interact with each other. When all neighbouring channels are turned on, FWM adds to SPM and XPM. Again Q measurements are performed as a function of per-channel power and received OSNR: the channels at the far end of the spectrum are turned on/off, leaving the neighbours on. By combining experimental results, it is possible to separately account for nonlinear effects. FWM penalty is found by subtracting SPM and XPM contributions from the total nonlinearity-due impairments.

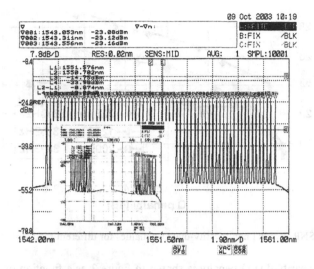

Figure 1. Power-equalized DWDM 64 CH-33 GHZ spaced spectrum and pump-probe measurement example (in the inset).

3. MEASUTEMENTS DISCUSSION

Figure 2 shows the total nonlinearity-due Q penalties for different system lengths and 33 GHz channel spacing. OSNR is chosen as reference parameter. A 1.5 dB Q penalty is found for 22.5 dB OSNR between 1257 and 2000 km propagation. For higher OSNR (higher per-channel powers) it is possible to compare all the three considered system lengths: total nonlinearity-due Q penalties grow more than linearly with total system length. When residual link dispersion is not compensated, performances worsen and a further Q penalty is added. Differently from FWM, XPM is influenced both by the dispersion compensation scheme [4,8] and by the residual dispersion value, because residual dispersion allows cumulated nonlinear phase to be further converted into intensity noise. Experiments carried out with a residual dispersion value of 557 ps/(nmkm) lead to an added Q penalty of 1 dB.

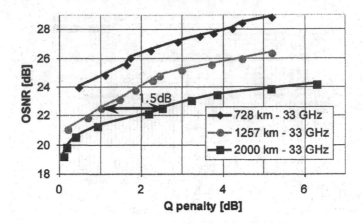

Figure 2. OSNR vs. total nonlinearity-due Q-penalties for different total system length.

The nonlinearity-due Q penalty is shown in figure 3 as a function of input per-channel power: XPM, FWM and SPM contributions are separately addressed. With 33 GHz channel spacing XPM-due impairment overweighs both SPM and FWM. When channel spacing increases, the amount of XPM, FWM and SPM becomes similar: in particular with 100 GHz channel spacing SPM represents the most important impairment in the 2000 km system, while XPM always produces a Q penalty higher than FWM.

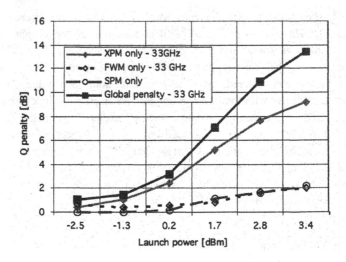

Figure3. Contribution of XPM, FWM and SPM to Q-penalties, for 2000 km system length.

The system under test is NZDS-fiber based and the dispersion map, typical of submarine systems, is designed to limit FWM penalties. Channels experience a significant walk-off before compensation, which is done every 6 spans, thus representing a good trade-off also for XPM penalties reduction [4,8]. At 33 GHz channel spacing, because of the relatively low fiber dispersion (D= −2.8 ps/(nmkm)), FWM still produces over 1 dB Q penalty, as high as SPM, for per-channel launch power over 1.7 dBm. XPM proves to be the main nonlinear system impairment when increasing the channel density (see Figure 3).

4. CONCLUSIONS

Nonlinearity-due system Q penalties in DWDM multispan systems are experimentally analysed as a function of system length and channel spacing. For the first time to our knowledge 33 GHz spacing is addressed for DWDM systems. SPM, FWM and XPM impairments are separately measured. At 33 GHz spacing XPM proves to be the most severe impairment, whereas at 66 and 100 GHz both XPM and FWM present inter-channel penalties <1 dB for relatively high powers, and the most limiting factor in increasing per-channel power and then span length is SPM, as reported in previous works [5].

REFERENCES

[1] S.Ten, K.M. Ennser, J.M. Grochocinski, S.P. Burtsev, V.L. daSilva, "Comparison of four-wave mixing and cross phase modulation penalties in dense WDM systems", in Proceedings of Optical Fiber Conference 1999, ThC4-1/43

[2] A.V.T. Cartaxo, "Cross-Phase Modulation in Intensity Modulation Direct Detection WDM Systems with Multiple Optical Amplifiers and Dispersion Compensators", J. Light. Tech. 17, 1999, pp. 178-190

[3] S. Betti and M. Giaconi, "Analysis of the Cross-Phase Modulation Effect in WDM Optical Systems", IEEE Phot. Tech. Lett. 13, 2001, pp. 43-45

[4] Bo Xu and Maité Brandt Pearce, "Comparison of FWM and XPM-Induced Crosstalk Using the Volterra Series Transfer Function Method", J. Light. Tech. 21, 2003, pp. 40-53

[5] Sebastien Bigo, Giovanni Bellotti and Michel W. Chbat, "Investigation of cross-phase modulation limitations on 10 Gbit/s transmission over various types of fiber infrastructures", IEEE Phot. Tech. Lett. 11, 1999, pp. 605-606

[6] H.J. Thiele, R. I. Killey and P. Bayvel, "Investigation of XPM Distortion in Transmission over Installed Fiber", IEEE Phot. Tech. Lett. 12, 2000, 669-671

[7] Livio Paradiso, Pierpaolo Boffi, Lucia Marazzi, Giandomenico Pozzi, Massimo Artiglia and Mario Martinelli, "Electrical measure of Cross-Phase Modulation impact in WDM optical Systems", in Proceedings of Conference in Laser and Electro Optics, 2003, CFJ4

[8]   C. Fuerst, C. Scheerer, G. Mohs, J. P. Elbers, C. Glingener, "Influence of the dispersion map on limitations due to cross-phase modulation in WDM multispan transmission systems", Proceedings of Optical Fiber Conference, 2001, MF4-1

[9]   C. Caspar, K. Habel, N. Heimes, M. Konitzer, M. Malach, H. Özdem, M.Rohde, F. Schmidt and E.-J. Bachus, "Penalties through XPM crosstalk in a switched long haul standard fiber WDM system based on normalized transmission sections", in Proceedings of Optical Fiber Conference, 2001, WI5-1

SUPPRESSION OF TRANSIENT GAIN EXCURSIONS IN EDFA'S

Comparison of Multiplicative and Additive Schemes for Combining Feedforward and Feedback Blocks

Mladen Males, Antonio Cantoni, John Tuthill
Western Australian Telecommunications Research Institute, 39 Fairway,CRAWLEY WA 6009, Australia, email: mmales@watri.org.au

Abstract: In this paper a comparative study of two pump control schemes for suppressing transient gain excursions in EDFA's is presented. The first control scheme is based on the traditional feedforward/feedback control, while the second scheme is new and uses multiplication to combine the feedforward and feedback blocks. The controller parameters are designed using linearised (small-signal) EDFA model derived from the nonlinear state-space model of Bononi and Rusch. When the controller design is applied to the nonlinear EDFA plant, the new scheme is shown to display some important performance improvements over the traditional scheme.

1.   INTRODUCTION

Erbium-doped fibre amplifiers (EDFA's) are widely used in multi-channel optical communications systems based on the wavelength-division multiplexing (WDM) technology. The EDFA's are usually operated in deep saturation so that high output powers are achieved. As a consequence, the gain provided by the EDFA to each channel is susceptible to changes in the total input signal power. Transient gain excursions in the surviving channels adversely affect the quality of service that the optical network operators can guarantee [1,2], and these unwanted gain excursions need to be mitigated. In literature, there is a large body of work reported on the transient control of EDFA's (see for example [3–8]). The approaches taken

320

generally fall into one of the following three groups: pump control [3-5], link control [6] and all-optical control [7,8]. In the work reported in this paper, two schemes for implementing the transient control of EDFA's based on the pump control approach are compared. Both schemes use the closed-loop control architecture of Figure 1 but the difference between the two schemes is in the type of the pump-control block in Figure 1. The two types of the pump-control block are:

- Multiplicative type, where $P_p^{in}(t) = U_{ff}(t)\left(1 + U_{fb}(t)\right)$

- Additive type, where $P_p^{in}(t) = U_{ff}(t) + U_{fb}(t)$

The additive pump-control block represents the traditional way of combining the feedforward and feedback blocks in the transient control of EDFA's (see for example [3]). The transient control of EDFA's using the closed-loop architecture with the multiplicative pump-control block is investigated for the first time in this paper.

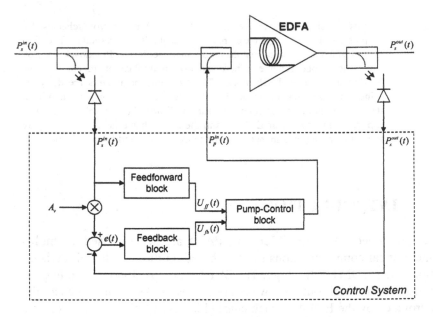

Figure 1. Using measurements of the total input signal power and the total output signal power, the control architecture produces the necessary pump power to maintain the EDFA gain constant. A_v is the desired gain, $P_s^{in}(t)$ and $P_s^{out}(t)$ are the total input and output signal powers, respectively, and $P_p^{in}(t)$ is the EDFA's input pump power.

The feedforward block in Figure 1 realises the following function:

$$U_{ff}(t) = K_{ff} P_s^{in}(t) + O_{ff} \tag{1}$$

For each input signal power there is a corresponding pump power $P_p^{in}(t) = K_{ff}^* P_s^{in}(t) + O_{ff}^*$ that maintains the EDFA gain constant (this fact was utilised in the early successful implementation of the feedforward approach in [5]). The drawback of the feedforward approach is that in practical implementations the exact feedforward parameters K_{ff}^* and O_{ff}^* depend on the EDFA's operating environment. The chosen feedforward parameters K_{ff} and O_{ff} are thus unlikely to produce the exact required pump power [3], and some form of feedback is necessary to provide a corrective action. The feedback block in Figure 1 contains a proportional-integral (PI) controller:

$$U_{fb}(t) = K_p e(t) + K_i \int_0^t e(\tau) d\tau \tag{2}$$

The PI controller reduces the error $e(t)$ to zero in steady state (this is achieved by the integral action [9]) and can be designed to minimise the transient overshoots and undershoots in the output signal power response.

The nonlinear blocks of the closed loop of Figure 1 are the feedforward block, pump-control block (when it is of the multiplicative type) and the EDFA itself. The controller parameters K_{ff}, O_{ff}, K_p and K_i are designed based on linear approximation of the closed-loop system of Figure 1. The controller design is then applied to the nonlinear closed-loop system and the performance of the two schemes is compared.

In the remainder of this report, the design approach is presented for the closed-loop system that includes the multiplicative pump-control block. A similar analysis can also be done for the system with the additive pump-control block but is not included here. Instead, important differences between the two systems are highlighted. An important assumption in the modelling of the EDFA is that the EDFA gain does not vary significantly with wavelength, and hence the total input and output signal power measurements yield sufficient information to keep the gain of each individual channel constant.

2.　　LINEARISATION OF LOOP COMPONENTS

Under the assumption that the spectral dependence of the EDFA gain is minimal, the EDFA can be modelled by the following nonlinear state-space model [10]:

$$\frac{dr}{dt} = -\frac{r}{\tau} + \left(1 - e^{B_s r - A_s}\right) P_s^{in}(t) + \left(1 - e^{B_p r - A_p}\right) P_p^{in}(t)$$

$$P_s^{out}(t) = P_s^{in}(t) e^{B_s r - A_s} \tag{3}$$

$P_s^{in}(t)$ and $P_s^{out}(t)$ are the total input and output signal powers, respectively, $P_p^{in}(t)$ is the input pump power, r and τ are the total number and the mean lifetime of excited erbium ions, respectively, and B_s, B_p, A_s and A_p are dimensionless constant parameters.

When forming linear approximations of the feedforward block, pump-control block and the EDFA, the input, state and output variables of these blocks are treated as small perturbations around their steady-state values (i.e. a variable $x(t)$ is treated as $x_0 + \delta x(t)$). The linearisation procedure follows that presented in [9], and involves keeping only the linear terms from the Taylor series expansion of nonlinear functions. The Laplace-domain linearised model for the nonlinear blocks of the closed loop of Figure 1 is given below.

$$\delta P_s^{out}(s) = \frac{P_{s0}^{in} e^{B_s r_0 - A_s} B_s \left(1 - e^{B_p r_0 - A_p}\right)}{s + 1/\tau + P_{s0}^{in} e^{B_s r_0 - A_s} B_s + P_{p0}^{in} e^{B_p r_0 - A_p} B_p} \delta P_p^{in}(s)$$

$$+ e^{B_s r_0 - A_s} \frac{s + 1/\tau + P_{s0}^{in} B_s + P_{p0}^{in} e^{B_p r_0 - A_p} B_p}{s + 1/\tau + P_{s0}^{in} e^{B_s r_0 - A_s} B_s + P_{p0}^{in} e^{B_p r_0 - A_p} B_p} \delta P_s^{in}(s)$$

$$\Box \frac{\alpha_0}{s + b_0} \delta P_p^{in}(s) + \frac{a_1 s + a_0}{s + b_0} \delta P_s^{in}(s)$$

$$\Box G_{sp}(s) \delta P_p^{in}(s) + G_{ss}(s) \delta P_s^{in}(s)$$

$$\delta U_{ff}(s) = K_{ff} \delta P_s^{in}(s)$$

$$\delta P_p^{in}(s) = U_{fb0} \delta U_{ff}(s) + U_{ff0} \delta U_{fb}(s)$$

$$= \left(\frac{K_{ff}^* P_{s0}^{in} + O_{ff}^*}{K_{ff} P_{s0}^{in} + O_{ff}} - 1\right) \delta U_{ff}(s) + \left(K_{ff} P_{s0}^{in} + O_{ff}\right) \delta U_{fb}(s) \tag{4}$$

3. CHARACTERISTICS OF LINEARISED CLOSED LOOP

The linear approximation of the closed loop in Figure 1 is shown in Figure 2, where the transfer function of the PI controller is $C(s) = K_p + K_i/s$.

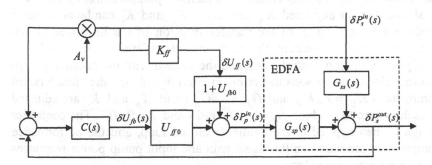

Figure 2. Linear approximation of the closed loop in Figure 1

The transfer function between the input signal perturbation and the output signal perturbation is:

$$T(s) = \frac{\delta P_s^{out}(s)}{\delta P_s^{in}(s)} = A_v \frac{s^2 + \left(b_0 + \dfrac{K_{ff}^* \alpha_0}{A_v}\left(\dfrac{\left(O_{ff}^*/K_{ff}^*\right) - \left(O_{ff}/K_{ff}\right)}{P_{s0}^{in} + \left(O_{ff}/K_{ff}\right)}\right)\right)s}{s^2 + \left(b_0 + \alpha_0 U_{ff0} K_p\right)s + \alpha_0 U_{ff0} K_i}$$

$$+ \frac{\alpha_0 U_{ff0} K_p s + \alpha_0 U_{ff0} K_i}{s^2 + \left(b_0 + \alpha_0 U_{ff0} K_p\right)s + \alpha_0 U_{ff0} K_i} \quad (5)$$

Whenever $(O_{ff}/K_{ff}) = (O_{ff}^*/K_{ff}^*)$, which occurs either when there is exact knowledge of the ideal feedforward parameters or when the actual feedforward parameters are in error by an equal proportion, the poles and zeros of $T(s)$ coincide and the transfer function reduces to $T(s) = A_v$. This leads to an interesting observation, albeit of limited practical value, that the perfect gain clamping can be achieved even if there are significant errors in the feed-forward parameters, as long as these parameters are erroneous by the same proportion. This is not the case when the pump-control block of Figure 1 is of additive type, as it can be shown that the perfect gain clamping is only obtained when the ideal feed-forward parameters are known exactly.

4. CONTROLLER DESIGN AND PERFORMANCE EVALUATION

For any fixed K_{ff} and O_{ff}, sufficient degrees of freedom exist in the controller to allow K_p and K_i to achieve any desired natural frequency ω_n and damping ξ of the second-order transfer function (5). Similarly, for the closed-loop system that contains the additive pump-control block, it can be shown that for any fixed K_{ff} and O_{ff}, K_p and K_i can be adjusted to produce any ω_n and ξ of the transfer function of the linearised closed loop. In order to compare the performances of the system with the multiplicative pump-control block and the system with the additive pump-control block, a reasonable error is introduced in the feedforward parameters ($K_{ff} = 2.2K_{ff}^*$ and $O_{ff} = 0.8O_{ff}^*$) and K_p and K_i are adjusted in each system to produce $\omega_n = 1 \times 10^{10}$ rad/s and $\xi = 0.707$. The controller design is then applied to the nonlinear closed loop, and the performance comparisons are based on the signal gain and input pump power responses of the nonlinear closed loop.

From Figure 3, the closed-loop system that uses the multiplicative pump-control block is seen to be more effective in minimising the transient gain excursions. The gain excursions are minimal in spite of reasonable input signal power changes and errors in the feedforward parameters.

From Figure 4, it can be seen that when the additive pump-control block is used, the controller can request a negative pump power during the transient period. As this is not physically possible, an additional element is added to the system that clamps the EDFA's input pump power at zero anytime the controller requests a negative pump power. A detrimental effect of the pump power clamping on the transient response is seen in Figure 3, as the largest overshoot/undershoot is observed for the case when the pump power is clamped at zero during the transient period. From Figure 4, it appears that the system with the multiplicative pump-control block is much less likely to produce a negative pump power requirement (compare the curves in Figure 4 for $P_s(\text{new}) = 0.3\text{mW}$).

In Figure 4, the system with the multiplicative pump-control block is also seen to produce smaller transients in the input pump power. This last property is very desirable in practical implementations, as the pump lasers may not be able to produce the high initial powers that the system with the additive pump-control block requests.

From Figure 3 and Figure 4, it can be seen that a close agreement is achieved between the responses of the nonlinear and linearised closed loops. This observation justifies the approach taken to design the controller by linearising the closed loop first.

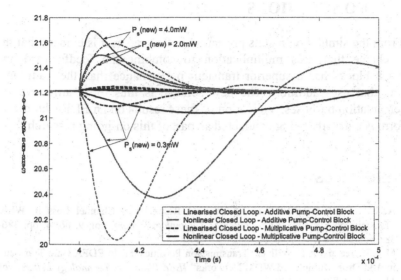

Figure 3. Transient gain excursions after the input signal power is abruptly switched from 1.0mW to a new value. When the pump-control block is multiplicative, the curves for $P_s(\text{new}) = 4.0\text{mW}$ and $P_s(\text{new}) = 2.0\text{mW}$ are almost indistinguishable on this plot.

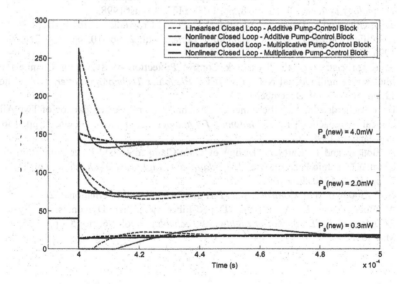

Figure 4. Transient input pump power adjustments after the input signal power is abruptly switched 1.0mW to a new value. When the pump-control block is additive, the curve for $P_s(\text{new}) = 0.3\text{mW}$ reveals that the controller requested a negative pump power (when this happened, an in-loop limiter clipped the input pump power at zero).

326

5.   CONCLUSIONS

From the simulation results presented in this paper, it is evident that the new scheme that uses multiplication to combine the feedforward and feedback blocks has a superior transient performance than the traditional scheme that uses addition. The new scheme is thus more attractive for implementation purposes. Validation of these results on an EDFA hardware platform is currently being conducted as part of this on-going research.

REFERENCES

[1]   A. K. Srivastava et al. "EDFA Transient Response to Channel Loss in WDM Transmission System" *IEEE Photonics Technology Letters*, vol. 9, no. 4, pp. 386–388, March 1997.

[2]   M. I. Hayee and A.E. Willner "Transmission Penalties due to EDFA Gain Transients in Add-Drop Multiplexed WDM Networks" *IEEE Photonics Technology Letters*, vol. 11, pp. 889–891, July 1999.

[3]   C. Tian and S. Kinoshita "Analysis and Control of Transient Dynamics of EDFA Pumped by 1480- and 980-nm Lasers" *IEEE Journal of Lightwave Technology*, vol. 21, no. 8, pp.1728–1734, August 2003.

[4]   M. Karasek and J. C. van der Plaats "Analysis of Dynamic Pump-Loss Controlled Gain-Locking System for Erbium-Doped Fiber Amplifiers", *IEEE Photonics Technology Letters*, vol. 10, no. 8, pp. 1171-1173, August 1998.

[5]   C. R. Giles and E. Desurvire, "Transient Gain and Crosstalk in Erbium-Doped Fibre Amplifiers" *IEEE Photonics Technology Letters*, vol. 2, no. 10, pp. 714–716, Oct. 1990.

[6]   A. K. Srivastava et al. "Fast-Link Control Protection of Surviving Channels in Multiwavelength Optical Networks" *IEEE Photonics Technology Letters*, vol. 9, no. 12, pp. 1667–1669, December 1997.

[7]   D. H. Richards et al. "Multichannel EDFA Chain Control: A Comparison of Two All-Optical Approaches", *IEEE Photonics Technology Letters*, vol. 10, no. 1, pp. 156–158, January 1998.

[8]   A. Bononi and L. Barbieri "Design of Gain-Clamped Doped-Fiber Amplifiers for Optimal Dynamic Performance" *IEEE Journal of Lightwave Technology*, vol. 17, no. 7, pp. 1229–1240, July 1999.

[9]   G. C. Goodwin et al. *Control System Design*. Prentice Hall International, 2001.

[10]   A. Bononi and L. A. Rusch, "Doped-Fiber Amplifier Dynamics: A System Perspective" *IEEE Journal of Lightwave Technology*, vol. 16, no. 5, pp. 945–956, May 1998.

Dynamic BER performance monitoring of WDM systems using a sum-of-Gaussian technique

B.Pal[1] and **R.Gangopadhy**ay[2]

[1] *Procyon Networks and Solutions, South Plainfield 07080*
email: biplab@procyonnet.com

[2] *Department of Electronics and Electrical Communication Engineering*
Indian Institute of Technology, Kharagpur, India
Tel:+913222-78028,Fax:+91 3222-55303
e-mail: ranjan@ece.iitkgp.ernet.in

Abstract: A sum-of-Gaussian (SGA) technique is presented for dynamic performance monitoring of WDM systems. The higher accuracy and usefulness of SGA technique over conventional Q-value analysis is established. It has been shown that Gaussian approximation (GA) overestimates the transmission penalty by a large amount.

1.   INTRODUCTION

With the advent of all optical networking, dynamic performance evaluation of optical digital transmission systems that are generally subjected to a large number of optical impairments is of extreme commercial interest because of automatic connection and fault management of the network [1]. In general, the power level, optical signal to noise ratio (OSNR) and the channel wavelength drift are measured at the input and output points of each network element like optical add/drop multiplexer, cross-connect etc. This kind of performance management, however, is inferior to the SONET layer physical performance monitoring of framer bytes A1/A2 and parity bytes B1/B2, because a large number of linear and non-linear effects in fiber that degrade the bit error rate (BER) of the system may not degrade the OSNR level at all. Hence Q value is normally proposed for dynamic channel monitoring [1]. However, forced usage of Q-value invites considerable inaccuracy in the prediction of system penalty. This is particularly true in situations handling inter-symbol interference (ISI), accumulated cross-talk, wavelength drift of optical channels, cascading effects of optical mux-demux etc. [2,3].

In this article, we propose a simple and fast numerical technique based on sum-of-Gaussian approximation (SGA) to represent accurately a non-Gaussian process of a time domain description of channel behavior available from optical channel monitor. We have developed an optimization algorithm to adapt this technique for BER evaluation. A Gaussian error index is also defined which quantifies the error in the estimation of BER when Q-value method is used.

2. THEORETICAL ANALYSIS

2.1 Sum of Gaussian approximation

Any non-Gaussian process defined by its probability density function (PDF) $W(x)$ can be expressed as a sum of N Gaussian PDFs each having a Gaussian PDF

$$W(x) = \sum_{i=1}^{N} c_i W_G^i(x, \sigma_i^2, \mu_i)$$ where c_i are the weights, $W_G^i(x, \sigma_i^2, \mu_i)$ is the 'i-th'

Gaussian PDF with variance σ_i^2 and mean μ_i. To estimate the parameters σ_i^2 and

μ_i, several stochastic methods and algorithms are available in practice based on

minimization of the likelihood function $J = E_x\left(\dfrac{W(x)}{W_A(x)}\right)$, where $E_x(.)$ denotes

expectation operation and $W_A(x)$ is the PDF of the approximated process.

The estimation of the set $[c_i, \sigma_i^2, \mu_i]$ requires solving p sets of non-linear

optimization equations by minimizing the norm of the error vector

$$E_{2p}(\mu_i, \sigma_i, c_i) = \frac{1}{p}\sqrt{\sum_{r=1}^{p} e_{2r}^2}$$ with each error vector element obtained from the

higher-order moments of the process: $e_{2r} = \sqrt[2p]{\dfrac{[M_{2p} - M_{2p}(c_i, \sigma_i)]}{(2p-1)!!\sigma^2}}$ where M_{2p} is

the actual 2p-th order moment of the process, $M_{2p}(c_i, \sigma_i)$ is the approximated 2p-th

order moment of the process and σ^2 is the actual variance of the process . The error

has been scaled by the Gaussian approximated 2p-th order moment of the process:

$M_{2p}(GA) = (2p-1)(2p-3)..3.1\sigma^2$.

2.2 The Algorithm

In a practical system, it is normally sufficient to consider the number of terms in the sum-of-Gaussian approximation to be equal to four since in most cases a bit is affected by two adjacent bits contributing to the ISI process (which is likely to be distorted by 4 Gaussian processes). Hence, for a starting condition of the optimization, each of the σ_i^2 is set to be the variance observed by the individual patterns (like 101,100 etc) and each pattern is considered to be equally likely. An iterative algorithm first optimizes the set $[c_i, \sigma_i^2, \mu_i]$ based on the second-order moments and then gradually uses the higher-order moments of the process for final determination of the set $[c_i, \sigma_i^2, \mu_i]$.

2.3 Gaussian Error Index (GEI)

We define a Gaussian error index (GEI) based on the calculation of *p-th order* moment for a non-Gaussian $W(x)$ as:

$$GEI(p) = \frac{1}{2p}[\sqrt{\sum_{r=1}^{p} \hat{e}_{2r}^2(1)} + \sqrt{\sum_{r=1}^{p} \hat{e}_{2r}^2(0)}] \tag{1}$$

where we define the error vector element for 1 as

$$e_{2r}(1) = \sqrt[2p]{\frac{[M_{2p}(1) - (2p-1)!!\sigma^2(1)}{(2p-1)!!\sigma^2(1)}}$$

with $M_{2p}(1)$ as the *2p-th* order moment of the detected values of '1' and $\sigma^2(1)$ is the variance of the process '1'. A high value of GEI is an indication that the process deviates from the Gaussian.

2.4 Bit Error Rate Evaluation

Once, c_i, σ_i^2 are computed, the probability of error in detecting '1' is determined by

$$P_e(1) = 0.5\sum_{i=1}^{N} c_i erfc(q_i) \text{ with } q_i = \frac{S_{th} - \mu_1(1)}{\sigma_1(1)}, \text{ where } S_{th} \text{ is determined by the}$$

point of intersection of the PDF curves corresponding to '1' and '0' respectively obtained from the SGA method. The quantity $P_e(0)$ can be similarly calculated.

Finally, the BER is calculated as $BER = 0.5 \times [P_e(1) + P_e(0)]$. The computation time with a 1GHz Pentium processor can be as low as below one second.

3.   RESULTS AND DISCUSSIONS

Experiments have been conducted to validate the SGA method. Two kinds of situations have been considered: In the first case, the degraded eyes due to SPM, GVD and ASE noise have been generated due to transmission of $2^8 - 1$ bits NRZ signal over 50 Km of SMF-28 at 9,10,12 and 14dBm launched power. In the second case, eyes are obtained when wavelength of the transmitter is drifted gradually from its ITU grid aligned position and the signal is obtained after it is passed through an equivalent 3-dB optical filter bandwidth of 20GHz centered at the transmitter ITU grid. In Figure 1, it can be seen that for both the cases, the sum-of-Gaussian method predicted BER is more close to the experimental BER and in the second case Q value method is inapplicable because of high ISI content in the error statistics. In Figure 2, the mismatch between Q, SGA predicted BER and actual BER is investigated with reference to GEI. It can be seen that at around 12 dBm launched power, the error statistics has highest GEI (maximum non-Gaussian in nature) and then again it tends to be Gaussian. This is due to the significant influence of Raleigh scattering after 12 dBm that drives the statistics to be Gaussian.

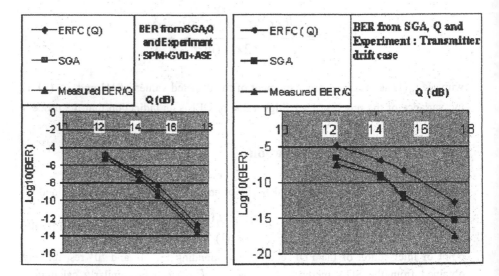

Figure1 A comparison of BER obtained from Q value and SGA method in GVD+SPM+ASE dominated and transmitter misalignment case

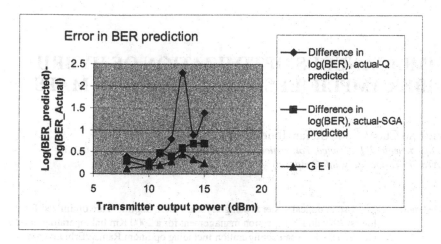

Figure 2 Error in predicting BER by SGA and Q value method for SPM+GVD+ASE noise case: Corresponding GEI values are also shown

4.   CONCLUSIONS

The sum-of-Gaussian technique provides an efficient way to simulate the true PDF of the receiver's output statistics affected by additive Gaussian noise and ISI. It has been observed that the error in system penalty can be very large if we use Gaussian approximation for the receiver output with a GEI more than 0.3. It is suggested that the Q-value technique may not be authentic for the receiver systems with a GEI more than 0.3.

REFERENCES

[1] I. Shake and H. Takara, "Transparent and flexible performance monitoring using amplitude histogram method", *OFC-2002*, TuE P.19
[2] S. Betti, G. DeMarchis and E. Innone, *Coherent Optical Communication System*. John Wiley and Sons, 1995
[3] L. F. Ribereiro, J. R. F. Da Rocha, "Performance evaluation of EDFA preamplified receivers taking into account inter-symbol interference", *IEEE J. of Lightwave Technol.*, vol.13 , no.9, pp.225-232, 1995

SIMULTANEOUS OPTIMIZATION OF HYBRID FIBER AMPLIFIERS AND DISPERSION MAPS

Vitorrio Curri[1,2] and Stefan Tenenbaum[1]

[1]*Dipartimento di Eletronica, Politecnico di Torino, Torino, Italy, optcom@polito.it*
[2]*ATS Srl, Torino, Italy, info@alps-telsoft.com*

Abstract: Multi-span optical systems using Hybrid Fiber Amplifiers are optimized for the lowest *BER* under dispersion management for a 2000 Km link operating at 42.7 Gbit/s. The best span configuration including optimum Raman/Erbium gain balance and ideal dispersion compensation degree is found. It is shown that hybrid amplifier configurations behave better than EDFA-only and Raman-only systems, accumulating less noise and nonlinearities.

1.   INTRODUCTION

Hybrid Raman/Erbium-Doped Fiber amplifiers (HFA) are an excellent choice for in-line optical amplification in multi-span links thanks to the improvement of system Optical Signal-to-Noise Ratio (*OSNR*) and broader bandwidth with respect to systems based on pure Erbium-Doped Fiber Amplifiers (EDFA). Furthermore, dispersion compensation and Raman gain can be integrated in a single unit [1-3].

In previous theoretical developments [4,5] the expression for the *OSNR* at the receiver was maximized varying the balance between Raman and EDFA gains, but there was no design of dispersion map since dispersion was assumed to be completely compensated at each span. In this work, we add to the previous analyses the simultaneous optimization of the dispersion map, i.e., the amount of in-line – length of the dispersion compensating fiber (DCF) inserted after each span - and total dispersion compensation – dispersion of the fiber grating (FG) inserted before the receiver.

The optimization is performed using a semi-analytical approach, then results are verified *a posteriori* using the optical system simulator OptSim® [6] in order to verify the impairments induced by propagation effects (chromatic dispersion and nonlinearities).

2.   SYSTEM CONFIGURATION

The system configuration we considered is presented in Fig. 1. It is a multi-span amplified optical link with N_{SPAN} periods. Each period is composed of a transmission fiber span (whose length is L_{SPAN}) backward pumped in order to get Raman amplification (RA), a first EDFA, a dispersion compensating fiber (DCF) span whose length L_{DCF} defines the degree of in-line dispersion compensation (and consequently the amount of in-line residual dispersion: $D_{res,IL}$), a gain flattening filter (GFF) of 4 dB loss and a second EDFA.

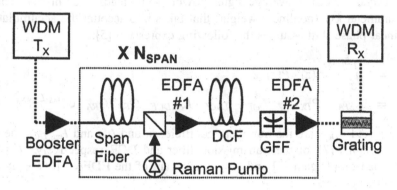

Figure 1.   System setup used for the optimization.

After the N_{SPAN} periods, we considered to insert an ideal fiber grating (FG) that contributes to the overall dispersion compensation and defines the total dispersion residue ($D_{res,TOT}$). Contrary to the DCF, the FG does not introduce nonlinear effects. Here are the analytical expressions for $D_{res,IL}$ and $D_{res,TOT}$:

$$\begin{aligned} D_{res,IL} &= D_{TF} \cdot L_{SPAN} + D_{DCF} \cdot L_{DCF} & \text{[ps/nm]} \\ D_{res,TOT} &= N \cdot D_{res,IL} + D_{FG} & \text{[ps/nm]} \end{aligned} \quad (1)$$

where D_{TF} and D_{DCF} are the dispersion coefficients, expressed in ps/nm/Km, for the transmission fiber and for the DCF, respectively, and D_{FG} is the amount of dispersion, expressed in ps/nm, introduced by the FG. The amplifiers completely recover the overall losses of each link period (losses of fibers + GFF loss).

Our analysis is applied to two system scenarios based on different transmission fibers: one based on SMF and the other based on non-zero dispersion shifted fiber (NZ-DSF). See Table 1 for transmission parameters of the considered fibers. We analyzed a link length of 2000 km subdivided in periods of L_{SPAN} = 50 km and L_{SPAN} = 80 km. We assumed to use standard IM-DD NRZ modulation and a bit-rate R_B = 42.7 Gbit/s (40 Gbit/s + FEC overhead).

The purpose of the work was to maximize the system *OSNR* (and consequently minimize the *BER*) varying the balance between Raman and EDFA gains. With respect to previous works [4,5], we added to the analysis the simultaneous optimization of the amount of in-line and total dispersion compensation (DCF length and FG dispersion) exploring the possible dispersion maps with $D_{res,IL}$ varying from −30 to 30 ps/nm and $D_{res,TOT}$ from −100 to 100 ps/nm.

3. THE ANALYSIS

Using RAs, the impact of nonlinearities is stronger than in EDFA-only systems, because the span average signal power profile tends to be higher. Thus, the parameter k_{NL} (nonlinear weight) that takes into account the accumulated nonlinear phase shift assumes the following expression [5]:

$$
\begin{aligned}
k_{NL} &= \int_0^{L_{LINK}} \gamma(z) \cdot P(z)dz \\
&= N_{SPAN} \cdot P_{TX} \cdot [\gamma \cdot L_{eff} + \gamma_{DCF} \cdot L_{eff,DCF} \cdot G_{E1} \cdot G_{RA} \cdot e^{-\alpha_S \cdot L_{SPAN}}] \quad (2)
\end{aligned}
$$

where γ and γ_{DCF} are the nonlinear coefficients, and L_{eff} and $L_{eff,DCF}$ the effective lengths [7], of the transmission fiber and DCF, respectively. G_{RA} is the Raman on-off gain [5] and G_{E1} is the gain of the EDFA #1. P_{TX} is the transmitted power.

As defined in [4], the noise accumulated on the link is $N_{TOT} = N_{SPAN} \cdot N_1$, where N_1 is noise power generated after a single span. Being P_1 the power launched to obtain $k_{NL} = 1$ with $N_{SPAN} = 1$, we can express the launched power with the help of Eq. 2 as $P_{TX} = k_{NL} \cdot P_1 / N_{SPAN}$. Using k_{NL}, the system $OSNR$ measured over a noise bandwidth equal to R_B after N_{SPAN} spans is:

$$
OSNR = \frac{P_{TX}}{N_{SPAN} N_1} = \frac{k_{NL} P_1}{N_{SPAN}^2 N_1} = \frac{k_{NL}}{N_{SPAN}^2} OSNR_1 , \quad (3)
$$

with $OSNR_1 = P_1/N_1$. The analytical expression for $OSNR_1$ defined in [4], is:

$$
OSNR_1 = \frac{P_1}{hf R_B \left[\left(n_{eq,RA} + n_{sp,E1} \frac{G_{E1}-1}{G_{RA} \cdot G_{E1}} \right) e^{\alpha_S \cdot L_S} + n_{sp,E2} (G_{E2} - 1) \right]} , \quad (4)
$$

where hf is the photon energy, $n_{sp,E1}$ and $n_{sp,E2}$ are the *spontaneous emission factors* [8] of the EDFAs and $n_{eq,RA}$ is the *equivalent input noise factor* [5] for the RA [5]. The optimal HFA setup (Raman/EDFA gain balancing) is the one that maximizes $OSNR_1$ for each L_{SPAN} and for each L_{DCF} (and consequently each value of $D_{res,IL}$). Therefore, the optimization is done on $OSNR_1$ [4] independently of N_{SPAN}, thus, for each L_{SPAN} and $D_{res,IL}$, there exists an optimal HFA configuration (amount of gain of the RA, EDFA #1 and EDFA #2) that gives the optimal $OSNR_1$ (function of L_{SPAN} and $D_{res,IL}$).

After deriving the optimal HFA for each L_{SPAN} and $D_{res,IL}$, we defined the $OSNR_{TARGET}$ as the value of $OSNR$ corresponding to the required system bit-error rate (BER_{TARGET}), remembering that

$$BER_{TARGET} \simeq \frac{1}{2} \cdot e^{-0.98 \cdot OSNR_{TARGET}} \tag{5}$$

We considered $OSNR_{TARGET} = 16$ dB corresponding to a $BER_{TARGET} \approx 10^{-16}$. Using Eq. 3, we were able to express the nonlinear weight as

$$k_{NL} = N_{SPAN}^2 \frac{OSNR_{TARGET}}{OSNR_1}. \tag{6}$$

Defining the total link length (2000 km) and the span length (we considered $L_{SPAN} = 50$ and $L_{SPAN} = 80$ km), N_{SPAN} is consequently defined ($N_{SPAN} = 40$ and $N_{SPAN} = 25$), and $OSNR_{TARGET}$ is fixed by imposing the target BER. Therefore, in Eq. 6, k_{NL} becomes dependent on the optimal $OSNR_1$ (that varies with $D_{res,IL}$), defining for each $D_{res,IL}$, the corresponding nonlinear weight.

At this point of the process, we had at our disposal, for each of the two considered span lengths, a set of optimal HFA configurations, each corresponding to a different amount of in-line dispersion compensation ($L_{DCF} \rightarrow D_{res,IL}$). $D_{res,TOT}$ does not influence the HFA configuration because the FG does not introduce nonlinearities, and consequently to each optimal HFA may correspond the overall range of $D_{res,TOT}$.

In order to evaluate the optimal dispersion map ($D_{res,IL}$, $D_{res,TOT}$), we needed to evaluate system performances for the set of possible system configurations. Therefore, we simulated the propagation for all the system setups using the optical system simulator OptSim® [6], deriving a set of values of the Q factor ($Q = 20 \cdot \log_{10}\{erfc^{-1}(2 \cdot BER)\}$ dB). Each derived Q value corresponds to a point of the explored plane ($D_{res,IL}$, $D_{res,TOT}$), where points with different values of $D_{res,IL}$ refer to different optimal HFA configurations. Therefore, final results are surfaces of Q factor in the ($D_{res,IL}$, $D_{res,TOT}$) plane: one for each considered L_{SPAN} (50 or 80 km) and type of transmission fiber (SMF or NZ-DSF). From the contour plots of these surfaces, the optimal dispersion maps can be deduced, i.e., the areas of the ($D_{res,IL}$, $D_{res,TOT}$) plane where the Q factor exceeds an established threshold. Besides being the optimal dispersion map areas, these correspond to optimal HFA configuration as well, yielding the simultaneous optimization of HFA and dispersion map.

4.   RESULTS

Even assuming to use a FEC, we considered to operate with an $OSNR_{TARGET}$ as high as 16 dB, corresponding to $BER \approx 10^{-16}$, i.e., $Q \approx 18.4$ dB. Of course, these values refer to the absence of propagation impairments. Using such ref-

Table 1. Transmission parameters of the considered fibers.

	α_S [dB/Km]	α_P [dB/Km]	D [ps/nm]	A_{eff} [μm^2]	γ [1/W/Km]	C_R [1/W/Km]
SMF	0.2	0.3	+ 16	80	1.27	0.4
DCF	0.5	-	- 100	25	4.1	-
NZ-DSF	0.2	0.3	+ 5	55	1.85	0.6

erence values, and plotting Eq. 6 in the following form:

$$k_{NL} = \left(\frac{L_{TOT}}{L_{SPAN}}\right)^2 \cdot \frac{OSNR_{TARGET}}{OSNR_1 (L_{SPAN}, D_{res,IL})}, \tag{7}$$

where $OSNR_1(L_{SPAN}, D_{res,IL})$ means $OSNR_1$ as function of L_{SPAN} and $D_{res,IL}$, we obtained the plots of Figs. 2 for the k_{NL} level curves in the plane $(L_{SPAN}, D_{res,IL})$. Fig. 2a and Fig. 2b refer to the use of SMF and NZ-DSF as transmission fiber, respectively. Each vertical line of the explored plane refers to a different HFA configuration. Highlighted horizontal lines refer to L_{SPAN} = 50 and 80 km, i.e., to the span lengths considered for the following analysis. The points crossing such values refer to the optimal HFA setups for each $D_{res,IL}$. It can be observed that increasing the amount of $D_{res,IL}$, k_{NL} increases because of the presence of a longer DCF span.

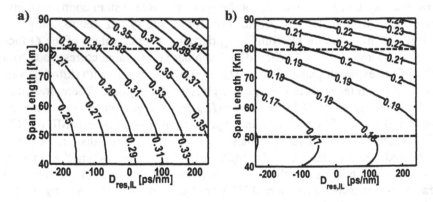

Figure 2. k_{NL} in the $(D_{res,IL}, L_{SPAN})$ plane for the case of using (a) SMF and (b) NZ-DSF.

After deriving the optimal HFA configurations for each considered scenario, we simulated the signal propagation for the corresponding system setup, varying $D_{res,TOT}$ beside varying $D_{res,IL}$. As previously described, variations of $D_{res,TOT}$ do not influence the optimal HFA configuration that depends only on

L_{SPAN} and $D_{res,IL}$. In order to define the optimal dispersion maps we swept $D_{res,IL}$ within [-250;+250] ps/nm and $D_{res,TOT}$ within [-100;+100] ps/nm.

Resulting contour plots of Q surfaces are presented in Fig. 3 for the following scenarios: SMF and L_{SPAN} = 50 km (Fig. 3a), SMF and L_{SPAN} = 80 km (Fig. 3b), NZ-DSF and L_{SPAN} = 50 km (Fig. 3c) and NZ-DSF and L_{SPAN} = 50 km with pure Erbium amplification (Fig. 3d) as comparison.

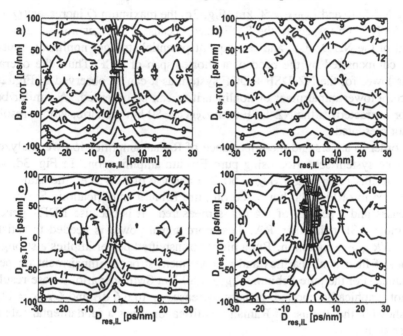

Figure 3. Q level curves for (a) SMF and L_{SPAN} = 50 km, (b) SMF and L_{SPAN} = 80 km, (c) NZ-DSF and L_{SPAN} = 50 km and (d) SMF EDFA-only and L_{SPAN} = 50 km, as comparison.

As can be observed on all graphs, the highest Q value is found off the central point of the ($D_{res,IL}$, $D_{res,TOT}$) plane, meaning that for the best transmission performance, dispersion should not be completely compensated (some in-line and total dispersion residue was left). This is a well known behavior due to the presence of nonlinearities — in particular, XPM — that is excited at the maximum if dispersion is totally compensated at each span [9].

In Fig. 3a it can be observed that using SMF as transmission fiber and L_{SPAN} = 50 Km a Q_{MAX} = 14 dB was obtained with 4.4 dB of Q penalty due to propagation impairments, while for L_{SPAN} = 80 km, the Q_{MAX} decreases to 13 dB (Fig. 3b). Increasing L_{SPAN} from 50 km to 80 Km, system performance decreases due to the enhancement of the nonlinear effects because:

- longer L_{SPAN} implies longer DCF span that contributes to increase the k_{NL} as we can see in Fig. 2a-b;

- the longer is the span of transmission fiber the larger is the loss per span, therefore a larger amount of amplification is needed to completely recover the attenuation. It implies a higher noise power added to the signal, which consequently requires a higher amount of transmission power to obtain the target *OSNR*. Thus, the impact of nonlinearity is stronger and induces a larger penalty.

Using NZ-DSF and L_{SPAN} = 50 km (Fig. 3c) the maximum Q increases (Q_{MAX} = 14.5 dB) with respect to the SMF case, because the larger nonlinear coefficient of the NZ-DSF — and consequent potential stronger nonlinear impact — is compensated by the need of a shorter span of DCF. Thus, the overall k_{NL} is lower for the NZ-DSF + DCF system as it can be observed in Fig. 2c. Hence, a lower impairment of nonlinearities can be observed. A similar behavior characterizes the NZ-DSF with L_{SPAN} = 80 km scenario, whose results are not presented.

In order to understand the importance of Raman amplification we analyzed the same system scenarios using pure Erbium amplification. In Fig. 3d, we report results referred to L_{SPAN} = 50 Km and use of SMF, i.e., the same system scenario of Fig. 3a, but based on pure Erbium amplification. System Q decreases and a penalty of 5.4 dB is measured. It is because EDFA generates more ASE noise than RA, thus more signal power is needed to satisfy the $OSNR_{TARGET}$ condition. But a higher transmitted power implies a stronger impact of fiber nonlinearities and a consequent stronger impairment on performance. For the pure Erbium, L_{SPAN} = 80 Km SMF scenario, whose results are not graphically reported, penalty increases up to 6.6 dB and Q_{MAX} = 11.8 dB, showing that the use of Raman amplification plays a fundamental role in longer spans.

5. CONCLUSION

We presented an innovative method that allows the simultaneous optimization of the HFA configuration and dispersion map for multi-span systems. We applied such method to show the difference in using SMF or NZ-DSF as transmission fibers for span lengths of 50 and 80 km. Furthermore, we compared the results of the HFA with EDFA-only: we conclude that HFA systems lead to higher OSNR and are less susceptible to nonlinearities than pure Erbium systems. By increasing the span length from 50 to 80 Km poorer performances were achieved in all studied cases. The use of low dispersion fibers seems to be preferable because it requires a lower amount of DCF.

ACKNOWLEDGMENTS

The authors would like to thank RSoft Design Group, Inc. for supplying the system simulation tool OptSim®.

REFERENCES

[1] P. B. Hansen, G. Jacobovitz-Veselka, L. Gruner-Nielsen, A. J. Stentz, "Loss compensation in dispersion compensating fiber modules by Raman amplification," in *Proc. Optical Fiber Communication Conference (OFC '98)*, (San Jose, USA), pp. 20-21, paper Tu.D.1, Feb. 1998.

[2] S. N. Knudsen, "Design and manufacture of dispersion compensating fibers and their performance in systems," in *Proc. Optical Fiber Communication Conference (OFC '2002)*, (Anaheim, USA), paper WU3, pp. 330-332, Mar. 2002.

[3] R. Hainberger, T. Hoshida, T. Terahara, and H. Onaka, "Comparison of span configurations of Raman-amplified dispersion-managed fibers," *IEEE Photonics Technology Letters*, vol. 14, pp. 471-473, April 2002.

[4] A. Carena, V. Curri, and P. Poggiolini, "On the optimization of hybrid Raman/Erbium-Doped Fiber Amplifiers," *IEEE Photonics Technology Letters*, vol. 13, pp. 1170-1172, November 2001.

[5] V. Curri, "System advantages of Raman Amplifiers," in *Proc. NFOEC 2000*, (Denver, USA), vol. 1, paper B1.1, pp 35-46, August 2000.

[6] www.rsoftdesign.com/products/system_simulation/OptSim40

[7] G. P. Agrawal, *Nonlinear Fiber Optics*, 2nd ed. New York, Academic Press, 2001.

[8] E. Desurvire, *Erbium-Doped Fiber Amplifiers*, John Wiley & Sons, 1994.

[9] G. Bosco, A. Carena, V. Curri, R. Gaudino and P. Poggiolini, "Modulation Formats Suitable for Ultra High Spectral Efficient WDM Systems," to be published on *IEEE Journal of Selected Topics on Quantum Electronics*.

ACCURATE BIT ERROR RATE EVALUATION IN OPTICALLY PREAMPLIFIED DIRECT-DETECTION

P. Martelli[1,2], S. M. Pietralunga[1], D. Nicodemi[1], and M. Martinelli[1,2]
[1]*CoreCom, Via G. Colombo 81, 20133 Milano, Italy, martelli@corecom.it*
[2]*Politecnico di Milano, Dipartimento di Elettronica e Informazione, Via G. Ponzio 34/5, 20133 Milano, Italy*

Abstract: We present an accurate method for evaluating the bit error rate in optically preamplified direct-detection, by accounting for both the intersymbol-interference and the exact Laguerre photon-count statistics. A quantum-limited sensitivity of 33.9 photons/bit is derived.

1. INTRODUCTION

We propose a novel method to evaluate the bit error rate (BER) in optically preamplified intensity-modulated direct-detection (IM-DD) systems, based on the Laguerre photon-count statistics predicted by the theory of the photodetection [1,2]. The previously published works [3,4] which use the Laguerre distribution to calculate the BER in on-off-keying (OOK) IM-DD systems, assume a measurement time $T_m = 1/R$, where R is the bit rate, and do not consider the signal distortion due to the optical filtering, so that the intersymbol-interference (ISI) is neglected. Besides, they assume a number of modes M of polarised ASE noise equal to the product between the optical bandwidth and the measurement time. These approximations can be considered as accurate whenever the optical bandwidth is much greater than R. Unfortunately, in dense wavelength-division multiplexing (DWDM) systems this condition is typically not verified, as one usually deals with optical and electrical bandwidths of comparable extent.

The method for evaluating the BER presented in this paper calculates the photon-count statistics of the directly detected noisy amplified signal, while accounting for both the effect of the ISI and the dependence of the ASE mode number on the optical filter shape and on the measurement time. The effects of optical filtering on the coherent part of the amplified signal and on the noise are separately considered.

2. BIT ERROR RATE EVALUATION

The scheme of the optically preamplified direct-detection receiver is shown in Fig. 1. An optical amplifier (OA) is followed by the cascade of a polariser (P), which eliminates the ASE component with state of polarisation orthogonal to the signal, an optical bandpass filter (F), a photodetector (PD) and an electrical integrator (I) over T_m. At the quantum-limit the electrical signal obtained at the output of the photodetector is directly proportional to the photon-counts over T_m.

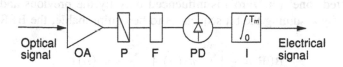

Figure 1. Scheme of an optically preamplified direct-detection receiver.

We consider an optical amplifier of gain G, with an input optical signal characterised by optical power $P_{in}(t)$, non-return to zero (NRZ) modulation format and rectangular intensity pulses. The optical field at the photodetector is given by the ASE noise superposed to a modulated coherent signal of power

$$P_s(t) = \left| \sqrt{GP_{in}(t)} * h(t) \right|^2 \qquad (1)$$

where $h(t)$ is the baseband field impulse response of the optical filter and $*$ represents a convolution. In this way we account for the signal distortion due to the optical filter. According to the theory of the photodetection, the photon-count probability is described by the following Laguerre distribution:

$$p(n) \;=\; \frac{(n_{ASE}/M)^n}{(1+n_{ASE}/M)^{n+M}} \cdot \exp\left(-\frac{n_s}{1+n_{ASE}/M}\right) \cdot$$
$$\cdot L_n^{(M-1)}\left\{-\frac{n_s}{(1+n_{ASE}/M) \cdot n_{ASE}/M}\right\} \qquad (2)$$

where $L_n^{(M-1)}$ is a generalised Laguerre polynomial [1]. The mean number of noise photon-counts is $n_{ASE} = n_{sp}(G-1)B_{eq}T_m$, where B_{eq} is the noise equivalent bandwidth of the filter and n_{sp} is the spontaneous emission factor [1]. The mean number n_s of signal photon-counts, if a photodetector with unitary quantum efficiency is assumed, is calculated by integrating the optical power $P_s(t)$ over the measurement interval and dividing by the photon energy $h\nu$, in order to account for the effect of the ISI in terms of photon-counts. The exact number of modes M is then given by the following integral [5]:

$$M = \left\{ \frac{1}{T_m} \int_{-T_m}^{T_m} \left(1 - \frac{|\tau|}{T_m} \right) \cdot \left| g^{(1)}(\tau) \right|^2 d\tau \right\}^{-1} \tag{3}$$

The degree of first-order coherence $g^{(1)}(\tau)$ of the filtered ASE is the normalised autocorrelation of $h(t)$.

From the knowledge of both the optical filter spectral shape and T_m, it is possible to calculate the Laguerre photon-count distribution of Eq. (2) for any bit sequence. In the case of rectangular impulse response of the optical filter, with the condition $B_{eq} \geq R$, the photon-count probability in correspondence of a transmitted 'one' (or 'zero') is influenced only by the previous and subsequent bit. By assuming the bit sequences to be equiprobable, the BER results in:

$$\text{BER} = \frac{1}{2} \left[\sum_{n=0}^{n_{th}} \pi_1(n) + \sum_{n=n_{th}}^{\infty} \pi_0(n) \right] \tag{4}$$

where n_{th} is the threshold level in terms of photon-counts and $\pi_0(n)$, $\pi_1(n)$ are the photon-count probability distributions, in correspondence to a transmitted reference bit being respectively 'zero' and 'one'. These distributions are obtained by summing with same weighting factor the Laguerre photon-count distributions calculated in correspondence of all the bit sequences with respectively 'zero' and 'one' transmitted reference bit. After the calculation of the Laguerre photon-count probability distribution functions p_{ijk} for each of three-bit sequences ijk (with $i,j,k = 0,1$), the distributions π_0, π_1 can be determined as:

$$\pi_0 = \frac{1}{4} \sum_{i,k=0}^{1} p_{i0k}$$

$$\tag{5}$$

$$\pi_1 = \frac{1}{4} \sum_{i,k=0}^{1} p_{i1k}$$

The optimal threshold level is then given by the conditions $\pi_1(n_{th}) \geq \pi_0(n_{th})$ and $\pi_1(n_{th}-1) \leq \pi_0(n_{th}-1)$. The BER obtained in such a way is a function of the noise equivalent optical bandwidth, of the measurement time and of the mean optical power at the amplifier input.

3. NUMERICAL RESULTS

By using the method previously described, we have numerically evaluated the BER at the quantum-limit in case of OOK-NRZ intensity modulation and

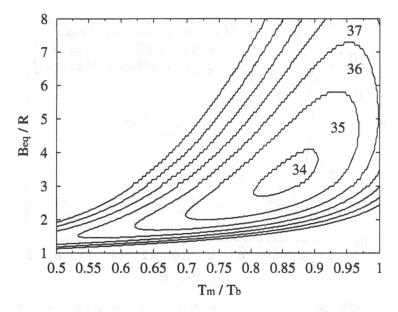

Figure 2. Contour lines of the quantum-limited sensitivity at BER 10^{-9} with rectangular $h(t)$, $G = 30$ dB and $n_{sp} = 1$.

optically preamplified direct detection, with $G = 30$ dB and $n_{sp} = 1$. The impulse response $h(t)$ of the optical filter is assumed to be rectangular.

Figure 2 shows the contour lines of the sensitivity calculated as a function of the noise equivalent bandwidth B_{eq}, normalised to the bit rate R, and of the measurement time T_m, normalised to the bit time $T_b = 1/R$. This sensitivity is defined as the mean photon number per bit at the amplifier input which gives a BER of 10^{-9}. The optimal sensitivity is 33.9 photons per bit in correspondence of $B_{eq}/R = 3.3$ and $T_m/T_b = 0.86$.

In Figure 3 it is plotted the BER as a function of the ratio B_{eq}/R, in case of a mean vale of 34 photons/bit at the amplifier input and $T_m/T_b = 0.86$. The continuous line represents the calculation with Laguerre distribution and exact value of M, given by Eq. (3). The dashed line corresponds to a Gaussian approximation of the photon-count distribution with exact M, while the dotted line corresponds to the case of Laguerre distribution with M approximated by the product $B_{eq}T_m$. In all these cases, the effect of the ISI has been considered. The pronounced minimum of the BER as function of B_{eq}/R is due to the competition between the reduction of ASE noise and the increase of the ISI, as B_{eq}/R is reduced. It can be seen that within the Gaussian approximation, the BER is higher than the BER calculated by considering the Laguerre photon-count distribution; the discrepancy is appreciable near the optimum value for optical bandwidth that minimises the BER. Furthermore it is evident that the

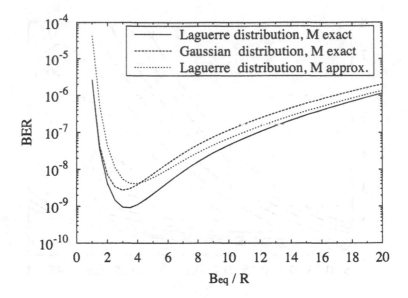

Figure 3. BER calculated as function of B_{eq}/R for $T_m/T_b = 0.86$ and 34 photons/bit at the amplifier input. Optical filter with rectangular $h(t)$, $G = 30$ dB and $n_{sp} = 1$.

approximation $M \simeq B_{eq}T_m$ gives a value of BER close to the exact one only at optical bandwidths much greater than the bit rate.

4.   CONCLUSIONS

We have proposed a simple and accurate method to evaluate the performances of optically preamplified direct detection optical communications system in terms of BER calculation. Such a method is based on the Laguerre photon statistics and accounts for the exact noise mode number M and for the effect of the ISI. Therefore, it is particularly suitable to model accurately the case of narrowband optical filtering, typical of DWDM systems, in which optical bandwidth and bit rate become comparable. In particular, in correspondence to optimised filtering condition, a quantum-limited optimal sensitivity of 33.9 photons per bit has been evaluated.

REFERENCES

[1] E. Desurvire, *Erbium-doped fiber amplifiers - Principles and applications*, J. Wiley & Sons, 1994.

[2] B. E. A. Saleh, *Photoelectron statistics*, Springer-Verlag, 1978.

[3] T. Li and M. C. Teich, "Bit-error rate for a lightwave communication system incorporating an erbium-doped fibre amplifier," *Electronics Letters*, vol. 27, pp. 598–599, 1991.

[4] H. A. Haus, *Electromagnetic noise and quantum optical measurements*, Springer-Verlag, 2000.

[5] J. W. Goodman, *Statistical optics*, J. Wiley & Sons, 1985.

TECHNO-ECONOMIC ANALYSIS OF DISPERSION-TOLERANT TRANSMISSION TECHNIQUES FOR 10GB/S DWDM SYSTEMS

Cornelius Fürst, Helmut Griesser, Jörg-Peter Elbers, and Christoph Glingener
Marconi, Stuttgarter Str. 139, 71522 Backnang, Germany,[*]
{cornelius.fuerst | helmut.griesser | joerg-peter.elbers | christoph.glingener}@marconi.com

Abstract: We analyse the prospects of dispersion-tolerant transmission techniques such as electronic distortion compensation and optical duobinary modulation. Our investigation of both technical and economical aspects shows that these techniques are predominantly beneficial for Metro DWDM transmission.

1.   INTRODUCTION

In the past years, robust transmission technologies have become a hot topic for optical core and metro networks. Whilst a number of dispersion-tolerant techniques have been proposed, the technical discussion mainly focused on two approaches. At the transmit side, optical duobinary transmission has been shown to provide superior tolerance to residual chromatic dispersion, allowing dispersion un-compensated transmission over 200km of standard fibre [1]. At the receive side, electrical distortion compensation (EDC) has been investigated [2], compensating for chromatic dispersion of several thousands of ps/nm as well as several tens of ps of polarisation mode dispersion (PMD).

---

[*] The work reported in this paper has been supported in part by the German Ministry of Education and Research (BMBF) under contract number 01 BP 260. The authors are responsible for the content of this paper.

Up to now there is no clear picture, under which conditions these technologies really exhibit their claimed technical benefits and, more importantly, do generate measurable cost savings in core and metro networks.

In this paper, we carry out a techno-economic analysis which indicates the network scenarios where techniques as duobinary transmission and EDC are beneficial for 10Gb/s DWDM systems. From the technical point of view, we investigate the dispersion tolerance provided by the these techniques for Metro and (Ultra) Long Haul systems. The economical aspect is illustrated by calculating the relative system cost savings that may be realised when a certain cost premium for each transmitter/receiver pair is assumed.

2.   DUOBINARY TRANSMISSION

Duobinary encoding is implemented by quarter-rate filtering (10th order Bessel filter) of an NRZ electrical signal. The signal then drives a Mach-Zehnder modulator at an amplitude of 2Vpi, generating a three-level signal for the electrical field (two levels for the intensity) with narrower optical bandwidth compared to conventional NRZ coding [1,3-5].

Assuming that the width of the optical filters (e.g. multiplexers) is much larger than the signal bandwidth, duobinary transmission shows a degradation of the back-to-back performance compared to the optimum performance (Q value), obtained for residual dispersion of approx. +/-2000ps/nm at a bit rate of 10.7 Gb/s (see inset of *Figure 1*). Approximately 1.5 dB of OSNR margin has to be allowed to obtain a dispersion tolerance of +3000 ps/nm (180 km of standard fibre). Spending another 2 dB extends the tolerance to +3500ps/nm.

That behaviour changes fundamentally after transmission of the signal over long-haul and ultra-long haul distances. *Figure 1* shows simulation results of a single channel propagated over 25 spans of each 80km standard fibre. Double-stage amplifiers compensate the fibre loss (20dB) and contain a dispersion compensating fibre (DCF) as an interstage device. Fibre input power levels of -3dBm to +5dBm into the standard fibre and -2dBm into the DCFs are chosen. The performance does not show anymore the "M"-shape. The dispersion tolerance is substantially reduced, even for moderate power levels. *Figure 1b* compares the results for duobinary and NRZ showing that the behaviour of both formats practically converges under the influence of fibre nonlinearities.

We have analysed the dispersion tolerance under various conditions, especially when varying the dispersion map (precompensation and slope of the accumulated dispersion). *Figure 2* shows the −1dB width of the Q-vs-dispersion curves for 25 spans of DWDM transmission at 50 GHz channel spacing.

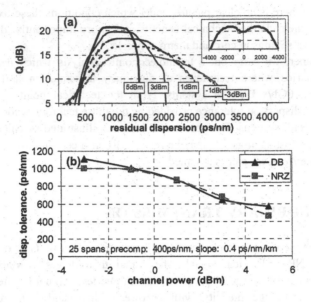

Figure 1. Dispersion tolerance of optical duobinary signals after transmission over 25 spans: (a) Simulated Q for different channel launch power levels into the standard fibre spans. Inset: Linear transmission showing the 'M'-shape, symmetric to zero dispersion. (b) Simulated nonlinear dispersion tolerance compared to NRZ.

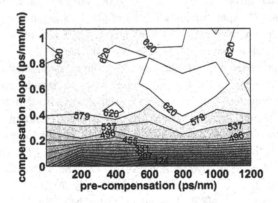

Figure 2. Behaviour of the dispersion tolerance (-1dB width of the Q-vs-dispersion curves) in ps/nm in dependence of the dispersion map parameters.

For a slope of the accumulated dispersion of > 0.3ps/nm/km, a low dispersion tolerance of 600 ps/nm is found, similar to the behaviour of NRZ. Low slope of the accumulated dispersion (< 0.2 ps/nm/km) has to be avoided to prevent resonant accumulation of distortion due to cross-phase modulation (XPM). The only improvement found with respect to NRZ is a slightly better tolerance of variations of the dispersion map (e.g. practically no influence of the precompensation).

We can conclude that whilst optical duobinary modulation is facilitator for dispersion-compensating fibre (DCF) free transmission in the Metro range (up to 200km) there is very limited technical benefit for LH and ULH scenarios.

3. ELECTRONIC DISTORTION EQUALISATION

Our assessment of electronic distortion equalisers is concentrated on maximum likelihood sequence estimation (MLSE) [6] due to its powerful distortion mitigation capabilities, but, for comparison, also includes simpler equalisers schemes such as feed-forward (FFE) and decision-feedback equalisers (DFE).

The optically pre-amplified receiver front-end consists of an EDFA and a 50 GHz super-Gaussian filter followed by a photodiode and a 7.5 GHz 5th order Bessel filter. Only optical amplifier noise is considered.

The simulation model for the MLSE is based on a moderate 4 state trellis (with corresponds to a channel memory of 2) and a 4 bit signal quantisation with two-fold oversampling [7]. The simulation for the FFE assumes 6 taps, fractionally spaced by T/2 (where T is the bit time), and for the DFE 4 feed-forward and 2 feedback taps, also with a spacing of T/2, are considered.

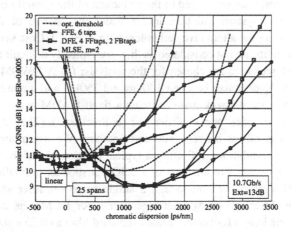

Figure 3. Dispersion tolerance of EDC receivers after a linear single span and a 25 span transmission of a single 10.7 Gb/s channel.

The system performance is evaluated by means of Monte-Carlo simulations. It is expressed as required optical signal to noise ratio (OSNR) in 0.1 nm bandwidth) to reach a bit error rate (BER) of 5e-4, sufficient for error-free operation with enhanced FEC.

Similar to the previous chapter, we compare the dispersion tolerance for the linear case (distance < 300km, low channel power levels) and for transmission after 25 spans of standard fibre with a launch power of 1dBm. In the linear case, we find a dispersion tolerance of 3000 ps/nm when spending an OSNR margin of ~4 dB (*Figure 3*). The figure also presents results for the ULH transmission, with MLSE providing a clear improvement with respect to the conventional optimum threshold receiver. Also the FFE and DFE equalisers enhance the dispersion tolerance, with some reduced gain for higher values of the residual dispersion. From these results, we can conclude that EDC technologies keep their technical benefit for a wide range of network applications.

4.   ECONOMICAL ANALYSIS

In this section, we provide an general investigation of the cost benefits that can be realised by the implementation of dispersion tolerant transmission techniques. The analysis treats the dispersion tolerance, the required OSNR margin and the cost premium (with respect to standard NRZ transmitters and receivers) as parameters, therefore providing results which are independent of the technique actually employed.

We consider cost changes related to the reduction of the quantity of equipment needed to set up a system and take into account the change of the costs of components by using the alternative technologies (e.g. EDC). Cost savings related to operational benefits are not considered as they are typically difficult to quantify.

Two network scenarios are chosen for the analysis: (1) A DWDM Metro ring network, comprising distances between 100km and 250km and a number of nodes between 5 and 8. Each network node contains a fixed OADM and, depending on the loss between the nodes, one amplifier pair (bidirectional). (2) A DWDM Long Haul Network at distances between 500 and 1500km. We use a simplified cost basis used for the calculations, reflecting a typical relationship of the costs of the main system components. For the costs of a 2.5G transponder normalised to 1, we assume the 10G transponder cost as 2.5, a single stage amplifier as 1.25 and a double stage amplifier (providing interstage access for DCMs) as 2. The DCM costs scale with the length L of compensated standard fibre as $0.25 + 0.01 \times L$.

4.1   Metro scenario

Most traffic today is a maximum 2.5G traffic intrinsic bandwidth. The introduction of 10G traffic needs handling of the chromatic dispersion, predominantly in standard G.652 networks. In order to allow in-service upgrade, dispersion compensation needs to be implemented during the installation of the system and therefore heavily affects the first-in costs of the system. Especially here, tolerant transmission techniques help to achieve acceptable economics for 10G upgrade traffic by saving capital expenditures (CAPEX). Actual cost savings are expressed by two items: (1) Partial or full replacement of DCM modules. (2) Potential saving of a single EDFA due to reduced network loss when DCMs are removed.

Figure 4 presents savings of first-in costs including 8x2.5G traffic. Rather than concentrating on an arbitrary length of the network, the results are averaged over system lengths from 100 to 250km. Cost savings of 15% are realised for an average network.

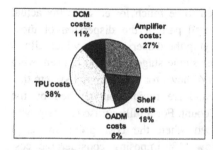 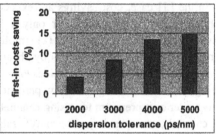

Figure 4. Assumed cost contribution and saving of first-in costs in a 8x2.5G Metro network.

Figure 5 presents the cost saving relative to the total system costs after upgrading to 8x10G vs. the 10G transponder cost premium and the achieved dispersion tolerance. For a dispersion tolerance of 3000 ps/nm, we find the break even point at a cost premium of 8% of the 10G transponders. The same calculation for 24 channels at 10G shows a break even point at 4%.

If the technology needs some extra OSNR margin (e.g. EDC needs 4dB for 3000ps/nm, duobinary 2dB for 3000ps/nm), cost savings are reduced since the saving of an amplifier becomes improbable. From further calculations we see that that the averaged savings are 0.5% lower than in the previous case at low OSNR penalty. Note that the obtained figures are averaged over a range of network extensions, and therefore individual first-in cost savings may be considerably higher.

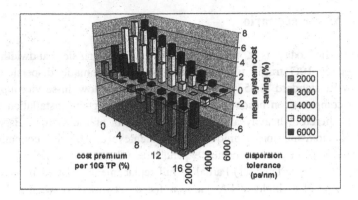

Figure 5. Savings in a Metro network loaded by 8x 2.5G and 8x10G traffic.

4.2   Long-Haul scenario

In the LH case, cost savings are expected to be much lower since the actual technologies can compensate for only a small part of the dispersion of these networks. Savings are realised by (1) simplifying the dispersion map which allows to replace double-stage amplifiers by cheaper single stage ones and (2) removing the terminal DCFs at the receive side. *Figure 6* shows for a 8-channnel system that savings are in the range of few percent and are rapidly outweighed by the transponder cost premium for rising channel count. For a fully utilised system with 40 channels, cost savings are only realised when the cost premium for a 3000 ps/nm transponder technology is below 3%, imposing considerable cost pressure onto any technique for enhanced chromatic dispersion tolerance.

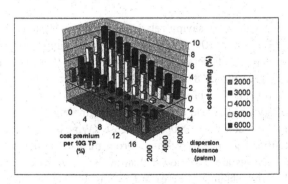

Figure 6. Cost savings relative to the total system costs for LH networks at distances between 500 and 1200km and a load of 8x10Gb/s.

5.   CONCLUSIONS

Duobinary modulation and electronic distortion compensation are promising technologies to realise cost savings in future DWDM Metro networks at 10Gb/s line rates. Both technologies provide a dispersion tolerance that enables the seamless upgrade of 2.5Gb/s systems to 10Gb/s for < 200km links, without provoking excessive first-in costs. By employing these techniques, first-in cost savings around 15% can be realised compared to systems with conventional dispersion compensation. In order to prevent cost penalties at higher channel population, the cost premium should be below about 4% of the original transponder costs when the technology is capable of tolerating 3000 ps/nm of dispersion.

For Long-haul systems, the situation is different. Due to impact of fibre nonlinearities, duobinary transmission does not provide superior dispersion tolerance anymore. EDC still increases the dispersion tolerance for the long-haul case, albeit the benefit is reduced compared to the linear transmission. Potential cost savings for any dispersion tolerant technology are clearly lower than in a Metro network and the acceptable transponder cost premium to avoid cost penalties at higher channel counts is even lower.

The cost estimations do not include some potential savings, which are difficult to take into account. There may be operational benefits such as a simpler installation procedure and design benefits such as relaxed specifications for DCMs in ULH networks. Additionally, dispersion-tolerant technology can simplify the setup of optical transparent networks, where the dispersion management is more difficult due to the diversity of optical paths.

REFERENCES

[1] S. Walklin et al., IEEE Photon. Technol. Lett., Vol. 9 (1997), page 1005.
[2] Haunstein et al., OFC 2001, paper WAA4-1.
[3] D. Penninckx et al., IEEE Photon. Technol. Lett., Vol. 9 (1997), page 259.
[4] H. Bissessur et al., OFC 2001, paper WDD36.
[5] W. Kaiser et al., ECOC 2000, paper 7.2.2.
[6] Benedetto et al., Digital Transmission Theory, Prentice Hall, 1987.
[7] Fludger et al., OFC 2004, paper WM7.

REFERENCES

PART B2:

MODULATION FORMATS

2.5 GBPS 2-PSK ULTRA-DENSE WDM HOMODYNE COHERENT DETECTION USING A SUB-CARRIER BASED OPTICAL PHASE-LOCKED LOOP

S. Camatel, V. Ferrero, R. Gaudino and P. Poggiolini
PhotonLab, Dipartimento di Elettronica, Politecnico di Torino, Corso Duca degli Abruzzi 24, 10129 Torino, Italy

Abstract: We present an optical phase locked loop based on sub-carrier modulation and designed using commercial optical components. In principle the output of a continuous-wave laser is subcarrier modulated by a common LiNbO3 MachZehender driven by an electrical voltage controlled oscillator. The proposed architecture is used as optical receiver for homodyne coherent detection of ultra-dense WDM 2.5 Gbps 2-PSK signals with 6.25 GHz spacing. This method offers the potential for providing many closely spaced multigigabit channels and enables coherent lightwave technology to become commercially viable.

1.   INTRODUCTION

The development of Ultra-Dense WDM (UDWDM) systems is currently under investigation for increasing the global capacity per single fiber. For example in [1], 10 Gbit/s transmission at 25 GHz channel spacing is demonstrated and studied in details. Anyway, it is shown in this and other works that optical filters for UDWDM have very tight requirements in pass-band shape and frequency stability. As described in [2] these requirements can be avoided using coherent communication systems which let transmission of a large number of WDM optical channels with very narrow frequency separations. Another advantage of coherent transmission is the ability to select any particular channel by simply tuning a local oscillator. An optical phase-locked loop (OPLL) has then to be implemented in order to obtain a tunable local oscillator.

OPLLs for homodyne or heterodyne coherent detection received a great deal of attention at the beginning of the 1990s [3] in order to increase the receiver sensitivity, but they never found practical applications, first because of their complexity, secondly due to the introduction of EDFAs, that greatly leveraged sensitivity issues. We proposed in [4-6] a novel and much simpler OPLL architecture based on commercial off-the-shelf optoelectronic components and without fast direct laser frequency tuning. Frequency tuning is obtained in our system through optical sub-carrier generation and tuning, and will be indicated as SC-OPLL.

This paper presents a possible application of our SC-OPLL for demultiplexing and detection of UDWDM signals. We propose and demonstrate a receiver setup capable of detecting 2-PSK signals at 2.5Gbit/s and 6.25 GHz channel separation. Our setup does not require narrow optical filters, and can be an enabling technology for future optical networks.

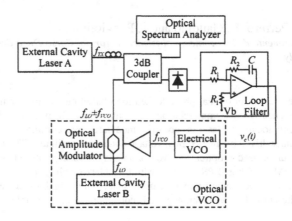

Figure 1. Schematic diagram of SC-OPLL.

2. SC-OPLL ARCHITECTURE

The SC-OPLL architecture is shown in Fig. 1. The Optical Voltage Controlled Oscillator (OVCO) is the key-element and is based on a commercial Continuous-Wave (CW) external cavity tunable laser at frequency f_{LO} that is externally amplitude modulated by the signal coming from an electrical VCO at frequency f_{VCO}. By biasing the external Mach-Zehnder (MZ) amplitude modulator at a null of its transfer function, a sinusoidal carrier-suppressed modulation is obtained. The resulting spectrum at the output of the OVCO is shown in Fig. 2. Two main sub-carriers at frequency $f_{LO} \pm f_{VCO}$ are generated, with spurious optical tones at f_{LO} and $f_{LO} \pm 2 \cdot f_{VCO}$ due respectively to a limited extinction ratio and nonlinearities of the amplitude modulator. Considering one of the two main sub-carriers, for example the one at $f_{LO} + f_{VCO}$, we are able to tune an optical frequency by simply changing the voltage applied to the electrical VCO, thus implementing an OVCO. Actually, this is the key issue of

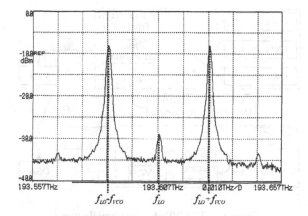

Figure 2. Spectrum at the output of the optical VCO.

the proposed architecture, that allows obtaining optical fine frequency tuning with the speed and stability of an electrical VCO, and thus to re-use typical RF PLL set-ups. In our case, a standard second-order PLL control circuit allows to lock $f_{LO}+f_{VCO}$ to the transmitted signal at frequency f_{TX} (laser A in Fig. 1). When the OPLL is locked to $f_{LO}+f_{VCO} = f_{TX}$, optical homodyne is obtained, allowing to track the incoming optical signal frequency and phase.

When the incoming signal is modulated, its optical spectrum is translated to base-band at the photodiode output. In principle, due to beating with the other sub-carriers, copies of this signal appears also around frequencies f_{VCO} and $2 \cdot f_{VCO}$, but they can be filtered out by the receiver filter if f_{VCO} is larger than the signal spectral width (or bandwidth). Other details of the setup can be found in Fig. 1. A polarization controller matches the polarization of TX and LO signals before being combined by a 3dB coupler and sent to an amplified photodiode. The resulting electrical signal is processed by a single-pole active filter, in order to obtain a second-order PLL with natural frequency f_{loop} and the damping factor ξ [5], which both depend on the loop filter parameters $\tau_1 = R_1 C$ and $\tau_2 = R_2 C$.

3. ULTRA-DENSE WDM COHERENT DETECTION

An ultra-dense WDM optical transmission system has been experimentally set up in order to evaluate the performance of our SC-OPLL. The description of the system's design and the obtained results follow.

3.1 System setup

The system experimental setup is shown in Fig. 3, and is an UDWDM extension of the one presented in [4]. Three CW lasers at frequency f_0 (central

360

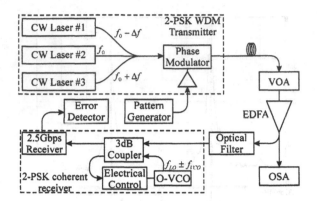

Figure 3. UD-WDM experimental setup.

channel) and $f_0 \pm \Delta f$ (adjacent channels) were optically multiplexed and sent to a LiNbO$_3$ external phase modulator, driven by an electrical NRZ 2.5Gbit/s PRBS signal to obtain a 2-PSK modulation. We used external cavity tunable lasers in order to be able to freely set the UDWDM channel separation Δf. The resulting optical signal, whose spectrum is shown in the upper part of Fig. 4, is sent to a variable optical attenuator (VOA) and then to an optical preamplifier. The resulting signal is then filtered by a 0.6 nm optical filter, which is used to reduce ASE noise, but is large enough to let the three channels pass through undistorted, i.e. this filter does not perform WDM demodulation. The coherent receiver is based on our SC-OPLL. The local oscillator frequency f_{LO} has been

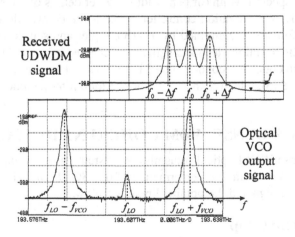

Figure 4. Spectra (0.01 nm resolution bandwidth) of the received UD-WDM signal (a) and of the OVCO output.

set such that the main sub-carrier $f_{LO}+f_{VCO}$ is able to lock the transmitted central channel frequency f_0. The spectrum at the output of the OVCO has been represented in the lower part of Fig. 4 which graphically represents the desired locking condition. The 2.5 Gbps receiver is a typical amplified photodiode and shifts the received WDM spectrum to base-band. Indeed, when the SC-OPLL is locked, the spectrum at the output of the photodetector includes a spectral component centered on zero frequency generated by the central channel and the contribution around Δf given by the adjacent two channels. UD-WDM channel demultiplexing and demodulation is directly obtained through the receiver electrical filters. In fact, a standard receiver filter such as SDH 4-poles Bessel filter at 1.8 GHz proved more than adequate to reject the two adjacent channels efficiently. In contrast, a standard DWDM receiver would demultiplex through optical filtering which is quite impractical at 6.25 GHz channel spacing.

In our experiment, we used a 2-PSK modulation with residual carrier, so that the SC-OPLL automatically locks to the 2-PSK residual carrier. For further details on this principle see [8]. The O-VCO contains an electrical VCO with f_{VCO} = 20 GHz and 5 GHz dynamic range, followed by a 40 GHz bandwidth LiNbO$_3$ amplitude modulator.

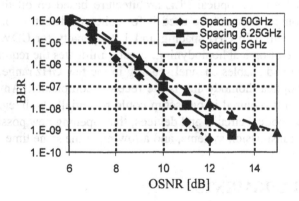

Figure 5. BER vs. received OSNR (0.1 nm resolution bandwidth) for 50, 6.25 and 5 GHz channel spacing values.

3.2 Experimental Results

The performances of our setup were tested by measuring the BER as a function of the OSNR (measured over 0.1nm resolution) for different channel spacing. The results are shown in Fig. 5 for Δf equal to 50, 6.25 and 5 GHz. The curve for Δf = 50 GHz can be taken as a reference for a situation without WDM crosstalk. In fact, we verified that it does not show any penalty with respect to single channel transmission. The curve for $\Delta f = 6.25$ GHz shows a penalty smaller than 1 dB at BER=10^{-9}. This result proves the feasibility of the proposed setup, even for 6.25 GHz channel spacing. The penalty becomes larger (more than 3 dB) for $\Delta f = 5$ GHz, where the UD-WDM channel spectra significantly overlap, thus giving rise to an intrinsic, receiver independent, channel crosstalk. Please note that we do not use any particular signal shaping at the transmitter [7], so that $\Delta f = 5$ GHz, i.e., twice the bit rate, is close to the theoretical limits (regarding bandwidth occupation).

In our experiment, due to hardware limitation, we used only three wavelengths at the transmitter side, which are anyway sufficient as a "proof-of-concept" of our technique. In a practical setup with many UD-WDM channels, when $f_{LO}+f_{VCO} = f_0$, the other SC-OPLL subcarrier at $f_{LO}-f_{VCO}$ could beat with another WDM channel. Anyway, this problem can be easily solved by using an optical filter with a passband of the order of $2f_{VCO}$, which could be significantly greater than Δf. In our experiment, we had 20 GHz, thus envisioning the use of a quite common 40 GHz bandwidth optical filter (tunable, if required by the network architecture).

4.   CONCLUSIONS

We proposed a novel optical PLL architecture based on off-the-shelf optoelectronic components and we used it for coherent detection of 2.5 Gbit/s 2-PSK signals with 6.25 GHz spacing and 1 dB penalty for UDWDM applications. The use of optical homodyning, greatly mitigates the requirements on optical filtering and enables channel spacing in the few GHz range. The price to be paid for optical homodyning is the receiver complexity. Anyway, most of the components required in our setup could potentially be integrated using next-generation optical circuits and devices, thus opening new possibilities for future optical transmission systems, and allowing at the same time a reduction in the costs.

ACKNOWLEDGMENTS

This project was partially funded by CISCO, University Research Program (URP). The authors would like to thanks S. Morasca (Avanex) for his invaluable support to the experiment.

REFERENCES

[1] I. Lyubomirsky, T. Qui, J. Roman, M. Nayfeh, M. Y. Frankel and M. G. Taylor, "Interplay of Fiber Nonlinearity and Optical Filtering in Ultradense WDM," *IEEE Photonics Technology Letters*, vol. 15, no. 1, pp. 147-149, Jan. 2003.

[2] R. E. Wagner, N. K. Cheung and P. Kaiser, "Coherent Lightwave Systems for Interoffice and Loop-Feeder Applications," *Journal of Lightwave Technology*, vol. LT-5, no. 4, April 1987.

[3] L. G. Kazovsky and D. A. Atlas, "A 1320-nm experimental optical phase-locked loop: performance investigation and PSK homodyne experiments at 140 Mbps and 2 Gbps", *Journal of Lightwave Technology*, vol. 8, no. 9, pp. 1414–1425, Sept. 1990.

[4] S. Camatel, V. Ferrero, R. Gaudino and P. Poggiolini, "10 Gbit/s 2-PSK Transmission and Homodyne Coherent Detection using Commercial Optical Components," *ECOC 2003*, paper We.P.122.

[5] V. Ferrero, S. Camatel, R. Gaudino and P. Poggiolini, "A novel Optical Phase Locked Loop architecture based on Sub-Carrier modulation," OFC 2004, paper FN6.

[6] S. Camatel, V. Ferrero, R. Gaudino and P. Poggiolini, "Optical phase-locked loop for coherent detection optical receiver," IEE Electronics Letters, vol. 40, no. 6, pp. 384-385, March 2004.

[7] L. G. Kazovsky, S. Benedetto and A. Willner, *Optical fiber communication systems*, chap. 4, Artech House, 1996.

[8] L. Kazovsky, "Balanced Phase-Locked Loops for Optical Homodyne Receivers: Performance Analysis, Design Considerations, and Laser Linewidth Requirements," Journal of Lightwave Technology, vol. LT-4, no. 2, pp. 182-195, Feb. 1986.

INFLUENCE OF OPTICAL FILTERS ON THE PERFORMANCE OF FSK/IM TRANSMISSION SCHEME

J. J. Vegas Olmos[1], I. Tafur Monroy[1], E. Tangdiongga[1], J.P.A. van Berkel[1], A.M.J. Koonen[1] and J. Prat[2]
[1]COBRA Research Institute, Eindhoven University of Technology, The Netherlands. E-mail j.j.vegas@tue.nl
[2] GCO Optical Communications Group, Technical University of Catalonia, Spain

Abstract: Influence of the filter shape on the performance of a single-wavelength combined FSK/IM scheme is investigated by simulations and experiments. For 156 Mbit/s FSK label and 10 Gbit/s IM payload, central wavelength misalignment of the signal and the optical filter can be tolerated up to 30 GHz without noticeable penalties. Results of the simulation agree very well with the experiments.

1.   INTRODUCTION

All-optical label switching (OLS) is a promising technique for switching internet protocol (IP) packet and for forwarding optical functions over Wavelength Division Multiplexing (WDM) networks [1]. By using short fixed-length labels the core nodes of the network can forward/switch packets quickly and efficiently while keeping the payload data entirely in the optical domain. A combined frequency-shift keying/intensity modulation (FSK/IM) scheme is a strong candidate for such an optical data router because of the simple label swapping mechanism and the scalability to higher data rates [2]. In this orthogonal modulation scheme, the payload is intensity modulated and the label is frequency modulated. When reaching an optical node, the label is separated from the payload and is subsequently processed. A new label will be created and re-inserted together with the payload to form a complete data packet. During the label processing, the payload remains unchanged in the optical domain. Recently, the generation of an FSK signal by using an agile tuneable laser is reported in [3]. A 10 Gbit/s FSK signal generated by a phase modulator is demonstrated in [4].

When an optical network is considered which consists of a number of optical nodes, one of the main network impairments is the spectral misalignment between the central wavelength of the modulated signal and the wavelength selective element of the nodes, such as arrayed-waveguide (AWG)-based optical filters [5,6]. This spectral misalignment will inevitably increase the network susceptibility to power loss and optical crosstalk in case of WDM transmission.

This destructive effect becomes stronger in the case of the combined FSK/IM scheme, where the broadening of the spectrum due to FSK modulation will not only disturb the signal-filter alignment but it can also shift a portion of the signal spectrum outside the filter bandwidth, possibly leading to a serious signal deformation.

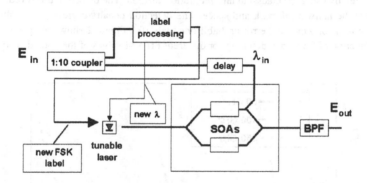

Figure 1. Generic node architecture for all optical packet switched. BPF: Band pass filter

Figure 1 shows a generic node structure and a label swapper architecture of the combined FSK/IM scheme. Ideally, the signal passes through an optical filter, for instance an optical (de)multiplexer or an optical (tuneable) band pass filter. This filter is in general needed in optically amplified WDM systems for selecting a desired wavelength channel and removing the outband amplified spontaneous emission (ASE) noise. To cope with signal-filter misalignment in an optical network, it is necessary to give extra margin in the power budget allocation. The FSK label with a sufficient optical power is locally generated in each node. Consequently, the critical constrain comes from the IM payload signal which is generated from a distant node and is kept in the optical domain in the nodes. We present in this paper a theoretical and experimental study on the performance of the combined FSK/IM scheme impaired by the filter shape.

1.1 Simulations

A continuous wave (CW) laser source, a chirp-free IM modulator, a FSK modulator, and a third order Gaussian optical bandpass filter (BPF) are assumed in the simulation. Both modulators are set to be lossless. The data format of the IM payload and FSK label signal is 2^7-1 pseudorandom nonreturn-to-zero (NRZ). The payload signal is coded at 10 Gbit/s and the label signal 156.25 Mbit/s. The label bitrate is set to be a subrate of the payload bitrate. The FSK labeling of the IM payload gives rise to a frequency deviation of 15 GHz. The optical receiver is modeled as a normal photodiode followed by a Bessel third order electrical low pass filter with a full-width half maximum (FWHM) bandwidth of 0.7 times the payload bit rate. The CW laser center frequency is set to 192.3 THz and the BPF center frequency is swept from 192.30 THz to 192.36 THz. Error probability is calculated by using the Gaussian tail integration method. The detection threshold is fixed at the halfway of mark and space. The detection penalty corresponds to the input power for which the error probability equals 10^{-9}. Figure 2 shows the penalty as a function of frequency detuning for different FHWM values of the optical BPF.

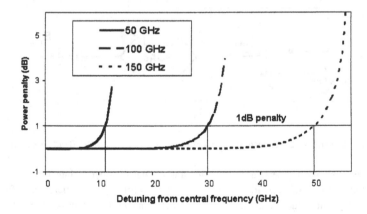

Figure 2. Penalty of the IM signal for the combined FSK/IM signal versus the detuning from the center wavelength of filters with different bandwidths

We take a penalty of 1 dB as the reference for performance evaluation. An optical filter with 50 GHz bandwidth is more sensitive to the frequency detuning than filters with larger bandwidths. For 1-dB penalty we can observe that the tolerance to the frequency misalignment increases with approximately 20 GHz as the bandwidth is enhanced with 50 GHz. The signal-filter frequency detuning must be kept under 12 GHz for a 50 GHz optical filter with penalties under 1 dB.

2.   EXPERIMENTAL SETUP

FSK modulation on the signal is obtained by modulating the current to the phase section of a tuneable laser source of type single grating assisted coupler sampled reflector (GCSR) [3]. The label format is again pseudorandom 2^7-1 NRZ at 100 Mbit/s bit-rate. The current to the other laser sections (coupler, reflector, and gain) is used for coarse and fine-tuning to a target wavelength [7], which is in the range of 1529.551561.42 nm. For this experiment the laser wavelength is 1558.92 nm or 192.30 THz with a frequency deviation of 15 GHz due to the FSK modulation format. The experimental setup is schematically depicted in Fig. 3.

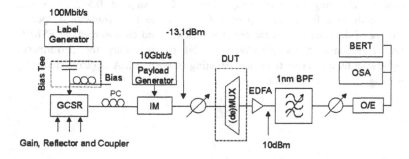

Figure 3. Experimental setup. DUT: Device under test. BPF: Band pass filter

After the labeling section, the signal is intensity modulated by using a Mach-Zehnder amplitude modulator with the same data format as the label but at 10 Gbit/s bit-rate. The combined FSK/IM signal has an extinction ratio of 6.2 dB and a Q factor of 8.5. This signal is launched to the device under test (DUT), which is an 8-channel AWG (de)multiplexer. Figure 4 shows the transmittance of the channel we used in the experimental assessment. The spectral response can be sufficiently modelled as a first order Gaussian filter with a FWHM bandwidth of 100 GHz.

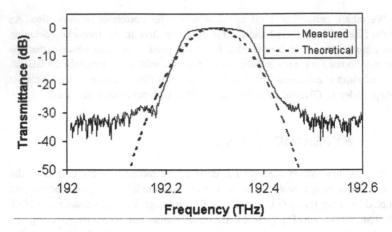

Figure 4. Theoretical and measured filter profile of 100-GHz bandwidth at 192.3 THz

Although the measured filter center frequency deviates 10 GHz from the theoretical one, the passband shape looks fairly identical. The laser frequency is tuned by changing the laser temperature with a step of 0.3°C. This step corresponds to a frequency step of 1 GHz. Due to the symmetrical spectral response of the filter around the center wavelength and the symmetrical FSK/IM spectral response to the pseudorandom bit patterns, only the performance degradation by a positive frequency detuning is evaluated. A negative frequency detuning gives a similar result.

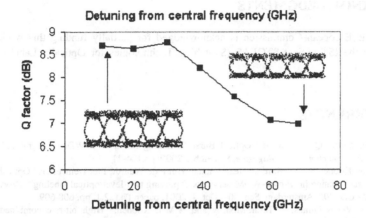

Figure 5. Performance of filtered FSK/IM signal as a function of frequency detuning: (a) received optical spectra and (b) Q value

Figure 5 shows the Q factor when the laser center frequency is swept within the filter bandwidth. For clarity, we displayed eye patterns at two extreme detuning frequencies. As observed in Figure 5, the Q factor is preserved for the detuning frequencies until 30 GHz and drops considerably by almost 2 dB at 70-GHz detuning. This is caused by the fact that one of the FSK peaks is converted into amplitude fluctuation, which in turn impairs the IM payload signal. The experimental values are confirmed by the simulations.

3. CONCLUSION

We have presented a comprehensive study on the impact of the filter shape on the performance of the combined scheme FSK/IM in a single channel for several filter bandwidths. From simulations and experiments, we observed that the frequency misalignments up to 30 GHz of the signal and the Gaussian filter can be accepted without noticeable degradation of the IM signal in systems operating at 10 Gbit/s payload. For OBS networks where the signal is regenerated locally at every node by the wavelength conversion, no significant performance degradation is expected.

ACKNOWLEDGMENTS

The European Commission is acknowledged for partially funding this work within the IST project STOLAS (Switching Technologies for Optically Labeled Signals).

REFERENCES

[1] Chunming Qiao, "Labeled Optical Burst Switching for IP-over-WDM Integration", *IEEE Communication Magazine*, September 2000, pp.104-114.

[2] Ton Koonen, Geert Morthier, Jean Jennen, Huug de Waardt, Piet Demeester, "Optical packet routing in IP-over-WDM networks deploying two-level optical labeling", *Proc. of ECOC'01*, Amsterdam, Sep. 30 – Oct. 4, 2001, paper Th.L.2.1, pp. 608-609.

[3] J.J. Vegas Olmos, I. Tafur Monroy and A.M.J. Koonen, "High bit-rate combined FSK/IM modulated optical signal generation by using GSCR tunable laser sources," *Optics Express*, Vol. 11, 3136-3140 (2003).

[4] Tetsuya Kawanishi, Takahide Sakamoto, Satoshi Shinada, Masayuki Izutsu, Kaoru Higuma, Takahisa Fujita, Junichiro Ichikawa, "High-speed optical FSK modulator for optical packet labeling," *Postdeadline paper in OFC 2004*, Paper PDP16, Los Angeles, 2004.

[5] N. Khrais, A. Elrefaie, R. Wagner and S. Ahmed, "Performance degradation of multiwavelength optical networks due to laser and (de)multiplexer misalignments," *IEEE Photonic Technology Letters*, vol. 7, pp. 1348-1350, Nov. 1995.

[6] C. Caspar, H. Foisel, R. Freund, U. Kruger and B. Strebel, "Cascadability of arrayed-waveguide grating (de)multiplexers in transparent optical networks," *Proc. of OFC'97*, 1997, Paper TUE2, pp.19-20.

[7] O. Lavrova and D.J.Blumenthal, "Detailed transfer matrix based dynamic model for multisection widely-tunable GSCR lasers," *J. Lightwave Technol.*, vol. 18, pp. 1274-1283, Sept. 2000.

CARRIER RESHAPING AND MUX-DEMUX FILTERING IN 0.8 BIT/S/HZ WDM RZ-DPSK TRANSMISSION

Ranjeet S Bhamber,[1] Sergei K Turitsyn,[1] and Vladimir Mezentsev [1]

[1]*Aston University, Bitmingham B4 7ET, United Kingdom*

Abstract: Numerical optimization of the ultra high dense WDM RZ-DPSK transmission has been performed by means of fine tuning the duty cycle and shape, carrier reshaping, and pre- and post-filtering parameters.

1.   INTRODUCTION

High spectral efficiency is an ultimate target in modern long haul fiber communications. New signal formats and RZ-DPSK in particular have become key enabling factors in increasing spectral efficiency [1–3]. However, experimental studies of applications of DPSK format in optical transmission have recently outpaced theoretical and numerical analysis. Such important resources for reaching the highest system performance as carrier reshaping and channel pre- and post-filtering have not been systematically studied in case of new formats. Asymmetric pre-filtering has been thoroughly studied for on-off keying formats, see e.g. [4] and has recently been used for CS-RZ DPSK [5].

In this paper we have systematically explored the optimal duty cycle, carrier reshaping and channel pre- and post- filtering on FEC free system performance by means of numerical optimization. Our main purpose is to fine tune the system and signal parameters corresponding to the highest experimentally available transoceanic transmission at spectral efficiency of 0.8 bit/s/Hz recently reported in [6].

2.   SYSTEM CONFIGURATION

The chosen system is essentially equivalent to that reported in [6].

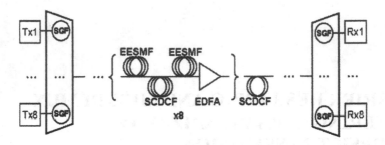

Figure 1. Scheme of the modelled system. The same super-Gaussian filters (SGF) are used for pre- and post-filtering.

The transmission line comprises periodically allocated dispersion map including eight symmetrically deployed pairs of Enlarged Effective area SMF (EESMF) with slope compensating DCF (SCDCF) between the spans of EESMF. Fiber parameters are summarized in Table 1 and the scheme of the modelled system is shown in Fig 1.

Each symmetric map is followed by EDFA to compensate for losses. The last section is followed by the short span of 3.75 km of SCDCF to equalize the residual chromatic dispersion over eight symmetric sections.

Table 1. Fibre parameters

Fibre	EESMF	SCDCF
Dispersion, $ps/nm/km$	20	-40
Dispersion slope, $ps/nm^2/km$	0.06	-0.12
Effective area μm^2	110	30
Length per section, km	28	15

Following [6], the choice of simple components was made to demonstrate superior ultra high density, up to 80% spectral efficient, transmission by means of fine tuning of system and signal parameters. No polarization division multiplexing have been used that could be considered as additional resort to improve the presented results.

3. MODELLING

We have performed numerical modelling of 8 channel RZ-DPSK transmission. Each channel had been pre-filtered before multiplexing in order to reduce inter-channel cross-talk. The same filters were used in demultiplexor. The following parameters were tuned during the optimization: i) MUX/DEMUX filter detuning, bandwidth, and shape. Super-Gaussian filters of different order with

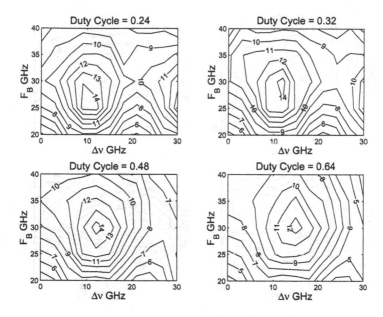

Figure 2. Contour plots of linear Q-factor versus filter bandwidth and detuning for different duty cycles.

filter transfer function $H(f) = exp\{-[(f - f_0)/B]^{N/2}\}$ have been tried to find the optimal filter steepness parameter N as well as filter bandwidth B and detuning f_0. ii) carrier duty cycle and carrier shapes varied from Gaussian to super-Gaussian of different orders.

A multi-stage optimization strategy has been adopted due to a very large number of optimization parameters. First, back to back optimization has been performed of the carrier duty cycle and shape as well as pre- and post-filtering parameters.

4.   RESULTS

The results of this optimization are presented in Figures 2 ,3 and 4. Fig. 2 shows the contours of the linear Q-factor versus filtering parameters (filter bandwidth and detuning) for different pulsewidth. Filter shapes were kept Gaussian. It was found that pulsewidth of 6ps provides the best performance.

Fig 3 shows the contours of the linear Q-factor versus the same filter parameters but for different filter shapes for the fixed duty cycle. It was found that steeper filters provide better discrimination between the channels and practically eliminate inter-channel crosstalk from N=12. In all the following simu-

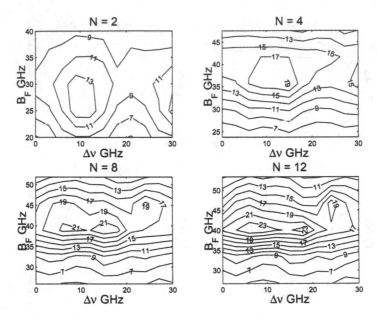

Figure 3. Contour plots of linear q-factor versus filter bandwidth and detuning for different filter shapes: Gaussian: (a) N=2 and Super-Gaussian (b) N=4, (c) N=8, (d) N=12.

lations this parameter was set to 16 as steeper filters do not gain performance and could be difficult to manufacture.

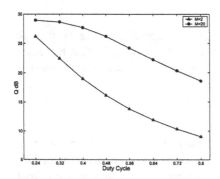

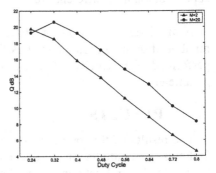

Figure 4. Back to back Q-Factor versus duty cycle for different carrier shapes: Gaussian M=2 (triangles) and super-Gaussian M=20 (circles)

Figure 5. Q-Factor (after 2000 km) versus duty cycle for different carrier shapes: Gaussian M=2 (triangles) and super-Gaussian M=20 (circles)

Figures 4 and 5 shows the influence of the carrier shape on system performance. Fig 4 shows back to back performance gain by using a super-Gaussian carrier versus duty cycle. It is seen that the performance gain achieved by simply reshaping the carrier can be as much as 10 dB whereas after 2000 km the performance gain is still respectable 4 dB in a wide range of the duty cycle as shown in Fig 5.

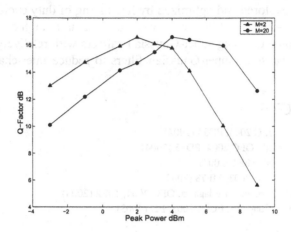

Figure 6. Q-Factor versus signal peak power cycle for different carriers: M=2 (triangles) and M=20 (circles) after 2000 km.

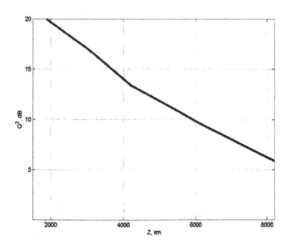

Figure 7. Q-Factor versus transmission distance for optimized transmission.

Figure 6 illustrates another benefit of using super-Gaussian carrier with steep shape index M=20. It shows the system performance after 2000 km versus signal peak power. It is seen that super-Gaussian signals perform better at higher power by approximately 2 dB which results in a noticeably better signal to noise ratio.

5. CONCLUSIONS

System performance at 0.8 bit/s/Hz CS-RZ DPSK transmission have been systematically explored and optimized by fine tuning of duty cycle, carrier re-shaping, and pre- and post filtering. It was found that ultra high density system gains performance by using super-Gaussian carriers with relatively short duty cycle of 0.24 and steep super-Gaussian filters to reduce inter-channel cross-talk.

REFERENCES

[1] J.-X. Cai et al., OFC 2004, PD34 (2004)
[2] A. H. Gnauck et al., OFC 2004, PD35 (2004)
[3] B. Zhu, OFC 2004, ThE1 (2004)
[4] A. Agata et al., OFC 2003, MF78 (2003)
[5] K. Tanaka, I.Morita, and N.Edagawa, OFC 2004, TuF2 (2004)
[6] I. Morita and N. Edagawa, ECOC2003, PD (2003)

EFFECT OF OPTICAL FILTERING ON 20-GBIT/S RZ-DQPSK TRANSMISSION OVER 2000 KM IN A 64-CHANNEL DWDM SYSTEM

Pierpaolo Boffi[1], Lucia Marazzi[1], Paolo Martelli[1], Livio Paradiso[1], Paola Parolari[1],
Aldo Righetti[1], Rocco Siano[1], and Mario Martinelli[1,2]
[1]*CoreCom, via G. Colombo, 81, 20133 Milano, Italy boffi@corecom.it*
[2]*Politecnico di Milano, piazza L. da Vinci 32, 20133 Milano,*

Abstract: We analyze 20-Gb/s RZ-DQPSK transmission in a 33-GHz spaced 64-channel DWDM system over a 2000-km laboratory straight line. In particular filtering impact on DQPSK modulation format due to optical filter and a de-interleaver at the receiver and pre-filter at the transmitter are experimentally evaluated and commented.

1. INTRODUCTION

Nowadays there is a growing effort to upgrade the transmission capacity of optical systems by increasing the data transmission spectral efficiency. For this purpose modulation formats alternative to usual IMDD are explored [1,2]. Optical differential quadrature phase shift keying (DQPSK) format appears very promising, as it combines increased information rate due to multi-level modulation with the simplicity of direct detection. Recently a few papers have presented experimental applications of this format, showing also DQPSK robustness toward fiber nonlinear effects [3,4,5] and DQPSK employment in submarine systems designed for standard IMDD transmission [6].

It is also known that optical communication channel performance is strongly influenced by the receiver optical filter characteristics as well as by the choice of the optical pre-filter. Of course this effect varies with the signal modulation format [7, 8].

In this contribution we experimentally analyze the impact of the filtering apparatus on the BER performance in a transmission of a RZ-DQPSK signal at 10 Gsymbol/s (20 Gb/s equivalent) in a 64-channel DWDM system with 33-GHz spacing over 2000 km. In particular at the receiving unit we analyze the combined effect of a 33/66GHz de-interleaver followed by an optical tunable band pass filter, while at the transmission side we study the effect on DWDM long-haul propagation of pre-filtering. The preliminary experimental results obtained with non-optimized filters clearly show the importance of an accurate project for filter shapes and bandwidths.

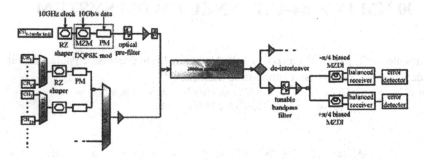

Figure 1. Experimental setup.

2.   EXPERIMENTAL SET UP

The experimental setup is presented in Figure 1. The 64 channels, 33-GHz spaced from 1543.03 nm to 1559.71 nm, are multiplexed in a 33-GHz grid. A tunable laser generates the DQPSK channel under test. RZ-DQPSK modulation is obtained by the cascade of a first RZ-shaper, a Mach-Zehnder modulator (MZM) and a phase modulator (PM). The RZ-shaping is performed by a Mach-Zehnder modulator (MZM) driven by the 10 GHz clock signal, obtaining a duty-cycle of 50% and an extinction ratio of 13 dB. The second MZM operates in push-pull mode and is driven by a 10-Gb/s PRBS of length 2^7-1 achieving a π–modulation depth. The last PM, driven by the complementary PRBS signal suitably delayed in order to obtain uncorrelation, performs an additional $\pi/2$-phase modulation. An RF phase shifter synchronizes data at the two phase modulators. The test channel is inserted after the multiplexer.

Setup allows RZ shaping of the other 63 channels, which are 0-π binary phase shift keyed (BPSK) by a single PM. Due to the same RZ format and to the BPSK modulation with π-depth, the obtained 63 channels optical spectra are very similar to the RZ-DQPSK channel under test, as demonstrated in [6]. The RZ-DQPSK channel under test can be pre-filtered before propagation by means of a 22 GHz filter. Actually employed pre-filter is a 66/33 interleaver, whose optical transfer function is shown in Figure 2. Receiver de-interleaver has the same transfer function and demultiplexes to 66 GHz the 33 GHz DWDM grid. A tunable band pass filter (FWHM of 33 GHz) selects the DQPSK channel under test for BER measurements. The in-phase and in-quadrature phases are directly detected by a pair of Mach-Zehnder delay interferometers (MZDI) biased at $+\pi/4$ and $-\pi/4$, respectively [9].

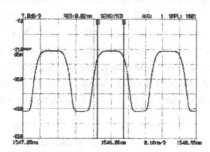

Figure 2. Measured optical transfer function of the employed optical pre-filter, which is an interleaver with the same spectral characteristics of the receiver interleaver.

We employ a pair of originally designed MZDIs [10] realized by SiON technology with a high index contrast between the guide core and upper cladding/substrate, allowing minimum bending radius with negligible losses in the 100 ps delay arm. In order to set the differential optical phase between the interferometer arms respectively to $+\pi/4$ and $-\pi/4$, the integrated SiON chips are thermally controlled by a heater inside a metallic case. The two outputs of each MZDI are detected by a commercially available balanced receiver (14-GHz bandwidth), mechanically controlling the optical path lengths of the balanced receiver. Then the BER-tester is programmed with the expected data sequence of 2^7-1 bits.

The DWDM channels copropagate over a 1994-km straight line constituted by 33 NZD fiber spans of 55 km compensated, every 6 spans, by SMR fibers. NZD fiber has a negative dispersion of -2.82 ps/nm km at 1548 nm, the link average dispersion at 1548 nm is 0.0488 ps/nm km. In the straight line 39 EDFAs with 17-nm flat optical bandwidth, 12-dBm saturation output power and 4.2-dB noise figure are employed. This laboratory link was originally designed for an IMDD submarine transmission by Pirelli Submarine Telecom Systems.

3. EXPERIMENTAL RESULTS AND DISCUSSION

DQPSK performance are evaluated in the above-described 2000 km DWDM system in order to understand design complexity and limitations due to filtering specifications over propagation impairments. In the following figures we show BER measurements relative to $+\pi/4$ biased MZDI receiver; similar results are obtained for $-\pi/4$ biased MZDI.

In particular we investigate BER performances of the RZ-DQPSK channel at 10 Gsymbol/s as a function of system optical signal-to-noise ratio (OSNR), at $\lambda=1548$ nm. First of all, Figure 3 shows curves comparison between unfiltered (black diamonds) and narrow filtered (gray squares) RZ-DQPSK. It can be noticed that RZ-DQPSK signal appears robust towards narrow-band filtering induced by the de-interleaver at the receiver. Moreover, the penalty due to the detuning of the signal wavelength with respect to the de-interleaver optical transfer function remains less than 1 dB for detuning within +/-7.5 GHz, as shown in Figure 4.

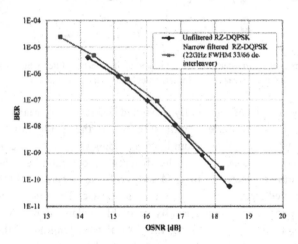

Figure 3. BER versus system OSNR for unfiltered (black diamonds) and narrow filtered (gray squares) RZ-DQPSK. Filtering is performed at the receiver by a 33/66 GHz de-interleaver. No propagation.

As shown in Figure 5, when all the 64 33-GHz-spaced DWDM channels are present, a penalty arises, depending on the relative state of polarization between the two closest channels. This penalty can be cancelled in case of orthogonal relative States Of Polarization (SOPs).

RZ-DQPSK signal performance over 2000 km in a 33-GHz DWDM environment is shown in Figure 6. 1548 nm RZ-DQPSK channel is polarization-scrambled. Back-to-back data (blank square curve) are compared with BER after 2000-km propagation (black square curve). As for previous measures, at the receiver channel is isolated from 64 DWDM copropagating channels by the cascade of the de-interleaver and optical band pass filter. Penalties due to propagation are evidenced: the phase modulation on the neighbor channels impairs propagation, owing to linear inter-channel cross talk due to PM spectral broadening and owing to non-linear interactions. As shown in figure, the experimented system, designed for standard IMDD propagation, is intrinsically limited in maximum achievable OSNR by the amplifier saturation output power. After 2000 km the maximum OSNR for the 64 DWDM channels is 17.5 dB.

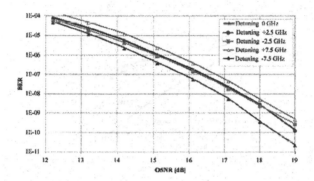

Figure 4. BER versus system OSNR for filtered RZ-DQPSK for signal wavelength detuning within ± 7.5 GHz with respect to the de-interleaver optical transfer function. No propagation.

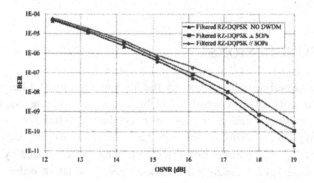

Figure 5. BER versus system OSNR for filtered RZ-DQPSK. When all 64 CH are presents (gray squares and diamonds) penalty varies with relative SOPs of closest channels. No propagation.

Pre-filtering impact is evaluated as well by employing, as a tight pre-filter, an interleaver with the same spectral characteristics of the de-interleaver. Comparison between the two un-propagated curves (blank square curve without pre-filter and blank circle with pre-filtering) shows about 2 dB penalties. It is noteworthy that total filtering transfer function is non-optimized for DQPSK format. Gray circle curve represents 2000-km propagation performance with pre-filtering. With respect to back-to-back curves, after 2000-km propagation lower penalties (about 0.8 dB) are found between pre-filtered (gray circles) and non-pre-filtered (black squares). Pre-filtering in fact limits the spectrum of DWDM channels broadened by phase modulation before propagation, thus reducing inter-channel cross talk.

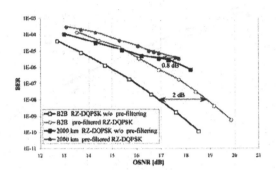

Figure 6. BER versus DQPSK channel OSNR in presence of the de-interleaver and of the 64 DWDM channels for back-to-back (blank squares) and after 2000 km (black squares); in presence of the de-interlaver, the 64 DWDM channels and the interleaver before propagation for back-to-back (blank circles) and after 2000 km (gray circles).

4. CONCLUSIONS

We have experimentally presented the impact of optical filtering on the transmission of a RZ-DQPSK signal at 20 Gbit/s equivalent bit rate in a 64-channel DWDM system. Propagation over 2000 km is obtained in a link not optimised for this type of modulation format. From these preliminary results, we expect that RZ-DQPSK degradation in ultra-long-haul transmission can be reduced by a suitable pre-filtering optimised for RZ-DQPSK modulation format in order to confine optical spectrum before propagation limiting non-linear interactions among channels.

ACKNOWLEDGMENTS

The authors thank CoreCom Optical Technologies and Integrated Optics Labs for the project and the realization of the MZDI in collaboration with Pirelli Labs – Milan, Italy, and Pirelli Submarine Telecom Systems – Milan, Italy for support during experiments.

REFERENCES

[1] T. Ono, Y. Iano, "Key technologies for Terabit/second WDM systems with high spectral efficiency of over 1 bit/s/Hz", *IEEE J. Lightwave Technol.*, vol. 34, pp. 2080-2088, 1998.

[2] T. Hoshida, O. Vassilieva, K. Yamada, S. Choudhary, R. Piecqueur, H. Kuwahara, "Optimal 40 Gb/s modulation formats for spectrally efficient long-haul DWDM systems", *J. Lightwave Technol.*, vol. 20, pp. 1989-1995, 2002.

[3] C. Wree, J. Leibrich, J. Eick, W. Rosenkranz, "Experimental investigation of receiver sensitivity of RZ-DQPSK modulation format using balanced detection", in *Proc. Optical Fiber Conference (OFC 2003)*, pp. 456-457, 2003.

[4] P.S. Cho, V.S. Grigoryan, Y.A. Godin, A. Salamon, Y. Achiam, "Transmission of 25-Gb/s RZ-DQPSK signals with 25-GHz channel spacing over 1000 km of SMF-28 fiber", *IEEE Phot. Tech. Lett.*, vol.15, pp. 473-475, 2003.

[5] H. Kim, R.J. Essiambre, "Transmission of 8x20 Gb/s DQPSK signals over 310-km SMF with 0.8-b/s/Hz spectral efficiency", *IEEE Phot. Tech. Lett.*, vol. 15, pp. 769-761, 2003.

[6] P. Boffi, L. Marazzi, L. Paradiso, P. Parolari, A. Righetti, D. Setti, R. Siano, R. Cigliutti, D. Mottarella, P. Franco, M. Martinelli "20 Gb/s differential quadrature phase-shift keying transmission over 2000 km in a 64-channel WDM system" *Optics Communications* to be pubblished

[7] P. J. Winzer, M. Pfennigbauer, M. M. Strasser, W. R. Leeb "Optimum Filter Bandwidths for Optically Preamplified NRZ Receivers" *J. Lightwave Technol.*, vol.19, no.9, pp. 1263-1273, 2001.

[8] P. J. Winzer, S. Chandrasekhar, H. Kim "Impact of Filtering on RZ-DPSK Reception" *IEEE Photon. Technol. Lett.*, vol.15, no.6, pp. 840-842, 2003.

[9] R.A. Griffin, A.C. Carter, "Optical differential quadrature phase-shift key (oDQPSK) for high capacity optical transmission", in Proceedings of Optical Fiber Conference 2002, WX6, 367-368, 2002.

[10] F. Morichetti, R.Costa, A.Cabas, M.Ferè, M.C.Ubaldi, A.Melloni, M.Martinelli "Integrated optical receiver for RZ-DQPSK transmission systems", Proceedings of Optical Fiber Conference 2004, FC8 (2004).

COMBINED (SYMBOL AND CLASSICAL) DWDM DATA TRANSMISSION

A. O. Nekuchaev [1], and U. Yusupaliev [2]

[1]*Telecom Transport, of. 27-30, bld. B2-2, Profsoyuznaya st., 84/32, Moscow, GSP-7, 117997, Russia, an@tt.ru*
[2]*Moscow State University, Physics Department, Leninskie Gori, Moscow, 119992, Russia, unirus@phys.msu.ru*

Abstract: Symbol DWDM transmission: every wavelength is symbol, carrying Log_2N bits, where N – number of wavelengths; one wavelength in fiber every moment. Combined transmission: several wavelengths (symbols) in fiber every moment. Application: ultra long haul.

1.　　INTRODUCTION

The essence of symbol technology is that standard DWDM equipment (for example, with 16 carriers λ_1-λ_{16}) is used not for simultaneous modulation of all carriers but in serial operating mode. That means the original data is divided into 4-bit clusters (i. e. 100010100010 = 1000 1010 0010), and every carrier is assigned to a single cluster (from 0000 to 1111). Each 4-bit cluster is transmitted in one time-slot, the first part of that time-slot is taken by corresponding carrier λ_i (pilot) and the second part is taken by carrier λ_{i+1} (co-pilot). The co-pilot for λ_{16} is λ_1. This is first FEC1. Under such modulation scheme (1 - turned on, 0 - turned off) the transmission rate of the pair "pilot - co-pilot" - i. e. 4-bit cluster - is equal to clock rate. Every five information clusters are accompanied by redundant sixth cluster with bit-by-bit control sum of theirs. This is second FEC2. So to transmit 10 Gbit/sec stream ir requires to transmit 10/4*(6/5) = 3 Gsymbols/sec. This case is shown on fig.1.

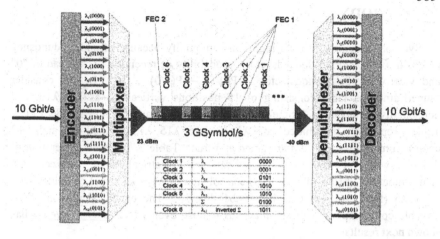

Fig. 1. Symbol DWDM data transmission.

An evolution of the symbol DWDM transmission idea gives rise to a combined DWDM transmission concept. Since modern production DWDM equipment supports up to 80 channels in a single fiber in 1530-1565 nm range, let us divide this range into 5 "tubes" with 16 carriers in every tube, fig. 2. Using the symbol transmission technology in every tube and applying 3 Gsymbol/sec lasers, it's possible to organize 10 Gbit/sec transmission in every tube or 50 Gbit/sec transmission over single fiber.

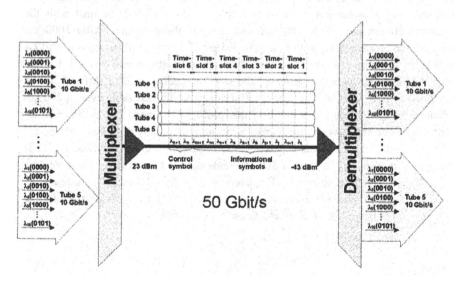

Fig. 2. Combined DWDM data transmission.

2.   BODY

Symbol transmission has a high noise immunity because of code redundancy (140% in our example). Indeed, let the probability of receiving "1" instead of "0" and vice versa on photodetector is $P(0/1) = P(1/0) \sim 10^{-3}$. Let us consider "superframe" as table consisting of 16 positions vertically , as $\lambda 1- \lambda 16$, and horisontally of 6, that is 5 time-slots as payload data and 6th time-slot as FEC2. Table 1 represents transmitted data $\lambda 1, \lambda 2, \lambda 16, \lambda 15, \lambda 15$ payload data and $\lambda 12$ control sum. "1"s and "0"s are green and true. Table 2 represents operation of decoder with mistakes due to optical and electrical noise at receivers. "Unfortunately" (probability is 1/16) control sum of $\lambda 1 + \lambda 2$ (true) coincides with $\lambda 7 + \lambda 5$ (false), correct data has probability ½. The exact calculation of all possible combinations with simultaneous 4 false events (see www.tt.ru/eng) has shown next results:

FEC input	10^{-3}	10^{-4}	10^{-5}
FEC output	10^{-9}	10^{-13}	10^{-17}

But additional employment of in-band triple-error-correcting Bose-Chaundhuri-Hocquenghem BCH-3 code (with was documented in an October 2000 revision to the ITU-T G.707 standard) can improve FEC1,2 significantly since the location of mistakes has been already known.

One of significant problems in symbol transmission is the chromatic dispersion compensation. Indeed, the dispersion factor in SMF is 17 ps/nm*km. Signals at neighbouring wavelengths in third spectral window (1530-1565 nm) with the interval between carriers 50 GHz are distanced at about 0,4 nm. After 1000 km they will come at receiver with time shift 8,5 ns. The time shift between λ_1 and λ_{16} will be 16 times more - about 136 ns. If the symbol length on one carrier is 0,17 ns then it will be necessary to process time-slots at receiver with a shift of 50-800 symbols (at 1000 km distance already). But this is hardly realizable even at 200 km distance. Nethertheless the solution of this problem has been found. Since ultra-long-haul fiber links (which the symbol technology is mainly designed for) operate in SDH mode, the SDH data format can be used for receiver synchronization at different wavelengths. In fact, in SDH systems data is transmitted by consecutive STM-1 frames with frequence 8 KHz. Every STM-1 frame consists of heading (81 bytes) and informational fields which sizes are fixed. Let us divide the heading data in 4-bit clusters (for 16 carriers symbol transmission).

Table 1. Transmitted values.

	Time-slot 1	Time-slot 2	Time-slot 3	Time-slot 4	Time-slot 5	Time-slot 6	Cluster
$\lambda 1$	1		1				0000
$\lambda 2$	1	1					0001
$\lambda 3$		1					0010
$\lambda 4$							0100
$\lambda 5$							1000
$\lambda 6$							0110
$\lambda 7$							1001
$\lambda 8$							1110
$\lambda 9$							1101
$\lambda 10$							0111
$\lambda 11$							1111
$\lambda 12$					1		1011
$\lambda 13$						1	0011
$\lambda 14$							1100
$\lambda 15$				1	1		1010
$\lambda 16$			1	1	1		0101

Table 2. Received values.

	Time-slot 1	Time-slot 2	Time-slot 3	Time-slot 4	Time-slot 5	Time-slot 6	Cluster
$\lambda 1$	0		1				0000
$\lambda 2$	1	0					0001
$\lambda 3$		1					0010
$\lambda 4$							0100
$\lambda 5$		0					1000
$\lambda 6$		1					0110
$\lambda 7$	0						1001
$\lambda 8$	1						1110
$\lambda 9$							1101
$\lambda 10$							0111
$\lambda 11$							1111
$\lambda 12$					1		1011
$\lambda 13$						1	0011
$\lambda 14$							1100
$\lambda 15$				1	1		1010
$\lambda 16$			1	1	1		0101

Let us define the first 4-bit cluster as "pilot - co-pilot" pair - λ_{16} and λ_1, the second cluster - as λ_8 and λ_9, the third cluster - as λ_{12} and λ_{13} and so on (so that every posterior pair take place in the middle of already appointed pairs). As a result, a bit succession which is specific for heading, known beforehand and repeated with 8 kHz frequency will be formed at every wavelength. Since the length of every frame and transmission rate are known, it is easy to determine which group of bits the received signal belongs to and to recover symbol spreaded in time. In fact, the buffer is necessary which size is determined by scattering of signals at λ_1 and λ_{16}. DWDM management system which operates at 1510 nm carrier must transmit all necessary information for that. It is worthy to remark that in the scheme of synchronization considered above pilot and co-pilot signals may be placed not at neighboring wavelengths but arbitrarily. This fact has a great significance because frequency response function of EDFA transparent section is not linear and has irregular nature. It results in the fact that signals at some wavelengths are transmitted better than at others. But choosing "good" co-pilots for "bad" pilots (and vice versa) it's possible to achieve nearly equal BER for every symbol, and so essentially mitigate and simplify complex and expensive procedure of gain equalization and tilt control of EDFA spectrum (which inevitably reduces OSNR after attenuation of signal power, especially when number of wavelength increases). Note that regeneration station may operate in 2R mode (reshaping, reamplification) or in 3R mode (2R+retiming). The latter means that STM frames recognition, error correction and data recovery take place. It is evident that during symbol transmission regeneration station operates in 3R mode. Combined method has significant competitive advantage for security of data - in classical DWDM intruder can copy digital information from one photodetector, but here he will be forced to copy even not 16 , but all 80 channels together (since inside one logic tube every wavelengh among 16 may be arbitary among 80) with general clock!

3. CONCLUSIONS

Our competitive advantages for target – 50 Ggbit/sec over SMF 4000 km regeneration distance – are as follows:

- Sensitivity +6 db
- Chromatic dispersion +2 db
- Polarisation dispersion +2 db
- Nonlinearities about +2 db (for example, Double Relay Scattering (DBS) puts upper limit to Raman amplification essentially, but we have every wavelength 8 times rarely – 16 pilots and 16 co-pilots – so can increase Raman pump power significantly)
- FEC is approximately the same as usual Reed Solomon
- 3R and SDH framing exist
- Security of data is extremely high

As a result the span (distance between optical amplifiers (Raman and EDFA combined) is at least 50 km longer than usual, so the main conclusion of this project is:

SDH nodes become "heavier" (or more expensive) and fiber links become "easier" (or cheaper), in the long distance we gain in cost significantly.

REFERENCES

The patent address of this idea is http://www.fips.ru, patent 2161374.

INTEGRATED DIRECT-MODULATION BASED QUANTUM CRYPTOGRAPHY SYSTEM

Johann Cussey[1], Matthieu Bloch[2], Jean-Marc Merolla[1] and Steven.W McLaughlin[2]

[1] *GTL-CNRS Telecom UMR 6174, 2-3 rue Marconi 57070 Metz, France,*
jcussey@georgiatech-metz.fr

[2] *Electrical and Computer Engineering, Georgia Institute of Technology, Atlanta, Georgia,*
USA,
mbloch@georgiatech-metz.fr

Abstract: We report a new quantum key distribution scheme using direct-modulation method associated with single sideband detection (SSB). Experiments were carried out at 850 nm using standard electronic and optical components.

1. INTRODUCTION

Quantum cryptography or quantum key distribution (QKD) has known an increasing interest because it offers higher security than public-key based key transfer systems [1]. Several systems [1-4] have been developed to exchange quantum keys via optical fibres, achieving key distribution up to 100km [5]. In a recent experiment [6], a 23km key transmission was performed in free space, hence demonstrating the feasibility of QKD for ground-to-ground or space application. However, only one method based on polarization-coded quantum states has been explored to realize free-space key distribution.

We report a new free-space transmission quantum key distribution method, using a direct-modulation technique associated to a single sideband detection method [7]. The use of directly modulated laser diodes and standard electronical components at the emitter enables suitable integration and thus offers potential satellite-to-ground applications. The experimental prototype operates at 850nm using off-the-shelf components.

2. PRINCIPLE

Figure 1 shows the proposed system.

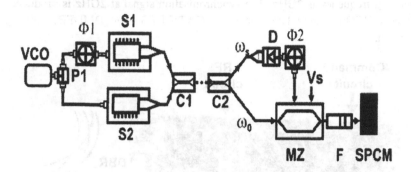

Figure 1. Schematic diagram of the direct modulation scheme.

The source (S1) is an attenuated laser diode operating at optical frequency ω_0 (quantum signal). S1 is directly modulated at $\Omega << \omega_0$ with a modulation depth $m<<1$. The modulating signal is produced by a voltage controlled oscillator (VCO) that drives simultaneously a second laser diode S2 operating at optical frequency ω_s. Both optical signals are transmitted thanks to a WDM coupler (C1). Their optical spectra are composed of a central peak and two sidebands at frequencies $\omega_0 \pm \Omega$ ($\omega_s \pm \Omega$) with phase Φ_1 (0) relative to the central peak. The phase Φ_1 is introduced by a phase shifter. At the receiver, a WDM demultiplexer (C2) separates the transmitted signals. The synchronisation signal is converted by a detector (D) that generates an electrical signal at frequency Ω The amplitude of the electrical signal is matched to the emitter modulation depth m and drives an unbalanced integrated Mach-Zehnder modulator (MZ). An additional phase shift Φ_2 is introduced thanks to a second phase shifter. The bias voltage Vs is matched to the chirp value of the source S1 such that the probability P_1 and P_2 of detecting one photon in the lower and the upper sidebands of the quantum signal are respectively governed by a sine-squared and a cosine-squared function of the phase difference ($\Phi_1 - \Phi_2$). The sidebands are separated by an optical filter (F) and photons are counted by a single photon counting module (SPCM). The BB84 protocol can then be implemented with only one detector as shown in [8].

3.   EXPERIMENT

The experimentals circuits are shown in figure 2 and 3. The actual prototype size emitter (fig.2) is 15x10 cm .The quantum signal is generated by a 1mW, 852nm DBR laser diode. The modulation amplitude is set to $m \approx 0.3$V thanks to an integrated VCO and amplifier (RF circuit). Calibrated attenuators attenuate the light so that with the chosen value of m, the average photon number sent in the fibre is approximately 0.21 photon per pulse in each modulation sideband. The operating frequency is 2GHz. The synchronisation signal at 2GHz is produced by the same VCO and modulates a standard CATV DFB laser (DMDFB) emitting at 1310 nm.

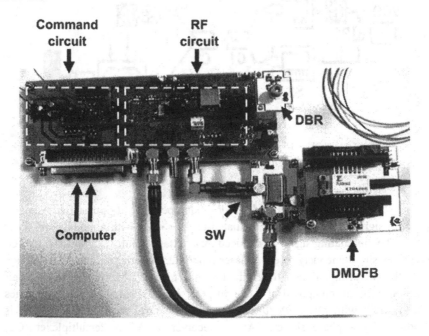

Figure 2. Prototype of the emitter.

Its average power was set to 10 µW to avoid crosstalk with the quantum signal. At the emitter a computer controls a quadrature phase-shift-keying (QPSK) modulator (including in the RF circuit) to introduce a phase variation Φ_l , randomly chosen among four possible values. An additional square-modulation at 1MHz is mixed with the synchronization signal to be used as a clock thanks to an external RF switch (SW). The command circuit converts the digital signal generated by the computer into the required analog signal.

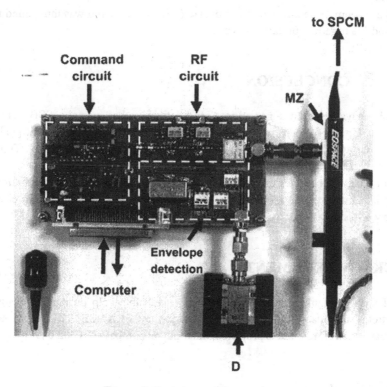

Figure 3. Prototype of the receiver.

At the receiver (fig.3), the quantum and the synchronization signal are separated by a 30dB isolating WDM demultiplexer (not shown in fig 3). The synchronisation signal is first converted into an electrical signal by a standard detector (D). The resulting electrical signal is then split and filtered in two different ways to obtain in one case the plain 2GHz frequency and in the other case the 1MHz square envelope. This 1MHz clock is used to switch randomly and synchrounously the phase Φ_2 among four possible values and generates a 50ns duration gate that allows the EGG single photon counting module to be gated. In these conditions the quantum efficiency and the dark count per gate are 50% and 1.10^{-6}. We use a customized LiNbO$_3$ intensity modulator (MZ) and fibre Fabry-Pérot interferometer (composed of two bragg gratings, not shown in fig 3) with a FSR of 10GHz and a finesse of 100 to select only one sideband. Quantum signal was detected by a SPCM. Key distributions have been performed with a 100m single mode fibre at 850nm. The reconciliation process [9,10] uses a LAN connection as the public channel. The global attenuation of the receiver was around

6dB at 850nm. The quantum bit error rate (QBER) measured was thus found to be around 1% for a raw bit rate of 5Kc/s.

4.   CONCLUSION

We have reported a new direct modulation method using a SSB detection scheme. Unlike a recent free space polarization based QKD experiment [11], the synchronisation and quantum signal used the same transmission channel. The first results obtained show the possible used of our method in free space transmission. Finally, the laser diodes can be replaced by VCSEL sources allowing future on-chip integration of the emitter with hybrid CMOS VLSI [12] electronic circuits. This work is under progress.

ACKNOWLEDGMENTS

I wish to thank Samuel Moec for his invaluable help during the design of electronic circuits. This work was supported by FRANCE TELECOM under contrat N°: 991B489 and is protected under patent number : WO 02/049267 A1.

REFERENCES

[1]   C.H.Bennett, F.Bessette, G.Brassard, L.Salvail and J.Smolin, "Experimental quantum cryptography", *Journal of Cryptology*, 5 (1992), 3.

[2]   A. Muller, H. Zbinden and N. Gisin, "Quantum cryptography over 23 km in installed under-lake telecom fibre", *Europhysics Letters*,33 (1996), 335-339.

[3]   P.D. Townsend, J.G. Rarity and P.R. Tapster, "Single photon interference in 10 km long optical fibre interferometer", *Electronics Letters*, 29 (1993), 634-635.

[4]   R.J. Hughes, G.L. Morgan and C.G. Peterson, "Quantum key distribution over a 48-km optical fiber network," *Journal of Modern Optics*, 47 (2000), 533-547.

[5]   H.Kosaka, A.Tomita, Y.Nambu, T.Kimura and K.Nakamura, "Single-photon interference experiment over 100 Km for quantum cryptography system using balanced gated-mode photon detector", *Electronics letters*, 39 (2003), 1199-1201.

[6]   C. Kurtsiefer, P. Zarda, M. Halder, P.M. Gorman, P.R. Tapster, J.G. Rarity and H. Weinfurter, "Long distance free-space quantum cryptography", *Proceedings of the SPIE*, 4917 (2002), 25-31.

[7]   O.L. Guerreau, J.-M. Merolla, A. Soujaeff, F. Patois, J.-P. Goedgebeur, F.J. Malassenet, "Long-distance QKD transmission using single-sideband detection scheme With WDM synchronization", *Journal of Selected Topics in Quantum Electronics*, 9 (2003), 1533-1540.

[8]   P. Moller, C. Schori, J.L. Sorensen, L. Salvail, I. Damgard and E. Polzik, " Experimental quantum key distribution with proven security against realistic attacks", *Journal of Modern Optics*, 48 (2001), 1921-1942.

[9]   G. Gilbert and M. Hamrick, " Secrecy, Computational Loads and Rates in Practical Quantum Cryptography " , *Algorithmica*, 34 (2002), 314.

[10] D. Gottesman and H.-K. Lo, "Proof of security of quantum key distribution with two-way classical communications", *IEEE Transactions on Information Theory*, 49 (2003), 457.

[11] J.C. Bienfang, A.J. Gross, A. Mink, B.J. Hershman, A. Nakassis, X.Tang, R.Lu, D.H. Su, C.W.Clark, C.J. Williams, E.W. Hagley, J. Wen "Quantum key distribution with 1.25 Gbps clock synchronization", *Optics Express*, 12 (2004), 2011-2016.

[12] A.V. Krishnamoorthy and D.A.B. Miller, "Scaling optoelectronic-VLSI circuit into the 21st century: a technologiy roadmap" , *Journal of Selected Topics in Quantum Electronics*, 2 (1996), 55-76.

DPSK OVER INVERSE-RZ OPTICAL PULSES FOR 2-BIT PER SYMBOL TRANSMISSION

Tetsuya Miyazaki, Fumito Kubota
National Institute of Information and Communications Technology, 4-2-1 Nukuikita, Koganei, 184-8795 Tokyo, Japan: tmiyazaki@nict.go.jp

Abstract: We successfully demonstrated superimposing DPSK over inverse-RZ optical pulses using SOA-XGM for 2-bit per symbol modulation/demodulation at 20-Gbit/s to simply double spectral efficiency. Error free operation less than 10^{-12} was achieved for both RZ and DPSK signals.

1.   INTRODUCTION

Multi-state per symbol modulation format such as differential quadrature phase-shift keying (DQPSK) has been investigated to enhance spectral efficiency and tolerance to chromatic dispersion and polarization-mode-dispersion [1,2]. A combination of amplitude-shift keying (ASK) and differential phase-shift keying (DPSK) is another scheme without using complex encoder and decoder[3,4]. However precise adjustment of the extinction ratio has been required in the ASK-DPSK scheme. On the other hand inverse-RZ format with more than 100% spectral efficiency was proposed without using multi-state per symbol modulation scheme[5]. In this study, we demonstrated superimposing of DPSK over inverse-RZ as a simple solution for doubling the spectral efficiency.

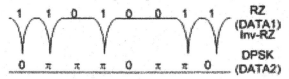

Figure 1 Timing chart of the overwriting of DPSK over inverse-RZ signal.

2. EXPERIMENTS AND RESULTS

Figure 1 shows the timing chart of the overwriting of DPSK over inverse-RZ signal. Inverting of RZ optical signal allows both "0" and "1" pulses to have finite pulse energy in the time slot without adjusting the extinction ratio.

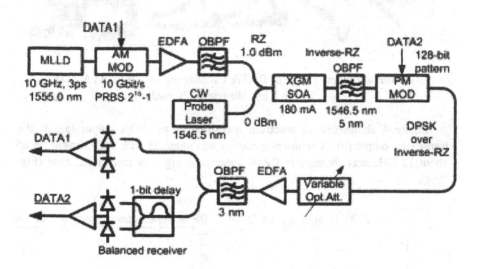

Figure 2 Experimental setup.

Figure 2 shows the setup. 10-GHz (9.95328 GHz), 3.0 ps optical pulses from a mode-locked laser diode (MLLD, 1555 nm) was modulated at 10 Gbit/s with a PRBS of 2^{15}-1 (DATA1) by an intensity modulator for RZ modulation. For inverse-RZ generation, cross-gain modulation (XGM) in a semiconductor optical amplifier (SOA, biased at 180 mA) was adopted by injecting continuous wave (CW) probe (1546.5 nm, 4.0 dBm) and the RZ pulses (-1 dBm). Previously reported technique for inverse-RZ generation in electrical domain using push-pull type Mach-Zehnder modulator[5] is also applicable with an encoder to compensate alternative phase inversion. Optical band-pass filter (OBPF, 5nm) was employed to select converted inverse-RZ optical pulses. Then DPSK modulation with 128-bit pattern length (DATA2) was superimposed. In the receiver side, optical pre-amplifier (EDFA) followed by an OBPF (3 nm) and a balanced receiver was used for both RZ and DPSK detection. Single-end direct detection was used to convert polarity from inverse-RZ into RZ pulses. We set the expected differential data pattern to an error detector.

Figure 3 shows optical waveform of DPSK with inverse-RZ measured at transmitter output (Fig. 3(a)), and detected eye diagrams of DATA 1 RZ signal

398

(Fig. 3(b)), and DATA 2 DPSK signal (Fig. 3(c)). Thanks to the pulse inversion, clear eye opening was obtained in detected DPSK signal (Fig.3 (C)). There was no prominent eye opening degradation in RZ detection by superimposing DPSK modulation (Fig.3 (b)).

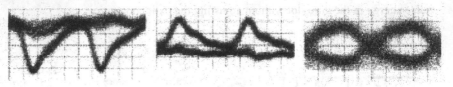

(a)transmitter output   (b) DATA 1 RZ signal   (c) DATA 2 DPSK

Figure 3 Eye diagrams (50 ps/div.)

Figure 4 shows optical spectrum measured at the SOA output (a), at the transmitter output (b). A compact spectrum was obtained with 3-dB bandwidth of about 12 GHz was obtained in DPSK superimposing over inverse-RZ case (Fig. 2(b)).

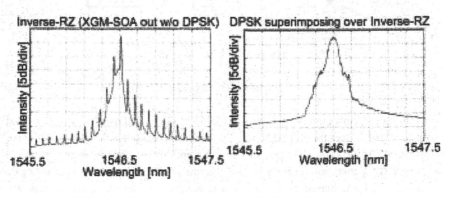

(a) SOA output   (b) Transmitter output

Figure 4 Optical spectrum (Span 2 nm, resolution 0.01 nm)

Figure 5 shows the bit error rate (BER) characteristics for DPSK only (open circles), DATA 1: RZ (closed squares), DATA 2 : DPSK with inverse-RZ (closed circles). The received optical power was defined at the optical pre-amplifier input. It should be noted that there is only 2-dB receiver sensitivity penalty at 10-9 in DPSK with inverse-RZ from DPSK only. There was no error floor and error free less than 10-12 was achieved for both DPSK with inverse-RZ and RZ with DPSK.

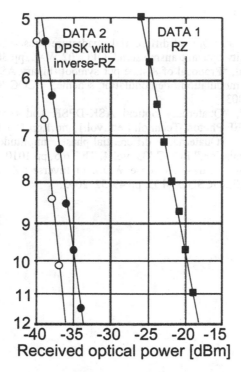

Figure 5 Bit error rate (BER) characteristics
open circles : DPSK only, closed circles : DPSK with inverse-RZ,
closed squares : RZ

3. CONCLUSION

We demonstrated DPSK over inverse-RZ format for 2-bit per symbol modulation/demodulation at 20 Gbit/s to simply double spectral efficiency. High sensitivity of DPSK data signal is preserved within 2-dB penalty from its original sensitivity by superimposing DPSK data onto inverse-RZ optical signal.

ACKNOWLEDGMENTS

The authors wish to thank Dr. Y. Matsushima for his continuous encouragement.

REFERENCES

[1] R.A.Griffin et al, "Optical differential quadrature phase-shift key (oDQPSK) for high capacity optical transmission," OFC '02,WX6, pp.367-368, 2002.

[2] S.Hayase, et al, "Proposal of 8-state per symbol (binary ASK and QPSK) 30-Gbit/s optical modulation / demodulation scheme," ECOC '03, vol.2, Tu1.6.6, pp.204-205,2003.

[3] M.Ohm et al, "Quaternary optical ASK-DPSK and receivers with direct detection," IEEE Photon. Technol. Lett.,vol.15,no.1,pp.159-161,2003.

[4] X.Liu et al., "Quaternary differential-phase amplitude-shift-keying for DWDM transmission," ECOC '03, vol.4, Th 2.6.5, pp.1010-1011, 2003.

[5] M.Ogusu et al., "Ultra-high dense WDM of Inverse-RZ signals," Technical Report of IEICE, OCS2003-105, pp.49-52, 2003.

PART B3:

ALL-OPTICAL PROCESSING

A 40 GHZ POLARIZATION MAINTAINING PICOSECOND MODELOCKED FIBER LASER EMPLOYING PHOTONIC CRYSTAL FIBER

Kazi S. Abedin[1] and Fumito Kubota[2]

[1] National Institute of information and Communications Technology, 4-2-1, Nukui-Kita, Koganei-Shi,Tokyo 184-8795 JAPAN, abedin@nict.go.jp

[2] National Institute of information and Communications Technology, 4-2-1, Nukui-Kita, Koganei-Shi,Tokyo 184-8795 JAPAN, kubota@nict.go.jp

Abstract: We demonstrate a harmonically mode-locked dispersion-managed polarization-maintaining erbium fiber laser that uses photonic crystal fiber for nonlinear pulse compression. The high nonlinearity and large anomalous dispersion of the PCF resulted in significant reduction in the cavity length and increased the long-term stability. The laser cavity, only 36-m-long, yielded stable 1.3 ps pulses at a 40-GHz repetition-rate, with supermode noise suppression of over 60 dB.

1.   INTRODUCTION

Compact tunable sources providing picosecond pulses with tens of GHz repetition rates and low timing jitter have high demands in many application such as, high speed optical communication systems, optical analog-to-digital conversion and so on. Harmonically active modelocking of fiber lasers is a very useful technique for generating picosecond pulses from fiber lasers with repetition rates of tens to even over a hundred GHz [1-7]. Mode-locked fiber lasers using the erbium-doped fibers (EDF) as the gain medium have been studied most extensively as they produce pulses in the 1.55 μm communication window where the silica fibers exhibit the lowest loss. Technically, short pulse generation at GHz-repetition-rate in erbium doped fiber laser is conveniently realized by taking advantage of the high saturation power and the availability of optical fibers with both normal and anomalous dispersion characteristic. One can appropriately manage the cavity-dispersion to favor intracavity nonlinear processes, such as the formation of optical solitons. Operation of the laser in the soliton regime not only yields pulses shorter than the Kuizenga-Siegman limit [8], but also ensures pulse-dropout which is particularly important for application in error-free optical communication.

In the high-repetition-rate (10 GHz or above), picosecond pulses can be generated by soliton-effect compression in a long segment of suitable anomalous dispersion fiber placed inside the cavity. Using dispersion shifted fiber (DSF) of 190 m long inside a laser cavity, solitons with widths of 2.7 ps at 10 GHz-rate [2] and 0.9 ps at 40-GHz rate were generated [6]. Recently, highly nonlinear DSF fibers were also deployed with an aim to reduce the length of nonlinear fiber, and indeed 1.2 ps pulses at 40 GHz-rate were obtained from an actively mode-locked fiber laser that used 100 m of highly nonlinear DSF fiber [4].

Alternatively, picosecond pulses can also be produced from a dispersion managed (DM) laser cavity, which has fiber segments with large local dispersion, but a small anomalous path-averaged dispersion. In a DM laser a pulse spends much of its time in a stretched state and experiences a lower effective nonlinearity than a fundamental soliton pulse in a comparable uniform dispersion cavity [9]. This allows pulses to circulate with higher energy in the cavity, yielding higher output power. Pulses with a width of 1.3 ps, timing jitters of only 10 fs, and pulse dropout ratio smaller than 10^{-14} were generated at 10 GHz rate from a DM sigma fiber laser [3, 10].

We have recently demonstrated a mode-locked erbium fiber laser which uses a polarization maintaining PCF (PM-PCF) with high nonlinearity and anomalous dispersion for nonlinear pulse compression [11]. The high nonlinearity and large anomalous dispersion of the PCF made nonlinear pulse compression achievable in fiber only 10-m long. Highly-stable pulses with 1-ps-width and 10-GHz repetition-rate were obtained over a range of 1535-1560 nm. In this paper, we demonstrate successful operation of PCF-based compact fiber laser at 40-GHz-repetition-rate. We have produced ~1.3 ps pulses at 40-GHz repetition rate with supermode noise suppression better than 60 dB and output power of over 14 mw.

2. EXPERIMENTAL SETUP

The experimental setup of the mode-locked fiber ring laser is shown in Fig. 1. The cavity consisted of a 20-m long Er-doped PANDA fiber, a Mach-Zehnder modulator, an optical isolator, a tunable bandpass filter (3-dB-bandwidth: 8 nm), a 30% output coupler, and a 10-m-long polarization maintaining PCF (PM-PCF) section. The Er-fiber had an un-pumped absorption coefficient of 3.28 dB/m and dispersion of -54 ± 5 ps/nm/km at 1550 nm. The PM-PCF, manufactured by Mitsubishi Cable Industries, had a Ge-doped core that was surrounded by a hexagonal array of holes. The cross section of the PM-PCF is shown in the inset of Fig 1. The fiber had a zero-dispersion wavelength at 876 nm, and exhibited a dispersion parameter of 104 and 126 ps/nm/km, for the slow and fast axes respectively. The PM-PCF had a nonlinear coefficient of 39.5 W^{-1}km^{-1} and a loss coefficient of 16 dB/km (at 1550 nm). To facilitate optical coupling, both ends of the PM-PCF were fusion-spliced to standard PANDA fiber through a mode-field converter, which yielded a loss/splice of only 1.6 dB. The cavity also employed a fused fiber polarization beam splitter (PBS) that restricted oscillation to slow axes in the cavity. The doped fiber and PM-PCF within the laser cavity mapped a dispersion-managed (DM) periodic system with large dispersion variation. The dispersion map of the cavity, which mostly consisted of the erbium fiber and the PCF is shown in Fig. 2. The average value of dispersion in the 36-m-long cavity was 1.4 ps/nm/km, and the fundamental cavity repetition rate was 5.6 MHz.

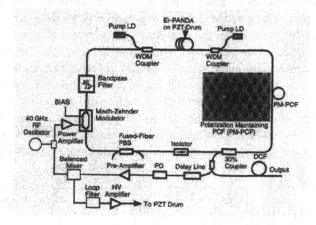

Figure 1. Experimental setup

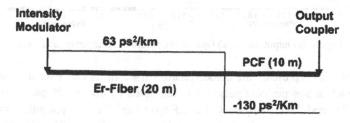

Figure 2. Dispersion map of the laser.

For modelocking at a 40-GHz-repetition-rate, we directly drove the modulator using an RF oscillator and a narrow band RF power amplifier. We also stabilized the cavity by controlling the length of the Er-PANDA fiber that was wound around a cylindrical piezoelectric transducer (PZT). The error signal necessary for active feedback was obtained from the phase of the output pulses relative to the clock signal available at the IF output port of the double balanced mixer. The error signal was processed with proportional and integration control circuitry, amplified and applied to the PZT [12].

3.   RESULTS

Modelocking of the laser was achieved at a repetition rate of ~40 GHz by carefully adjusting the oscillator frequency to match a harmonic of the fundamental cavity-repetition-rate and the bias voltage of the modulator. The optical spectrum of the mode-locked pulses is shown in Fig. 3(a), which shows the optical comb of frequencies separated by a frequency equal to pulse repetition rate. The ASE noise-floor was over 10^{-4} times lower than the peak of spectrum. The envelope of the optical spectrum could be fitted by a Gaussian-like envelope with an FWHM bandwidth of 2.69 nm. In soliton systems that has largely differing dispersions [3, 10], optical spectrum with the Gaussian rather than sech-profile are commonly seen. In our experiment, we obtained output pulses with nonzero chirp that was accounted for by external chirp-compensation using dispersion compensating fiber (DCF) of about a 4-m-long. The autocorrelation trace of the shortest pulse thus obtained is shown in Fig. 3(b), which yielded an FWHM pulsewidth of 1.29 ps.

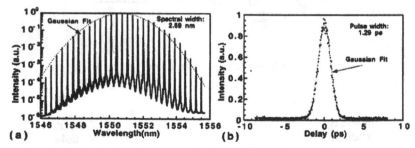

Figure 3. Output pulses. a) Optical spectrum. b) Autocorrelation trace.

Figure 4(a) represents the autocorrelation trace measured over a longer time scale, which shows periodic pulse trains separated by about ~25 ps, in consistent with the RF modulation frequency. The RF spectrum of the output pulses showed a frequency component equal to pulse repetition-rate of 39.463 GHz. The supermode noise was suppressed by more than 60 dB. The average output power was 14.4 mW, which was limited by the power of the pump LDs. The wavelength of the output could be changed through the use of the tunable bandpass filter.

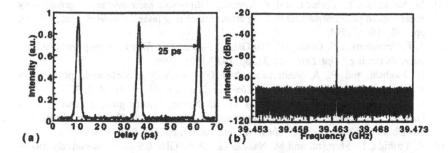

(a)

(b)

Figure 4. Output pulses. a) Auto/cross correlation trace, b) RF spectrum.

It is known that in a fiber system with dispersion maps that have large deviations of the local dispersion from the average-dispersion, the pulses have enhanced energy relative to solitons in a system with uniform dispersion that is equal to the path-averaged dispersion of the map [13]. The dispersion map strength factor $\gamma_s = 2 \sum_n |(\beta_{2n} - \beta_{2avg})_n l_n| / \tau_{FWHM}^2$ of the DM soliton laser, where β_{2n} and l_n is the group velocity dispersion and length of the nth fiber segment forming the cavity, τ_{FWHM} is the shortest FWHM pulse duration, β_{2avg} is the average dispersion, can be determined using a value of 1.29 ps for τ_{FWHM} and the parameters of the PCF and Er-PANDA. We obtain a value for γ_s of ~3.0 for the 36-m-long dispersion map, while it was 4.5 for the same laser used for producing 1.07 ps pulse at 10 GHz, repetition rate [11]. Higher value of γ and, correspondingly shorter pulses with larger output power are expected by using pump LDs with higher power and further optimization of the dispersion map of the cavity.

4.   CONCLUSIONS

In this paper we have demonstrated the application of highly nonlinear and anomalously dispersive PCF in compact pulse sources that generate tunable picosecond pulse at high repetition rate suitable for optical communication. The use of PCF has helped reduction in the cavity length by an order of magnitude. We successfully produced 1.3 ps pulses at a 40 GHz repletion rate, and the supermode noise was suppressed by more than 60 dB.

408

REFERENCES

[1] J. D. Kafka, T. Baer, and D. W. Hall, "Mode-locked erbium-doped fiber laser with soliton pulse shaping," *Opt. Lett.*, vol. 14, pp. 1269-1271, 1989.

[2] M. Nakazawa, E. Yoshida, and Y. Kimura, "Ultrastable harmonic and regeneratively mode-locked polarization-maintaining erbium fiber ring laser," *Electron. Lett.*, vol. 30, pp. 1603-1605, 1994.

[3] T. F. Carruthers, I.N. Duling III, "10-GHz, 1.3-ps erbium fiber laser employing soliton pulse shortening," *Opt. Lett.*, vol. 21, pp. 1927-1929, 1996.

[4] B. Bakhshi and P. A. Andrekson, "40 GHz actively modelocked, polarization maintaining erbium fiber ring laser," *Electron. Lett.*, vol. 36, pp. 411-413, 2000.

[5] D. Ellis, R. J. Manning, I. D. Phillips, and D. Nesset, "1.6 ps pulse generation at 40 GHz in phase locked ring laser incorporating highly nonlinear fiber for application to 160 Gbit/s OTDM networks," *Electron Lett.*, vol. 35, pp. 645-646, 1999.

[6] E. Yoshida, N. Shimizu, and M. Nakazawa, "A 40-GHz 0.9-ps regeneratively mode-locked fiber laser with a tuning range of 1530-1560 nm," IEEE Photon. Technol. Lett., vol. 11., pp. 1587-1589, 1999.

[7] K. S. Abedin, M. Hyodo and N. Onodera and, "Active stabilization of a higher-order mode-locked fiber laser operating at a pulse repetition-rate of 154 GHz," *Opt. Lett.*, vol. 26, pp. 151-153, 2001.

[8] D. J. Kuizenga and A. E. and Siegman, "FM and AM mode locking of the homogeneous laser — Part I: Theory," *IEEE J. Quantum Electron.*, vol. 6, pp. 694-708, 1970.

[9] K. Tamura, E. P. Ippen, H. A. Haus, and L. E. Nelson, "77-fs pulse generation from a stretched-pulse mode-locked all-fiber ring laser," Opt. Lett., vol.18., pp. 1080-1082, 1993.

[10] T. Carruthers, Irl N. Duling, M. Horowitz, and C. R. Menyuk, "Dispersion management in harmonically mode-locked soliton laser," *Opt. Lett.*, vol. 25, pp. 153-155, 2000.

[11] K. S. Abedin and F. Kubota, "A 10-GHz, 1-ps regeneratively modelocked fibre laser incorporating a highly nonlinear and dispersive photonic crystal fibre for intracavity nonlinear pulse compression," Electron. Lett., 40, pp. 58-60, 2004.

[12] X. Shan, D. Cleland, A. Ellis, "Stabilizing Er fiber soliton laser with pulse phase locking," Electron. Lett., vol. 28, pp. 182-184, 1992.

[13] V. S. Grigoryan, T. Yu, E. A. Golovchenkom, C. R. Menyuk and A. N. Pilipetskii, "Dispersion-managed soliton dynamics," *Opt.Lett.*, vol. 22, pp. 1609-1611, 1997.

40 GHZ ADIABATIC SOLITON GENERATION FROM A DUAL FREQUENCY BEAT SIGNAL USING DISPERSION DECREASING FIBER BASED RAMAN AMPLIFICATION

Ju Han Lee[1], Taichi Kogure[2], Young-Geun Han[1], Sang Hyuck Kim, Sang Bae Lee[1], and David. J. Richardson[3]

[1] Photonics Research Center, Korea Institute of Science and Technology (KIST)
39-1 Hawolgok-Dong, Seongbook-Gu, Seoul, Zip:136-791, Republic of Korea
TEL: +82-2-958-5719, FAX:+82-2-958-5709, Email: j.h.lee@ieee.org

[2] Information Technology R&D Center, Mitsubishi Electric Corporation
Kamakura 247-8501, Japan

[3] Optoelectronics Research Centre, University of Southampton
Highfield, Southampton, SO17 1BJ, United Kingdom

Abstract: We demonstrate the adiabatic soliton compression of a dual frequency beat signal using distributed Raman amplification (DRA) in a dispersion decreasing fiber (DDF). A 40 GHz beat signal generated from a LiNbO$_3$ modulator at a driving RF frequency of 20 GHz is compressed into ~ 2.2 ps soliton pulses using DRA in a 20 km DDF. The generation of high quality of soliton pulses from the 40 GHz sinusoidal beat signal is readily achieved with a significantly enhanced compression efficiency using DDF based DRA, compared to the case of using a DDF without DRA or a DSF with DRA.

1.   INTRODUCTION

All-optical high speed time division multiplexing (OTDM) systems operating at data rates in excess of 40 Gbit/s requires the development of stable sources of high repetition rate ultrashort optical pulses operating in the 1550 nm telecommunication band [1]. Among many other methods one very simple and cost-effective approach to high-repetition rate short pulse generation, is to use adiabatic soliton pulse compression of a dual frequency beat signal. In this approach the dual-frequency beat signal is usually obtained by either two separate distributed feedback (DFB) lasers [2] or a combination of a single DFB laser and a LiNbO$_3$ Mach-Zehnder modulator [3]. The Mach-Zehnder modulator based dual frequency generation method is likely to be a more reliable and simple approach since the generated two optical carriers are phase-synchronized and the corresponding beat signal has a very low phase noise [4]. In order to achieve high quality output pulses using adiabatic soliton compression of the frequency beat signal it is essential to ensure that the compression process is as adiabatic as possible. In a simple scheme based on dispersion decreasing fiber (DDF) the dispersion lengths associated with such sinusoidal beat signal dictate the use of multi-tens of km length scale DDF's in order to ensure that sufficiently adiabatic compression is achieved throughout the system. For such device lengths background fiber loss can have a significant impact on the compression process and limit the degree of compression that can be achieved for physically reasonable ranges of dispersion variation. In a distributed Raman amplification (DRA) based scheme multi-tens of km of fiber is also typically required to obtain the net on-off gain needed for practical values of pump power. Recently several research groups suggested a novel concept of the use of DRA in a DDF to achieve high quality adiabatic soliton compression [5, 6]. This approach enhances the compression factor for which high quality adiabatic compression is achieved for seed pulse sources carved by an electro-absorption modulator (EAM), and reduces the pump power requirements relative to those required for pure DRA compression.

In this paper we experimentally demonstrate the use of distributed Raman amplification in a dispersion decreasing fiber for the efficiency enhancement of adiabatic soliton compression of a dual frequency beat signal. We compress a 40 GHz beat signal generated from a LiNbO$_3$ modulator at a driving RF frequency of 20 GHz into ~ 2.2 ps soliton pulses using DRA in a 20 km DDF. The output pulses are then compared to those generated from the DDF without DRA in terms of pulse pedestal level. The same experiments are also performed with a 20 km DSF for the overall performance comparison. The generation of high quality of soliton pulses from the 40 GHz sinusoidal beat signal is readily achieved with an enhanced efficiency using distributed Raman amplification in a DDF.

2.   EXPERIMENTAL SETUP AND RESULTS

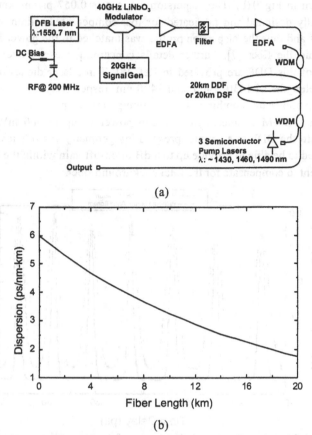

(a)

(b)

Figure 1. (a) Experimental setup. (b) Dispersion profile of the dispersion decreasing fiber (DDF) used in this experiment

The schematic of our experiment is shown in Fig. 1(a). A continuous wave (CW) DFB laser was first phase-modulated at a RF frequency of 200 MHz to suppress stimulated Brillouin scattering (SBS) in the dispersion decreasing fiber used in this experiment. The output beam from the DBF laser was then externally modulated using a LiNbO₃ Mach-Zehnder modulator which was biased at a null transmission point and driven at a RF frequency of 20 GHz to obtain a 40 GHz sinusoidal optical beat signal. A low-noise erbium doped fiber amplifier (EDFA) followed by a bandpass filter was used to amplify the modulated output beam and another high power EDFA was then employed to boost the optical power of the beat signal sufficient to form the first-order soliton at the input end of our DDF. The DDF used in this experiment had a length of 20 km and its dispersion followed an exponentially tapered profile at 1550 nm along the length from 6 to 1.75 ps/nm-km as shown in Fig. 1(b). The dispersion slope was 0.057 ps/nm²-km. The fiber was originally designed and fabricated for loss-compensated soliton transmission applications and thus the dispersion profile was matched to the power loss of 0.27 dB/km within the fiber [7]. Further details concerning the characterization and fabrication of this DDF are provided in Ref. [8]. Three laser diodes operating at center wavelengths of 1430, 1460, and 1490 nm, respectively were employed for the Raman pump. After combing the three pump laser outputs using a fiber based 14XX/C-band WDM coupler, a total pump power of up to 500 mW could be launched into the DDF in a counter-propagating geometry and this level of pump was observed to be able to provide up to 8 dB of on-off gain within the fiber which was sufficient to compensate for the total background fiber.

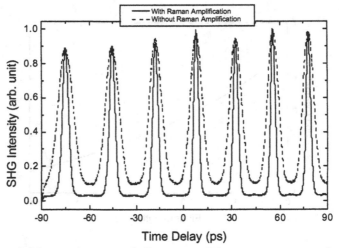

Figure 2. Measured autocorrelation traces of the output compressed pulses after the DDF for both the case with and without distributed Raman amplification.

Fig. 2 shows the measured second harmonic generation (SHG) autocorrelation traces of the compressed pulses after the DDF for both the case with and without DRA. High-quality, low pedestal soliton pulses were obtained with DRA in the DDF whilst both high pedestal level and low compression factor were observed without DRA. Note that the broad background pedestals in the autocorrelation traces are mainly due to the poor extinction ratio of the LiNbO₃ modulator used in this experiment and the non-uniform pulse intensity distribution can be attributed to the residual 20 GHz frequency components of the beat signal caused by the imperfect DC bias setting of the modulator as shown in the optical spectrum in Fig. 3(b). The temporal shape of the output compressed pulses with DRA in the DDF was observed to be a hyperbolic secant with a full width at half maximum (FWHM) of 2.2 ps as shown in Fig. 3(a). The corresponding optical spectrum of the pulses is shown in Fig. 3(b). The solid line over the measured spectrum shows a least squares sech2 fit, from which the spectral bandwidth was estimated to be about 1.1 nm. The time -bandwidth product of the pulses is 0.31, indicating that the output compressed pulses are almost transform-limited solitons.

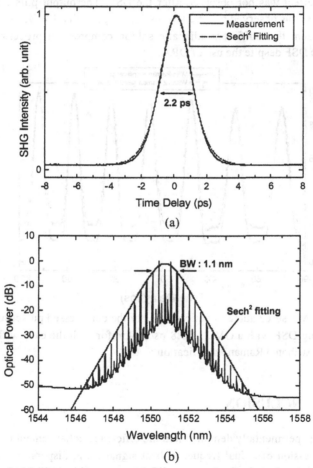

(a)

(b)

Figure 3. (a) Measured autocorrelation trace of a single pulse compressed with distributed Raman amplification in the DDF. (b) The corresponding optical spectrum of the puls e.

In order to confirm the enhanced performance of adiabatic soliton compression using the proposed scheme we performed the same experiment with a 20 km long dispersion shifted fiber (DSF) which was specially chosen considering the average value of the GVD of the DDF. The GVD of this DSF was 3.8 ps/nm-km. These results are summarized in Fig. 4. Although distributed Raman amplification along the DSF initiated an adiabatic soliton train formation in some degree, significant pulse compression process was not observed in terms of compression factor and pulse pedestal compared to the case of DRA in the DDF. Without DRA any pulse compression effect was not observed after the DSF. The output pulses after the DSF with DRA was observed to have a Gaussian shape with a FWHM of 5.8 ps. It is clearly evident that sufficient adiabatic soliton compression process was not induced in the DSF despite the use of DRA.

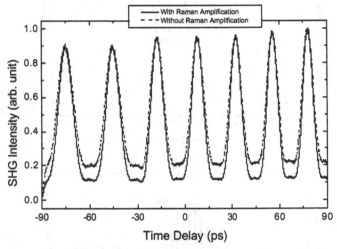

Figure 4. Measured autocorrelation traces of the compressed pulses after a 20 km long DSF with a GVD of 3.8 ps/nm-km for both the case with and without distributed Raman amplification.

3.   CONCLUSION

We have experimentally demonstrated the efficiency enhancement of adiabatic soliton compression of a dual frequency beat signal using dispersion decreasing fiber based distributed Raman amplification. Highly efficient adiabatic compression of a 40 GHz beat signal generated from a Mach-Zehnder modulator into a ~ 2.2 ps soliton pulse train, was achieved using DRA in a 20 km of DDF. The use of a DDF with an optimized dispersion profile and a high extinction ratio

LiNbO$_3$ modulator should allow for providing further improvements in terms of compression efficiency and background pedestals.

REFERENCES

[1]   E. Ciaramella, G. Contestabile, A, D'Errico, C. Loiacono, and M. Presi, 'High-power widely tunable 40-GHz pulse source for 160-Gb/s OTDM systems based on nonlinear fiber effects," *IEEE Photon. Technol. Lett.*, vol. 16, pp.753-755, 2004.

[2]   A. V. Shipulin, E. M. Dianov, D. J. Richardson, and D. N. Payne, "40 GHz soliton train generation through multisoliton pulse propagation in a dispersion varying optical fiber circuit," *IEEE Photon. Technol. Lett.*, vol. 6, pp.1380 – 1382, 1994.

[3]   A. D. Ellis, W. A. Pender, T. Widdowson, D. J. Richardson, R. P. Chamberlin, and L. Dong, "All-optical modulation of 40-GHz beat frequency conversion soliton source," *Electron. Lett.*, vol. 31, pp.1362-1364, 1995.

[4]   E. A. Swanson, and S. R. Chinn, "40-GHz pulse train generation using soliton compression of a Mach-Zehnder modulator output," *IEEE Photon. Technol. Lett.*, vol. 7, pp.114-116, 1995.

[5]   I. Morita, N. Edagawa, M. Suzuki, S. Yamamoto and S. Akiba, "Adiabatic soliton pulse compression by dispersion decreasing fiber with Raman amplification," in *Proc. Opto-Electronics & Communications Conference*, 17P-16, 1996.

[6]   T. Kogure, J. H. Lee, and D. J. Richardson, "Wavelength and duration tunable 10 GHz, 1.3 ps pulse source using dispersion decreasing fiber based distributed Raman amplification," *IEEE Photon. Technol. Lett.*, vol. 16, pp.1167-1169, 2004.

[7]   D. J. Richardson, R. P. Chamberlin, L. Dong and D. N. Payne, "High quality soliton loss-compensation in 38km dispersion-decreasing fibre," *Electron. Lett.*, vol. 31, pp.1681-1682, 1995.

[8]   N. G. R. Broderick, D. J. Richardson, and L. Dong, "Distributed dispersion measurements and control within continuously varying dispersion tapered fibers," *IEEE Photon. Technol. Lett.*, vol.9, pp.1511-1513, 1997.

ALL-OPTICAL NONLINEAR SIGNAL PROCESSING AT A RZ RECEIVER

Sonia Boscolo, Sergei K. Turitsyn, and Keith J. Blow
Photonics Research Group, School of Engineering and Applied Science, Aston University, Birmingham B4 7ET, United Kingdom

Abstract:　　We propose a novel all-optical signal processor for use at a return-to-zero receiver utilising loop mirror intensity filtering and nonlinear pulse broadening in normal dispersion fibre. The device offers reamplification and cleaning up of the optical signals, and phase margin improvement. The efficiency of the technique is demonstrated by application to 40 Gbit/s data transmission.

1.　　INTRODUCTION

In optical fibre communication systems using return-to-zero (RZ) modulation format, an electrical low-pass filter is normally employed after direct detection at the optical receiver. This filter limits the receiver bandwidth in order to reduce the noise power mainly due to accumulated amplified spontaneous emission noise generated by amplifier repeaters. Also, the expansion of the pulse width given by the filter reduces the influence of the fluctuation of pulse position in time caused by timing jitter [1]. But there exists a trade-off in the cut-off frequency of the filter among inter-symbol interference, signal amplitude, and signal-to-noise ratio (SNR). In this paper, we present a novel scheme of all-optical nonlinear signal processing for application to a RZ optical receiver front-end, which combines the intensity filtering action of a nonlinear optical loop mirror (NOLM) for reamplification and cleaning up of the optical signals [2, 3] with the nonlinear pulse broadening in a normal dispersion fibre (NDF) for phase margin improvement [4, 5]. The efficiency of the proposed technique is numerically demonstrated by application to 40 Gbit/s RZ data transmission.

2.   OPERATION PRINCIPLE AND CONFIGURATION

The proposed pulse processor consists of an optical amplifier, a section of NDF, and an unbalanced NOLM (see Fig. 1). Qualitatively, the idea for the method is as follows. An input pulse to the pulse processor is amplified to the preferred power level of the device by the optical amplifier. During transmission along the NDF, the temporal waveform of the pulse is changed to a rectangular-like profile by the combined action of group-velocity dispersion and Kerr nonlinearity [6]. As a result, the pulse width is broadened and the centre portion of the pulse is changed to be flat. By utilising this property, the phase margin of a RZ pulse train can be improved [4, 5]. The phase margin improvement enables reduction of the influence of the displacement of pulse position at the receiver caused by timing jitter. Following the NDF, the pulse enters the NOLM. The unbalanced NOLM acts as a saturable absorber [7] and, hence, filters out low-intensity noise and dispersive waves from the higher-power pulse. This allows for restoration of the pulse amplitude and cleaning up of the distorted pulse. Also, whenever the NOLM operates in the region just after the peak of its switching curve, it enables stabilisation of amplitude fluctuations. In the case of a pulse train, the noise and radiative background in the zero timing slots is suppressed by the saturable absorption action of the NOLM, and the amplitude jitter of ones is also reduced [2, 3].

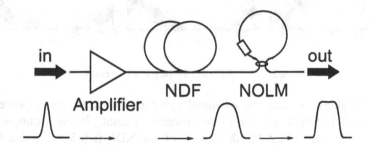

Figure 1.   Schematic diagram of the pulse processor.

In the sample system used for demonstration of the technique, the amplifier is an erbium-doped fibre amplifier (EDFA) with a noise figure of 4.5 dB. The NDF has a dispersion of $-20\,\mathrm{ps/(nm\,km)}$, an effective area of $30\,\mu\mathrm{m}^2$, and an attenuation of 0.24 dB/km. The NOLM incorporates a 50:50 coupler, and a 1.5 km-long loop of dispersion-shifted fibre with zero dispersion, an effective area of $25\,\mu\mathrm{m}^2$, and an attenuation of 0.3 dB/km. Unbalancing of the NOLM is achieved with an optical attenuator asymmetrically placed in the loop, and the loss of the loop attenuator is -27.1 dB. The NOLM is preferably operated in the stable region of its switching curve.

3.   MODELLING RESULTS

To demonstrate the efficiency of the proposed pulse processor, without loss of generality, the following model simulations are run. $2^7 - 1$ pseudorandom RZ single-channel pulse trains are transmitted at 40 Gbit/s in a dispersion-managed system whose transmission performance is severely degraded by intra-channel nonlinear effects when regenerators are not used (see [8] for details). In such a system, the periodical deployment of in-line NOLMs effectively sta-bilises the accumulation of amplitude noise and pulse distortion mainly driven by the intra-channel four-wave mixing, and the transmission distance is lim-ited by intra-channel cross-phase modulation-induced timing jitter [8]. This presents a good model situation to demonstrate the action of the proposed de-vice. The pulses after 20000 km transmission are used as the input for the pulse processor. The input full-width at half-maximum (FWHM) pulse width and energy are approximately 7 ps and 0.011 pJ, respectively. The signal qual-ity is evaluated in terms of the standard (Gaussian-based) Q-factor.

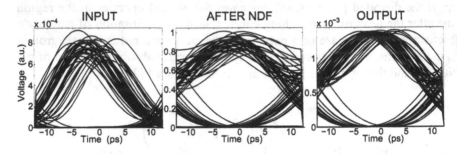

Figure 2.   Eye-diagrams in the pulse processor.

Figure 2 shows an example of signal eye-diagrams at the pulse processor input, the NDF output, and the pulse processor output. In this example, the gain of the EDFA is 33.5 dB, the length of the NDF is 0.5 km, and a fifth-order Bessel filter with a cut-off frequency of 30 GHz is used as a receiver low-pass filter. It can be seen that the input eye is closed mainly due to a significant timing jitter of the optical pulses. Dispersion and nonlinearity in the NDF broaden the pulse duration and simultaneously flatten the pulse shape. In this example, the FWHM pulse width is broadened to approximately 25 ps. Consequently, the eye opening at the NDF output is wider than at the pulse processor input. It is also seen that the amplitude jitter of pulses at the centre of the bit slot is slightly smaller. The additional widening of eye opening that can be observed at the pulse processor output is given by a significant reduction of the amplitude jitter provided by the NOLM.

The improvement of the signal quality that can be achieved in the nonlinear pulse processor-modified receiver with respect to the conventional receiver is

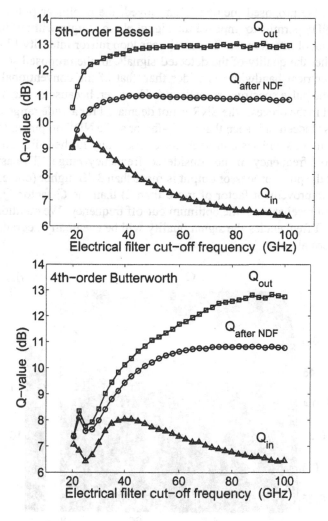

Figure 3. Q-values versus cut-off frequency of the receiver electrical filter.

evident from Fig. 3, where the signal Q-factor at the pulse processor input, Q_{in}, the NDF output, $Q_{after\,NDF}$, and the pulse processor output, Q_{out}, is plotted as a function of the cut-off frequency of the receiver electrical filter for two commonly encountered filter types. Here, the EDFA gain and the NDF length are the same as those used in Fig. 2. For the conventional receiver (see Q_{in}-curve), when the cut-off frequency is too low, the signal quality is degraded by the increase of inter-symbol interference and the decrease of pulse amplitude. Large cut-off frequency leads to a decrease of the SNR in the detected electrical signals, and this also results in degradation of the signal quality. On the other hand, when the proposed method is employed, the nonlinear pulse broadening in the NDF permits to improve the signal phase margin without increasing the inter-symbol interference. The nonlinear loop mirror intensity filtering improves further the quality of the detected signals. In the proposed scheme, the required electrical bandwidth is wider than that of the conventional receiver since also the pulse spectrum spreads out. However, because of the nonlinearity involved in the process, the SNR is not decreased for a wide range of cut-off frequencies. Indeed, it is seen that the Q-factor at the NDF output and the pulse processor output stabilises and does not degrade any further after some value of the cut-off frequency in the considered frequency range. The asymptotic Q-factor at the pulse processor output is more than 3 dB higher (corresponding to a linear improvement factor of more than 2) than the Q-factor Q_{in} for the conventional receiver with the optimum cut-off frequency. We mention that for larger cut-off frequencies the signal quality will be eventually degraded by the receiver thermal noise.

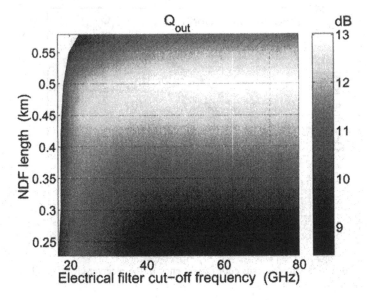

Figure 4. Q-factor at the pulse processor output versus receiver electrical cut-off frequency and NDF length.

A key parameter to be tuned in the proposed pulse processor is the length of the NDF. Figure 4 shows the Q-factor at the pulse processor output as a function of the cut-off frequency of the receiver electrical filter and the NDF length. Here, a fifth-order Bessel filter is used, and the conditions $Q_{out} \geq Q_{after\,NDF} \geq Q_{in}$ are required to be met. For each value of the NDF length, the gain of the EDFA is adjusted so as to provide both adequate enhancement of the nonlinearity in the NDF and adequate power at the NOLM input. It is seen that the pulse processor works with NDF lengths between approximately 0.2 and 0.6 km, and the optimum NDF length is 0.5 km. The decrease of Q_{out} for lengths shorter/longer than the optimum one at a fixed cut-off frequency is due to the fact that the power level of pulses at the NDF output differs from the correct power level for the NOLM to operate in the region just after the switching peak. It can also be seen that for each allowed NDF length, the Q-factor asymptotically stabilises at some value with increase of the cut-off frequency.

4. SUMMARY

We have described a novel all-optical pulse processing technique for use at a RZ optical receiver that exploits the intensity filtering action of a NOLM for reamplification and cleaning up of the optical signals and the Kerr effect in a NDF for improvement of the signal phase margin. The efficiency of the proposed pulse processor has been demonstrated by application to jitter-limited 40 Gbit/s RZ data transmission. The estimated Q-factor has been improved by more than 3 dB compared to the conventional receiver.

REFERENCES

[1] B. Bakhshi, P. A. Andrekson, M. Karlsson, and K. Bertilsson, "Soliton interaction penalty reduction by receiver filtering," *IEEE Photon. Technol. Lett.*, vol. 10, pp. 1042–1044, 1998.

[2] S. Boscolo, S. K. Turitsyn, and K. J. Blow, "Study of the operating regime for all-optical passive 2R regeneration of dispersion-managed RZ data at 40 Gbit/s using in-line NOLMs," *IEEE Photon. Technol. Lett.*, vol. 14, pp. 30–32, 2002.

[3] S. Boscolo, S. K. Turitsyn, and K. J. Blow, "All-optical passive 2R regeneration for $N \times$ 40 Gbit/s WDM transmission using NOLM and novel filtering technique," *Opt. Commun.*, vol. 217, pp. 227–232, 2003.

[4] M. Suzuki, H. Toda, A. H. Liang, and A. Hasegawa, "Improvement of amplitude and phase margins in an RZ optical receiver using Kerr nonlinearity in normal dispersion fiber," *IEEE Photon. Technol. Lett.*, vol. 13, pp. 1248–1250, 2001.

[5] M. Suzuki and H. Toda, "Q-factor improvement in a jitter limited optical RZ system using nonlinearity of normal dispersion fiber placed at receiver," in *Tech. Dig. OFC 2001*, Anaheim, California, WH3, March 17–22, 2001.

[6] H. Nakatsuka, D. Grischkowsky, and A. C. Balant, "Nonlinear picosecond-pulse propagating through optical fibers with positive group velocity dispersion," *Phys. Rev. Lett.*, vol. 47, pp. 910–913, 1981.

[7] N. J. Smith and N. J. Doran, "Picosecond soliton transmission using concatenated nonlinear optical loop-mirror intensity filters," *J. Opt. Soc. Am. B*, vol. 12, pp. 1117–1125, 1995.

[8] S. Boscolo, S. K. Turitsyn, and K. J. Blow, "All-optical passive quasi-regeneration in transoceanic 40 Gbit/s return-to-zero transmission systems with strong dispersion management," *Opt. Commun.*, vol. 205, pp. 277–280, 2002.

MODIFICATION OF DECODER FOR 2-D WAVELENGTH/TIME OPTICAL CDMA SYSTEM BY OPTICAL HARD-LIMITERS

Jozef Chovan[1,2], František Uherek[1,2]

[1] *International Laser Center, Ilkovičova 3, 812 19 Bratislava, Slovak Republic,*
e-mail: chovan@ilc.sk
[2] *Slovak University of Technology, Faculty of Electrical Engineering and Information Technology, Department of Microelectronics, Ilkovičova 3, 812 19 Bratislava, Slovak Republic, e-mail: uherek@elf.stuba.sk*

Abstract: We present the modification of decoder for 2-D Wavelength/Time (W/T) Optical Code Division Multiple Access (O-CDMA) system by optical hard-limiters. The performance improvement of the system using proposed decoder modification is shown by numerical simulation of the average Signal to interference difference and Bit error probability as a function of active user number and weight of codeword for the different topology of the decoder. The influence of non-ideal properties of the used optical hard-limiters upon the bit error probability in system is studied.

1. INTRODUCTION

Optical Code Division Multiple Access (O-CDMA) system is advance all-optical multi-channel scheme for fiber optic networks, which allows asynchronous transmission mode and accession to the optical network simultaneously. One of the most important problems for O-CDMA systems is the Multiple-access interference (MAI), which limits the number of active users in the system. We proposed the modification of 2-D W/T decoder [1] by placing the second optical hard-limiter (O-HL) after its last optical coupler (O-HL2 in Fig.1). The O-HL is a nonlinear all-optical threshold device. It has two valued output optical intensity, which depends on the optical intensity at the input [2]. The topologies

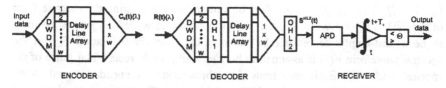

Figure 1. 2-D W/T O-CDMA encoder and modified decoder with O-HLs

of 2-D W/T encoder and modified 2-D W/T decoder are shown in Fig.1. The first O-HL (O-HL1) is placed into the each delay lines of delay line array of the decoder, which clips back the amplitudes of the unwanted created optical pulses at the same wavelength. The second O-HL (O-HL2) selects the autocorrelation pulse from the output optical pulses sequence, which determines the transmitted information data bits.

In the performance evaluation of the 2-D W/T O-CDMA system we assume an implemented passive O-HLs proposed in [2] by L. Brzozowski and E. H Sargent. The proposed passive O-HL is a periodic structure consisting of alternating layers of materials possessing different optical Kerr nonlinearity [2]. We used the approximation their properties by two ideality parameters IP1 and IP2 made in [3]. These two parameters characterize the measure of O-HL non-ideal properties. An intensity modulation and direct detection by APD optical receiver are assumed. Bit error probability (BEP) is used for performance evaluation of the system. BEP is calculated by noise model of APD optical receiver published in [4]. This model contains the excess noise of APD, thermal noise and MAI noise. In this paper we have used the average signal-to-interference difference (SID) as the MAI merits calculated by the following proposed numerical model of the 2-D W/T O-CDMA system.

2.   SYSTEM MODEL

We consider an incoherent optical O-CDMA system with 2-D W/T codewords. N users share the same optical medium usually, but not exclusively, in a star topology. Each k-th user's information bit form is encoded into a 2-D W/T codeword

$$C_k(t)(\lambda) = \sum_{i=1}^{m} \sum_{j=1}^{n} c_k[j][i] \, p_j(t - jT_c) \, v_i(\lambda - i\lambda_c), \tag{1}$$

where n is the length of the codeword in the time domain, $T=nT_c$ is the time duration of the information data bit, m is number of the available wavelength, $\Delta\lambda = m\lambda_c$ is the spectral width of optical information data bit and $C_k=[j][i]\epsilon\{0,1\}$, for $1 \leq j < n$, $1 \leq i < m$, is the j, i-th chip pulse of the k-th user's codeword. Let

$C_k=\{c_k[1][1], c_k[1][1], ... ,c_k[m][n]\}$ be a matrices representing the discrete form of the 2-D W/T codeword. Further, the chip signaling waveform $p_j(t)$ is assumed to be a unit-amplitude rectangular pulse of the chip time duration T_c and the spectral waveform $v_i(\lambda)$ is assumed to be a unit-amplitude rectangular pulse of the spectral width λ_c. Each user transmitter broadcasts its encoded signal to all receivers in the system by star topology of the system. The received signal is a sum of all active N users' transmitted signals

$$R(t)(\lambda) = \sum_{k=1}^{N} b_k C_k (t - \tau_k)(\lambda),$$ (2)

where $b_k \in \{0,1\}$ is the k-th user's information bit and $0 \leq \tau < T$ is the time delay for $k=1, ..., N$. The decoder performs the correlation function of the sequence of incoming optical chip pulses. The output signal of desired k-user's optical decoder is thus

$$S_k(t) = C_k(t)(\lambda)R(t)(\lambda).$$ (3)

If the second O-HL is applied to the output signal of optical decoder, then the output signal of the optical decoder can be express as

$$S_k^{HL2}(t) = I_{HL2}(S_k(t)),$$ (4)

where I_{HL2} is transmission function of O-HL [3]. The output signal of k-th user decoder is sequence of optical chip pulses with different amplitudes. It can be expressed as

$$S_k(t) = \sum_{i=1}^{2n} a_k^l[i]p_i(t - iT_c),$$ (5)

where a_k^l is a normalized amplitude of the optical chip pulse of i-chip time, $l \in \{0,1\}$ for $b_k=1$ and $b_k=0$ transmitted data bit, respectively. Let $A_k^l=\{a_k^l [1], a_k^l [2], ..., a_k^l [n]\}$ be a vector representing the discrete form of the output chip pulse sequence of k-th user decoder. Let

$$a_{k,max}^0 = \max\{a_k^0[1], a_k^0[2], ...a_k^0[n]\},$$ (6)

$$a_{k,max}^1 = \max\{a_k^1[1], a_k^1[2], ...a_k^1[n]\}.$$ (7)

In this paper the average SID is defined as

$$SID = \frac{1}{N}\sum_{k=1}^{N} a_{k,max}^1 - a_{k,max}^0.$$ (8)

The probability that a specified number of photons are absorbed from an incident optical field by an optical detector over a chip interval T_c is given by a Poisson distribution [4]. The photon absorption rate λ_s is given by

$$\lambda_s = \frac{\eta \, P_r \, \lambda}{h \, c}, \tag{9}$$

where η is the APD quantum efficiency, P_r is received laser power, λ is central wavelength and h is Plank's constant. Hence, due to "one" and "zero" data bit transmission, the mean $\mu_{b(1,0)}$ and variance $\sigma^2_{b(1,0)}$ of number of received photons [4] can be written as

$$\mu_{b(1)} = M_{APD} T_C \left[w\lambda_s + \left(\lambda_0 + \frac{I_b}{e} \right) \right] + T_C \frac{I_s}{e}, \tag{10}$$

$$\sigma^2_{b(1)} = M^2_{APD} F_e T_C \left[w\lambda_s + \left(\lambda_0 + \frac{I_b}{e} \right) \right] + \left(T_C \frac{I_s}{e} + \sigma^2_{th} \right), \tag{11}$$

$$\mu_{b(0)} = M_{APD} T_C \left[(w - SID)\lambda_s + \left(\lambda_0 + \frac{I_b}{e} \right) \right] + T_C \frac{I_s}{e}, \tag{12}$$

$$\sigma^2_{b(0)} = M^2_{APD} F_e T_C \left[(w - SID)\lambda_s + \left(\lambda_0 + \frac{I_b}{e} \right) \right] + \left(T_C \frac{I_s}{e} + \sigma^2_{th} \right), \tag{13}$$

where T_C is a time chip interval, n is a length of codewords and λ_0 is photon absorption rate due to a background light. The values of the parameters of APD detector is used the same as in [4]. In this performance evaluation we assume Gaussian approximation at the APD output. An optimum receiver uses the value of threshold [4], which minimizes the overall error probability. The average BEP is given by

$$BEP = \frac{1}{2} \left[erfc\left(\frac{\mu_{b(1)} - \theta_{opt}}{\sigma_{b(1)}} \right) + erfc\left(\frac{\theta_{opt} - \mu_{b(0)}}{\sigma_{b(0)}} \right) \right]. \tag{14}$$

3.   NUMERICAL RESULTS

The obtained numerical result of BEP variance as a function of number active users and weight of codewords for different values of IP2 and topology of decoder is calculated by PC software. PC software calculates the average value of BEP by equations (1-14). At the beginning of the BEP calculation, 2-D generalized multi-wavelength prime codewords (GMWPC) are generated for each user of the 2-D W/T O-CDMA system by algorithm proposed in [5]. The independent input parameters of this algorithm are weight of codewords w and a set of prime $p_k \geq p_{k-1} \geq p_{k-2...} \geq p_1 \geq w$. The output of the algorithm is $p_1 p_2... \, p_k$ GMWPC with these parameters $(w \times p_1 p_2 p_k, \, w, \, 0, \, 1)$, zero autocorrelation side lobes, cross-correlation functions of at most 1. The choosing of the active users is random generated up to required number. The time delay between each user's codewords, τ_c in equation (2), is random generated by Gaussian distribution. After that it calculates the output optical pulse sequence of each active user 2-D W/T O-CDMA decoder for "one" and "zero" data bit transmission and the average value of SIR and BEP in 2-D W/T O-CDMA system with specified parameters.

The two periods of input and three periods of output signals of the 101 -th user decoder in the O-CDMA system with 150 active users and (7 x 343, 7, 0, 1) GMWPC are shown in Figs. 2 - 5. Figure 2 shows the input and output signal of this decoder without any O-HLs for "one" data bit transmission in time domain. The input signal is a sum all active N users' transmitted codewords (Eq. 2) and it contain optical chip pulses with normalized amplitudes more grater than one. It can be seen that, when O-CDMA decoder is without any O-HL, the amplitude of the created autocorrelation optical chip pulse does not have the defined normalized amplitude by used GMWPC and some other optical chip pulses have the same amplitude as the autocorrelation optical chip pulse. Thus in this case it is not possible to detect transmitted data bit by threshold device at the end of optical receiver, which is determined by autocorrelation pulse.

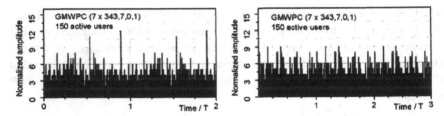

Figure 2. Input and output signals of the decoder without any O-HL

The left part of Fig.3 shows the first delay line signal of decoder, which contains the unwanted optical pulses created by MAI at the same wavelength with normalized amplitude greater than 1. Therefore the first O-HL1 is placed into the each delay lines, which clips back the normalized amplitudes of the unwanted created optical pulses to 1 (the right part of Fig.3).

Output signals of the decoder with only first O-HL1s for "one" data bit transmission are shown in the left part of Fig.4. In this case the amplitude differences between autocorrelation and the other optical pulses are not significant. In the case, when decoder is only with the second O-HL2, the decoder output signal contains the unwanted optical pulse with the same normalized amplitude than autocorrelation pulses (right part of Fig. 4).

Output signals of the decoder with both O-HLs for "one" and "zero" data bit transmission are shown in Fig. 5. In this case of the decoder topology it is possible to detect the transmitted information data bit by threshold device of optical receiver.

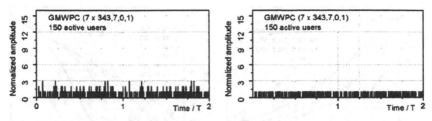

Figure 3. Signal of the decoder at first wavelength without any and with first O-HL1

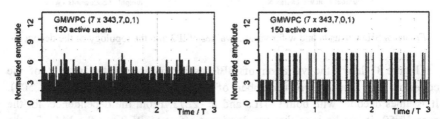

Figure 4. Output signals of the decoder with only first O-HL1 and with only second O-HL2

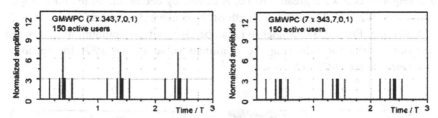

Figure 5. Output signals of the decoder with both O-HLs for "one" and "zero" data bit transmission

The obtained numerical result of average value of SID (Eq. 8) variance as a function of number active users and weight of codewords for different values of IP2 and the topology of the decoder can be seen in Fig. 6. The improvement of SID value by using decoder with both O-HL can be seen.

Figure 7 shows the dependence BEP versus number of the active users in the O-CDMA system with the decoder with only first O-HL (HL1) and with both O-HLs (HL1_HL2) for the fixed weight of codewords $w=9$ and fixed photon absorption rate $\lambda_s=1E14$. This dependence is calculated for three different value of IP2 (0.25, 0.5, and 0.75). Figure 7 depicts the improvement of BEP by using decoder with both O-HLs. The value of BEP for number of active users $N=200$ and $IP2=0.75$ is changed from 10E-1 to 10E-8 by using the second O-HL. Figure 7 also shows the influence of non-ideal properties of the used O-HL to the BEP. The influence of the second O-HL to the BEP is greater when its properties convergent to ideal.

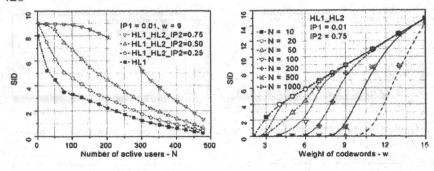

Figure 6. SID versus N and w for various values of IP2 and the topology of the decoder

Figure 8 shows the variation of BEP as a function of the weight of the codeword for four different number of active user (50, 100, 200, and 500). It assumes decoder with both O-HLs and $\lambda_s=2E13$. From the dependence, for the case $N=50$, it can be seen that the dominant effect to *BEP* has the MAI for the weight of the codeword approximately up to 6. The increasing of value of BEP, for the weight of codeword greater than 6, is caused by decreasing of the mean signal photon count per T_c, because the length of the codeword is increasing and thus T_c is decreasing. From this dependence it can be seen that the optimum weight of the codewords exits for achieving the minimum value of the *BEP* for specified parameters of the O-CDMA system.

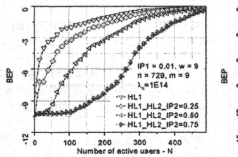

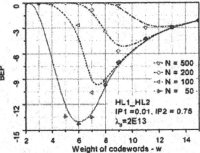

Figure 7. BEP versus N for the decoder with first one and both O-HLs and for various value of IP2

Figure 8. BEP versus w with decoder with both O-HLs for different number of active users N

4. CONCLUSIONS

The performance improvement of 2-D W/T O-CDMA system by using decoder with two O-HLs and the influence of O-HL properties have been shown by numerical modeling of the average Signal to interference difference and Bit error probability as a function of active users number and weight of codewords for different decoder topology.

The performance evaluation has shown that the dependence of bit error probability versus the weight of codewords has a local minimum, what allows choosing an optimum weight of the codewords for achieving the minimum bit error probability in 2-D W/T O-CDMA system for given system parameters.

ACKNOWLEDGMENTS

The work at the International Laser Center was supported by the project No.1/0130/03 and at the Department of Microelectronics of Faculty of Electrical Engineering and Information Technology by the project No.1/0152/03 both from the Slovak Ministry of Education.

REFERENCES

[1] J. Mendez, et al., "Design and Performance Analysis of Wavelength/Time (W/T) Matrix Codes for Optical CDMA," J. Lightwave Technol., vol. 21, pp. 2524-2533, Nov. 2003.

[2] L. Brzozowski, E. H. Sargent, "Nonlinear distributed-feedback structures as passive optical limiters," J. Opt. Soc. Am. B., vol.17, pp.1360-1368, 2000.

[3] J. Chovan, F. Uherek, P. Hábovčík, "Performance analysis of an optical CDMA system with non-ideal optical hard-limiters," Optics Communication, vol. 216, pp. 289-297, 2003.

[4] A.Srivastava, et al., "Performance evaluation of PIN+OA and APD receiver in multi-wavelength CDMA and WaCDMA networks," Optics Communication, vol. 191, pp. 55-66, May 2001.

[5] G. C. Yang, W.C.Kwong, "Performance Comparison of Multiwavelength CDMA and WDMA – CDMA for Fiber-Optic Networks," IEEE Trans. Communication, vol. 45, pp. 1426-1434, Nov. 1997.

THE INTEGRATION OF THE ALL-OPTICAL ANALOG-TO-DIGITAL CONVERTER BY USE OF SELF-FREQUENCY SHIFTING IN FIBER AND A PULSE-SHAPING TECHNIQUE

Takashi NISHITANI, Tsuyoshi KONISHI, and Kazuyoshi ITOH
Graduate School of Engineering, Osaka University, 2-1, Yamadaoka, Suita, Osaka 565-0871, Japan

Abstract: Integration of the all-optical analog-to-digital (A/D) converter using self-frequency shifting in fiber and a pulse-shaping technique is described. Optical A/D conversion has attracted much attention, in order to realize high-speed and high-throughput system for photonic networks. A/D conversion generally consists of sampling, quantization and coding. Whereas various optical sampling techniques have been proposed, there are few investigations of the optical quantization and optical coding. Previously, we have proposed an all-optical A/D converter. In this paper, we aim at the integration of the proposed all-optical A/D converter using high nonlinear fiber, arrayed waveguide grating, and variable optical attenuator, and demonstrate its operation.

1. INTRODUCTION

Analog-to-digital (A/D) converters have been widely investigated as connecting continuous analog signals in nature to discrete digital signals for signal processing and transmission. Because of the difficulty in achieving high-speed A/D converter by use of current electronic technology, the A/D converter has been and continues to be a bottleneck of the realization of high-speed, high-throughput system. Recent advance in communication markets has renewed interest in pursuit of high-speed A/D converters. For realization of high speed A/D converter, optical approach has attracted much attention recently [1].

A/D conversion consists of three operation parts: sampling, quantization, and coding. In the most of previously proposed optical A/D conversion system, quantization and coding are realized by electrical processing technique after the optical sampling process [2–4]. Because of the difficulty in realizing optical multilevel thresholding, which is essential to optical quantization, there are few investigations of the optical quantization and optical coding technique [5–7]. Nevertheless, all optical A/D conversion, which consists of optical sampling, optical quantization and optical coding, is indispensable for high speed A/D converter.

Previously, we have proposed the all-optical A/D converter for realization of optical quantization and optical coding by use of self-frequency shifting in a fiber and a pulse-shaping technique [8]. The proposed system is composed of dispersion shifted fiber (DSF) and a bulk system including gratings and lenses. To make it more stable one which can be used in the actual field, it would be one promising approach to integrate a whole of system by use of planar lightwave circuit technique. Besides, we have to make fiber much shorter for generating self-frequency shift. Because of the low nonlinearity of DSF, we need to propagate the ultra-short pulse in a long DSF for generating self-frequency shift.

In this paper, we aim at the integration of the proposed all-optical A/D converter by examining the above-mentioned points. We consider the use of high nonlinear fiber (HNLF), arrayed waveguide grating (AWG) and variable optical attenuator (VOA) for the system integration. And we demonstrate the integrated all-optical A/D conversion.

2. PRINCIPLE OF ALL OPTICAL A/D CONVERTER

2.1 Theoretical Background

The proposed A/D converter is composed of two parts: optical quantization and optical coding. In optical quantization the center wavelength of an output signal is shifted as a function of the power of an analog input signal. By using the difference of the center wavelength, we can achieve optical quantization of an input signal. In optical coding an output signal after optical quantization is shaped to an arbitrary digitized signal by use of the pulse-shaping technique.

In optical quantization, we use self-frequency shifting in a fiber [9, 10] and AWG. The Raman self-frequency shift in a fiber, given by

$$\frac{d\kappa}{dZ} = \frac{8}{15}\sigma_R\eta^4 \tag{1}$$

where κ is the center frequency of a soliton pulse, Z is the propagation distance in a fiber, σ_R is a coefficient of the self-induced Raman effect, and η is the amplitude of an input pump pulse [11]. Equation (1) suggests that the amount of frequency shift increases in proportion to the fourth of the amplitude of an input pulse. On the other hand, some experimental results show that the wavelength shift of a dispersed pulse increases in proportion to the twice of the amplitude of an input pulse [11, 12]. Although the mechanism of wavelength shift of a dispersed pulse is still under investigation, this power-proportional feature is suitable for quantization processes.

In optical coding, we use the pulse-shaping technique [13]. The pulse-shaping technique enables to generate high bit rate digitized signals by modulating a seed ultrashort pulse in frequency domain.

The pulse-shaping operation can be derived and described by

$$e_{out}(t) = \int E_i(\omega)H(\omega)exp(-i\omega t)d\omega \qquad (2)$$

where $e_{out}(t)$, $E_i(\omega)$, and $H(\omega)$ are the modulated temporal signal, the complex spectral amplitude of the original ultrashort pulse, and the Fourier transform function of the frequency-filter function. If we use an adequate frequency filter $H(\omega)$ in a pulse-shaping system and modulate the spectra of the original ultrashort pulse, we can synthesize an arbitrary-shaped pulse from a cross correlation between the original seed ultrashort pulse and the frequency filter function.

To provide a digitized signal corresponding to each power of an analog input signal, we use this pulse-shaping technique. Since the center wavelength of an ultrashort pulse is changed after self-frequency shifting, we can obtain an arbitrary digitized signal by preparing a different frequency filter for each center wavelength.

2.2 Integrated System Configuration

Figure 1 shows the schematic diagram of the integrated all-optical A/D converter by using a fiber and PLC devices.

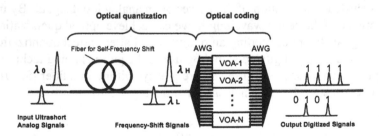

Figure 1. Schematic diagram of the integrated all-optical A/D converter

We compose the all-optical A/D converter by use of HNLF, AWG and VOA for integration of the system. The self-frequency shifted signals by use of short HNLF are generated more effectively and lower noise than by use of long DSF, and enable the integration of the system. The pulse-shaping technique by use of AWG and VOA enable to compose the compact system and achieve the high stability. On the other hand, a fiber for generating self-frequency shift is still long to integrate a whole of system.

If we can make it much shorter, we can integrate the system further. To do so, we need to optimize various parameters of fiber.

2.3 Simulation of self-frequency shift for integration of A/D converter

For the integration of the all-optical A/D converter, we have to make a fiber much shorter for generating the self-frequency shift. Because a high nonlinear fiber can be generated the low noise frequency shifted signal by propagating in a short fiber, we try to use a HNLF. To verify the characteristic of the shift of the center wavelength, we have a simulation of the pulse propagation in 5-m HNLF.

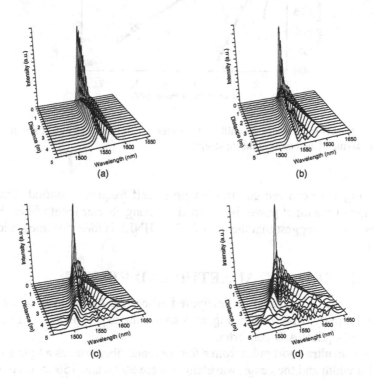

Figure 2. Simulation result of ultra short pulse propagation in 5m-HNLF: Peak power of input pulses (a) 100W (b) 200W (c) 300W (d) 400W

Figure.2 shows the simulation result of pulse propagation in 5-m HNLF by Split Step Fourier method [14]. The pulse width and the center wavelength of an input Fourier transform-limited pulse were 300fs and 1560nm, respectively.

From Fig.2 we can obtain the self-frequency shifted signal respected to the input pulse power by propagating in a short fiber. However, as the power of an input signal is higher, the noise of the self-frequency shifted signal is increased. It is necessary to achieve the low noise self-frequency shifted signal at wide range of the input power for the improvement of the quantization bit rate.

Figure.3 shows the simulation result of variation of the shift of the center wavelength in 2-m, 5-m, and 10-m HNLF

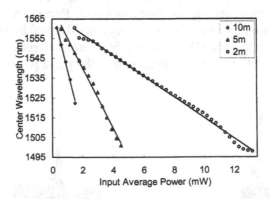

Figure 3. Simulation result of the shift of the center wavelength of an output signal form a HNLF by varying of the power of the input signal.

From Fig.3 we can see that the low noise self-frequency shifted signal at wide range of the input power is obtained by using shorter length fiber. From these results, we suggest that the use of a 2-m HNLF is ideal for integration of the system.

3.   EXPERIMENTAL SETUP AND RESULT

To verify the operation of the integrated all-optical A/D converter, we executed preliminary experiments. Figure 5 shows the experimental setup of the proposed all-optical A/D converter.

We used an ultra-short pulse from a femtosecond fiber laser as a light source. The pulse width and the center wavelength were 300fs and 1560nm. As input analog signals, we prepare three average power for the input pulses of 9.8mW, 10.1mW and 10.3mW adjusting the power of the input pulse by using a variable attenuator. To generate the self-frequency shift, signals propagated in a 2-m HNLF. Figure 4 shows the experimental result of variation of the shift of the center wavelength in a 2-m HNLF. From Fig.4 we verify the shift of the center wavelength in proportion to the power of input pulse.

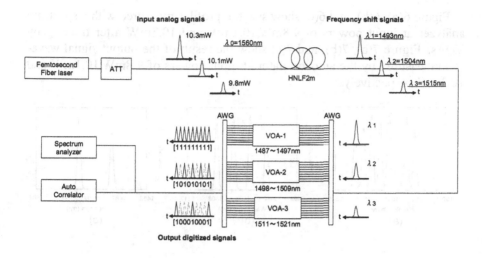

Figure 4. Experimental setup of the composed all-optical A/D converter

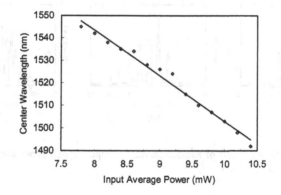

Figure 5. Experimental result of the shift of the center wavelength of an output signal from a 2m-HNLF by varying of the power of the input signal.

From this result, the center wavelengths of each analog signal changed to 1515nm, 1504nm and 1493nm, respectively. The wavelength shifted signals input to a pulse-shaping system composed of two AWGs and three VOAs. Each signal inputs to VOA-1, VOA-2, VOA-3 after by branching by AWG, and then is filtered in the frequency domain, as a result output three different digitized signals in the time domain. To measure the temporal profile of a generated signal, we used an interferometric autocorrelator.

Figure 6(a), 6(b), and 6(c) show spectral profiles measured with a spectrum analyzer at input powers of 9.8mW,10.1mW and 10.3mW after filtering by VOAs. Figure 7(a), 7(b), and 7(c) show the result of the output signal waveform measured by the autocorrelator at input powers of 9.8mW,10.1mW and 10.3mW, respectively.

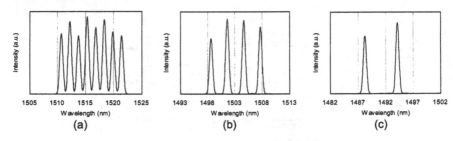

(a)   (b)   (c)

Figure 6. Experimental result of the output spectrum after filtering by VOA: Input average power (a) 9.8mW (b) 10.1mW (c) 10.3mW

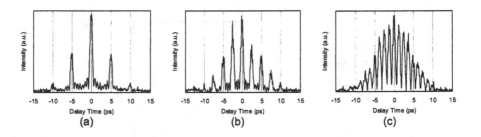

(a)   (b)   (c)

Figure 7. Experimental result of autocorrelation of output signal: Input average power (a) 9.8mW (b) 10.1mW (c) 10.3mW

From Figs 6 and 7, we can confirm that the integrated all-optical A/D converter successfully operates and output three different digitized signals, [100010001], [101010101] and [111111111] at three input power 9.8mW, 10.3mW and 10.5mW, respectively.

4. CONCLUSION

We integrate the proposed all-optical A/D converter by using HNLF, AWG and VOA. For integration of a fiber used, we simulate the pulse propagation in a HNLF. From simulation result we verify using HNLF is very effective for the integration of the system.

Preliminary experimental results show that 3-level different digitized signals from three average power of an ultrashort analog input pulse can be generated and the operation of the integrated system can be verified.

REFERENCES

[1] B. L. Shoop, *Photonic Analog-to-Digital Conversion*. Berlin, Germany: Springer-Verlag, 2001.

[2] S. Diez, R. Ludwig, C. Schmidt, U. Feiste, and H. G. Wiber, "160-Gb/s Optical Sampling by Gain-Transparent Four-Wave Mixing in a Semiconductor Optical Amplifier," *IEEE Photon. Technol. Lett.*, Vol.11, No.11, pp.1402-1404, November. 1999.

[3] M. Shirane, Y. Hashimoto, H. Yamada, and H. Yokoyama, "A Compact Optical Sampling Measurement System Using Mode-Locked Laser-Diode Modules," *IEEE Photon. Technol. Lett.*, Vol.12, No.11, pp.1537-1539, November. 2000.

[4] I. Kang, KF. Dreyer, "Sensitive 320 Gbit/s eye diagram measurements via optical sampling with semiconductor optical amplifier-ultrafast nonlinear interferometer," *Electron. Lett.*, Vol.39, No.14, pp.1081-1083, July. 2003.

[5] P. P. Ho, Q. Z. Wang, J. Chen, Q. D. Liu, and R. R. Alfano, "Ultrafast optical pulse digitization with unary spectrally encoded cross-phase moduration," *Appl. Opt.*, Vol.36, No.15, pp.3425-3429, May. 1997.

[6] L. Brzozowski and E. H. Sargent, "All-optical analog-to-digital converters, hard limiters, and logic gates," *J. Lightwave Technol.*, Vol.11, No.1, pp.114-119, January. 2001.

[7] S. Oda, A. Maruta, and K. Kitayama, "All-Optical Quantization Scheme Based on Fiber Nonlinearity," *IEEE Photon.Technol. Lett.*, Vol.16, No.2, pp.587-589, February. 2004.

[8] T. Konishi, K. Tanimura, K. Asano, Y. Oshita, and Y. Ichioka, "All-optical analog-to-digital converter by use of self-frequency shifting in fiber and pulse-shaping technique," *J. Opt. Soc. Am. B*, VOl.19, No.11, pp.2817-2823, November. 2002.

[9] F. M. Mitschke and L. F. Mollenauer, "Discovery of the soliton self-frequency shift," *Opt. Lett.*, Vol.11, No.10, pp.659-661, October. 1986.

[10] J. P. Gordon, "Theory of the soliton self-frequency shift," *Opt. Lett.*, Vol.11, No.10, pp.662-664, October. 1986.

[11] A. Hasegawa and Y. Kodama, *Solitons in Optical Communications*. Oxford U. Press, New York, 1995.

[12] N. Nishizawa, R. Okamura, and T.Goto, "Analysis of widely wavelength tunable femtosecond soliton pulse generation using optical fibre," *Jpn. J. Appl. Phys.*, Vol.38, No.8, pp.4768-4771, August. 1999.

[13] A. M. Weiner, "Femtosecond Fourier optics: Shaping and Processing of Ultrashort Optical Pulses," in *Trends in Optics and Photonics*, T.Asakura, ed., Berlin, Germany: Springer-Verlag, 1999.

[14] G. P. Agrawal, *Nonlinear Fiber Optics*. New York, Academic Press, 2001.

BER IMPROVEMENT USING A 2-R REGENERATOR BASED ON AN ASYMMETRIC NONLINEAR OPTICAL LOOP MIRROR

Markus Meissner,[1] Klaus Sponsel,[1], Kristian Cvecek,[1] Andreas Benz,[2] Stefan Weisser,[2] Bernhard Schmauss,[3] and Gerd Leuchs[1]

[1]Insitute of Optics, Information and Photonics, Max-Planck Research Group, University of Erlangen-Nuernberg, Guenther Scharowskys Str.1, D-91058 Erlangen, Germany [2]Lucent Technologies, Nurnberg, Germany; [3]Electrical Engineering Departement, University of Applied Science Regensburg, Germany

Abstract: BER improvement of 15 decades is observed using an asymmetric NOLM at a bitrate of 40Gbit/s and an input OSNR of 28.7dB. The BER improvement can be converted into 3.9dB of OSNR gain. Both were achieved by optimizing NOLM input power and splitting ratio. This allows for longer spans and thus reduces the over all amount of amplifiers or allows for an increased system reach.

1. INTRODUCTION

Signal regeneration is a key element for future data transmission at high bit rates. 3-R and 2-R regenerators are currently under investigation using all optical techniques. If no retiming is required NOLM based 2-R regenerator are an interesting alternative to 3-R regenerators, as they are less complex but have high potential for increasing the system reach [1, 2]. Their set-up uses only a few components and integration in an existing optical transmission system is simple [1, 2]. We use a 2-R regenerator based on an asymmetric NOLM suggested by Smith and Doran [3]. With the nonlinear power transfer characteristic of the device (see Fig. 2) it is possible to suppress noise and thus to prohibit the formation of new bit errors. The feasibility of such a setup concerning noise reduction and BER improvement was experimentally shown [4, 5].

We present for the first time to our knowledge a detailed analysis of the NOLM's impact on the BER. The BER improvement in dependence of the NOLM input power and splitting ratio is investigated, and the OSNR performance with optimized NOLM parameters is measured.

2. SIGNAL REGENERATION WITH AN ASYMMETRIC NOLM

Due to the nonlinear transmission characteristic of the NOLM (e.g. Fig.2) it is possible to reduce amplitude noise imposed on the "1" bits of the signal. In addition the noise on the "0" bits is reduced to some degree. By this noise reduction the probability density functions (pdf) of "1" and "0" bits are compressed, compensating for the broadening in the preceding EDFA. Therefore the formation of new bit errors due to an overlap of both pdfs is reduced. The NOLM itself is a passive device and thus cannot reverse any already existing overlap between the two pdfs, being responsible for the BER. Thus the usage of NOLMs as in-line regenerators has best impact on the signals evolution at the beginning of a transmission-link as done in our experimental setup. Implementation of the NOLM at the end of the link has only little impact on the BER. Furthermore the NOLM should be used at good OSNR around 30dB, as maximum noise reduction is achieved under these conditions [6].

3. EXPERIMENTAL SETUP

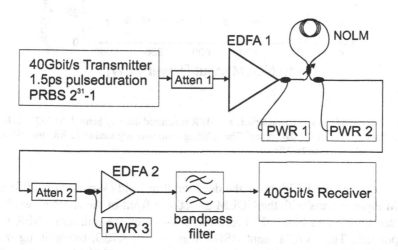

Figure 1. Experimental Setup. The setup emulates a 2 span transmission system in which fiber is emulated by attenuators.

The experimental setup is depicted in Fig.1. A 1.5ps laser with 10GHz repetition rate is multiplexed to a 40GB/s pulse train which is modulated with PRBS $2^{31} - 1$ data. With attenuator1 and the following high power EDFA1 the NOLM input OSNR and the NOLM input power are adjusted. The NOLM consists of a variable fiber coupler and 1840m of NZDSF fiber. After the NOLM the signal is attenuated and amplified again by EDFA2. This emulates a fiber span following the NOLM. The input power of EDFA2 is kept constant to -24dBm with attenuator2, thus adding always a constant noise to the NOLM output signal. Then the data signal is band pass filtered and detected.

4. EXPERIMENTAL RESULTS

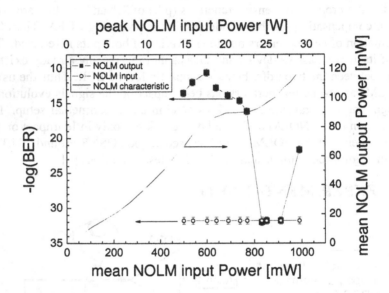

Figure 2. NOLM transfer characteristic and BER measured directly behind the NOLM. BER measured behind EDFA1 is shown, too. The splitting ratio was adjusted to 12:88, the OSNR at the output of EDFA1 was 28.7dB.

We measured the BER versus the decision threshold (V-curves) at different NOLM input powers with the NOLM directly in front of the receiver, excluding attenuator 2 and EDFA 2. From these V-curves the minimum BER was extrapolated. The NOLM input OSNR was set to 28.7dB, the splitting ratio was 12:88. As shown in Fig.2, the BER taken at the NOLM output, varies with the NOLM input power. For comparison the BER without the NOLM is also shown. It is clearly visible that the NOLM can only degrade the BER in this application, but never improve it. Only at the upper end of the plateau the NOLM induces no BER degradation, and the same level as without the NOLM is reached. Thus, implementing a NOLM at the end of a transmission link, directly in front of the receiver, is shown to be ineffective. Broaden-

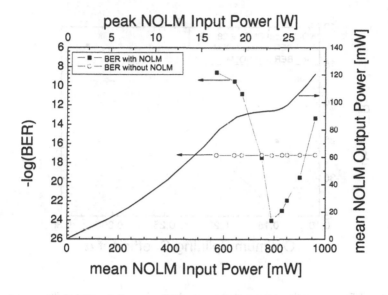

Figure 3. NOLM transfer characteristic, BER with and without NOLM measured at the output port of bandpass filter. The splitting ratio of the NOLM was set to 14:86, the OSNR at the output of EDFA1 was set to 28.7dB

ing the probability density functions of "1" and "0"bits with the subsequent EDFA2, by adding noise to the signal after the NOLM, BER improvement can be demonstrated as shown in Fig. 3. The receiver input OSNR was set to 25.7dB. As expected from the measurement without adding noise, the optimum operation point is the upper end of the plateau region. In these measurements, too, the BER is extrapolated from V-curves. The BER is reduced from 10^{-17} to 10^{-25}. Changing the NOLM input power the BER improvement reduces, and for powers below and above the plateau region even BER degradation may be observed. Varying the splitting ratio the BER can be optimized further, as shown in Fig.4. Best results are achieved for a splitting ratio of 16:84. For more symmetric splitting ratios the BER improvement reduces and for splitting ratios more symmetric than 28:72 the NOLM only degrades the signal. Measuring the OSNR performance of the NOLM leads to Fig. 5. This figure shows the BER with and without NOLM as a function of the output OSNR of EDFA1. Additionally this figure shows the OSNR gain achieved with the NOLM on its right hand side. As expected from noise measurements [6] and Fig.4 adaption of the NOLM splitting ratio on the NOLM input OSNR was necessary to optimize the results. The BER with NOLM was taken at the upper end of the plateau, as shown in Fig.3 for splitting ratios of 12:88 and

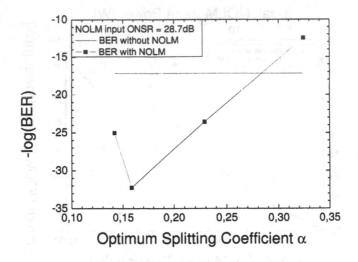

Figure 4. BER taken at the output port of the bandpass filter for different splitting ratios. The BER without NOLM is given for reference. The BER was taken at the upper end of the plateau.

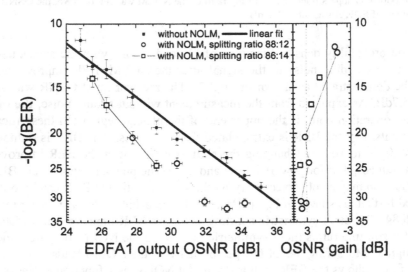

Figure 5. OSNR performance without and with NOLM for different OSNR, taken at the output of EDFA1, and different splitting ratios.

14:86. For NOLM input ONSRs between 27dB and 32dB, corresponding to BERs between 10^{-21} and 10^{-32}, more than 3dB of OSNR gain are achieved. Best OSNR gain of 3.9dB is achieved for a BER of 10^{-24}. Measurement of data for lower OSNR and higher BER could not be performed due to a lack of power, provided by EDFA1. At low OSNR a large part of EDFA1's pump power generates ASE noise, thus reducing the power available for signal amplification. Currently the power needed to operate the NOLM is very high. But with the development of modern photonic crystal fibers, fibers will be available which have a high nonlinearity and thus can reduce the power requirement to a few mW.

5. CONCLUSIONS

BER improvement by a 2-R regenerator based on an asymmetric NOLM has been studied. 15 decades of BER improvement at a NOLM input OSNR of 28.7dB were demonstrated after optimizing the splitting ratio to 16:84. For optimum operation the NOLM input power has to reach the upper end of the plateau. More than 3dB gain of OSNR were observed for NOLM input OSNR between 27dB and 32dB by implementing the NOLM into a transmission link. A maximum OSNR gain of 3.9dB was observed for NOLM splitting ratio of 86:14 and a BER of 10^{-24}. Thus by using the NOLM as an in-line regenerator the system reach can be more than doubled. Alternatively the length of a span can be elongated by up to 19km, as up to 3.9dB of extra attenuation can be tolerated. This in turn can be used to reduce the amount of amplifiers needed for a transmission system. The amount of amplifiers used for example in the FLAG-Atlantic-1 cable can be reduced by 31% when using NOLMs.

ACKNOWLEDGMENTS

We gratefully acknowledge the financial support of the Deutsche Forschungsgemeinschaft (DFG) and the Bayrische Staatsregierung.

REFERENCES

[1] Z. Huang, A. Gray, Y. W. A. Lee, I. Khrushchev and Ian Bennion: "40GB/S Transmission over 4000Km of standard fibre using in-line nonlinear optical loop mirror", *Proc. ECOC*, Rimini, Italy, Vol. 1, Mo4.6.3, 2003

[2] F. Seguineau, B. Lavigne, D. Rouvillain, P. Brindel, L. Pierre, and O. Leclerc: "Experimental demonstration of simple NOLM based 2-R regenerator for 42.66 GBit/s WDM long-haul transmissions", *Proc. OFC*, Los Angeles, USA, WN4, 2003

[3] N.J. Smith and N.J. Doran:"Picosecond soliton transmission using concatenated nonlinear optical loop-mirror intensity filters" *J. Opt. Soc. Am. B*, Vol. 12, pp. 1117–1125, 1995.

[4] M. Meissner, M. Roesch, B. Schmauss, G. Leuchs:"12dB of noise reduction by a NOLM based 2-R-regenerator",*Photonics Technology Letters*, Vol. 15, pp. 1297–1299, 2003

[5] R. Ludwig, A. Sizmann, U. Feiste, C. Schubert, M. Kroh, C.M. Weinert and H.G. Weber :"Experimental Verification of Noise Squeezing by an Optical Intensity Filter in High-Speed Transmission",*Proc. ECOC*, Amsterdam, Netherlands, 2001

[6] M. Meissner, M. Roesch, B. Schmauss and G. Leuchs: "Noise reduction performance of a NOLM based 2-R-regenerator in dependence on the OSNR", *Proc. ECOC*, Rimini, Italy, Vol. 3, P3.08, 2003

Calculate BER Improvement due to Nonlinear Regenerators

F.G. Sun, Z.G. Lu, G.Z. Xiao, and C.P. Grover
Photonic Systems Group, Institute for National Measurement Standards
National Research Council, M-50 1200 Montreal Road, Ottawa, Canada K1A 0R6
Email: fengguo.sun@nrc.ca

Abstract *Use the method we developed recently we calculate the bit-error-rate (BER) improvement as a function of transmitter extinction ratio and the optical link noise parameter.*

Introduction

Various optical regeneration techniques have been proposed and demonstrated [1]-[3] in order to eliminate noise, crosstalk, and signal distortion. All-optical 2R regeneration based on polarization rotation induced by nonlinear birefringence in a semiconductor optical amplifier was recently demonstrated [4] with an improved extinction ratio of 15dB for an input extinction of 5dB. The operating principle of such regenerators relies on the nonlinear input-output transfer characteristic. Recently we proposed a new method to evaluate the performance of a regenerator [5]. With this method in this paper we calculate the bit-error-rate (BER) improvement as a function of transmitter extinction ratio and the optical link noise parameter.

Calculation

We consider an optical transmission link of length L. The transmitter in the system is assumed to have a finite extinction ratio. A nonlinear regenerator is set at position l between the transmitter and the receiver. The system model is illustrated in Fig. 1.

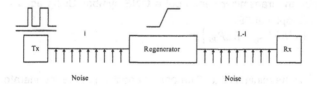

Fig. 1. System model.

The regenerator transforms the input signal x into an output $f(x)$

$$\bar{x} = f(x).$$

(1)

Because of the noise accumulation, the probability that a signal and noise will appear at a given level x is a function of the propagation length. Let $P_0(x,l)$ ($P_1(x,l)$) be the probability of getting a signal at a level x in the position l when the symbol ZERO (ONE) is sent from a transmitter, and let $P^N(y,l)$ be the probability of finding additional noise at a level y after the signal has travelled over a distance l. Assuming that the ZERO and ONE symbols are equally probable, in the absence of a regenerator in the optical link, the BER can be represented by

$$BER_N = \frac{1}{2}\int_D^\infty P_0(x,L)dx + \frac{1}{2}\int_{-\infty}^D P_1(x,L)dx .$$

(2)

The first (second) term is the contribution of the ZERO (ONE) rail. D is the decision level. When a regenerator is used, from the probability theory, the BER contribution of the ZERO rail becomes

$$BER_{R0} = \frac{1}{2}\int_{-\infty}^\infty \tilde{P}_0[g(\bar{x}),l]\left|g'(\bar{x})\right| \times \int_{D-\bar{x}}^\infty P^N(y,L-l)dyd\bar{x} .$$

(3)

where $\tilde{P}_0[g(\bar{x}),l]\left|g'(\bar{x})\right|$ is the probability of finding the output of the regenerator at a level $\bar{x}$ when the transmitter sends out a ZERO symbol, $P^N(y,L-l)$ is the probability of finding an additional noise in the second interval $L-l$ at a level y, and $g(\bar{x})$ is the inverse function of the nonlinear transfer function. Similarly, the BER contribution of the ONE rail can be written as follows

$$BER_{R1} = \frac{1}{2}\int_{-\infty}^\infty \tilde{P}_1[g(\bar{x}),l]\left|g'(\bar{x})\right| \times \int_{-\infty}^{D-\bar{x}} P^N(y,L-l)dyd\bar{x} ,$$

(4)

where $\tilde{P}_1[g(\bar{x}),l]\left|g'(\bar{x})\right|$ is the probability of finding the output of the regenerator at a level $\bar{x}$ when the transmitter sends out a ONE symbol. Using equations (3) and (4), we obtain an optimal BER

$$\min_{0\le l\le L}\{BER_{R0} + BER_{R1}\},$$

(5)

where $\min\{x\}$ is the minima of x. The optimal position of the regenerator is l_o and it provides the best BER value given by equation (5). The BER improvement attributable to the regenerator is

$$\log(BER_N) - \log(\min_{0\le l\le L}\{BER_{R0} + BER_{R1}\})$$

(6)

The method described here can be generalized to a situation in which more than one nonlinear regenerator is placed in the optical link. However, the calculation is much more complicated than if a single regenerator is used. In the following part of the paper we will limit ourselves to a single regenerator.

We assume that these probability distribution functions have a Gaussian form, its standard deviation σ is a function of l

$$\sigma = a\sqrt{l},$$

$$(7)$$

and a is a constant. We also assume that the nonlinear transfer function has the form [4]

$$\bar{x} = f(x) = \begin{cases} \gamma \cdot x & x < 1/2 \\ \gamma \cdot (x-1)+1, & x > 1/2 \end{cases},$$

$$(8)$$

Using the method we proposed recently, we obtain
The BER improvement is defined by

$$\Delta \log(BER) = \log(BER_{N0} + BER_{N1}) - \log(BER_{R0} + BER_{R1}).$$

$$(9)$$

In Fig. 1 we show the BER improvement, $\Delta \log(BER)$, as a function of the extinction ratio when the standard deviation is set at 0.1. This noise level is set to give a BER of 7.42×10^{-7} without regenerator for an extinction ratio equals 20dB. When a regenerator is used, the BER is improved to 5.18×10^{-12}. Then, we set $l/L = 0.5$ to calculate $\Delta \log(BER)$ as a function of $a\sqrt{L}$. The result is shown in Fig. 2.

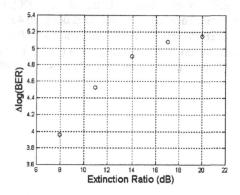

Fig. 1. $\Delta \log(BER)$ versus transmitter extinction ratio when the regenerator is located in its optimal position; $\sigma = 0.1$.

448

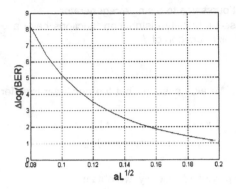

Fig. 2. Calculated $\Delta \log(BER)$ as a function of $a\sqrt{L}$ for $l/L = 0.5$ and an extinction ratio of 20dB.

Conclusions

In summary, we calculate the bit-error-rate (BER) improvement as a function of transmitter extinction ratio and the optical link noise parameter with a new method we proposed recently.

References

1　Geert Morthier et al IEEE Photon. Technol. Lett., vol. 12 (2000), pp.1516-1518
2　T. Otani et al IEEE Photon. Technol. Lett., vol. 12 (2000), pp.431- 433
3　Mingshan Zhao et al IEEE Photon. Technol. Lett., vol. 15 (2003), pp.305-307
4　Peter Öhlén et al IEEE Photon. Technol. Lett., vol. 9 (1997), pp.1011-1013
5　F.G. Sun et al IEEE Photon. Technol. Lett. Vol. 16 (2004), pp.1406-1408

PART B4:

PHOTONIC DEVICES

BRAGG GRATINGS PHOTOIMPRINTED IN INTEGRATED OPTICAL COMPONENTS: IMPROVING OF APODIZATION PROFILES

Lech Wosinski[1], Romano Setzu[2], and Matteo Dainese[3]
Laboratory of Photonics and Microwave Engineering, Royal Institute of Technology (KTH),
Electrum 229, S-16440 Kista, Sweden
[1]lech@imit.kth.se, [2]itex_ros@it.kth.se, [3]matteo@imit.kth.se

Abstract: Apodized waveguide Bragg gratings are written using Sagnac type interferometer. As the quality of these gratings is not enough for DWDM devices, a procedure for shaping of filter profiles was developed. Obtained 10 mm long grating gives line width of 0.4 nm and almost completely suppressed side lobes.

1.   INTRODUCTION

Silica-based planar lightwave circuits (PLCs) on a silicon platform will play a key role for fabrication of multifunctional devices for WDM network systems. Silica waveguides and waveguide devices will serve as communication paths between different elements on the platform as well as fill wavelength selective functions. Silica-on-silicon technology has shown the ability to keep high performance even for devices with high levels of integration, such as arrayed waveguide gratings.

Change of the refractive index of germanium doped silica under UV irradiation is commonly used for fabrication of UV imprinted fiber Bragg gratings. UV-induced changes of the refractive index in Ge-doped planar structures allow for adding new functionality to these components.

Imprinting of UV generated Bragg gratings in silica waveguides allows for building demultiplexers and add/drop filters [1] and can also be used for external cavity lasers to stabilize output wavelength [2]. UV processing of silica-based integrated devices also includes post-fabrication phase adjustment and wavelength tuning [3], direct UV-writing of waveguides [4], birefringence compensation by UV illumination [5] and UV-assisted poling [6]. WDM communication networks based on multifunctional integrated systems are a major step toward fully optical networking.

Bragg gratings with periodic modulation of the refractive index along the light guiding media were discovered in 1978 by Hill and co-workers in germanosilica optical fibers using illumination by an Ar-ion laser at 488 nm [7]. The use of UV-light at 244 nm was much more effective and the method proposed in 1989 by Meltz *et al.* [8] based on transverse holographic projection allowed to write gratings with any period by changing the angle between interfering beams. The experiment of Meltz and co-workers was a starting point for a dynamically grown research area of fiber Bragg gratings and later, waveguide gratings and related components.

Although it has gone more than 25 years since Hill's discovery, UV induced refractive index changes in germanosilicates are still not fully understood. Three main mechanisms that have been suggested by different research groups to explain the phenomena responsible for refractive index change in germanosilicate fiber cores due to UV illumination are: color-center model [9, 10], stress-relief or structural relaxation model [11, 12] and densification or compaction model [13, 14].

2.   MATERIAL ANALYSIS

The photosensitivity of planar waveguides usually consists of contributions from the three mentioned processes and the net result is determined by their relative importance that depends on a number of factors.

A large compressive stress at the germanosilica – silicon interface plays here an important roll, but due to complicated microstructure the relaxation of this stress upon UV irradiation can be accompanied by localized compaction and positive index change.

Similarly to standard communication fibers the UV-induced refractive index change in germanosilicate films is usually not sufficient and hydrogen loading is necessary prior to UV-illumination. As hydrogen loading is not very convenient, especially in the case of planar components, increasing of photosensitivity by boron codoping has been used here as an interesting alternative. Boron doping has an opposite behavior on the refractive index, with respect to Ge. This means that B addition has a twofold advantage regarding UV photosensitivity, allowing higher Ge content for the same designed Δ as well as contributing itself to material densification upon UV exposure [15].

To test the boron influence on the UV photosensitivity of Ge doped silica films, three types of samples have been compared: a purely Ge doped PECVD deposited film as a reference, and two PECVD films with increasing boron content, 2 at% and 5 at%. The results are shown in Figure 1.

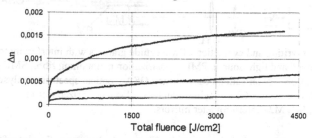

Figure 1. Variation of the refractive index in the germanium doped, boron codoped waveguide core upon UV exposure at 248nm. Pulse energy density 30mJ/cm². Curves present from bottom: pure Ge-doped material, 2 at% B codoped, 5 at% B codoped.

The maximum obtained change of the refractive index due to UV illumination exceeded 1.5×10^{-3}, which is the result comparable to hydrogen loading. For further experiments we have chosen the film with higher boron codoping.

3. EXPERIMENTAL SETUP

For writing Bragg gratings in fibers a simple interferometrical method based on Sagnac interferometer is often used [16]. A similar system consisting of a continuous wave (CW) UV laser, an interferometer and a monitoring unit has been used here for writing gratings in waveguides (Figure 2).

A frequency doubled argon ion laser Coherent Innova 90 Fred provides a high quality laser beam at 244 nm. The output power at this wavelength is 100 mW and the beam diameter is 0.6 – 0.8 mm. A cylindrical lens system focuses the beam in the waveguide plane as a sharp 20µm x 0.8 mm line. Such a formed beam is reflected from a motorized mirror MM and then split mainly to the –1 and +1 diffraction orders at the phase mask. Zero- and higher-order diffracted beams have quite low efficiency and fall out of the system. The phase mask is used here as a beam splitter. The waveguide is placed just above the phase mask. The two beams go in opposite directions through mirrors M1 and M2 and then are recombined on the waveguide aside from the upcoming beams by turning the mirrors M1 and M2 in such a way that down-going beams are slightly tilted out of the interferometer plane. This tilt is very small and does not introduce visible distortions in the interference plane, when using the system for fibers.

By moving the motorized mirror MM along the phase mask, the interference pattern is moved along the waveguide allowing writing gratings with maximum

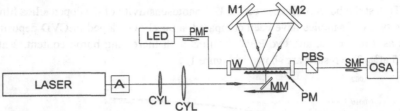

Figure 2. Setup for writing and evaluation of waveguide gratings with interferometrical method; LED – broadband light source, PMF – polarization maintaining fibre, SMF – single mode fiber , PBS – polarisation beam splitter, OSA – optical spectrum analyser, A – programmable attenuator, CYL – cylindrical lens, MM – motorized mirror, PM – phase mask, M1, M2 –interferometer mirrors.

length equal to the length of the phase mask. The angle between the interfering beams at the waveguide plane determines the written grating period and depends on the phase mask period (diffraction angle) and tilt of mirrors M1 and M2. When the system is symmetric, the waveguide is situated at the same plane as the phase mask and the imprinting angle is equal the diffraction angle as in standard phase mask technique.

The spectral response of the grating is measured during the writing of the grating using a broadband light source and an optical spectrum analyzer.

The main drawback of this system in application to waveguides is the fact that for writing gratings in waveguides situated in larger distance from the sample edge, a considerable out of plane tilt of mirrors M1 and M2 must be introduced. The analytical study of the interferometer geometry and the subsequent computer simulations of the setup allowed to follow the beam path through the interferometer to the interference plane. To introduce minimum distortion the out of plane tilt of the interfering beams should be introduced by tilting of the mirror MM. This tilt causes that the interfering beams (focal lines) are crossing each other. The larger the tilt is, the crossing angle increases. This angle changes also when motorized mirror MM moves along the phase mask during grating writing. In result the illumination intensity as well as the length of the interference pattern responsible for writing gratings change also during the movement of the motorized mirror MM. As it will be shown later the gratings imprinted in waveguides suffer from the more significant side lobes on the shorter part of the spectrum in comparison to gratings imprinted in fibers.

4.   GRATING APODIZATION

The spectral response of the grating is a Fourier transform of the envelope of the index modulation Δn along the grating. In the simplest case, when a finite-length grating has a uniform modulation of refractive index, Δn=*const*, as seen in Figure 3.a), the main peak or Bragg resonance in the spectral response is accompanied by a series of side lobes at adjacent wavelengths (sinc function). In WDM applications in which high rejection of adjacent channels is required it is important to lower the reflectivity of the side lobes, or to apodize the reflection spectrum of the grating.

A known method of suppressing the side lobes is to apodize the index profile such that towards the edges of the grating the index modulation approaches zero (Gaussian apodization) as seen in Figure 3.b). The gratings of this index profile show a significant suppression of side lobes, but on the short wavelength side residual peaks remain. Direct use of an expanded laser beam with a natural Gaussian intensity profile allows for the exposition of Gaussian apodized Bragg gratings.

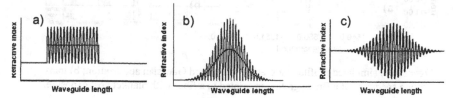

Figure 3. a): finite-length grating with uniform modulation of refractive index; b): self Gaussian apodization and c): raised Gaussian apodization of the laser intensity modulation for writing Bragg gratings. Line in the middle shows an average index profile. The period of the grating has been exaggerated for illustrative purposes.

Figure 4.a) and b) presents the simulated profile with Gaussian apodization and the spectral response of the grating obtained experimentally. Both profiles show the same behavior that can be explained by a non-uniform average index profile along the grating. Sections with a lower refractive index on both sides of the grating contribute to reflection of light with a shorter wavelength than the center part of the grating. Additionally these two reflections form Fabry-Perot resonances

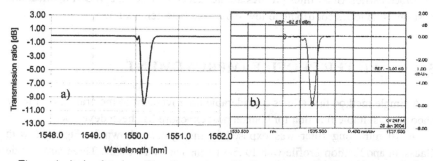

Figure 4. a) simulated profile of a grating with Gaussian apodization; b) transmission spectrum of a Bragg grating imprinted in a fiber with help of frequency doubled argon ion laser with natural Gaussian beam profile.

that give additional peaks at the shorter wavelengths.

To suppress the side lobes caused by the described effect it is necessary to raise the average index of refraction to be constant along the grating length. An apodization using this index profile, the raised Gaussian apodization is shown in Figure 3.c) and the simulated filter profile as well as experimental transmission spectra of such a grating are shown in Figure 5.a) and b), respectively.

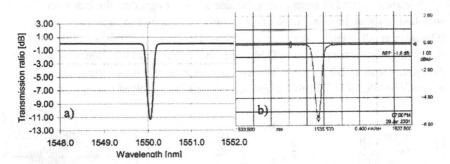

Figure 5. a) simulated profile of a grating with raised Gaussian apodization; b) transmission spectrum of a Bragg grating imprinted in a fiber with blanket post exposure.

To implement this apodization as well as other even more complicated apodization functions, programmable double exposure is necessary. Using a programmable tunable attenuator A and motorized mirror MM (shown in Figure 1) one can expose an arbitrarily chosen intensity profile along the grating to get an appropriate Δn profile. Prior to this exposure or after it an "inverted" blanket exposure should be done, which gives an opposite profile of the mean refractive index change. Then, when overlapped with the modulated one a uniform average index profile along the grating is obtained.

In practice grating profiles usually obtained in planar waveguides contain additional deformations due to non-perfect alignment and out of plane distortions described earlier. Such filter profiles are not acceptable for the WDM applications as strong side lobes contribute to low separation between channels.

5.   FILTER PROFILE IMPROVEMENT

A simple method that allows considerable improvement of the grating profile by choosing a suitable function for blanket post exposure has been developed.

A 10 mm long filter was exposed in the system shown in Figure 1 with Gaussian apodization profile with 10 mm length and FWHM 0.27. Three strong side lobes are clearly visible on the short wavelength side of the filter; one is of the same order of magnitude as the main peak (Figure 6.a)). To improve the shape of this filter, several post-exposures were performed over the same filter area. The five subsequent exposures have been done with inverted Gaussian profiles of different FWHM. The results are shown in Figure 6.b) – f).

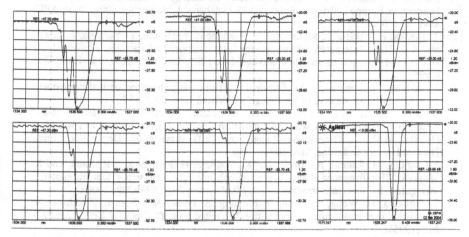

Figure 6. a) result after grating imprinted with Gaussian apodization profile with 10 mm length and FWHM 0.27; b) – f): consecutive phases of apodization after blanket exposures with inverted Gaussian profile and FWHM: b) 0.27, c) 0.21, d) 0.21, e) 0.162 and f) 0.162.

Figure 6.f) shows the final result after post exposure completed. The raising of the refractive index at the edges of the grating by choosing suitable intensity profile has been geometrically adjusted to the side lobes characteristics. The side lobes on the shorter wavelength side have been almost completely erased, while the formation of longer wavelength side-lobes has been avoided. The improving process has been stopped when the right-side lobe started to be comparable with the left side one which means that the mean refractive index started to be higher on the edges than in the filter center. The obtained filter bandwidth (FWHM) decreased to 0.4 nm and grating strength reached 17.5 dB, whereas side lobes observed 40 GHz from the center of the filter decreased to the value below 0.5 dB.

6.   CONCLUSIONS

In this paper we have demonstrated a new technique for improving of apodization profiles of Bragg gratings imprinted in planar waveguides. This method is simple to implement and does not require any additional equipment beyond that for standard apodization. Suppressed side lobes below 0.5 dB for 17.5 dB grating have been demonstrated. Higher suppression of side lobes can be also achieved by changing the geometry of the interferometer to diminish out of plane distortions of the interferometric pattern.

458

REFERENCES

[1] L. Wosinski, M. Dainese, H. Fernando and T. Augustsson, "Grating-assisted add-drop multiplexer realized in silica-on-silicon technology", Conference "Photonics Fabrication Europe", Brugge, Belgium, 28 October - 1 November 2002, Proc. SPIE **4941**, 43 (2002).

[2] T. Tanaka, H. Takahashi, M. Oguma, T. Hashimoto, Y. Hibino, Y. Yamada, J. Albert, and K.O. Hill, "Integrated external cavity laser composed of spot-size converted LD and UV written grating in silica waveguide on Si", Electron. Lett., **32**, 1202-1203 (1996).

[3] M. Åslund, J. Canning and M. Bazylenko, "Tuning of integrated optical component using UV-induced negative index change", Electron. Lett., **35**, 236-237 (1999).

[4] D.Zauner, K. Kulstad, J. Rathje and M. Svalgaard, "Directly UV-written silica-on-silicon planar waveguides with low insertion loss", Electron. Lett., **34**, 1582-1584 (1998).J. Williams, "Narrow-band analyzer," Ph.D. dissertation, Dept. Elect. Eng., Harvard Univ., Cambridge, MA, 1993.

[5] J. Canning, M. Åslund, A. Ankiewicz, M. Dainese, H. Fernando, J. K. Sahu, and L. Wosinski, "Birefringence control in plasma-enhanced chemical vapor deposition planar waveguides by ultraviolet irradiation", *Applied Optics*, **39**, 4296-4299 (2000).

[6] T. Fujiwara, D. Wong, Y. Zhao, S. Fleming, S. Poole, M. Sceats, "Electro-optic modulation in germanosilicate fibre with UV-excited poling", Electron. Lett. , **31**, 573-575 (1995).

[7] K. O. Hill, Y. Fujii, D. C. Johnson, and B. S. Kawasaki, "Photosensitivity in optical fiber waveguides: Application to reflection filter fabrication", Appl. Phys. Lett., **32**, 647-649 (1978).

[8] G. Meltz, W. W. Morey, and W. H. Glenn, "Formation of Bragg gratings in optical fibers by a transverse holographic method", Opt. Lett., **14**, 823-825 (1989).

[9] D. L.Williams, S. T. Davey, R. Kashyap, J. R. Armitage, and B. J. Ainslie, "Direct observation of UV induced bleaching of 240 nm absorption band in photosensitive germanosilicate glass fibres," Electron. Lett. **28**, 369–370 (1992).

[10] R. M. Atkins, V. Mizrahi, and T. Erdogan, "248 nm induced vacuum UV spectral changes in optical fibre preform cores: support for a colour centre model of photosensitivity," Electron. Lett. **29**, 385–386 (1993).

[11] M. D. Sceats, S. B. Poole, "Stress-Relief – The Mechanism of Photorefractive Index Control in Fiber Cores", Proc. 16th Australian Conference on Optical Fiber Technology", 302-305 (1991).

[12] M. D. Sceats, G. R. Atkins, and S. B. Poole, "Photolytic index changes in optical fibres," Ann. Rev. Mater. Sci. **23**, 381–410 (1993).

[13] C. Fiori, R:A:B. Devine, "Ultraviolet irradiation induced compaction and photoetching in amorphous , thermal SiO2", Mat. Res. Soc., **61**, 187-195 (1986).

[14] B. Poumellec, P. Guenot, I. Riant, P. Sansonetti, P. Niay, P. Bernage and J.F. Bayon, "UV induced densification during Bragg grating inscription in Ge:SiO2 preforms", Opt. Mater., **4**, 441-445 (1995).

[15] M. Douay, W. Xie, T. Taunay, P. Bernage, P. Niay, P. Cordier, B. Poumellec, L. Dong, J.F. Bayon, H. Poignant and E. Delevaque, "Densification involved in the UV-based photosensitivity of silica glasses and optical fibers", *J. Lightwave Technol.* **15**, 8, 1329–1341 (1997).

[16] P.-Y. Cortés, H. Fathallah, S. laRochelle, L. A. Rusch and P. Loiselle, "Writing of Bragg gratings with wavelength flexibility using a Sagnac type interferometer and application to FH-CDMA", ECOC'98, 411 (1998).

EXTERNAL OPTICAL MODULATOR USING A LOW-COST FABRY- PEROT LASER DIODE FOR OPTICAL ACCESS NETWORKS

H. J. Lee[1], H. Yoo[2], Y. D. Jeong[2], and Y. H. Won[2]
[1]*Division of Information & Communication Engineering, Kyungnam University, 449 Wolyoung-dong, Masan, 631-701, KOREA   hyuek@kyungnam.ac.kr*
[2] *Optical Internet Research Center (OIRC), Information and Communications University (ICU), P .O. BOX 77, Yusong-Gu, Daejeon, 305-600, KOREA   yhwon@icu.ac.kr*

Abstract: We propose and demonstrate an external optical modulation method based on TE/TM-mode absorption nulls in a Multiple Quantum Well (MQW) Fabry-Perot laser diode (FP-LD). The center wavelength of the absorption nulls is rapidly shifted to short-wavelength by the small current change ($\sim$ 1mA) in the FP-LD, which can modulate optical signal with more than 10 dB of extinction ratio (ER). The shift of the center wavelength comes from the refractive index change due to anomalous dispersion and plasma effect in MQW FP-LD waveguide. Non-inverting and inverting signals are made by TE- and TM-mode absorption nulls at 155.52 Mbps and BERs for the signals are measured, respectively.

1.   INTRODUCTION

Recently, Wavelength Division Multiplexing (WDM)-Passive Optical Network (PON) for optical access networks has received a great deal of an attention. WDM-PON has a high security, a protocol transparency, and wide bandwidth compared with Time Division Multiplexing (TDM)-PON. However, WDM-PON typically requires high-cost DFB-LDs, which becomes a big bottleneck for a commercial deployment. Therefore, many researchers have been intensively trying to find out cost-effective methods for WDM-PON [1-3]. One of the methods is a spectrum-sliced WDM source by optically filtering of an incoherent broadband Amplified Spontaneous Emission (ASE) light such as a LED or an EDFA [1]. The others are based on injection-locked Fabry-Perot laser diode (FP-LD) sources by external coherent [2] or incoherent [3] light. However, the injection-locked FP-LD source may have a demerit of potentially unstable operation by unexpected behaviours inside the laser diode [4]. On the other hand, the LED light source may be a good candidate for WDM-PON, but its output power is not enough for the purpose. In case of the EDFA, the spectrum-sliced source has relatively much higher output power than the LED, but expensive external optical modulators [1] or SOAs [5] for signal modulation are needed.

In this paper, we propose and demonstrate an external optical modulator employing a low-cost FP-LD, which can be effectively used for WDM-PON as an optical access network application. The modulation is based on the shift of TE/TM-mode absorption nulls (by refractive index change) due to anomalous dispersion and plasma effect in a FP-LD. The optical modulated signal with high extinction ratio (> 10dB) at 155.52 Mbps can be achieved. Based on the proposed optical modulator, a cost-effective WDM-PON architecture is proposed.

2.   PRINCIPLE

When a multiple-quantum-well (MQW) type FP-LD is driven over a threshold current, the FP-LD makes light by lasing on TE-mode. On the other hand, TM-mode in the FP-LD shows only absorption nulls not lasing owing to very small TM-gain inside the Fabry-Perot mirror (resonator). If no current is applied to the FP-LD, both modes (TE- and TM-mode) have absorption nulls. As the current increases from 0 mA to the threshold current, all the absorption nulls are moved to short-wavelength without any lasing on both modes. Even though light due to spontaneous emission occurs, the light at small current level can be ignored in comparison to externally incident light for optical modulation. The wavelength of the incident light is spectrally aligned with the center wavelength of one among the absorption nulls. Thus, optical modulation can be achieved by vibrating the center wavelength of the null due to current signal injection into a PF-LD.

The shift of the absorption null comes from the change of the refractive index in an MQW FP-LD waveguide, which is due to anomalous dispersion, plasma effect, and bandgap shrinkage [6]. However, the bandgap shrinkage can be neglected when an input light wavelength is longer than a bandgap wavelength of the waveguide in the MQW FP-LD [7]. In this case, the anomalous dispersion and the plasma effect become dominant for the refractive index change. The refractive index change due to the plasma effect in a bulky waveguide can be written as follow [6,8].

$$\Delta n = \frac{e^2 N}{2\omega^2 \varepsilon_0 \varepsilon_r m_c^*} n \qquad (1)$$

where, N is a carrier density, n is the refractive index in the absence of the plasma effect, $\varepsilon_0 \varepsilon_r$ is the dielectric constant of the active region, and m_c^* is effective mass of the carrier. The total refractive index change due to plasma effect in a MQW waveguide can be easily obtained by summing refractive index changes in wells and barriers [6]. In (1), because $\omega = 2\pi / \lambda$, the $\sqrt{\Delta n}$ is proportional to an input light wavelength λ. On the contrary, a refractive index change due to the anomalous dispersion according to the wavelength λ is reduced. Therefore, as the input light wavelength increases, the plasma effect is more dominant than the anomalous dispersion. As a result, the total refractive index change becomes almost constant for a wide range of wavelength (> 100nm) [6].

Let us consider a modulation speed for the proposed PF-LD modulator. To operate a PF-LD as an optical modulator, the injection current J with less than a threshold current J_{th} must be applied, i.e. carrier density N < threshold density N_{th}. As a consequence, a photon density N_{ph} becomes almost zero, which means a stimulated emission can be neglected. Thus, the rate equation for the carrier density N can be expressed as [9]

$$\frac{dN}{dt} = \frac{J}{qd} - \frac{N}{\tau_e(N)} \qquad (2)$$

where, q and d are the charge constant and the thickness of the active layer, respectively. And, $1/\tau_e(N)$ is the effective recombination rate, which $1/\tau_e(N) = A + BN + CN^2$, where AN is the nonradiative recombination rate, BN^2 is the radiative spontaneous rate, and CN^3 is the Auger recombination rate. Because of the carrier dependence term $\tau_e(N)$, Eq. (2) is nonlinear and can be solved numerically or approximately. Assuming τ_e is constant, i.e. $B = C = 0$, we can know easily that the carrier density $N(t)$ increases logarithmically and decays exponentially with the time constant τ_e. Thus, the speed for the proposed modulator strongly depends on τ_e. By the calculation from typical parameter values (A= 5×10^8, B= 1×10^{-10}, C= 3×10^{-29}, $N_0 = 1 \times 10^{18}$, and $N_{th} = 2.61 \times 10^{18}$) in [9], $\tau_e(N_0) \approx 1.58$ nsec and $\tau_e(N_{th}) \approx 1.04$ nsec, which mean roughly 630 Mbps ~ 1 Gbps modulation speed.

3. EXPERIMENTAL RESULTS

Fig. 1 shows the experimental result for absorption nulls with respect to the injection current in a FP-LD, which the laser diode used in the experiment is

462

InGaAsP MQW type. The absorption nulls are made by being kept and dissipated in a FP-LD cavity for the input light with the wavelength that the phase after each round trip has to be an integral multiple of 2π. In other words, $\lambda_p = 2nL / p$, where λ_p is the wavelength of the pth cavity mode, L is the cavity length. When a current is injected, all the absorption nulls are continuously shifted to short-wavelength by negative refractive index change. But, for over the threshold current (I_{TH} = ~11mA), the nulls start to move in the opposite direction (to long-wavelength) due to the thermal effect by lasing [10]. Here, it is worthy noting that the nulls are rapidly shifted to short-wavelength by a small amount of current from 0 to 1 mA as shown in Fig. 2. The center wavelength of the nulls is simultaneously moved by ~0.25 nm (~30GHz) due to only 1 mA current, which is enough to modulate optical signal. The absorption null has the notch filter characteristic of ~0.1 nm bandwidth at -3 dB and ~19 dB attenuation at the center wavelength. For the small current level such as 1 mA, the spontaneous emission noise generated by the FP-LD itself is negligible compared to the input light.

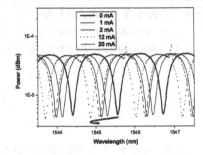
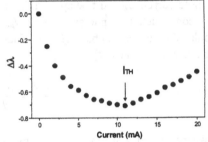

Figure 1 TM-mode absorption spectra of Fabry-Perot laser-diode (FP-LD) with the current change from 0 to 20 mA.

Figure 2 Center wavelength displacement of the absorption null versus injection current change.

Fig. 3 shows the experimental setup for an external optical modulation using a low-cost FP-LD. Laser light from the tunable-LD is incident into the FP-LD through PC (Polarization controller)1, OC (Optical Circulator), PBS (Polarization Beam Splitter) and PC2. By the PC2, the polarization (TE or TM) of the incident light into the FP-LD can be determined. The reflected light from the FP-LD comes out of 3 port of the OC through the PBS. The output signal may have an inverted or a non-inverted data format, which depends on which side (short-wavelength side -> inverting, long-wavelength side -> non-inverting) of absorption nulls is used. BERs for the inverted and the non-inverted signal on TE- and TM-mode absorption null at 155.52 Mbps (2^{31}-1 PRBS) are measured and shown in Fig. 4, respectively. Non-inverted signal that is made on long-wavelength side of the TM-mode absorption null, shows the best BER performance. The worst case, i.e. inverted data signal on the short-wavelength side of the TE-mode absorption null, has ~2.3

dB power penalty (@10^{-9}). The minimum and the maximum insertion loss of the proposed modulator are 10.2 dB and 13.5 dB for the non-inverted signal on the TM-mode null and for the inverted signal on the TE-mode null, respectively. ERs for all the data show more than 10 dB.

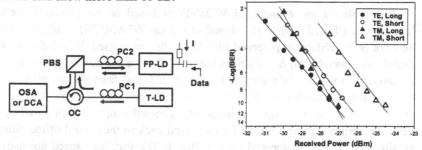

Figure 3 The experimental setup for the proposed external modulator based on the absorption null modulation of a FP-LD.

Figure 4 Measured BER curves and the corresponding eye diagrams for (●) non-inverted, (○) inverted data signal by a TE-mode null and (▲) non-inverted, (△) inverted data signal by a TM-mode null.

The MQW FP-LD used in the experiment has a cavity length of ~300 um and then shows that the wavelength difference between nulls as shown in Fig. 1 is around 1.17 nm. For the WDM signal with the ITU-T grid, the FP-LD has to be designed to get the optimal cavity length. In order to know the wavelength sensitivity for the proposed FP-LD modulator, we measured ERs and insertion losses from 1520 nm to 1630 nm at 155.52 Mbps for non-inverting signal on TM-mode nulls as shown in Fig. 5. The proposed FP-LD modulator shows good performances (ERs > 11.5dB and insertion losses < -13.4 dB) for very wide wavelength range (~100nm). This result is in very good agreement with the result in [6].

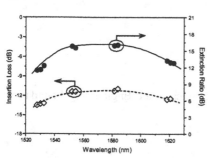

Figure 5 Measured Extinction Ratio (dB) and Insertion Loss (dB) corresponding to Wavelength.

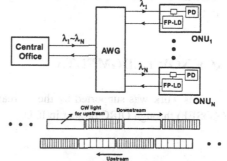

Figure 6 WDM-PON architecture with the proposed FP-LD modulator.

464

4. APPLICATION TO WDM-PON

Fig. 6 shows one example of WDM-PON based on the proposed method. Instead of a DFB-LD source (employed in typical WDM-PON) inside the optical network unit (ONU), the proposed FP-LD modulator is used. WDM downstream signals with wavelength $\lambda_1 \sim \lambda_N$ from the central office are wavelength-routed and transmitted to ONUs. As shown in Fig. 6, each signal from the central office is composed of the modulated downstream data for half and the CW light for the other half. Here, the CW light is used for the generation of upstream signal by the FP-LD modulator at the ONU and then routed back to the central office. Similar architectures had been proposed [2, 11]. But, in [2], very high-speed downstream signal is required to make the single mode operation of a FP-LD by injection-locking, which may be potentially unstable. Also, in [11], high-cost optical modulators are needed. On the contrary, the proposed modulator is based on the absorption nulls in a PF-LD and thus shows very stable operation in a cost-effective way.

5. CONCLUSIONS

In this paper, new optical modulation method using a low-cost FP-LD has proposed and experimentally demonstrated. Due to the cost-effectiveness, the proposed optical modulator has many applications in optical access networks, optical signal processing, and so on. Even though the modulation speed in the experiment is 155.52 Mbps with more than 10 dB insertion loss, ~1 Gbps modulation and low (~6 dB) insertion loss will be possible with the small modification of a FP-LD such as a length and a reflectance control for the FP cavity.

ACKNOWLEDGMENTS

This work was supported by the Korean Science and Engineering Foundation (KOSEF) through OIRC project in ICU.

REFERENCES

[1] J.S. Lee, Y.C. Chung and D.J. DiGiovanni, "Spectrum-sliced fiber amplifier light source for multichannel WDM applications," *IEEE Photon. Technol. Lett.*, vol. 5, pp. 1458-1461, 1993.

[2] L.Y. Chan, C.K. Chan, D.T.K. Tong, F. Tong and L.K. Chen, "Upstream traffic transmitter using injection-locked Fabry-Perot laser diode as modulator for WDM access networks," *Electron. Lett.*, vol. 38, pp. 43-45, 2002.

[3] H.D. Kim, S.G. Kang and C.H. Lee, "A low-cost WDM source with an ASE injected Fabry-Perot semiconductor laser," *IEEE Photon. Technol. Lett.*, vol. 12, pp. 1067-1069, 2000.

[4] G. Yabre, H. Waardt, H.P.A. Boom and G.-D. Khoe, "Noise characteristics of single-mode semiconductor lasers under external light injection," IEEE *J. Quantum Electron.*, vol. 36, pp. 385-393, 2000.

[5] P. Healey, P. Townsend, C. Ford, L. Johnston, P. Townley, I. Lealman, L. Rivers, S. Perrin and R. Moore, "Spectral slicing WDM-PON using wavelength-seeded reflective SOAs," *Electron. Lett.*, vol. 37, pp. 1181-1182, 2001.

[6] J.I. Shim, M. Yamaguchi and M. Kitamura, "Refractive index and loss changes produced by current injection in InGaAs(p)-InGaAsP Multiple Quantum-Well (MQW) waveguides," *IEEE J. Select. Topics Quantum Electron.*, vol. 1, pp. 408-415, 1995.

[7] B.R. Bennett, R.A. Soref and J.A. Del Alamo, "Carrier-induced change in refractive index of InP, GaAs, and InGaAsP," *IEEE J. Quantum Electron.*, vol 29, pp. 113-122, 1990.

[8] K. Iizuka, *Elements of Photonics*, Volume II, Chapter 14, Wiley & Sons, New York, N.Y., 2002.

[9] M.M.-K., Liu, *Principles and applications of optical communications*, Chapter 12, IRWIN, Times Mirror Higher Education Group, Inc., 1996.

[10] T. Higashi, T. Yamamoto, S. Ogita and M. Kobayashi, "Experimental analysis of temperature dependence of oscillation wavelength in quantum-well FP semiconductor lasers," *IEEE J. Quantum Electron.*, vol. 34, pp. 1680-1689, 1998.

[11] N.J. Frigo, P.P. Iannone, P.D. Magill, T.E. Darcie, M.M. Downs, B.N. Desai, U. Koren, T.L. Kock, C. Dragone, H.M. Presby and G.E. Bodeep, "A wavelength-division multiplexed passive optical network with cost-shared components," *IEEE Photon. Technol. Lett.*, vol. 6, pp. 1365-1367, 1994.

AN ACCURATE MODEL OF A FULL OPTICAL ENCODER/DECODER IN A WGR CONFIGURATION

Gabriella Cincotti[1], N. Wada[2], and K. Kitayama[3]
[1]*Department of Applied Electronics, University of Roma Tre, via della Vasca Navale 84, I-00146 Rome, Italy, g.cincotti@uniroma3.it*
[2]*National Institute of Information and Communications Technology Tokio, Japan, wada@nict.go.jp*
[3]*Department of Electronic and Information Systems, Osaka University, 2-1 Yamadaoka, Suita, Osaka 565-0871, Japan, kitayama@comm.eng.osaka-u.ac.jp*

Abstract: We present a field model of a full Encoder/Decoder (E/D) in a Waveguide Grating Router (WGR) configuration that generates/processes a set of Phase-Shift Keyed (PSK) codes. We furnish the design procedure to synthesize the device, determining the optimal number of the ports and the tolerance in the differential path length as functions of the code detection parameter.

1.   INTRODUCTION

In Refs. [1-3], we presented innovative planar architectures for a full optical Encoder/Decoder (E/D), that is passive optical devices with a single input and N outputs to generate/process N PSK codes simultaneously. In particular, we showed that the standard Waveguide Grating Router (WGR) configuration of Fig. 1 can be synthesized as a full E/D: if a single optical pulse is directed into one of the device inputs, at the device outputs we obtain N PSK codes. On the other hand, if one PSK code is forwarded into the same device input, at the device outputs we obtain the correlation functions between the input code an all the PSK codes in the look-up table.

WGRs are recognized as key passive components in current optical communication systems and they have been proposed to implement a large variety of applications, as dispersion compensation, frequency demultiplexers, tuneable filters and optical signal processing [4-7]. The transmission function of WGR frequency demultiplexers has been largely investigated, performing the spectral analysis by means of the Fourier optics [8, 9].

But the mechanism to build a set of PSK codes can be more easily described analyzing the WGR configuration in the time domain: the input slab coupler generates N copies of the input pulse, with phases given by the Rowland circle configuration [4]. The optical pulses travel different paths in the Arrayed Waveguides (AW) grating and the output slab coupler recombines the pulses to built N codes at the device outputs; each PSK code is composed of N optical pulses, which are often referred to as chips. The differential path delay $\Delta\tau$ in the AW grating is chosen smaller than the input pulse width δ_i, so that the chips in the output codes do not overlap. It is evident that the design requirements of a full E/D and a standard demultiplexer are greatly different: in the present paper we furnish an accurate model for the full E/D, along with the corresponding design guidelines.

The remainder of the paper is organized as follows: in Sec. 2, we give a very simple but comprehensive field model of the E/D, and introduce the PSK codes detection parameter. In Sec. 3, we evaluate the optimal number of ports and the fabrication tolerance of the proposed device architecture, as functions of the detection parameter. Conclusions and hints for future works are given in the last section.

2.   FIELD MODEL

We refer to the device architecture of Fig. 1 and the parameters listed in Table I, and evaluate the impulse response from the i-th input to the k-th output as

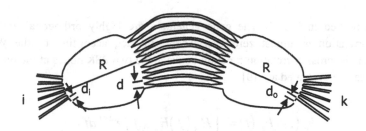

Figure 1. Geometry of the E/D in the WGR configuration.

Table 1. Device parameters

parameter	description	value
f_0	carrier frequency (λ_0=1550 nm)	193.55 THz
N	number of the AW grating arms	16
R	input/output slab focal length	20.85 mm
d	spacing in the AW grating	24.6 μm
d_i, d_o	waveguide spacing in the input/output grating	56.47 μm
L	shortest waveguide length	26 mm
ΔL	differential path length	1.0316 mm
n_e	refractive index in the AW grating	1.454
n_s	refractive index in the slab regions	1.451

$$h_{ik}(t) = \sum_{j=0}^{N-1} \delta\left(t - \frac{n_e(L+j\Delta L)}{c}\right) e^{-j2\pi j\left[\frac{n_s d}{\lambda_0 R}(id_i + kd_o)\right]} \quad , \quad (1)$$

where $\delta(t)$ is the delta function, j the imaginary unit and $\Delta\tau = n_e\Delta L/c$ is the chips interval. Assuming that the input and the output waveguide gratings have the same spacing, i.e. $d_i = d_o$, we set $\lambda_0 R /(dd_o n_s) = N$; in this case, the transfer function from the i-th input to the k-th output is

$$H_{ik}(f) = e^{-j2\pi\frac{n_e L}{c}f} \sum_{j=0}^{N-1} e^{-j2\pi j\left(\frac{i+k}{N}+\Delta\tau f\right)}$$

$$= e^{-j2\pi\frac{n_e L}{c}f} e^{-j\pi\left(\frac{i+k}{N}+\Delta\tau f\right)(N-1)} \frac{\sin\left[\pi(i+k+N\Delta\tau f)\right]}{\sin\left[\pi\left(\frac{i+k}{N}+\Delta\tau f\right)\right]}, \quad (2)$$

that is plotted in Fig. 2. The code generated are highly orthogonal and their crosscorrelation is almost zero everywhere; in fact, according to the Wiener theorem, the cross-correlation function between the two PSK codes at the outputs k and k' can be evaluated as [10]

$$h_{ik}(t) * h_{ik'}(t) = \int_{-\infty}^{\infty} H_{ik}(f)H_{ik'}(f)e^{i2\pi ft} df \quad . \quad (3)$$

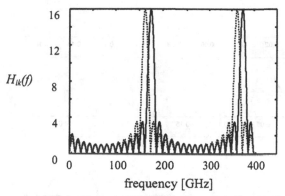

Figure 2. Transfer function at two adjacent outputs.

Since the transfer functions at two different outputs do not overlap (see Fig. 2), the Cross-Correlation Peak (CCP) is very small.

The Auto-Correlation Peak (ACP) detected at one of the device outputs reveals an exact match between the incoming PSK code and the corresponding code in the look up table; the parameter that characterizes the code detectability is the ratio r=ACP/CCP. It is quite immediate verify that the ACP=N^2, whereas the CCP depends on the two outputs k and k'. From an inspection of Eq. (2), it is evident that the CCP is higher for two PSK codes generated at two adjacent outputs, and in the followings we evaluate the r parameter in this worst case condition. We remark that lower values of the r parameter correspond to outputs that are spatially adjacent, not because the crosstalk, but only because the codes generated at adjacent outputs are more correlated.

To numerically investigate the performances of the proposed PSK codes set, we consider a Gaussian optical pulses of width δ_t=2 ps

$$p(t) = e^{-\frac{t^2}{2\delta_t^2}} e^{-i2\pi f_0 t} \quad , \quad (4)$$

at one of the device inputs, that generates N=16 PSK codes, each of them composed of N=16 chips. Setting the chip interval $\Delta\tau$=10ps, we obtain the PSK codes of Fig. 3, with ACP=256, CCP=26.27 and r=9.74.

Quite similar results are obtained decreasing the chip interval: in the case of $\Delta\tau$=5ps, we have ACP=296.43, CCP=30.9 and r=9.74, as shown in Fig. 4.

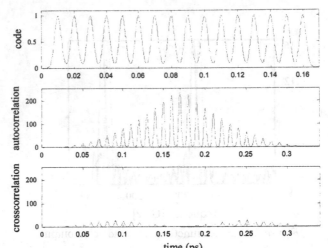

Figure 3: PSK code composed of *N=16* Gaussian chips, of width δ_i=2 *ps* and chip interval $\Delta\tau$=10*ps*.

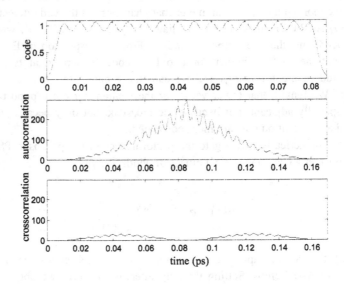

Figure 4: PSK code composed of *N=16* Gaussian chips, of width δ_i=2 *ps* and chip interval $\Delta\tau$=5*ps*

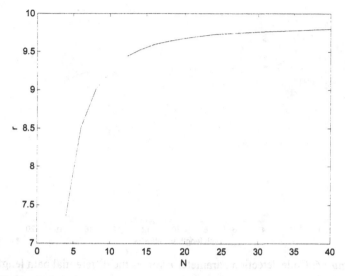

Figure 5: Label detection parameter r=ACP/CCP versus the number of the outputs N.

3.   DESIGN EQUATIONS

The number N of the device inputs/outputs coincides with both the number of the PSK codes and the code length, i.e. the number of the chips in every code. Therefore, we have to trade the code cardinality, i.e. the number of PSK codes that we can simultaneously generate and process, for the code processing time, i.e. the time necessary to perform the correlations that is $(N-1)\Delta\tau$ [11]. In the case $N=16$, the code processing speed is $1/(N-1)\Delta\tau=13.3\ 10^9$ pps.

The ACP coincides with N^2, and to increase the code detection parameter r=ACP/CCP, we should consider a high value of N. In Fig. 5 we plot the r parameter for two adjacent outputs as function of N: it is evident that the r parameter increases with N and that there is a saturation behaviour for $N>25$.

The main parameter of the WGR architecture is the differential path length ΔL, that determines the chip interval. The tolerance of this parameter depends on the input pulse width δ_t and the chip interval $\Delta\tau$. Fig. 6 shows the r parameter as a function of the differential path length variation, when all the path lengths in the AW grating increase (+) or decrease (-) of the same value.

The other device parameters has to be chosen so that $\lambda_0 R/(dd_o n_s)=N$, to set the phases of the chips. Therefore, any change in the device parameters d, d_o, R, n_s or the reference wavelength λ_0 causes a change in the code phases. However, their change slightly influences the device performances, as shown in Fig 7.

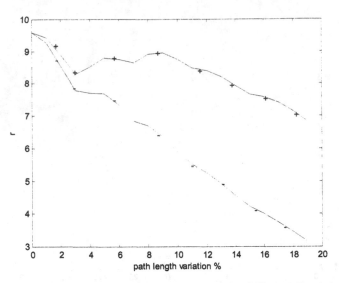

Figure 6. Code detection parameter *r* versus the differential path length variation.

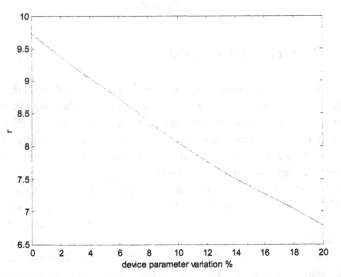

Figure 7. Label detection parameter *r* versus the parameter $dd_o n_s / (\lambda_0 R)$ variation.

4. CONCLUSIONS

A simple but accurate model for a full E/D that generates/processes *N* PSK codes simultaneously has been presented. We showed that a standard WGR configuration can be designed as a full E/D, and we furnish the design guidelines.

The code cardinality, i.e. the number of PSK codes generated has to be traded for the code processing time. In addition, we show that the code detection parameter increases with N and that it presents a saturation behavior for $N>25$. We also demonstrate that the proposed configuration is quite insensitive to the device parameter variation. A more complete model that takes the actual field distribution inside the waveguides will be the subject of a future work.

REFERENCES

[1] G. Cincotti, *"Full optical encoders/decoders for photonic IP routers"*, IEEE J. of Lightwave Technol., vol. 22, n. 2, pp. 337-342, 2004.

[2] G. Cincotti, *"Design of full optical encoders/decoders for MPLS network"*, IEEE J. of Lightwave Technol., at press June 2004.

[3] G. Cincotti, "Optical device to generate and process simultaneously optical codes," PCT Patent IT03/000879 filed by University of Roma Tre, December 30th 2003.

[4] C. K. Madsen, and J. H. Zhao, Optical *filter design and analysis. A signal processing approach*, New York: John Wiley and sons inc., 1999.

[5] C. Dragone, C. A. Edwards, and R. C. Kisler, *"Integrated optics NXN multiplexer on silicon,"* IEEE Photon. Technol. Lett., vol. 3, n. 10, pp. 896-899, 1991.

[6] H. Takahashi, S. Suzuki, K. Kato, and I. Nishi, *"Arrayed-waveguide grating for wavelength division multi/demultiplexer with nanometer resolution,"* Electron. Lett., vol. 26, pp. 87-88, 1990.

[7] R. Adar, H. Henry, C. Dragone, R. C. Kistler, and M. A. Milbrodt, *"Broad-band array multiplexers made with silica waveguides on silicon,"* IEEE J. of Lightwave Technol., vol. 11, n. 2, pp. 212-219, 1993.

[8] H. Takahashi, K. Oda, H. Toba, and Y. Inoue, *"Transmission characteristics of arrayed waveguide NXN wavelength multiplexer,"* IEEE J. of Lightwave Technol., vol. 13, pp. 447-455, 1995.

[9] P. Muñoz, D. Pastor and J. Capmany, *"Modeling and design of arrayed waveguide gratings,"* IEEE J. of Lightwave Technol., vol. 20, n. 4, pp. 661-674, 2002.

[10] K-i. Kitayama, N. Wada, H. Sotobayashi, *"Architectural considerations for photonic IP router based upon optical code correlation,"* IEEE J. of Lightwave Technol., vol. 18, n. 12, pp. 1834-1844, 2000.

[11] K-i. Kitayama, and N. Wada, H. Sotobayashi, *"Photonic IP routing,"* IEEE Photon. Technol. Lett., vol. 11, n. 12, pp. 1689-1691, 1999.

CHIRPED FIBER BRAGG GRATING AS ELECTRICALLY TUNABLE TRUE TIME DELAY LINE

Vincenzo Italia[1], Marco Pisco[1], Stefania Campopiano[1], Andrea Cusano[1] and Antonello Cutolo [1]
[1] Divisione di Optoelettronica – Dipartimento di Ingegneria, Università del Sannio, Corso Garibaldi 107 Benevento, Italy e-mail stefania.campopiano@unisannio.it

Abstract: A new optical time delay line based on a Chirped Fiber Bragg Grating is proposed. Numerical results show the time delay can be electronically varied by changing the grating temperature with a minimum step of 1ps up to 30GHz.

1.   INTRODUCTION

During the last years, optical true-time delay (OTTD) units have been investigated for wideband squint-free beamforming for phased array antennas [1]-[3]. The advantages of the optical beamforming networks are quite well know: low insertion loss, high phase stability, electromagnetic interference immunity and low mass and volume.

Recently, fiber Bragg gratings (FBG) have been used to realize OTTD units. The first approaches were based on a number of uniform FBG written at different positions on optical fibers and the distances between gratings determine the beampointing direction of the array antenna [4]. This system assures broad-band operation, but it only allows a discrete number of beampointing angles.

Subsequently, it has been demonstrated that linearly chirped fiber gratings (LCFG) can produce a linear phase delay of the modulating signal at microwave frequencies and the slope of the phase response can be continuously modified by tuning the wavelength of the optical carrier [5]. Broadband operation and continuous spatial scanning properties have been demonstrated in these systems. However these systems are very complicated as the need to use multiwavelength tunable laser source and tunable bandpass filter that should be tuned synchronously with wavelengths of the tunable laser source.

In this work, we propose and numerically analyze a novel variable OTTD based on a LCFG operating at a fixed optical wavelength that simplifies the architecture of the beamforming system. The operation principle, based on uniform temperature perturbation of a single LCFG, is illustrated in Section 2. As consequence of the temperature perturbation, the grating complex amplitude reflectivity moves rigidly, causing a delay's variation for the input optical carrier. The different time delays are obtained by changing the grating temperature. In Section 3 the numerical results of the proposed OTTD are presented.

2.   PRINCIPLE OF OPERATION

The proposed OTTD is illustrated in Figure 1. It's based on the use of one single LCFG uniformly perturbed in temperature.

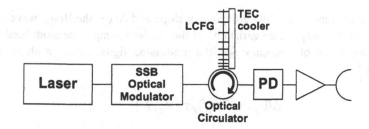

Figure 1. Single branch, OTTD based on a LCFG uniformly perturbed.

When a grating is uniformly perturbed in temperature its complex amplitude reflectivity is moved rigidly and the new Bragg wavelength is [8],

$$\lambda_B(T) = \lambda_{B_0} + k_T \lambda_{B_0}(T - T_0),$$ (1)

where T_0 is the room temperature, $(T-T_0)$ is the variation of the temperature introduced by the actuator, λ_{B_0} is the unperturbed Bragg wavelength and k_T is a constant determined by the characteristics of the grating. Figure 2 shows the

476

temperature induced reflectivity and group delay shift as a function of the optical wavelength.

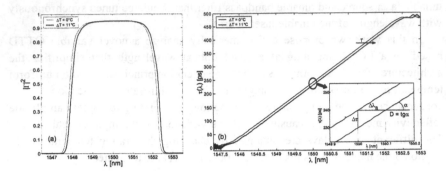

Figure 2. Reflectivity (a) and Time delay (b) responce for two values of temperature variation ΔT ($\Delta T=0°C$, $\Delta T=11°C$)

Assuming the wavelength of the optical carrier equals to the unperturbed Bragg wavelength, $\lambda_0 = \lambda_{B_0}$, the actual time delay as a function of the thermal changes, as shown in Figure 2, can be expressed as:

$$\tau(\lambda_B) = \tau(\lambda_B)\big|_{\Delta T=0} - D \cdot \Delta\lambda_B = \tau(\lambda_B)\big|_{\Delta T=0} - k_T \lambda_B D \cdot \Delta T \tag{2}$$

where D is the mean FBG group-delay slope and $\Delta\lambda_B$ is the Bragg wavelength shift due to the temperature variation. If the carrier is amplitude modulated by a microwave signal of frequency f_{RF}, the modulated signal suffers a phase delay given by [3]

$$\Delta\Phi_{RF} = 2\pi f_{RF} \tau(\lambda_0, T) \tag{3}$$

Hence, the grating produces a linear phase shift in the modulating signal and the phase slope can be continuously varied by changing the temperature.

The drawbacks in the use of the LCFG as a wide bandwidth OTTD unit can be identified in the spectral distortion and in the RF power degradation due to chromatic dispersion induced by the grating phase response. The first problem can be neglected in practical applications due to limited bandwidth of the RF modulation. For the second one optical losses can be significant if DSB amplitude modulation is used. As illustrated in detail in the following section, this problem can be solved by using SSB modulation [3].

3.   NUMERICAL RESULTS

In the modelling of the device, we used a 5-cm-long chirped grating with its bandpass centered at λ_{B0}=1550nm and a 3-dB bandwidth of 4.32nm. The mean group-delay slope is 100 ps/nm with a mean group delay ripple (GDR) of 0.1ps. The reflectivity and the time delay response of the grating are shown in Figure 3.

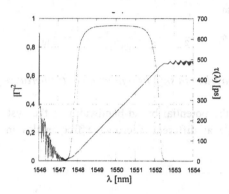

Figure 3. Grating reflectivity and time delay response of LCFG.

To obtain this GDR value an apodized LCFG with a positive hyperbolic-tangent apodization profile [6] has been used. An iterative method of GDR correction [7] has been used too. The apodization enabled a first and drastic reduction of the GDR while the iterative technique has been used to reduce the mean amplitude of the ripples under 0.1ps value. For thermal sensitivity of LCFG a value of 5.76×10^{-6} [9] is chosen for k_T. Grating complex reflectivity has been evaluated by using multilayer approach [10].

Delay line's simulations have been performed for temperature varying from room temperature T_0 to (T_0+11°C) with steps of 1.1°C to introduce variable delays from 1ps to 10ps with a step resolution of 1ps and SSB optical modulation have been used. Figure 4 shows the Bragg wavelength shift of the grating as the function of temperature perturbation. It presents a linear dependence on perturbation according to (1).

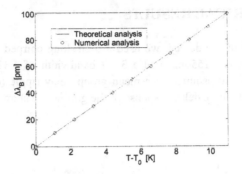

Figure 4. Bragg wavelength shift as a function of temperature perturbation.

In order to show the potential broad-bandwidth of the system the numerical simulations were made at different microwave frequencies in the range [1-30 GHz].

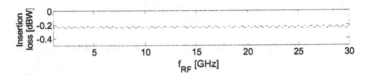

Figure 5. Insertion loss of the proposed OTTD as a function of detected frequency (f_{RF})

Figure 5 shows the insertion loss of the OTTD at different modulating frequencies obtained only considering the losses of the grating and neglecting the contributions of the other optical devices in the chain. The mean value of the evaluated insertion losses is due to the limited reflectivity of the grating.

In order to verify the time-delay behaviour described in eq. (2), the phase of the detected signal has been estimated as a function of RF frequency in the 1–30 GHz range for all values of perturbation.

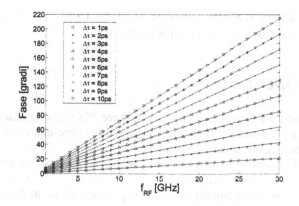

Figure 6. - Phase of the microwave signal at different RF frequencies after reflection by the LCFG for all values of perturbation.

Figure 6 shows the relative detected phase of the microwave signal after reflection in the grating. It presents a linear dependence on the frequency, with different slopes according to eq. (2).

In order to check the time-delay independence from detected frequency, the ripple around the ideally constant group delay has been calculated. The result is plotted in Figure 7 showing a worst-case standard deviation of 0.07 ps from 1 to 30 GHz.

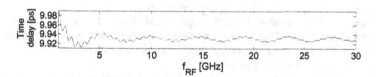

Figure 7. Time-delay ripple at against detected frequency (fRF) for temperature perturbation of 10°C.

The total system time-delay error should be estimated assuming two main causes of error: the temperature stability $\Delta T_{att.}$ and the FBG time-delay response deviation from the linear slope σ_{ripple}. The temperature stability can be translated to time precision by taking into account the grating time delay response

$$\sigma_{att.} = k_T \lambda_B D \cdot \Delta T_{att.} \tag{4}$$

A standard deviation of 0.07 ps would resume the time delay uncertainty due to the grating group-delay ripple. Taking into account both effects (temperature stability and the FBG group delay response) a total system time-delay precision,

480

for assuming a temperature stability of 0.1°C, is ~0.1 ps, which in turn implies a time-delay equivalent to a 3° phase error at 30 GHz.

4.   CONCLUSIONS

In this work, we have presented a new chirped fiber grating beamformer. The device operates at a fixed optical wavelength and the time delay is linearly tuned by simply changing the grating temperature.

Numerical results show that the time delays is linearly proportional to the temperature, according to the theoretical analysis. The proposed system can be used for wideband beamforming at radio frequencies up to 30 GHz with a 1-ps minimum time delay.

Inaccuracies due to the temperature stability and the linearity of the chirped grating group delay have also been considered, and their impact on the PAA performance is reported.

REFERENCES

[1] Istvàn Frigyes, *Senior Member, IEEE,* and A.J. Seeds, *Senior Member, IEEE*, "Optical Generated True-Time Delay in Phased-Array Antennas," *IEEE Trans. on Microwave Theory and Techniques*, vol. 43, no. 9, September 1995.

[2] J.L. Corral, J. Martì, *Member, IEEE,* S. Regidor, J.M. Fuster, R. Laming, M. J. Cole, "Continuously Variable True Time-Delay Optical Feeder for Phased-Array Antenna Employing Chirped Fiber Gratings," *IEEE Tran. on Microwave Theory and Techniques,* vol. 45, no. 8, August 1997.

[3] B. Ortega, J.L. Cruz, J. Capmany, M.V. Andrés, and Daniel Pastor, "Analysis of a Microwave Time Delay Based on a Perturbed Uniform Fiber Bragg Grating Operating at Constant Wavelength," *J. of Lightwave Technology*, vol. 18, no. 3, March 2000.

[4] A. Molony, C. Edge, and I. Bennion, "Fiber grating time delay for phased array antennas," *Electron. Lett.*, vol. 31, pp. 1485–1486, 1995.

[5] J. L. Cruz, B. Ortega, M.V. Andrés, B. Gimeno, D. Pastor, J. Capmany and L. Dong, "Chirped Fibre Bragg Gratings fr Phased Array Antennas," *Electron. Lett.*, vol. 33, no. 7, March 1997.

[6] K. Ennser, M. N. Zervas, and R.I. Laming, "Optimization of Apodized Linearly Chirped Fiber Gratings for Optical Communications," *IEEE J. of Quantum Elettron.* vol. 34, no. 5, May 1998.

[7] M. Sumetsky, P.I. Reyes, P.S. Westbrook, N.M. Litchinitser, and B.J. Eggleton, "Group Delay Ripple Correction in Chirped Fiber Bragg Gratings," *Opt. Lett.*, vol. 28, no. 10, May 2003.

[8] V. Bhatia, "Properties and Sensing Applications of Long-Period Gratings", *Ph.D. Thesis*, November 1996.

[9] N.P. Bansal et al., Handbook of Glass Properties, Academic Press, Florida, 1986.

[10] Raman Kashyap *Fiber Bragg Gratings*, Academic Press, 1999.

ANALYSIS OF TUNING TIME IN MULTIPLE-SECTION CURRENT-INJECTION TUNABLE LASER DIODES

Efraim Buimovich-Rotem[1], Dan Sadot[1]
[1]*Ben Gurion University, POB 653 Beer-Sheva, 84105 Israel. E-mail: buimov@ee.bgu.ac.il
Sadot@ee.bgu.ac.il*

Abstract: A generalized model for calculating the tuning time of multiple-section current-injection laser diodes is presented. The method is applied theoretically and experimentally to the Grating assisted co-directional Coupler with Sampled Reflector (GCSR) laser.

1.   INTRODUCTION

Optical packet switching is considered a key technology in the next generation optical communication networks. Different architectures rely on various photonic components such as tunable lasers, tunable filters, etc. to achieve fast optical routing of data. Fast tunable lasers are used to color packets of data according to their destination. Overhead and latency considerations dictate fast tuning time of the order of 50 nanoseconds for implementing efficient packet switching. [1]

Among the different types of tunable lasers those based on the free carrier plasma effect such as DBR, SG-DBR, GCSR, etc., offer the combination of wide tuning range and fast tuning. Such lasers have been implemented in various system demonstrations [2-4]. The inherent physical limitation on tuning time in these lasers is due to the carrier lifetime in the tuning sections, of the order of 1 nanosecond. Achieving fast tuning times requires fast electronic driving combined with accurate selection of the tuning currents (or operating points).

In this paper we present a general approach for calculation of the tuning time in these lasers based on the properties of the laser and the electronic driving. This model can be used to optimize the laser design and to calculate the inherent physical tuning time limitations (assuming ideal electronic driving). Conversely, the expected tuning time can be calculated for a particular electronic driving design and can be compared with its inherent physical limitations. We later apply this model to the GCSR laser and calculate the expected tuning time of a particular fast tunable transmitter. The calculations are verified experimentally.

2.   THEORY

We are interested in obtaining the temporal frequency evolution $f(t)$ of the laser following a change in the tuning currents designed to switch the laser's wavelength from a certain source wavelength λ_S to a destination wavelength λ_D. We define the laser's detuning as its frequency deviation relative to the destination wavelength using: $\Delta f(t) = f(t) - f_D$, where f_D is the laser frequency at the destination wavelength. A relevant problem is to find the tuning time t_{TX} for which $|\Delta f(t_{TX})| = f_{sys}$, where f_{sys} is the required frequency accuracy as defined for a system in which the laser is to be used. For practical reasons, the longest tuning time over the entire set of wavelengths defined for the laser will be regarded as the general tuning time.

We define for each of the laser's tuning sections a detuning coefficient $(\partial f / \partial N)_i$ where the index i indicates the i-*th* tuning section, as the change in the laser's cavity mode frequency caused by a change in the carrier concentration N at that section. We assume these coefficients are constant and verify that experimentally. A change ΔN_i in the carrier concentration will therefore cause a frequency change $\Delta f = (\partial f / \partial N)_i \Delta N_i$ provided the tuning is within the limits of continuous tuning and a mode-hop does not occur. While within the continuous tuning limits the detuning coefficient can be written as:

$$\left(\frac{\partial f}{\partial N}\right)_i = \left(\frac{\partial f}{\partial n'}\right)_i \cdot \left(\frac{\partial n'}{\partial N}\right)_i \tag{1}$$

where n' is the real part of the refractive index. The first term in the RHS is calculated by solving the amplitude and phase conditions of the cavity [6] and is dependant of the tuning section structure. The second term in the RHS is given by $\partial n' / \partial N = \beta_{pl}$, where β_{pl} is the free-carrier plasma coefficient [6] and is therefore equal for all the tuning sections.

We define $\Delta N(t) = N(t) - N_D$ as the change in the carrier concentration in the tuning section relative to the predefined steady state value N_D of the destination wavelength λ_D. The carrier concentration $N(t)$ is obtained by solving for each tuning section the rate equation: [6]

$$\frac{dN}{dt} = \frac{\eta i(t)}{q V_a} - (AN + BN^2 + CN^3) \qquad (2)$$

where η is the current confinement factor, $i(t)$ is the tuning current, V_a is the junction volume, q is the electron charge, A,B, and C are the non-radiative, bi-molecular and Auger recombination coefficients respectively. The tuning current $i(t)$ can be taken as a theoretical step function in order to investigate the physical limitations of the laser, or as the actual current pulse of the current driver used in a real laser transmitter. The effect of an improved current pulse such as a predistortion pulse can be explored in this manner.

Finally, we assume that within the boundaries of continuous tuning, where only one longitudinal mode exists, the tuning sections affect the laser frequency only through their phases. Therefore, the laser frequency detuning is given by summing the contributions of all individual tuning sections:

$$\Delta f(t) = \sum_i \left(\frac{\partial f}{\partial N} \right)_i \Delta N_i(t) \qquad (3)$$

Since the required frequency accuracy in practical systems is usually smaller than the entire continuous tuning range, this model is sufficient for the purpose of calculating the tuning time t_{TX}.

3.   TUNING TIME OF GCSR LASER

The GCSR laser consists of four sections: a multiple quantum well gain section, a co-directional coupler, a phase matching section, and a sampled Bragg reflector section. The tuning range is 1529-1561nm. A more detailed description of the laser structure is available in [5].

The reflection peaks of the sampled reflector are about 4nm apart. The coupler acts as a coarse filter, and the current applied to the coupler section is controlled to select one reflector peak. Fine tuning is achieved by tuning the reflector peak to a desired wavelength and adjusting the phase current to align the cavity mode comb to the selected wavelength.

In fig.1 a partial mode-plane map of the laser output power versus the coupler and reflector currents is presented. The boundaries of a single longitudinal mode are highlighted. The detuning coefficients are defined inside those boundaries and mode hops of ±0.2nm and ±4nm occur when the mode boundaries are crossed.

484

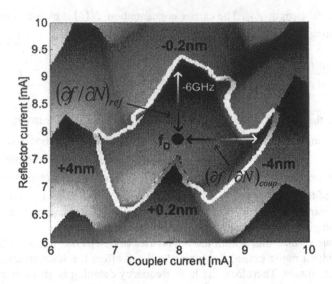

Figure 1. Map of output power versus coupler and reflector sections tuning currents of
GCSR laser showing lasing modes.

The detuning coefficients were measured empirically by introducing small
increments to the tuning currents and measuring the change in laser frequency. The
amount of change in the carrier concentration was then calculated by solving (2) in
steady state. The physical parameters of the laser are given in [7]. The detuning
coefficients measurement was repeated at different operating wavelengths, and the
coefficients were verified to be constant. The measured values of the tuning
coefficients $\{\partial f / \partial N\}_i$ were $1.64 \cdot 10^{-22}, 3.27 \cdot 10^{-23}, 4.07 \cdot 10^{-22} [GHz \cdot m^3]$ for
the coupler, phase and reflector sections respectively. The assumption expressed in
(3) was also verified experimentally with an accuracy of 10%.

We shall proceed to calculate the longest tuning time expected in a particular
GCSR laser transmitter [7].

We define I_{max}, I_{min} as the high and low limits of the tuning currents of each
section. We shall focus on the worst case where all tuning sections are switched
from I_{max} to I_{min} [8]. In fig.2 we present an example of the frequency evolution
calculation after the reflector current is switched. In fig.2(a) the current pulse $i(t)$
is shown. The current pulses supplied by the transmitter to the tuning sections of
the laser were measured using a digital scope and were used to solve (2). The
reflector current is switched from 25mA to 1mA, and the current settling time is
~30ns. In fig.2(b) the carrier concentration $N(t)$ in the reflector obtained by
solving (2) numerically is shown. In fig.2(c) the frequency detuning

$\Delta f(t) = (\partial f / \partial N) \Delta N(t)$ is presented. The frequency at the center of the mode, as shown in fig.1, is taken as the destination frequency f_D. As also shown in fig.1, beyond the vertical mode boundaries a cavity mode hop of 0.2nm occurs, which is associated with changing the reflector current beyond a certain value. While inside the mode boundaries, a frequency detuning of up to 6GHz can be observed between the mode center and the upper boundary. Consequently, fig.2(c) is valid only for $\Delta f \leq 6GHz$.

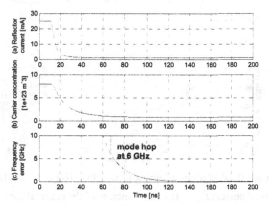

Figure 2. Tuning time calculation. (a) Reflector section current. (b) Reflector section carrier concentration. (c) Laser frequency detuning.

We investigated the sensitivity of the tuning time to I_{min} and I_{max} separately. In the first set of calculations we examined the effect of I_{max} on the tuning time. I_{min} was constant at 2mA in all the calculations, and I_{max} was gradually increased from 25mA to 50mA. Fig.3(a) presents the individual contributions of each tuning section (while keeping the other two fixed), and the overall tuning time while switching all three tuning currents from I_{max} to I_{min}.

In the second set of calculations we examined the effect of I_{min} on the tuning time. I_{max} was constant at 25mA, and I_{min} was gradually increased from 1mA to 15mA. Fig.3(b) presents the individual contributions of each tuning section assuming switching of only one section, and the overall tuning time assuming switching of all tuning sections from I_{max} to I_{min}. The required frequency accuracy f_{sys} was 6GHz in all cases.

We observe in Fig.3 that I_{min} has a strong influence on the tuning time while I_{max} has very little influence. Further we notice that the reflector section dominates the tuning time, due to its large tuning coefficient, and its large volume,

486

which increases the denominator in (2) and slows the tuning process. The dominant effect of the reflector section was also reported previously [8].

We therefore concluded that in a GCSR laser transmitter, the worst case tuning time is primarily a function of the reflector minimal current.

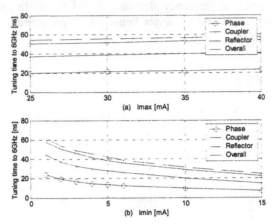

Figure 3. Theoretical calculations of tuning time with 6GHz accuracy. (a) As a function of the maximal current I_{max} . (b) As a function of the minimal current I_{min} .

Experimental verification

In this section we present the results of a set of experiments that was performed in order to verify the theoretical calculations of tuning time under the worst case conditions described above.

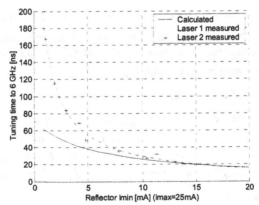

Figure 4. Theoretical and measured tuning time as a function of the reflector current minimal value I_{min}

As noted earlier the frequency at the mode upper boundary was measured to be approximately 6GHz less than the frequency at the mode center, throughout the laser tuning range. By using a 20GHz FWHM band-pass optical filter tuned to the frequency at the center of the destination mode, the mode hop into the destination mode, and consequently into filter pass band, produces a sharp increase in the detected signal, indicating the instant when the laser frequency obeys $\Delta f(t) = 6GHz$.

In Fig.4 we present the calculated and measured tuning time for two GCSR lasers. 15 operating points were selected with reflector currents ranging between 1mA and 18mA. For currents above 7mA there is a very good agreement between measured and calculated results.

When switching to operating points with reflector currents below 7mA the measured tuning times are increasingly longer than the calculated values. In order to achieve a tuning time of less than 50ns with a frequency accuracy of 6GHz the reflector current of the working points should be above 7mA.

In order to investigate the long tuning times observed while switching to low currents we measured the voltage across the reflector during the tuning process. The reflector voltage is a direct indicator of the carrier concentration [6] thus providing an important view into the laser dynamics.

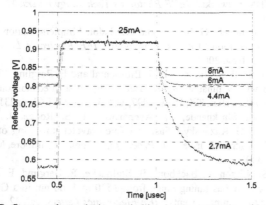

Figure 5. Reflector voltage during switching between 25mA and lower currents.

As shown in Fig.5, there is a strong correlation between the measured tuning times and the measured reflector voltage transients. As the final reflector current decreases the voltage settling time increases considerably.

Furthermore, the voltage-current curve of the reflector section was measured and its knee point was observed at 5mA. This indicates that most of the mismatch between the theoretical and measured tuning times occurs while the reflector current is below the diode cut-off current. We conclude that while the tuning section operates at low forward bias (2) does not provide an accurate model for the tuning section dynamics and a better model is required.

488

4.   CONCLUSIONS

We have presented a method for calculating the tuning time of multiple section current injection lasers. This method uses a system engineering approach that incorporates the electronic drive circuits together with basic semiconductor dynamics theory and the optical properties of the tuning sections which can be either calculated or easily measured.

The method was applied to the GCSR laser, and the limitations on tuning time in this laser were calculated theoretically and verified experimentally.

The theoretical calculation of tuning time which is based on the carrier concentration rate equation has proved to be accurate for currents above 7mA. Below this current we observed longer tuning times, which were also accompanied by a slow change in the reflector voltage. This indicates that at small forward bias a different dynamic model has to be considered.

REFERENCES

[1] D. Sadot, and I. Elhanany, "Optical switching speed requirements for Terabit/sec packet over WDM networks," *IEEE Photonics Technology letters*, 12, no., 4, pp. 440-442, April 2000

[2] S. Rubin, E. Buimovich, G. Ingber, and D. Sadot, "Implementation of an ultra-fast widely-tunable burst-mode 10Gbps transcevier," *IEE Electronic Letters*, vol. 38, no. 23, pp.1462-1463, Nov. 2002.

[3] P. J. Rigole, M. Shell, S. Nilsson , D. J. Blumental and E. Berglind, "Fast wavelength switching in a widely tunable GCSR laser using pulse predistortion technique," in *Proc. of OFC conf.*, Dallas, TX, USA, 1997, vol. 6, paper WL63, pp231-232.

[4] Y. Fukashiro, K. Shrikhande, M. Avenarius, M. S. Rogge, I. M. White, D. Wonglumsom, L. G. Kazovsky, "Fast and fine wavelength tuning of a GCSR laser using a digitally controlled driver," in *Proc. of OFC conf.*, Baltimore, MD, USA, 2000, vol.2, pp.338 –340.

[5] P. J. Rigole, S. Nilsson, L. Backbom, B. Stalnacke, E. Berglind, E. J. P. Weber, B. Stoltz, "Quasi-continuous tuning range from 1520 to 1560 nm in a GCSR laser, with high power and low tuning currents," *IEE Electronics Letters*, vol. 32, no.25, pp. 2352–2354, Dec. 1996.

[6] M.C.Amann J.Buus, *"Tunable Laser Diodes"*, Artech House pub. 1998.

[7] E. Buimovich, D. Sadot, "Physical limitations of tuning time and system considerations in implementing fast tuning of GCSR lasers", *Journal of Lightwave Technology*, **22**, no.2, pp. 582-588, February 2004.

[8] J. E. Simsarian, A. Bharwaj, K. Dreyer, J. Gripp, O. Laznicka, K. Sherman, Y. Su, C. Webb, L. Zhang, M. Zirngibl, "A widely tunable laser transmitter with fast accurate switching between all channel combinations," in *Proc. of ECOC conf.*, Copenhagen, Denmark, 2002, paper 3.3.6.

CHANGING RESONANCE WAVELENGTHS OF LONG-PERIOD FIBER GRATINGS BY THE GLASS STRUCTURE MODIFICATION

Katsumi Morishita[1] and Akihiro Kaino[2]
[1]*Osaka Electro-Communication University, Neyagawa, Osaka 572-8530 Japan, e-mail:*
morisita@isc.osakac.ac.jp
[2]*Osaka Electro-Communication University, Neyagawa, Osaka 572-8530 Japan, e-mail:*
m04105@isc.osakac.ac.jp

Abstract: Long-period fiber gratings written by arc discharge are heated at different temperatures, and the post-heating changes of transmission characteristics are investigated.   The resonance wavelengths are shifted to longer wavelengths by heating at the lower temperature than the structural temperature of the fiber, and they move more quickly with increasing heating temperature. The resonance wavelength shifts more largely for the loss peak generated by the higher cladding mode.   It becomes evident that the resonance wavelength can be changed and adjusted by heating temperature and heating time without significant degradation.

1.   INTRODUCTION

Modifying refractive index locally is a very important technique for fabricating optical devices and changing their optical properties.   Fiber gratings, which are commonly made by local index change, have been increasingly used in a wide variety of optical communication and sensing applications.   It is very valuable for practical use to change and adjust their transmission characteristics.   In this paper we have studied the possibility of changing resonance wavelengths of LPGs by the glass structure modification.   The glass structure change is a simple and widely applicable method to modify refractive index.   The index difference between the core and the cladding of dispersive fibers was controlled by the glass structure change generated by annealing [1, 2].   Mode-field transformers were made by the glass structure modification induced by heating locally [3]. Long-period gratings (LPGs) were written in a conventional silica fiber [4] and a pure silica holey fiber [5] by the rapid glass structure rearrangement induced by arc discharge.

The temperature sensitivity of LPGs written in silica fibers [6, 7] and a pure silica holey fiber [8] was investigated for temperatures up to about 1200 °C. The resonance wavelength shifted almost linearly with increasing temperature for up to about 800 °C for silica fibers and 900 °C for the holey fiber. For higher temperatures the resonance wavelength shift had a nonlinear dependence on temperatures. However the post-heating changes of the transmission characteristics of LPGs have not been examined yet.

In this paper, the LPGs are heated at different temperatures to change the glass structure, and the post-heating changes of resonance wavelengths and peak losses are examined against heating temperature and heating time. The mechanisms of the resonance wavelength shift and the peak loss change are investigated based on the glass structure change.

1. INDEX MODIFICATION BY THE GLASS STRUCTURE CHANGE

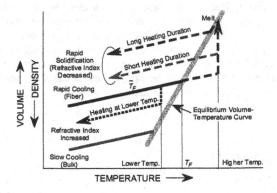

Figure 1. Schematic diagram of volume-temperature variation of a glass for heat treatments.

The index modification generated by the glass structure change results from the structural relaxation [9]. Figure 1 shows a schematic diagram of volume-temperature variation of a glass for heat treatments. When a glass is maintained at a constant temperature, the volume changes with time until it reaches a certain equilibrium glass structure. The temperature that corresponds to the equilibrium glass structure is called the structural temperature. The light gray line indicates an equilibrium volume-temperature curve. When temperature drops slowly, the glass structure changes with temperature, and the viscosity of the glass increases. Finally the glass cannot trace the equilibrium curve, and then the glass structure is frozen. The fiber glass is cooled faster than the bulk glass and the glass structure is frozen at the higher temperature T_F than that of the bulk glass. The density and the refractive index of the drawn fiber become lower than those of the bulk glass. Since the glass structure of the drawn fiber is the same as the equilibrium glass structure at T_F, the structural temperature of the fiber is expressed by $\overline{T}_F$.

In the LPG fabrication, a drawn fiber is heated locally to above the melting temperature, and then is cooled rapidly. The refractive index is decreased by rapid solidification as indicated by the broken lines. The index reduction can be adjusted by heating temperature and heating time. In case of heating at the lower temperature than T_F, the fiber glass approaches the equilibrium state, and the refractive index is increased as shown by the dotted line. In this paper, LPGs are written in a standard silica fiber (Corning SMF-28) by arc discharge with moving the discharge point periodically [4], and the LPGs are heated at the lower temperature than T_F to change the transmission characteristics.

1. SPECTRAL TRANSMISSION CHANGES INDUCED BY HEATING LONG-PERIOD FIBER GRATINGS

It was shown that the observed structural temperatures of the drawn fibers were in the range of 1150 – 1660 °C for single-mode silica fibers [10]. The fabricated LPGs, therefore, are heated at different temperatures below 1150 °C to increase the refractive index of the fiber, and the post-heating variations of the transmission characteristics are investigated.

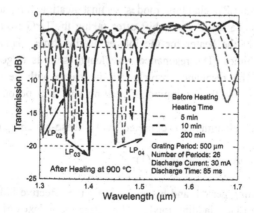

Figure 2. The transmission spectra of the LPG written with the discharge current and time of 30 mA and 85 ms before and after heating at 900 °C for 5, 10, and 200 minutes.

492

Figure 2 indicates the transmission spectra of the LPG with the grating period of 500 μm before and after heating at 900 °C for 5, 10, and 200 minutes. The LPG is written with the discharge current and time of 30 mA and 85 ms. The number of periods is 26. The LPG is heated by a ceramic heater. The spectral transmissions of the LPG are recorded after cooling to room temperature. The loss peaks are generated by coupling from the LP_{01} core mode to the LP_{02}, LP_{03}, and LP_{04} cladding modes, and are indicated by LP_{02}, LP_{03}, and LP_{04}, respectively. The resonance wavelengths are shifted to the longer wavelength region with heating time. The peak loss of the LP_{02} mode decreases with heating time, and those of the LP_{03} and LP_{04} modes decrease at the beginning, and begin to increase after that.

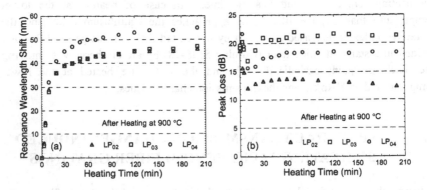

Figure 3. The changes of (a) the resonance wavelengths and (b) the peak losses of the LPG with the grating period of 500 μm against heating time at 900 °C. The discharge current and time are 30 mA and 85 ms, and the number of periods is 26.

Figure 3(a) shows the resonance wavelength shifts of the loss peaks generated by the LP_{02}, LP_{03}, and LP_{04} cladding modes against heating time at 900 °C. We measure the resonance wavelength shift after taking the LPG from the heater and then take it back in the heater for the longer heating time. Heating time is the accumulated total time. The resonance wavelength shift is larger for the higher cladding mode. The resonance wavelength increases with heating time. The increase rate is large at the beginning of heating and then becomes smaller with heating time.

The resonance wavelength λ_{res} is obtained by the phase-matching condition

$$\lambda_{res} = (n_{01} - n_{0m})\Lambda, \qquad (1)$$

where Λ is the grating period, and n_{01} and n_{0m} are the effective indexes of the LP_{01} core mode and the LP_{0m} cladding mode. The refractive indexes of the core and the cladding are raised by heating the LPG at the lower temperature than T_F. The effective index is increased more greatly for the lower mode because of its larger power fraction within the fiber. Therefore the effective index difference, $(n_{01} - n_{0m})$, is increased more largely for the higher cladding mode, and the resonance wavelength shift becomes larger for the higher cladding mode as shown in Figure 3(a).

When the fiber glass with the structural temperature of $\bar{T}_F$ is rapidly taken to the temperature T_H, the structural temperature $\bar{T}(t)$ changes as follows [9]:

$$\frac{d\bar{T}(t)}{dt} = A(\bar{T}_H - \bar{T}(t)), \qquad (2)$$

where A is the reciprocal of the viscosity, and $\bar{T}_H$ is the structural temperature corresponding to the equilibrium glass structure at the heating temperature T_H. The structural temperature $\bar{T}(t)$ is obtained by solving (2), and is shown by

$$\bar{T}(t) = \bar{T}_H + (\bar{T}_F - \bar{T}_H)\exp(-At). \quad (3)$$

The change of the structure temperature decreases exponentially with time, and the structural temperature $\bar{T}(t)$ approaches $\bar{T}_H$. The change becomes faster for the higher heating temperature T_H because of the lower viscosity. In case the heating temperature T_H is lower than the structural temperature of the drawn fiber $\bar{T}_F$, the refractive indexes of the core and the cladding increase and their increase rates decrease exponentially with heating time. Therefore the resonance wavelength grows quickly at the beginning of heating and then increases more slowly with heating time as shown in Figure 3(a).

Figure 3(b) shows peak losses generated by the LP_{02}, LP_{03}, and LP_{04} cladding modes against heating time at 900 °C. The peak losses decrease rapidly within about 10 minutes, and then gradually increase, and come to change just a little. The change of the peak loss is caused by the variation of the coupling coefficient between the core and the cladding modes, which is determined by the form of the index reduction generated by arc discharge along the fiber. The residual stress is released at the middle of the discharged area, but is not released completely around the discharged region. When the LPG is heated, the residual stress around the discharged part is released quickly, and the stress relaxation changes the coupling coefficients and the peak losses.

The residual stress relaxation in the undischarged part modifies the effective indexes and moves the resonance wavelengths. The resonance wavelength of a LPG written in SMF-28 shifted by 1.2 nm to short wavelengths owing to the stress relaxation induced by heating at 800 °C for 1 hour [7]. The residual stress of pure-silica-core/fluorine-doped-silica cladding fibers was almost removed by heating at 900 °C within 10 minutes [11]. Therefore we consider that the residual stress of a Ge-doped-silica core/pure-silica cladding fiber, SMF-28, is almost released by heating at 900 °C within 10 minutes and the peak losses change rapidly within 10 minutes. After the residual stress relaxation within 10 minutes, the glass structure change becomes a main factor of the peak loss variations. The glass structure changes more slowly with heating time, and the form of the index modification changes moderately, and the peak losses vary slowly as shown in Figure 3(b).

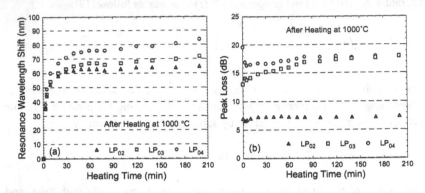

Figure 4. The changes of (a) the resonance wavelengths and (b) the peak losses of the LPG with the grating period of 500 μm against heating time at 1000 °C. The discharge current and time are 33 mA and 100 ms, and the number of periods is 30.

Figure 4(a) shows the resonance wavelength shifts of the LPG against heating time at 1000 °C. The resonance wavelength increases with heating time. The increase rate at 1000 °C is very great at the start of heating and larger than that at 900 °C, and then becomes smaller. Since the viscosity becomes lower at the higher temperature, the glass structure changes faster and reaches the equilibrium state in the shorter time as shown by (3). Therefore the resonance wavelength moves more quickly at 1000 °C than 900 °C and reaches a constant value earlier. The resonance wavelengths become almost constant at about 80 minutes, and after 80 minutes the resonance wavelengths begin to increase again. The increase after 80 minutes would be caused by elongation of the fiber.

Figure 4(b) shows peak losses generated by the LP_{02}, LP_{03}, and LP_{04} cladding modes against heating time at 1000 °C. The peak losses change rapidly within 5 minutes, and then gradually increase. We consider that the rapid changes of peak losses within 5 minutes result from the residual stress relaxation, because the residual stress is relaxed faster at 1000 °C than 900 °C. The peak losses are almost constant after 80 minutes, and it becomes evident that heating at 1000 °C does not degrade the LPG until 200 minutes.

Figure 5(a) shows the resonance wavelength shifts of the LPG against heating time at 1100 °C. The resonance wavelengths move rapidly to longer wavelengths within 3 minutes, and then shift slowly to longer and shorter wavelengths. The glass structure changes faster and reaches the equilibrium state in the shorter time at 1100 °C than 1000 °C. The glass structure in the undischarged area is thought almost to reach the equilibrium state in about 5 minutes. Therefore the effective indexes of the core and the cladding modes, n_{01} and n_{0m}, hardly increase after 5 minutes, and the resonance wavelengths change little. The resonance wavelength shifts after heating time of 20 minutes result from broadening of loss peaks, elongation of the fiber, and degradation of the LPG.

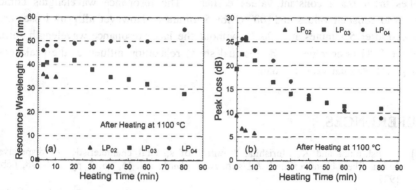

Figure 5. The changes of (a) the resonance wavelengths and (b) the peak losses of the LPG with the grating period of 500 μm against heating time at 1100 °C. The discharge current and time are 33 mA and 95 ms, and the number of periods is 32.

Figure 5(b) shows the peak losses generated by the LP_{02}, LP_{03}, and LP_{04} cladding modes against heating time at 1100 °C. All peak losses reduce for the longer heating time than 5 minutes. The glass structure in the undischarged part almost reaches the equilibrium state in about 5 minutes, but that in the discharged part does not become equilibrium yet owing to the higher structural temperature. Therefore the glass structure in the discharged part continue to approaches the equilibrium state after 5 minutes, and the index difference between the discharged and the undischarged parts decreases with heating time. That reduces the amplitude of the index modification along the LPG, and the loss peaks become shallower and broader with heating time. Since heating at 1100 °C makes the residual stress release very fast and the glass structure change very rapidly, it is difficult to observe the peak loss variations caused by the residual stress relaxation.

1.   CONCLUSIONS

Long-period fiber gratings written by arc discharge are heated at different temperatures, and the post-heating changes of resonance wavelengths and peak losses are investigated. It is shown that the resonance wavelengths can be changed and adjusted by heating temperature and heating time without significant degradation. The resonance wavelengths are shifted to longer wavelengths by heating LPGs at the lower temperature than the structural temperature of the fiber. The resonance wavelength shifts more greatly for the loss peak generated by the higher cladding mode because of its lower power fraction within the fiber. Heating at the higher temperature moves the resonance wavelengths more quickly and makes them reach constant values earlier. The resonance wavelengths come almost to constant values at 1100 °C for 5 minutes. However all peak losses are reduced after 5 minutes at 1100 °C without the large resonance wavelength shift, and the LPG is degraded. The residual stress relaxation influences the peak loss though the residual stress is small.

REFERENCES

[1] J. Nishimura and K. Morishita, "Control of spectral characteristics of dispersive optical fibers by annealing", *J. Lightwave Technol.*, vol. 15, no. 2, pp. 294-298, Feb. 1997.

[2] J. Nishimura and K. Morishita, "Changing multimode dispersive fibers into single-mode fibers by annealing and guided mode analysis of annealed fibers", *J. Lightwave Technol.*, vol. 16, no. 6, pp. 990-997, June 1998.

[3] J. Nishimura and K. Morishita, "Mode-field expansion and reduction in dispersive fibers by local heat treatments", *IEEE J. Select. Topics Quantum Electron.*, vol. 5, no. 5, pp. 1260-1265, Sept./Oct. 1999.

[4] K. Morishita, S. F. Yuan, Y. Miyake and T. Fujihara, "Refractive index variations and long-period fiber gratings made by the glass structure change", *IEICE Trans. Electron.*, vol. E86-C, no. 8, pp. 1749-1758, Aug. 2003.

[5] K. Morishita and Y. Miyake, "Fabrication and resonance wavelengths of long-period gratings written in a pure silica photonic crystal fiber by the glass structure change," *J. Lightwave Technol.*, vol. 22, no. 2, pp. 625-630, Feb. 2004.

[6] G. Rego, O. Okhotnikov, E. Dianov, and V. Sulimov, "High-temperature stability of long-period fiber gratings produced using an electric arc," *J. Lightwave Technol.*, vol. 19, no. 10, pp. 1574-1579, Oct. 2001.

[7] G. Humbert and A. Malki, "Electric-arc-induced gratings in non-hydrogenated fibres: fabrication and high-temperature characterizations," *J. Opt. A*, vol.4, no. 2, pp. 194-198, Feb. 2002.

[8] G. Humbert, A. Malki, S. Février, P. Roy, and D. Pagnoux, "Characterizations at high temperatures of long-period gratings written in germanium-free air-silica microstructure fiber," *Opt. Lett.*, vol. 29, no. 1, pp. 38- 40, Jan. 2004.

[9] T. S. Izumitani Optical Glass, American Institute of Physics, New York, 1986, ch. 1

[10] D. −L. Kim, M. Tomozawa, S. Dubois, and G. Orcel, "Fictive temperature measurement of single-mode optical-fiber core and cladding," *J. Lightwave Technol.*, vol. 19, no. 8, pp. 1155-1158, Aug. 2001.

[11] S. Ishikawa, H. Kanamori, T. Kohgo, M. Nishimura, and H. Yokota, "New mode-field conversion technique in optical fiber using thermal relaxation of residual stress," in *Tech. Dig. Conf. Optic. Fiber Commun./Int. Conf. Integrated Optics Optic. Fiber Commun. (OFC-IOOC)*, 1993, San Jose, USA, paper TuB4.

PART B5:

POLARIZATION MODE DISPERSION

ADAPTIVE ELECTRONIC PROCESSING IN OPTICAL PMD-IMPAIRED SYSTEMS

T. Foggi[1], G. Colavolpe[2], E. Forestieri[1,3], and G. Prati[1,3]
[1]*CNIT, Photonic Networks National Lab, Pisa, Italy*
[2]*Università di Parma, Dipartimento di Ingegneria dell'Informazione, Parma, Italy*
[3]*Scuola Superiore Sant'Anna, Pisa, Italy*
tommaso.foggi@cnit.it

Abstract: In high-speed optical transmission systems, one of the most challenging impairments is represented by the signal distortions produced by polarization-mode dispersion (PMD). In this paper, we analyze the limits of electronic compensation schemes, in the form of maximum-likelihood sequence detection receivers. With a constraint on the front-end processing after photo-detection, we compute the signal statistics necessary to implement this strategy. The receiver adaptivity is also discussed. The relevant performance is analyzed by means of computer simulations and accurate analytical performance bounds, and compared with the performance of all-optical compensators and decision-feedback equalizers. A significant performance loss with respect to optical compensation is observed for large values of the differential group delay (DGD) showing that, after the non reversible transformation operated by the photo-detector, it is not possible to effectively cope with the PMD-induced impairments.

1.   INTRODUCTION

In recent years, the application of electronic processing to optical preamplified transmission systems has been deeply investigated in order to combat linear and non-linear impairments of the optical channel. In fact, although optical compensation has demonstrated to be very efficient in recovering signals affected by heavy penalties induced by PMD (see [1] and references therein), these devices need advanced and cost-ineffective technologies [2], so that the use of well-known and popular electronic schemes is of great interest. In the first-order approximation, the PMD effects are represented by the introduction of a differential group delay (DGD) between the two principal states of polarization (PSPs), causing intersymbol interference (ISI). Classical methods of electronic processing of ISI-affected signals have been applied, such as linear equalization [3], and non-linear cancellation has also been used since the optical to electrical conversion causes a non-linear transformation of the signal [4]. Recent studies have focused on the comparison between these equalization techniques and optical compensation [5], evidencing qualities and lacks of both solutions.

More effective electronic techniques, based on maximum-likelihood sequence detection (MLSD) implemented through the Viterbi algorithm (VA) [6], have been proposed in [2, 3, 5]. In [3], the presence of optical amplifiers is not considered. As a consequence, the amplified spontaneous emission (ASE) noise is not present and the statistics of the received signal, necessary to compute the VA branch metrics are conditionally Gaussian due to shot and thermal noise. Since making a Gaussian assumption for noise statistics after photo-detection when ASE noise is dominant leads to inaccurate results in the computation of the receiver metrics [7], in [2, 5] the statistics of the received signal are approximately measured and updated in real-time during transmission and assuming no decision errors. This method has been compared with classical equalization schemes in [2, 5], showing that a better performance, as expected, can be obtained by using the MLSD strategy.

In this paper, we derive a practically optimal receiver, in the sense of MLSD, with the constraint of one sample per bit at the output of the postdetection filter. The exact statistics of the received signal are computed and the branch metrics of the VA expressed accordingly. A closed-form approximation of these branch metrics, which entails a negligible performance loss, is also given. Based on the exact branch metric computation, analytical bounds for the system performance are also provided, allowing to reach values of bit error rate (BER) below 10^{-12}. This analytical method also represents an interesting tool for optimizing the receiver parameters without resorting to time-consuming computer simulations. Since PMD is a time-varying phenomenon, the receiver has also to adaptively update some parameters. This aspect is also discussed. A comparison with commonly adopted electronic equalization and optical compensation techniques is also provided showing that after the irreversible transformation given by the photodetector, the receiver is not able to effectively cope with the PMD distortions.

2. SYSTEM MODEL

The considered transmission system is shown in Fig. 1, where double arrows denote two orthogonal states of polarization (SOPs) in the fiber. The optical signal generated by a laser modulator is launched in a single-mode fiber (SMF) operating in linear regime. At the receiver end, after a flat-gain optical amplifier (OA), which also introduces ASE noise, the received signal is opti-

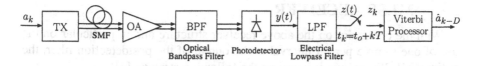

Figure 1. System model.

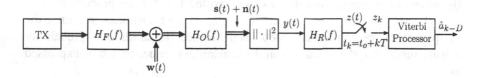

Figure 2. Low-pass equivalent of the considered system.

cally filtered prior photo-detection. The photo-detection process and the following electronic circuitry will add shot and thermal noise, respectively. The ASE noise is modeled as additive white Gaussian noise (AWGN). By using high-gain amplifiers we can assume that the other sources of noise are negligible. The detected signal $y(t)$ is then low-pass filtered and the resulting signal $z(t)$ is sampled at one sample per bit interval T. Finally, a Viterbi processor elaborates the received sequence $\mathbf{z} = \{z_k\}$ in order to take reliable decisions on the transmitted bit sequence $\mathbf{a} = \{a_k\}$.

The low-pass equivalent model of the system is shown in Fig. 2. At the optical filter output, the components of the two-dimensional complex vectors $\mathbf{s}(t) = (s_1(t), s_2(t))$ and $\mathbf{n}(t) = (n_1(t), n_2(t))$ represent the useful signal and noise components on two orthogonal SOPs, respectively. The noise components are Gaussian but not white, since they are obtained by filtering the low-pass equivalent AWGN $\mathbf{w}(t) = (w_1(t), w_2(t))$. In the following, we will assume that $w_1(t)$ and $w_2(t)$ have two-sided power spectral density N_0. The low-pass equivalent frequency response of fiber, optical and postdetection filters are denoted by $H_F(f)$, $H_O(f)$, and $H_R(f)$, respectively. At the output of the photodiode the detected signal can be described as the sum of two contributions, one for each SOP:

$$y(t) = \|\mathbf{s}(t) + \mathbf{n}(t)\|^2 = |s_1(t) + n_1(t)|^2 + |s_2(t) + n_2(t)|^2 . \qquad (1)$$

Clearly, after photo-detection the noise becomes signal-dependent and its statistics change. In the following, optical and postdetection filter parameters can be chosen arbitrarily, since the proposed receiver is independent of a particular choice of the filter shape or bandwidth.

3. MLSD RECEIVER

With the constraint on the above receiver structure and in particular on the use of one sample per bit interval at the output of the postdetection filter, the optimal MLSD receiver is based on the following strategy [6]

$$\hat{\mathbf{a}} = \arg \max_{\mathbf{a}} p(\mathbf{z}|\mathbf{a}). \tag{2}$$

By means of computer simulations, we verified that for the practically used optical and postdetection filters, conditioning on the transmitted sequence the samples z_k can be considered independent. Hence, the conditional probability density function (pdf) $p(\mathbf{z}|\mathbf{a})$ given the transmitted sequence can be expressed as

$$p(\mathbf{z}|\mathbf{a}) = \prod_k p(z_k|\mathbf{a}). \tag{3}$$

Assuming that the system is causal and with finite memory L, it will be $p(z_k|\mathbf{a}) = p(z_k|a_k, a_{k-1}, \ldots, a_{k-L})$. Therefore, the optimal MLSD algorithm can be efficiently implemented by means of the VA with the following branch metrics:

$$\lambda_k(a_k, \sigma_k) = \ln p(z_k|a_k, \sigma_k) \tag{4}$$

having defined the trellis state as $\sigma_k = (a_{k-1}, a_{k-2}, \ldots, a_{k-L})$. Hence, the number of states $S = 2^L$ increases exponentially with the channel memory L.

A closed-form expression for the $2S$ pdfs in (4), necessary to implement the optimal MLSD strategy, does not exist.[1] We have numerically evaluated all these necessary conditional pdfs by using the method described in [8] and stored them in a look-up table that will be addressed, in order to compute branch metrics, by the received samples properly quantized and by the considered trellis transition. Although from an implementation point of view this solution is not feasible, since entire pdfs must be stored, we have followed this approach in order to investigate the ultimate performance of the MLSD strategy. The obtained results also represent a benchmark for alternative solutions based on an approximate, possibly closed-form, branch metrics expression. In particular, we use the following non-central chi-square approximation [7,9]

$$p(z_k|a_k, \sigma_k) \simeq \frac{1}{N_0} \left(\frac{z_k}{s_R(a_k, \sigma_k)} \right)^{\frac{\nu-1}{2}} \exp\left(-\frac{z_k + s_R(a_k, \sigma_k)}{N_0} \right)$$
$$\cdot I_{\nu-1}\left(\sqrt{\frac{2 z_k s_R(a_k, \sigma_k)}{N_0}} \right) \tag{5}$$

---

[1]It is well known that conditioning on the transmitted sequence, the samples at the photo-detector output have non-central chi-square distribution [6]. However, the presence of the electrical filter modifies these statistics.

in which $I_{\nu-1}(x)$ is the modified Bessel function of first kind and order $\nu - 1$, $\nu = \frac{B_O}{B_R}$, where B_O and B_R are the optical and postdetection filter noise equivalent bandwidths, respectively, and $s_R(a_k, \sigma_k)$ is the output of filter $H_R(f)$, sampled at time t_k, in the absence of noise. This term obviously depends not only on the considered optical and postdetection filters but also on the PMD parameters. Although we verified that (5) is not suitable for BER evaluation in standard receivers with arbitrary filters (the BER would be heavily underestimated), its use to approximate the branch metrics expression produces a negligible performance loss with respect to the optimal solution, as it will be shown in the numerical results.

As already mentioned, the number of trellis states, and thus the complexity, depends exponentially on the channel memory L. For commonly used optical and electrical filters, and DGD values lower than a bit interval, we verified that $L \leq 3$. Hence the number of states is at most $S = 8$. In addition, the application of reduced-state sequence detection (RSSD) techniques [10] allows to substantially reduce the number of trellis states. In particular, a reduced state $\sigma'_k = (a_{k-1}, a_{k-2}, \ldots, a_{k-L'})$, with $L' < L$, may be defined. The resulting number of states is reduced to $2^{L'} < 2^L$. In order to compute the branch metrics (4) in a reduced trellis, the necessary symbols not included in the state definition may be found in the survivor history [10]. We note that, in the limiting case of $L' = 0$, the trellis diagram degenerates and symbol-by-symbol detection with decision feedback is performed. The resulting receiver can be considered as a non linear equalizer with decision feedback. Our numerical results, not shown here for a lack of space, show that the resulting performance loss is negligible.

Since PMD is a time-varying phenomenon, the receiver parameters should be adaptively updated. By using the above mentioned approximated closed-form expression of the branch metrics, when PMD changes the receiver has to simply adaptively identify the term $s_R(a_k, \sigma_k)$. This can be easily done by using a gradient adaptation algorithm and, as a cost function, the one defining the non linear branch metrics.

4. PERFORMANCE ANALYSIS

The classical union bound on the bit error probability P_b has expression [6]

$$P_b \leq \sum_{\mathbf{a}} P(\mathbf{a}) \sum_{\hat{\mathbf{a}} \neq \mathbf{a}} b(\mathbf{a}, \hat{\mathbf{a}}) P(\mathbf{a} \rightarrow \hat{\mathbf{a}}) \tag{6}$$

in which $\mathbf{a} = \{a_k\}$ and $\hat{\mathbf{a}} = \{\hat{a}_k\}$ denote the bit sequences corresponding to the correct and erroneous paths, respectively, $b(\mathbf{a}, \hat{\mathbf{a}})$ is the number of bit errors entailed by the error event $(\mathbf{a}, \hat{\mathbf{a}})$, $P(\mathbf{a} \rightarrow \hat{\mathbf{a}})$ is the pairwise error probability and $P(\mathbf{a})$ is the a priori probability of sequence $\mathbf{a}$. The pairwise error probability $P(\mathbf{a} \rightarrow \hat{\mathbf{a}})$ is the probability that the sum of the branch metrics relative to the erroneous path exceeds the sum of the branch metrics on the correct path. This probability may be easily computed by employing the same method described in [8] already used to compute the *exact* expression of the conditional density functions $p(z_k|\mathbf{a})$.

From the union bound, we derive a lower and an approximated upper bound for the bit error probability. The lower bound is simply obtained by considering the most likely error event. The approximated upper bound is obtained by truncating the union bound considering a few most frequently occurring error events only. The accuracy of these bounds will be shown in the numerical results.

5. NUMERICAL RESULTS

In this section, the performance of the considered detection schemes is assessed in terms of BER versus E_b/N_0,[2] E_b being the received signal energy per information bit. In all simulations, the optical filter is assumed to be a fourth order Butteworth filter with low-pass equivalent bandwidth equal to $0.95/T$, whereas the postdetection filter is a fifth order Bessel filter with bandwidth $0.75/T$.

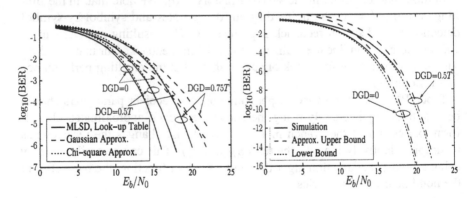

Figure 3. Performance of the VA with exact, chi-square, and Gaussian metrics.

Figure 4. Lower bound and and approximated upper bound.

In Fig. 3, we show the performance of the VA when the exact and approximated branch metrics are used. First-order PMD, with power splitting equal to 0.5 and different values of the DGD, is considered. It can be observed that the chi-square closed-form approximation (5) entails a negligible performance

---

[2]The ratio E_b/N_0 is proportional to the optical signal-to-noise ratio and represents the number of detected photons per bit at the input of the OA.

loss with respect to the optimal solution, especially for low BER values. On the contrary, when a Gaussian approximation for the branch metrics is used [3], the performance loss is remarkable (see Fig. 3).

The accuracy of the described lower and approximated upper bounds is shown in Fig. 4. In this case also, first-order PMD is considered with DGD values of 0 and $0.5T$. It is clear that it is not necessary to have recourse to time-consuming computer simulations since this tool predicts very well the receiver performance and can be also used to optimize the system parameters.

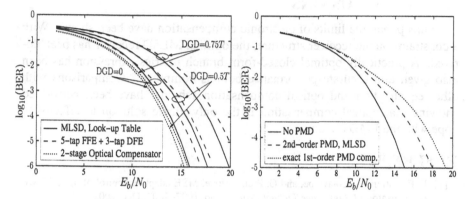

Figure 5. Comparison between optical and electronic compensation structures.

Figure 6. Performance in the case of second-order only PMD.

In Fig. 5, we compare the performance of the MLSD approach with that of optical and electronic compensators. The considered optical compensator consists of a cascade of some optical devices. The first optical device is a polarization controller (PC) which allows us to modify the polarization of the optical signal at its input. Then we have in this example two polarization maintaining fibers (PMF) separated by another PC. A PMF introduces a DGD between the components of the optical signal on the two orthogonal SOPs corresponding to its slow and fast axes. In the considered example, the DGD of the considered PMFs is $0.5T$. The other considered electronic compensation scheme is a $T/2$-spaced decision-feedback equalizer (DFE) with a feedforward filter with 5 taps and a feedback filter with 3 taps. No performance improvement has been observed by considering a more complex equalizer. In this case also, the presence of first-order PMD is considered. As it can be observed, the optical compensator is able to cope with the PMD effects almost perfectly. On the contrary, although MLSD performs better than equalization, electronic processing schemes, operating after the non-reversible transformation by the photodiode, entail a remarkable penalty.

Finally, in Fig. 6 we assume that the first-order effects have already been perfectly compensated and that the residual second-order parameters are the following: DGD derivative equal to $0.2T^2$ and PSP rotation rate equal to $0.4T$ (equal signal power splitting among the PSPs is also considered). The MLSD receiver is able to recover only 1 of the 4 dB penalty produced by the 2nd order PMD. Although we point out that only the outage probability can give a final answer, it seems that electronic compensation cannot be considered a viable solution not even joint with a simple optical scheme which perfectly compensates the effects of the first order PMD.

6. CONCLUSIONS

In this paper, the limits of electronic compensation have been shown. With a constraint on the receiver structure, the optimal MLSD receiver has been derived. A practically optimal closed-form branch metrics expression has been also given and accurate performance bounds computed. Comparisons with other equalization and optical compensation schemes have been performed showing that optical compensation is the only viable solution to effectively cope with the PMD effects.

REFERENCES

[1] E. Forestieri, G. Colavolpe, and G. Prati, "Novel MSE adaptive control of optical PMD compensator," *J. Lightwave Technol.*, vol. 20, pp. 1997–2003, Dec. 2002.

[2] H. F. Haunstein, W. Sauer-Greff, A. Dittrich, K. Sticht, and R. Urbansky, "Principles for electronic equalizationof polarization-mode dispersion," *J. Lightwave Technol.*, vol. 22, pp. 1169–1182, Apr. 2004.

[3] J. H. Winters and R. D. Gitlin, "Electrical signal processing techniques in long-haul fiber-optic systems," *IEEE Trans. on Commun.*, vol. 38, pp. 1439–1453, Sept. 1990.

[4] J. H. Winters and S. Kasturia, "Adaptive nonlinear cancellation for high-speed fiber-optic systems," *J. Lightwave Technol.*, vol. 10, pp. 971–977, July 1992.

[5] F. Buchali and H. Bulow, "Adaptive PMD compensation by electrical and optical techniques," *J. Lightwave Technol.*, vol. 22, pp. 1116–1126, Apr. 2004.

[6] J. Proakis, *Digital Communication*. New York: McGraw-Hill, 3rd ed., 1996.

[7] Y. Cai, J. M. Morris, T. Adali, and C. R. Menyuk, "On turbo code decoder performance in optical-fiber communication systems with dominating ASE noise," *J. Lightwave Technol.*, vol. 21, pp. 727–734, Mar. 2003.

[8] E. Forestieri and G. Prati, "Exact analytical evaluation of second-order PMD impact on the outage probability for a compensated system," *J. Lightwave Technol.*, vol. 22, pp. 988–996, Apr. 2004.

[9] P. A. Humblet and M. Azizoglu, "On the bit error rate of lightwave systems with optical amplifiers," *J. Lightwave Technol.*, vol. 9, pp. 1576–1582, Nov. 1991.

[10] M. V. Eyuboğlu and S. U. Qureshi, "Reduced-state sequence estimation with set partitioning and decision feedback," *IEEE Trans. on Commun.*, vol. 38, pp. 13–20, Jan. 1988.

PSO ALGORITHM USED FOR SEARCHING THE OPTIMUM OF AUTOMATIC PMD COMPENSATION

Xiaoguang Zhang, Yuan Zheng, Yu Shen, Jianzhong Zhang, and Bojun Yang
Department of Physics, School of Sience, Beijing University of Posts and Telecommunications, Beijing 100876, China

Abstract: The Particle Swarm Optimization (PSO) was introduced into automatic PMD compensation. It showed the merits of rapid convergence to the global optimum not being trapped in local sub-optima in searching process for automatic PMD compensation and robust to noise. In this paper we describe how implementing PSO as a control algorithm in automatic PMD compensation. The comparisons of performances between global version of PSO and local version of PSO were carried out theoretically and experimentally.

1.   INTRODUCTION

Adaptive compensation for polarization mode dispersion (PMD) may be one of the urgent tasks for next-generation high bit-rate optical fiber transmission systems. The control algorithm is critical in an adaptive PMD compensator. For adaptive PMD compensation using feedback scheme, many challenges such as rapid convergence to the global optimum without being trapped in local sub-optima, robust to noise etc. make it a hard work to find a practical feedback control algorithm. In most of related literatures, the adopted control algorithms have not been mentioned. Reference [1] and [2] reported the algorithm they used for controlling PMD compensators were gradient based peak search methods. We found that when the numbers of control parameters increased, gradient based algorithm had large numbers of opportunities to be locked into local sub-optima rather than the global-optimum. Besides, it would be less effective for a system with a relatively

508

high noise level in PMD monitor, because the gradient information between neighboring signals would be submerged in noise. For the fist time, we introduced Particle Swarm Optimization (PSO) into adaptive PMD compensation as the control algorithm, which .showed the good performance of rapid convergence to the global optimum without being trapped in local sub-optima and robust to noise.

2. THE ROLE OF CONTROL ALGORITHM IN ADAPTIVE PMD COMPENSATION

The configurations shown in Fig.1 and 2 are the typical schemes of optical post-compensation for PMD. It is widely believed that the one-stage compensators in Fig.1 are able to compensate the PMD to the first-order. They have 3 or 4 parameters (or degree of freedom, DOF) to be controlled depending on whether the differential group delay (DGD) line is fixed or varied. The Two-stage compensators in Fig. 2 can compensate the PMD up to the second-order [3][4]. They have 6 or 7 parameters (or DOF) to be controlled depending on whether the delay line is fixed or varied. (Here, the

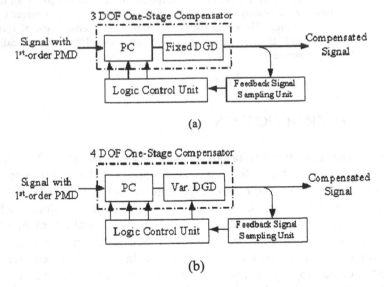

Figure 1. One-Stage Compensator

reason why we use 3 parameters instead of 2 to adjust polarization controllers (PC) is that, we found in the experiments, only adjusting at least

3 waveplates can a PC transform a fixed input state of polarization (SOP) into output states covering entire Poincaré sphere.)

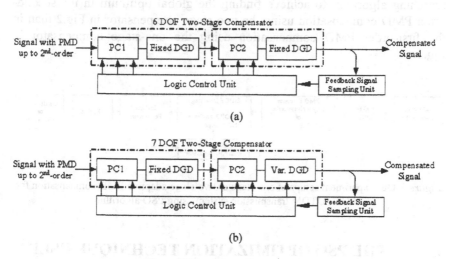

Figure 2. Two-Stage Compensator

The adaptive PMD compensation is a process for a control algorithm to find optimal combinations of control parameters, in order for the feedback signal to reach a global optimum, in an intelligent, fast, and reliable manner. In our experiment shown in Fig.3, the degree of polarization (DOP), obtained by an in-line polarimeter, was used as feedback signal. The optical pulses at the receiving end have DOP of 1 when there is no PMD in the fiber link, and DOP value decreases as PMD increases. The polarization controller used in compensation unit is the electrically controlled one which has four fiber-squeezer cells to be adjusted with voltage of 0-10V, out of which the three cells were used in the experiment. In this case, the problem of adaptive PMD compensation can be described as the problem of maximization of DOP in mathematics:

$$\underset{parameters}{MAX}\,(function) \tag{1}$$

where the *function* in bracket represents the DOP value in the experiment. The *parameters* here are the voltages for controlling PCs and varied delay line. The *function* in Eq.(1) is not simply predictable in the adaptive PMD compensation system. Therefore a good searching algorithm is required to solve problem (1), which is the problem of searching global maximum in D-dimensional hyperspace. The number of D depends on the number of DOF the compensator scheme chosen as shown in Fig.1 and 2. Generally, the more the DOFs are the more sub-maxima exist, which would increase the hard task of the searching algorithm. Therefore it is more difficult for a searching algorithm to achieve finding the global optimum in the second-order PMD compensation using the two-stage compensator in Fig.2 than in the first-order PMD compensation using the one-stage compensator in Fig.1.

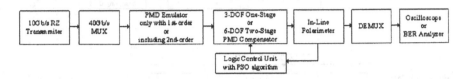

Figure 3. Our experiment setups of one-stage or two-stage adaptive PMD compensation for 40Gb/s OTDM transmission system using PSO algorithm

3.   THE PSO OPTIMIZATION TECHNIQUE USED AS THE CONTROL ALGORITHM

The PSO algorithm, proposed by Kennedy and Eberhart [5], has proved to be very effective in solving global optimization for multi-dimensional problems in static, noisy, and continuously changing environments [6]. We introduced for the first time the PSO technique into automatic PMD compensation in our previous works [7] and the latest experiment described in Fig.3, where it was shown to be effective.

The PSO is an optimization technique based on researches on swarms such as bird flocking. According to the research results for bird flocking, birds are finding food by flocking (not by each individual). It leads to the assumption that information is shared in flocking. Therefore it is easier for bird flocking to find only a food in a region than each individual.

At the beginning, the PSO algorithm randomly initializes a population (called swarm) of individuals (called particles). Each particle represents a single intersection of multi-dimensional hyperspace. The position of i-th particle is represented by the position vector $X = (x_{i1}, x_{i2}, \cdots, x_{iD})$. The particles evaluate their position relative to a goal at every iteration. In each iteration every particle adjusts its trajectory (by its velocity $V_i = (v_{i1}, v_{i2}, \cdots, v_{iD})$) toward its own previous best position, and toward the previous best position attained by any member of its topological neighborhood. Generally, there are two kinds of topological neighborhood structures: global neighborhood structure, corresponding to the global version of PSO (GPSO), and local neighborhood structure, corresponding to the local version of PSO (LPSO). For the global neighborhood structure the whole swarm is considered as the neighborhood, while for the local neighborhood structure some smaller number of adjacent members in sub-swarm is taken as the neighborhood. The two typical topological structures for global and local neighborhood structure are shown in Fig.4 [8].

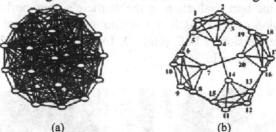

(a)   (b)

Figure 4. The topologic structure for global neighborhood (a) and one typical topologic structure for local neighborhood (b).

In the global version of PSO, each particle keeps track of its coordinates in hyperspace which are associated with the best solution (fitness) it has achieved so far. (The value of that fitness is also stored.) The value (and its location) is called *pbest*. At the same time, the overall best value, and its location, obtained so far by any particle in the population, is also tracked. This is called *gbest*. The particle swarm optimization concept consists of, at each time step, changing the velocity (accelerating) of each particle toward its *pbest* and *gbest*, at the same time changing its position according to following equations [9]:

$$v_{id} = v_{id} + c_1 \times rand() \times (pbest_{id} - x_{id}) + c_2 \times rand() \times (gbest_d - x_{id}) \qquad (2)$$

$$x_{id} = x_{id} + v_{id} \qquad (3)$$

where v_{id} is the i-th particle's velocity component, x_{id} is the current position component of the i-th particle, $pbest_{id}$ is the component of *pbest* of i-th particle along d axis, and $gbest_d$ is the component of *gbest* along d axis. The constants c_1 and c_2, termed learning rates, determine the relative influence of *pbest* and *gbest*. The rand() generates pseudo random numbers that are uniformly distributed in the interval of [0, 1]. If any particle's position is close enough to the goal function, it is considered as having found the global optimum and the recurrence is ended. In the global neighborhood structure, each particle's search is influenced by the best position found by any member of the entire population. In contrast, each particle in the local neighborhood structure is influenced only by parts of the adjacent members. In other words, for the local version of PSO (LPSO), particles have information only of their own and their neighbor's bests, rather than that of the entire group. Instead of moving toward the stochastic average of *pbest* and *gbest* (the best location of the entire group), particles move toward points defined by *pbest* and "*lbest*", which is the index of the particle with the best evaluation in the particle's neighborhood. Therefore the local version of PSO has less opportunity to be trapped in sub-optima than the global version of PSO (GPSO). In our experiment 20 particles are used either in GPSO or LPSO.

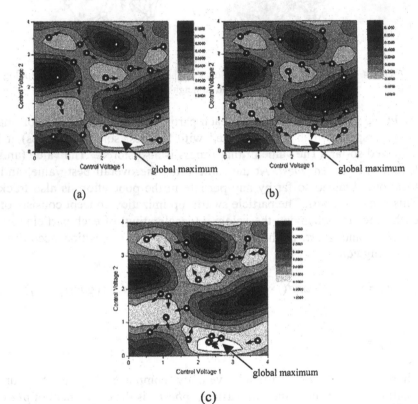

Figure 5. A visualized procedure for demonstrating a process of global maximum searching using PSO algorithm. (a)Initializing a swarm of particles at the beginning; (b)The particles in searching course according to Eq.(2) and Eq.(3); (c)The global DOP maximum has been founded.

Fig.5 is a visualized procedure for demonstrating a process of global maximum searching using PSO algorithm. For the sake of simplicity we choose a searching problem in 2-dimensional space---for example, finding global DOP maximum through controlling two voltages on a PC in the PMD compensator.

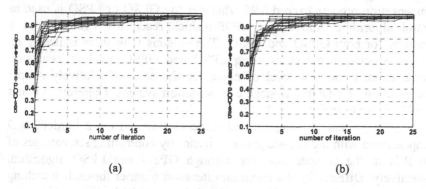

(a)   (b)

Figure 6. The best DOP vs. iteration recorded in 3-DOF 1st-order PMD compensation using LPSO (a) and GPSO algorithm (b).

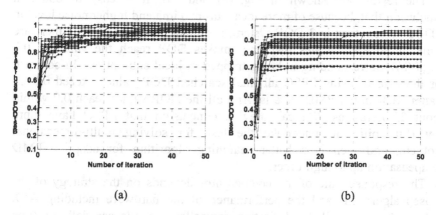

(a)   (b)

Figure 7. The best DOP vs. iteration recorded in 6-DOF 2nd-order PMD compensation using LPSO (a) and GPSO algorithm (b).

In the adaptive PMD compensation experiment shown in Fig.3, we made the comparison of effectiveness of both GPSO and LPSO, respectively either used in 3 DOF one-stage or in 6-DOF two-stage PMD compensation. At first, we made 18 times of the 1st-order PMD compensation experiments with the one-stage compensator by controlling the three voltages of the electrically controlled PC in the compensator through GPSO and LPSO algorithm, respectively. We made randomly the 18 kinds of different initial PMD states of the PMD emulator (corresponding to 18 different initial DOP values) for 18 times of experiments. In every process of global DOP maximum searching, we recorded variation of best DOP values in each iteration, and the maximum iteration number is set to 25, with the results shown in Fig.6. It can be seen that all the final searched DOP values in any compensation process exceed 0.95 whatever the GPSO or LPSO is used as control algorithm. And all the DOP values reach 0.9 within about 8 iterations for both GPSO and LPSO. The reasons why there is nearly no difference between the cases using LPSO and GPSO are less local sub-optima existed and low level of noise for the sake of relatively simple configuration of 3-DOF one-stage compensation system. Therefore it is not so difficult for both GPSO and LPSO to undertake.

Secondly, in comparison, we also made 18 times of the 2nd-order PMD compensation with the two-stage compensator by controlling six voltages of two PCs in the compensator also through GPSO and LPSO algorithm, respectively. Differently, the maximum iteration number for each searching process is set to 50, since we guess that it is more difficult searching task because of more complicate configuration or more DOF, and higher level of noise.

The results are shown in Fig.7(a) and (b). It is easy to obtain a conclusion that, because of more local sub-maxima and higher noise level in 6-DOF system than in 3-DOF system, for the case of using GPSO there are some initial PMD states that only makes DOP reach the value of 0.7 (Fig.8(b)), corresponding to being trapped in local sub-optima. In contrast, for the case of using LPSO all final searched DOP values exceed 0.9 no matter what initial PMD state is. And all the DOP values reach 0.9 within about 25 iterations. We can draw a conclusion that LPSO have more powerful ability to undertake the task for solving multi-dimensional problem, and then is a better searching algorithm for adaptive PMD compensation up to high-order.

The response time of the compensator depends on the strategy of the chosen algorithms and the performance of the hardware including A/D, D/A, voltage-controlled polarization controller, etc. We can define a time unit as the time used for one particle treatment in a PMD compensation loop. Many events happen in one time unit: (1) D/A converters writing multi-voltages to the voltage-controlled PCs, (2) waiting for the PCs to reach their steady states, (3) multiple A/D conversions, (4) processing the data in the computer processor with the PSO algorithm. Then the next D/A conversion begins. In our experiment one time unit was measured to be about 0.8ms for 3-DOF and 1.1ms for 6-DOF system. One iteration (containing 20 particles treatment) for searching is equivalent to 20 time units. If we set DOP value of 0.9 as the criterion which is considered to achieve the goal to complete the compensation, the compensation time used is:

compensation time=time unit × number of particles
×iterations used to reach the criterion

$$(4)$$

So for 1st-order compensation by 3-DOF one-stage compensator the compensation time is 0.8×20×8=128ms. And for 2nd-order compensation by 6 DOF two-stage compensator using LPSO algorithm the compensation time is 1.1×20×25=550ms. By analyzing time used by every part of hard wares and algorithm, we find the PSO algorithm only occupies fewer than 16% of the whole time. Therefore if we can afford to high speed hardwares the compensator with faster speed of real level of milliseconds can be achieved.

Furthermore, the PSO algorithm can also be used as the control algorithm for multi-channel PMD compensation like in WDM system. One of feasible strategy is trying to keep the worst compensated channel as good as possible, which can be described as another maximization problem as follows:

$$\underset{parameters}{MAX}\left(\underset{j}{MIN}\left(function_j \right) \right) \tag{5}$$

where j represents the j-th channel and the other parameters are similar to those in single channel.

4. CONCLUSION

In conclusion, for the first time we have introduced PSO algorithm into adaptive PMD compensation, which showed the good performance of rapid convergence to the global optimum without being trapped in local sub-optima and robust to noise. With comparison of using GPSO and LPSO in the one-stage and the two-stage adaptive PMD compensation in 40Gb/s OTDM transmission system, the LPSO algorithm is proved to be more suitable for solving with multi-DOF PMD compensation. With PSO algorithm, we have achieved the automatic PMD compensation with several hundreds of milliseconds. And it is shown that the compensator with PSO algorithm has deep potential to increase its response speed. We also made an expectation to use PSO for PMD compensation in WDM system.

ACKNOWLEDGMENT

This work was supported by the National "863" High Technology Projects No. 2001AA122041 and No. 2003AA311070, and the National Nature Science Foundation of China No. 60072042 and 60377026.

REFERENCES

[1] R. Noé, D. Sandel, M. Yoshida-Dierolf, S. Hinz, V. Mirvoda, A. Schöpflin, C. Glingener, E. Gottwald, C. Scheerer, G. Fischer, T. Weyrauch, and W. Haase, "Polarization mode dispersion compensation at 10, 20, and 40Gb/s with various optical equaliziers," J. Lightwave Technol., Vol. 17, pp1602-1616, 1999.

[2] J. C. Rasmussen, "Automatic PMD and chromatic dispersion compensation in high capacity transmission," 2003 Digest of the LEOS Summer Topical Meetings, pp47-48, 2003.

[3] J. Patscher and R. Eckhardt, "Component for second-order compensation of polarization-mode dispersion," Electron. Lett., Vol. 33, No.13, pp1157-1159, 1997.

[4] S. Kim, "Schemes for complete compensation for polarization mode dispersion up to second order," Opt. Lett. Vol 27, No 8, pp577-579, 2002.

[5] J. Kennedy and R. C. Eberhart, "Paticle Swarm Optimization," Proc. of IEEE International Conference on Neural Networks, Piscataway, NJ, USA, pp1942-1948, 1995.

[6] E. C. Laskari, K. E. Parsopoulos, and M. N. Vrahatis, "Particle Swarm Optimization for Minimax Problems," Proc. of the 2002 Congress on Evolutionary Computition, Vol.2, pp1576-1581, 2002.

[7] Xiaoguang Zhang, Li Yu, Yuan Zheng, Yu Shen, Guangtao Zhou, Bojun Yang, "Adaptive PMD compensation using PSO algorithm," OFC'2004, Los Angeles, CA, 2004, Paper ThF1.

[8] J. Kennedy and R. Mendes, "Population Structure and Particle Swarm Performance," Proc. of the 2002 Congress on Evolutionary Computition, Vol.2, pp1671-1676, 2002.

[9] R. Eberhart and J. Kennedy, "A New Optimizer Using Particle Swarm Theory," Proc. of the Sixth International Symposium on Micro Machine and Human Science, pp39-43, 1995.

DYNAMICAL LIMITATIONS OF SINGLE-STAGE PMD COMPENSATORS

Ernesto Ciaramella[1]

[1]Scuola Superiore Sant'Anna, V. Cisanello 145 (56124) Pisa ITALY;
email: ernesto.ciaramella@cnit.it

Abstract: Dynamical effects in single-stage polarization-mode dispersion compensators are investigated. A typical compensator can actually work in two different operating modes. Dynamic evolution of the input signal can require a continuous switch between these modes, producing significant performance degradation. We introduce a numerical procedure to simulate dynamical effects, which confirms the occurrence of critical configurations.

1. INTRODUCTION

Polarization Mode Dispersion (PMD) is the main limiting effect for future 40 Gbit/s optical systems [1]. In a fiber link, PMD impairments depend on the State Of Polarization (SOP) of the input signal and on the fiber parameters (to first order approximation: Differential Group Delay, DGD, and Principal States of Polarization, PSPs). As they are statistically varying [2], the Outage Probability (OP) concept [3] is used to evaluate system impairments.

The effectiveness of PMD-Compensators (PMDCs) is estimated by means of the Outage Probability (OP) as well [4]. To this aim, demanding numerical simulations are required, which first model the signal propagation, then optimise the PMDC (by setting it to an absolute or local maximum) and finally evaluate the penalty.

This quasi-static approach can be quite questionable. When we assume that the PMDC can be at a proper operating point at any time, we neglect what may happen during transients from one operating condition to the following. In

518

principle, any change of the input SOP (State Of Polarisation) and/or of the fibre PMD corresponds to a time-dependent trajectory of the PMDC variable parameters.

Therefore the quasi-static assumption would be correct only if these trajectories had no discontinuity. If they have, no practical PMDC could have such a short response time to reach the next operating point in one or few bits; hence a PMD-compensated system could manifest unexpected degradation, with higher error rate (whatever good the OP could be).

Here we show how dynamical effects can arise in feedback-based PMDCs, with one fixed-delay stage.

2. THEORETICAL MODEL

We use a simplified first order model to explain the critical PMDC dynamics. Let us concentrate on IM-DD optical signals and also assume that the signal SOP is varying very fast, compared to the fibre PSP's (Principal States of Polarisation) and DGD (Differential Group Delay), which is a realistic case.

As known, PMDCs can typically use one of two different compensation strategies [1]. Up to 1st order PMD approximation, system impairments only depend on $|\vec{\Omega} \times \vec{s}|$, where $\vec{s}$ and $\vec{\Omega}$ are the Stokes vectors of the signal SOP and overall 1st order PMD, respectively ($\vec{\Omega}$ is the vectorial combination of PMD of the fibre and of the PMDC, $\vec{\Omega}_F$ and $\vec{\Omega}_C$ respectively [4]). Hence a PMDC works effectively if either $\vec{\Omega}$ is parallel to $\vec{s}$ or if $|\vec{\Omega}|$ is simply minimised [4].

Let us now investigate one particular dynamical case, shown in Fig. 1. On the left, the signal is injected in one fibre PSP (i.e., $\vec{s}$ is parallel to

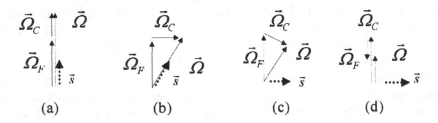

(a)   (b)   (c)   (d)

Figure 1 PMDC dynamics starting from input SOP in one PSP (a).
As SOP changes (b, c), the PMDC should switch to the minimisation mode (d).

$\vec{\Omega}_F$) and the PMDC aligns $\vec{\Omega}_C$ to both. However, this is a critical choice: if the signal SOP is rotating (b), it may reach an angle where this strategy cannot work anymore (c). In that configuration, the PMDC would suddenly have only one operating mode (shown in Fig. 1 d) and it should switch to the minimization strategy. However this switching cannot be done instantaneously: hence during a limited time window, the PMDC is not opti-mised and may produce eye closure and error bursts. We outline that the usual OP treatment neglects this effect (and possible others, which are not discussed), as it would never show configuration (c).

As can be seen, if $|\vec{\Omega}_F| < |\vec{\Omega}_C|$ (i.e. $DGD_F < DGD_C$) a PMDC might be able to use the parallel mode for any SOP rotation. Yet this is an unpractical choice. As $|\vec{\Omega}_F|$ statistically varies, $|\vec{\Omega}_C|$ should be very large, but this can produce a non-negligible 2nd order PMD, when combined with $\vec{\Omega}_F$. This was already shown by the static analysis, as DGD_C should be optimised to have the minimum OP, e.g., in a 40 Gb/s system for one-section and $0.3 < (<DGD>/T_{bit}) < 0.5$ the optimum is $DGD_C \approx 16$ ps [4].

Using this optimum DGD_C value, dynamical limitations might arise for $|\vec{\Omega}_F|$ greater than around 64% the bit time T_{bit}: this is, however, the DGD range where the PMDC beneficial effects would be really needed.

3.   NUMERICAL SIMULATIONS

We introduce here for the first time a numerical technique to simulate PMDC dynamics. We condsider a 40 Gbit/s chirp-free NRZ signal. Using the wave-plate model [6], a 64-bit sequence is propagated in a fibre having 8 ps ($\approx T_{bit}/3$) mean DGD. Among the various random wave-plate configurations, the software can select those having a chosen DGD. The PMDC has a one-stage with fixed delay (DGD_C) and a Polarisation Controller (PC), i.e. it basically has two Degrees of Freedom (DoF). The receiver is a fast photodiode followed by a Bessel filter (4th order, 28 GHz bandwidth). As the feedback signal, we use the eye opening. Note that, to simplify the analysis, no noise is added hence the eye opening is in principle the best possible feedback.

The dynamics is modelled as follows. First, we set an initial signal SOP and run an exhaustive search in all the PMDC configurations, optimising the two DoFs of the PMDC. After the starting configuration has been found, the signal SOP is varied by small steps, i.e., less than 0.02 rad/iteration on both azimuth and ellipticity [7], and the PC is re-optimised. Now, to simulate a continuous evolution, in any iteration both PC angles start from the previous condition and cannot change by more than a fixed amount (0.2 rad).

We first investigate the performance when $DGD_C=16$ ps, i.e. the optimum

520

value [5]. Obtained results show two situations. In a first (very common) case, the PMDC starts in a stable condition, the PC parameters do not change significantly for any SOP state and no significant penalty is observed. This suggests that if the PMDC starts in the minimisation mode, it locks firmly. However, in some cases the PMDC can start in an unstable condition, so that SOP rotations may produce severe eye closure (i.e., <-1.5 dB). After the eye closes, the PMDC may either recover to a stable condition or it may also reach another unstable condition (in which case, eye closure can occur again). Even allowing for the SOP to change at a slower rate, the same effects are obtained. As expected, these catastrophic events are likely obtained if DGD_F is high and the input SOP is very close to one PSP when the PMDC starts.

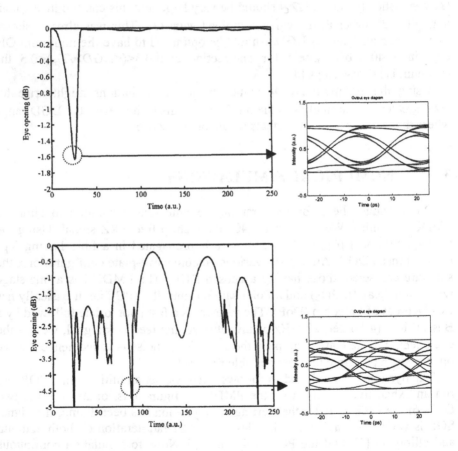

Figure 2 Eye closure evolution for a PMDC with DGD_C=16 ps. We report the eye closure versus time for two fiber cases (1st row: fibre DGD_F=15.3 ps; 2nd row: DGD_F=21.7 ps). Initial signal SOP is aligned to one PSP. Worst case eye diagrams are shown on the right.

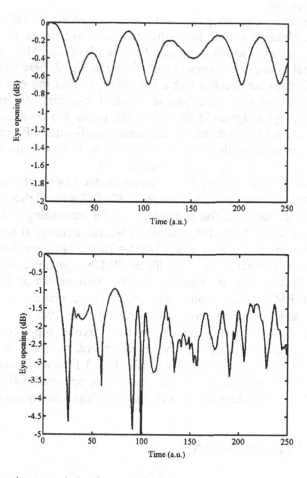

Figure 3 Eye closure evolution for a 1-stage PMDC with $DGD_C=26$ ps. Same other parameters as in Fig. 2.(upper row: fibre $DGD_F=15.3$ ps; lower row: $DGD_F=21.7$ ps).

Fig. 2 shows two evolution curves obtained for $DGD_C=16$ ps for two fiber configurations and input SOP close to one PSP. In the upper row, we report the case when the fiber has $DGD_F=15.3$ ps. Here, after the initial locking, the eye suddenly suffers from a significant penalty, meanwhile the PMDC switches to the minimisation mode, so that no other penalty is observed since then. Note that $DGD_F<DGD_C$, so the switching produces a significant penalty that was neither predicted by the 1st-order model presented heretofore.

On the right (same row), we show the eye diagram taken during this switching time, which indeed shows around 1.5 dB penalty. We note that, in the same fibre, an uncompensated system can suffer, in the worst case, from around the same penalty.

This result is suggesting possible limitations of the PMDCs. Still, the worst results are obtained when, due to the statistical variations, the fibre reaches a higher DGD_F value, which is exactly the case when the PMDC is really needed. An apparent example is shown in the 2nd row in Fig. 2: here DGD_F=21.7 ps and the PMDC does not reach a stable condition, so that the eye closure tends to oscillate repeatedly to unacceptable values ($\ll$-3 dB). An example of such high-penalty eye diagram is shown on the right. We outline that, again, this dynamics cannot be predicted by the usual quasi-static OP analysis. Indeed in this case, the quasi-static analyisis would show a very limited penalty, which cannot be attained in a dynamical environment.

Some of the previous results can be understood by means of the above first-order model. According to that model, a PMDC with higher DGD_C ($\gg$16 ps) might be more useful in the dynamical regime (producing a $\vec{\Omega}$ vector always parallel to $\vec{s}$). To check this possibility, we run dynamical simulations for the same above cases, but now using DGD_C=26 ps. Now we find that when the fibre DGD_F is comparable to T_{bit}, still the PMDC dynamics is strongly affected (see Fig. 3). This clearly indicates that for high values of DGD_F and DGD_C high-order PMD effects come into play. These effects cannot actually be neglected and again this eventually prevents the PMDC to attain a stable condition. This is clearly shown in both trends reported in Fig. 3 (for DGD_F=15.3 and 21.7, respectively). Furthermore, in this case some oscillating behaviour is observed also for low DGD_F (Fig. 3 1st row). These results seem to indicate that, as was found for the quasi-static approach [4] [5], a quite high DGD_C value can produce worse performance also in the dynamical regime.

4.   CONCLUSIONS

A first-order model indicates that a single-stage PMDC with feedback may produce relevant system impairments if it locks to an unstable condition and if DGD_F/DGD_C>1. Numerical simulations confirmed that PMDCs could suffer from this effect. Despite the first order model, higher length PMDCs could suffer even more due to 2nd order PMD.

These results might hold also for a single stage with variable length and for a double stage PMDC, but this issue is still being investigated. On the other hand, PMDCs using polarization scrambling [8] and PMDCs with no feedback (e.g., [9],[10]) could not suffer from these dynamical limitations.

The obtained results seriously indicate that the Outage Probability may not completely describe the performance of a PMDC: it provides indeed an upper bond, a bound that could not always be met.

Finally, in our scheme we assumed an initial startup of the PMDC, with an exhaustive search. In a practical environment, the PMDC may lock to an unstable condition either because of noise fluctuations, or the random evolution of fibre parameters.

ACKNOWLEDGMENTS

The author acknowledges stimulating discussions with E. Forestieri and the fruitful cooperation with the Optical Networks Group of Marconi Communications in Genoa, whose activity has stimulated and supported this work. This work was supported by Marconi Communications, under a grant.

REFERENCES

[1] H. Sunnerud, M. Karlsson, C. Xie, P.A. Andrekson, "Polarization Mode Dispersion in High Speed Fiber-Optic Transmission Systems" J. Light. Technol., 20, 12 2204-2219 (2002).

[2] F. Curti, B. Daino, G. De Marchis, and F. Matera, "Statistical treatment of the evolution of the principal states of polarization in single-mode fibers," J. Lightwave Technol., vol. 8, pp. 1162–1166, 1990.

[3] H. Bulow, "System outage probability due to first and second order PMD" IEEE Photon. Technol. Lett. 10, 696-699 (1998)

[4] H. Sunnerud, C. Xie, M. Karlsson, R. Samuelsson, P.A. Andrekson, "A comparison between different PMD compensation techniques" J. Light. Technol., 20, 3, 368-378 (2002).

[5] C. Xie, and H. Haunstein, "Optimum Length of One-Stage Polarization-Mode Dispersion Compensators With a Fixed Delay Lin", " IEEE Photon. Technol. Lett. 15, 9, 1228-1231 (2003)

[6] A. Galtarossa, L. Palmieri, M. Schiano, T. Tambosso, "Single-End Polarization Mode Dispersion Measurement Using Backreflected Spectra Through a Linear Polarizer", J. Lightwave Technol., 17, 10, 1835-1842 (1999).

[7] R. M. A. Azzam N.M. Bashara, "Ellipsometry and polarized light", Elsevier Science Ed., Amsterdam, 1999.

[8] H. Y. Pua, K. Peddanarppagari, B. Zhu, C. Allen, K. Demarest and R. Hui, "An adaptive First order Polarization Mode Dispersion compensation system aided by polarization scrambling: theory and demonstration", J. Light. Technol. 18, 6, 832-841 (2000).

[9] P. Chou, J. M. Fini, H. Haus, "Demonstration of a feed-forward PMD compensation technique" IEEE Photon. technol. Lett. 14, 2, 161-163 (2002).

[10] J. M. Fini, P. Chou, H. Haus, "Estimate of polarization dispersion parameters for compensation with reduced feedback" IEEE Photon. Technol. Lett. 14, 2, 161-163 (2002).

NEW APPROACH TO OPTICAL POLARISATION MODE DISPERSION MITIGATION: EXPERIMENTAL ANALYSIS OF THE DYNAMIC PERFORMANCES OF A COST-DRIVEN DEVICE

Raoul Fiorone, Aldo Perasso, Massimo Speciale, Marco Camera, Andrea Corti
MarconiCommunications, Via A. Negrone, 1/A – 16153 Genova, Italy
raoul.fiorone@marconi.com, aldo.perasso@marconi.com, massimo.speciale@marconi.com, marco.camera@maconi.com, andrea.corti@marconi.com

Abstract: An innovative approach to optical mitigation is proposed. We experimentally investigate the dynamic performance of a simple Polarisation Mode Dispersion optical mitigator, and demonstrate that a cost-effective solution can minimise the real criticality of optical structures.

1.   INTRODUCTION

Polarisation Mode Dispersion (PMD) can be the major limiting factor for future 40Gbit/s optical systems [1] and can constitute a headache in case of 10Gbit/s all-optical networks. An intense activity has been carried out in the last years by the scientific community [2], but to date no real solution to mitigate the effects of PMD is available for practical deployment. The development of a PMD compensator (PMDC) is not a usual engineering problem, given the high number of non-independent variables, the statistical behaviour of the physical phenomena, and the economical and technological constraints related to a practical use.

This paper describes the design and characterisation work behind the development of a PMDC, and analyses the behaviour of a cost-driven single-stage device, starting from where other solutions have failed so far, both cost-wise and performance-wise.

The performances of a simple, cost-effective and reliable optical PMDC come out to be quite attractive, even without possible supplementary aids like Forward Error Correction (FEC), adaptive Clock and Data Recovery and electronic equalisation.

2. THE DESIGN OF EXPERIMENT

A bunch of polarisation controllers (PC) and some metres of polarisation maintaining fibre can be easily used to open a PMD-affected eye diagram. However, it is not immediate to figure out the subtle troubles that the design of an optical PMDC involves. Our first hand experience has taught us that without a systematic Design of Experiment (DOE) approach [3,4] and properly chosen lab equipment, the chances to find the right recipe to design a good (cost sensible, effective and reliable) optical PMDC are very low [5].

A DOE is a structured, organised method to identify and determine the mutual relationship among the factors affecting a complex process. It is a sort of mathematical procedure to span the entire operating domain with a minimum set of test cases.

3. THE "COMPENSATION MODE DILEMMA"

The impairments due to PMD depend upon the signal's State of Polarisation (SOP) and the fibre parameters, Differential Group Delay (DGD) and Principal States of Polarisation (PSP) in a 1st order analysis. Given the statistical behaviour of all the parameters [6], the Outage Probability (OP) concept [7] is needed to evaluate the system performances, both without and with PMDC [8].

The obvious expectation is a compensated system showing an essentially lower OP on a PMD affected link, so that a known margin, whose probability to be exceeded over time is equal to the OP, can be allocated in the system design.

The OP is usually calculated and checked emulating the real world dynamic behaviour with a sequence of (quasi) static conditions. The two cases do not exactly coincide, and different behaviours can unsurprisingly be expected. In case of real dynamic characterisation of an optical PMDC, it is possible to find out that the transients from one working condition to the next are not always hitless, as one could expect [8]. A few solutions have been proposed with other rationales to constrain the PMDC operation mode, with the result of adding extra cost and complexity [9].

526

The assumption that, given a PC endless operation, for every continuous evolution of the signal SOP and fibre PMD, the optimum PMDC parameters evolve continuously, is not necessarily true, and in our DOE-like analysis we have, for the first time, experimentally observed and explained this potentially dangerous phenomenon. To do so, it is necessary to perform accurate PMDC state monitoring, long-term logging, and Bit Error Ratio (BER) floor and burst detection and monitoring, in order to identify, unequivocally recognise and finally understand the killer events.

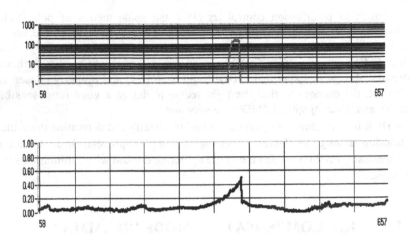

Figure 1. Experimental evidence of the "compensation mode dilemma" phenomenon

In Figure 1 the experimental evidence of the occurrence of the "compensation mode dilemma" is presented. What is important to notice is that the error burst occurred (error graph - top) without any reset condition of the PC's and with the feedback signal well aware of the problem (mean square error (MSE) graph - bottom). In this experiment the feedback signal was based on a conventional RF spectrum monitoring, and the PMDC was a two-section optical structure, with ample capability to mitigate the PMD values emulated.

The OP is not enough to fully describe the PMDC performances, and the reliability of the whole system, but it can usefully provide an upper bound limit.

4. THE OBJECTIVE

Our objective was to adopt a cost and integration driven approach to build a simple 1st order PMDC.

It is curious to notice that the best performing PMDC that we have tested in our lab (not completely exempt from the "compensation mode dilemma", anyhow) was too big and expensive for real deployment: it was a performance-driven device.

Our choice has been a single-stage optical PMDC, using RF spectrum based PMD monitoring. The target was to mitigate up to 0.6÷0.8 Unit Interval peak DGD with real world polarisation variations, which are normally slower than those emulated (up to Scan Rate (SR) 4 of the Agilent 11896A PC's used with the PMD Emulator - PMDE). No "compensation mode dilemma" criticality was a design objective and considered a real differentiator.

5. PERFORMANCE ANALYSIS

The design of the PMDC has been driven by the ability to fully assess the device's behaviour using a DOE-like approach and a programmable PMDE. The performances have been investigated to check, in the presence of different Optical Signal to Noise Ratio (OSNR) and Polarisation Dependent Loss values: compensation capability, dynamic behaviour in terms of response speed, endless operation, BER floor and "compensation mode dilemma".

In Figure 2 the sketch of the test bed is given.

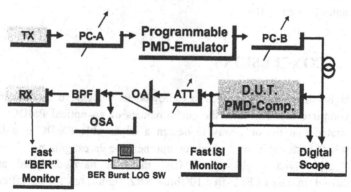

Figure 2. The test bed

In Figure 3 the system tolerance against DGD is shown in case of 10Gbit/s NRZ signal, with 19dB OSNR, no FEC, linear operation and practically no chromatic dispersion. Using a 2dB OSNR penalty threshold, the system resistance

528

against DGD (and SOP/PSP scanning) results increased from 30ps to more than 80ps, nearly a three-fold improvement.

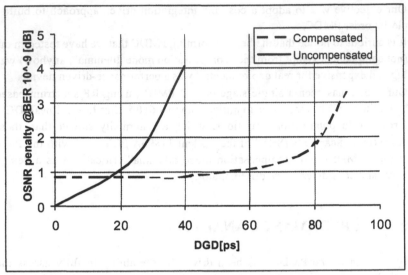

Figure 3. Long-term performance results against DGD

In our preliminary results, we have found no BER floor and no "compensation mode dilemma" events in a series of test sessions long at least 36 hours (SR 2 for PC-A and PC-B, Figure 2). It is worth noting that, in the same dynamic conditions, other PMDC prototypes have shown "compensation mode dilemma" events in a few minutes (typically 10÷30).

6. CONCLUSIONS

A highly innovative approach has allowed us to achieve for the first time a really complete assessment of the performances of an optical PMDC. We have demonstrated that the only way to design a deployable PMDC is a DOE-like method, with both static and dynamic comprehensive investigations.

The cost-driven single-stage PMDC that we have designed and fully characterised, in case of FEC-free 10Gbit/s NRZ signal, has shown efficacy up to more than 80ps DGD, with fast tracking capability, no error burst events and no BER floor.

REFERENCES

[1] H. Sunnerud, M. Karlsson, C. Xie, P.A. Andrekson, "Polarization Mode Dispersion in High Speed Fiber-Optic Transmission Systems" J. Lightwave Technol., 20, Dec 2002, pp. 2204-2219.

[2] Proceedings of Venice Summer School on PMD, June 2002

[3] R. A. Fisher, "The design of experiments" (9th ed.), 1971, New York, Hafner Publishing Co.

[4] J. J. Kennedy, A.J. Bush, "An introduction to the design and analysis of experiments", Lanham, MD, University Press of America Inc.

[5] A. Perasso, R. Fiorone, A. Corti, ECOC2003 Workshop on "PMD: Causes, Effects and Cures".

[6] F. Curti, B. Daino, G. De Marchis, and F. Matera, "Statistical treatment of the evolution of the principal states of polarization in single-mode fibers J. Lightwave Technol., vol. 8, 1990, pp.1162-1166.

[7] H. Bulow, "System outage probability due to first and second order PMD" IEEE Photon. Technol. Lett. Oct. 1998, pp. 696-699

[8] E. Ciaramella, "Theoretical Evidence of Dynamical Limitations in Practical Single-stage PMD Compensators", submitted to IEEE Photon. Technol. Lett.

[9] H. Y. Pua, K. Peddanarppagari, B. Zhu, C. Allen, K. Demarest and R. Hui, "An adaptive First order Polarization Mode Dispersion compensation system aided by polarization scrambling: theory and demonstration", J. Lightwave Technol, vol. 18, June 2000, pp. 832-841

NUMERICAL IMPLEMENTATION OF THE COARSE-STEP METHOD WITH A VARYING DIFFERENTIAL-GROUP DELAY

M. Eberhard[1] and C Braimiotis[1]
[1]*Photonics research group, Engineering department, Aston university, Birmingham, U.K*

Abstract: The effect of having a fixed differential-group delay term in the coarse-step method results in a periodic pattern in the autocorrelation function. We solve this problem by inserting a varying DGD term at each integration step, according to a Gaussian distribution. Simulation results are given to illustrate the phenomenon and provide some evidence,about its statistical nature.

1.   INTRODUCTION

As demonstrated in [2] the autocorrelation function (ACF) produced after the use of the coarse-step method, deviates from the analytical model. A repetitive pattern appears, owing to the additive effect of the convolution, in the time domain of the signal with a fixed differential group delay (DGD) term, at each integration step. Moreover it is possible, to minimise this effect by allowing the DGD coefficient to change at each step as a Gaussian variate.

2.   REVIEW OF THE COARSE-STEP METHOD

Following in the derivation of [1] the starting point is the coupled nonlinear Schroedinger equation (CNLS),

$$i\frac{\partial\Psi}{\partial z}+\tilde{\Sigma}\Psi+ib'\sigma_3\frac{\partial\Psi}{\partial t}-\frac{1}{2}\beta''\frac{\partial^2\Psi}{\partial t^2}+n_2k_0\left[\frac{5}{6}|\Psi|^2\Psi+\frac{1}{6}(\Psi^\dagger\sigma_3\Psi)\sigma_3\Psi+\frac{1}{3}N\right]=0 \tag{1}$$

Should be denoted that, $\Psi = R(z)A$ where $R(z)$ and $\tilde{\Sigma}$ are the following matrices,

$$R(z) = \begin{pmatrix} cos\alpha & sin\alpha \\ -sin\alpha & cos\alpha \end{pmatrix} \qquad (2)$$

$$\tilde{\Sigma} = \begin{pmatrix} b & -i\alpha_z \\ i\alpha_z & -b \end{pmatrix} \qquad (3)$$

Equations 2 and 3 follow the rapid evolution of the field along distance z, as the direction of the birefringence axes will be rapidly changing. Where the birefringence parameter $b = (\beta_1 - \beta_2)/2$ and the specific group delay per unit length $b' = (\beta'_1 - \beta'_2)/2$. In the CNLS n_2 is the Kerr coefficient while $k_0 = \frac{2\pi}{\lambda}$ is the wavenumber. As indicated in [1] if the birefringence axes are fixed then the angle of rotation equals zero, $\alpha_z = d\alpha_z/dz = 0$ and thus $N = (\Psi_1^*\Psi_2^2, \Psi_1^2\Psi_2^*)^t$ varies rapidly and can be dropped as observed in [3,4]. Equation 1 then becomes,

$$i\frac{\partial\Psi}{\partial z} + ib'\sigma_3\frac{\partial\Psi}{\partial t} - \frac{1}{2}\beta''\frac{\partial^2\Psi}{\partial t^2} + n_2 k_0\left[\frac{5}{6}|\Psi|^2\Psi + \frac{1}{6}(\Psi^\dagger\sigma_3\Psi)\sigma_3\Psi\right] = 0 \quad (4)$$

When $b' \neq 0$ the signal is subjected to polarization mode dispersion(PMD), while the third term causes chromatic dispersion.

Assuming that the step size is large enough so that the field has lost memory of its initial polarization, the solution of the CNLS can be multiplied by a scattering matrix S so that the polarization is randomly reorientated.

$$S = \begin{pmatrix} cos\alpha & sin\alpha exp(i\phi) \\ -sin\alpha exp(i\phi) & cos\alpha \end{pmatrix} \qquad (5)$$

The multiplication of the signal at different frequencies with a fixed DGD term induces a periodicity in the autocorrelation function as it is shown through our results. The ACF thus deviates from the theoretical model. This effect is eliminated including a varying DGD from step-to-step and consequently averaging out the unwanted peaks that are present in the ACF.

3. AUTOCORRELATION FUNCTION

The simulations performed on a system having the following characteristics $D_{PMD} = 3ps/sqrt(km)$, correlation length $L_c = 100m$, integration step 1 km while the optical bandwidth of the simulation is 4 THz. The ACF was compared with the following function [5,6], where $\Delta\tau$ is the mean DGD,

$$ACF_{analytical} = \frac{3}{\Delta\tau^2(\omega - \omega_0)^2}[1 - exp(\frac{-\Delta\tau^2(\omega - \omega_0)^2}{3})] \qquad (6)$$

and calculated from the simulations according to the following formula,

$$ACF_{analytical} = |\frac{\langle\Omega(\omega)\rangle\langle\Omega(\omega_0)\rangle}{\langle\Omega(\omega_0)\rangle\langle\Omega(\omega_0)\rangle}| \qquad (7)$$

$\Omega(\omega)$ is the polarization dispersion vector calculated as in [2]. As can be shown in figure 1, using a fixed value DGD we are getting the periodic pattern in the ACF of the coarse-step method.

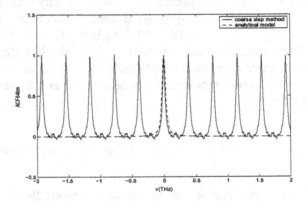

Figure 1. ACF of coarse step method with a fixed DGD of 3ps for 64km.

Allowing the DGD to vary as a random Gaussian variate with a standard deviation σ ranging from 0.009 - 1 ps minimizes the problem.

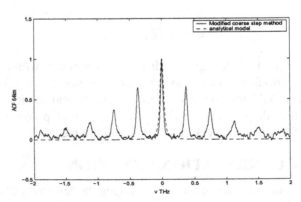

Figure 2. ACF of coarse step method with DGD varying according to a Gaussian distribution of $\sigma = 0.09ps$ and mean $\mu = 3ps$ for 64km.

It is obtained through the simulation results that the harmonics of the ACF gradually diminish as the standard deviation of the distribution is getting larger.

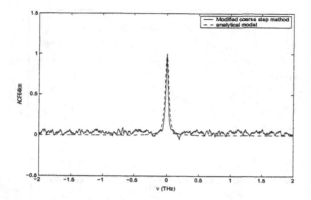

Figure 3. ACF of coarse step method with DGD varying according to a Gaussian distribution of $\sigma = 1ps$ and mean $\mu = 3ps$ for 64km.

It is worth noting that the PDF of the DGD does not change when we modify the coarse-step method, as it is shown in figure 4.

Moreover we can follow the evolution of σ, through our simulations by following the difference ΔA between the centre peak amplitude and the first harmonic amplitude as presented in figure 5. From the graph we can didact that at a value of $\sigma = 0.6ps$, the peak of the first harmonic drops at 10 percent, of its original value.

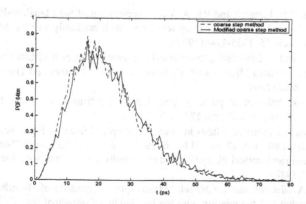

Figure 4. comparison of the PDF of the coarse-step method with a fixed DGD with a modified coarse-step method using a DGD varying according to a Gaussian distribution of $\sigma = 0.09ps$ and mean $\mu = 3ps$ for 64km.

534

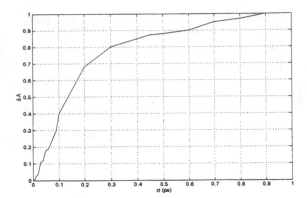

Figure 5. Evolution of the difference in amplitude between the fundamental frequency and the first harmonic for different values of σ.

4. CONCLUSION

Instead of using a fixed DGD we inserted a Gaussian variate, so that the peaks of ACF of the coarse-step method average out and the numerical implementation outcome agrees with the theoretical results. Moreover we provided simulation results that illustrate the decrease in the amplitude of the side peaks of ACF and give some evidence about a future analytical treatment of the phenomenon.

REFERENCES

[1] D. Marcuse, C.R. Menyuk and P.K.A. Wai, "Application of the Manakov-PMD equation to studies of signal propagation in optical fibers with randomly varying birefringence", J.Lightwave Tech. 15, 1735-1746(1997).

[2] M. Eberhard and C. Braimiotis, "Numerical implementation of the Manakov-PMD equation with precomputed M(ω) matrices", Nonlinear Guided Waves and Their Applications proceedings,MC3(2004).

[3] C.R. Menyuk, "Nonlinear pulse propagation in birefringence optical fibers", IEEE J.Quantum Electron., vol. 23, pp. 174 - 176(1987).

[4] C.R. Menyuk, "Stability of soltons in birefringent optical fibers. I: Equal propagation amplitudes", Opt. Lett., vol. 12, pp. 614-616, 1987; see also, C.R.Menyuk,"Stability of solitons in birefringent optical fibers.II:Arbitrary amplitudes" J. Opt. Soc. Amer. B., vol. 5, pp.392 -402(1988).

[5] M. Shtaif, A. Mecozzi and J.A.Nagel, "Mean-square magnitude of all-order polarization mode dispersion and the relation with the bandwidth of proncipal states", IEEE Photon. Tech. Lett.12, 53-55(2000).

[6] M. Karlsson and J. Brentel, "Autocorrelation function of the polarization mode dispersion vector", Opt. Lett. 24, 497-469(1999).

POSTER SESSION

ALL OPTICAL 3R REGENERATION AND WAVELENGTH CONVERTION

Davide M. Forin, Franco Curti, Giorgio M. Tosi-Beleffi, Francesco Matera[1],
Andrea Reale, Silvello Betti, Simone Monterosso, Alessandro Fiorelli[2],
Michele Guglielmucci, Sergio Cascelli[3]

[1] Ugo Bordoni Foundation, Via Baldassarre Castglione. No. 59,00142Rome, Italy;
Tel: (+39)0654802235, Fax: (+39)0654804402, e-mail: dforin@fub.it
[2] Department of Electronic Engineering, University of Tor Vergata, Via del Politecnico 1,
00133 Rome, Italy
[3] Istituto Superiore delle Comunicazioni e Tecnologie dell'Informazione, Viale America
201,Rome, Italy

Abstract: We experimentally demonstrate the possibility to realize an all-optical 3R
regenerator wavelength converter, based on a three-stage cascade
configuration. A multi-wave-mixing based on the phase modulation of an
auxiliary optical carrier induced by a two-signals interaction is used to
perform simultaneously reshaping, re-amplification and wavelength
conversion of a propagated signal. A fiber ring based unit is used to perform
the all-optical timing extraction while an orthodox four-wave-mixing process
inside a dispersion-shifted fiber is adopted for the retiming process. The
work is patent pending.

1. INTRODUCTION

Optical devices and subsystems like wavelength converter for optical cross
connects and 2R-3R optical regenerators are under development because of their
fundamental role in the future optical communication systems.

3R regeneration preserves data quality and allows for improved transmission
distances, thus enhancing transparency, scalability, and flexibility of optical
networks. In this work we present results concerning the three main cascaded

function to obtain a 3R regenerator: re-amplification and reshaping, timing extraction and re-sampling.

The Re-shaping and the Re-amplification unit (2RU) is studied via the simulations and the experimental results, at 10 Gbit/s, regarding a new technique (patent pending) based on the adoption of an auxiliary continuous-wave incoming optical carrier. The characteristic non-linearity of a particular dispersion shifted optical fiber has been Exploited [1]. In that cases the envelope of the beating between a strong pump and a modulated signal, near the zero dispersion area (around 1539 nm), determined a phase modulation of the whole incoming optical field; this beating generate a modulation of the refractive index of the fiber with frequency equal to frequency difference between pump and signal. Then any other carrier in the fiber is phase modulated. Hence the phase-modulated new carrier's spectrum presents a series of replicas typical of the phase modulation, each one reproducing the high data rate amplitude modulation of the original incoming signal enhancing, in this manner, the overall device wavelength conversion range. Furthermore, the in-out power-transfer characteristic of the n-th order replica is alike the correspondent order Bessel function, a reshaping function occurs too.

The Clock Recovery unit (CRU) is based on a fiber ring laser, which contains a single semiconductor optical amplifier (SOA), as active device, which provides both gain and modulation. The ring circuit was tested with both pseudorandom data sequences and periodic data patterns up to 10 Gb/s generating clock pulse trains across a 30 nm tuning range. The operation of the present clock recovery ring based structure relies on the fast gain saturation of the SOA, induced by the incoming data stream, that generates the modulation in the cavity and determines the injection locking of the fiber laser.

The standard mode-locking SOA ring based technique has been extended introducing a Fabry-Perot cavity to select the right oscillating frequency [4]. Preliminary results, not reported here, shown that this configuration is able to works at 40 Gbit/s too. Desirable features of the CRU unit are the wide repetition frequency and wavelength locking ranges and the broad wavelength tuning range of recovered signal that it may provide.

The Re-Sampling unit (RU), is based on the usual FWM non-linear effect. The incoming modulated signal, in fact, is coupled with the pulsed optical clock signal, recovered by the second stage, and injected into a 6 km long standard dispersion shifted fiber (DS). The effect is a re-sampled optical replica of the input modulated signal.

In order to improve the performances of the non-linear effect we used a clock signal obtained by cross-gain modulation effect (XGM) into a semiconductor optical amplifier (SOA) in order to narrow the optical spectrum of the clock.

2. DESCRIPTION OF THE THREE STAGE SET-UP

The key element of the reshaping section is a 10 km long DS fiber, with loss _=0.2 dB/km, nonlinear coefficient _=2.2 [W*km]-1, chromatic dispersion D=0.19 ps/nmKm (at the pump wavelength 1542.14 nm), dispersion slope SO=0.0634 ps/nm2km and differential group delay (PMD factor) DGD=0.13 ps[2].

Two DFB (DFB Source and DFB pump) laser sources used to generate respectively a signal (S), at λ=1542.9 nm, and a cw pump (CWP1), at λ=1542.1 nm, are mixed together, in order to induce the requested non-linearity, as reported in Fig.1.

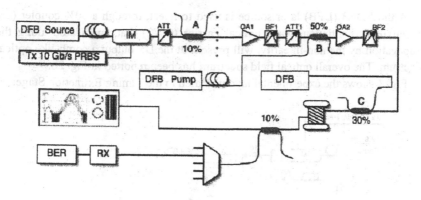

Figure 1. 2R unit set- up

In order to restrict the power depletion, due to Brillouin scattering phenomena, signal and pump are directly phase modulated. The signal, PRBS 10 Gbit/s NRZ, passes through a 10% coupler (A), preceded by an attenuator (ATT), used to measure the real optical input at the entrance of wavelength converter reshaper (WCR). After the 10% coupler there is a block made up of an EDFA (OA1) and a band pass filter (BF1) followed by an attenuator (ATT1) that allow us to set the value of signal power for optimum reshaping function. Combination of signal and pump, through 50% (B)coupler, is processed by a block made up of an erbium-doped power amplifier (OA2) and a band pass filter (BF2). This block increases the power of the pump and the signal up to the power needed in the non-linear processes.

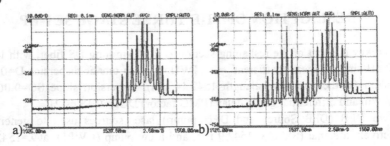

Figure 2. a)spectrum at the DS output of the beating between a cow pump and a modulated signal, b)spectrum at the DS output of the sum of beating and the third carrier

A third DFB (DFB) laser source is used to inject, through a 30% coupler (C), the auxiliary continuous-wave optical carrier (CW2) that, modulated along the propagation by the non-linearity, will present at the DS output a comb-like optical spectrum. The overall optical field spectrum has been reported in Fig. 2b.

Fig. 3 shows the experimental layout related to the Timing Extraction Stage.

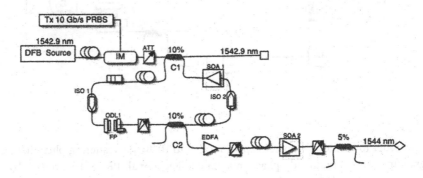

Figure 3. Clock Recovery Unit set-up

The test data patterns were produced modulating a DFB laser source which operates at 1542.9 nm, with a lithium niobate external modulator driven by a 10 Gbit/s PRBS pattern generator. This incoming non-return-to-zero (NRZ) data pattern, injected into the loop circuit with a 10% optical fiber coupler (C1), goes into an Alcatel M128 bulk model SOA1, with a 0.5 mm cavity length, 20 dB gain for small signal and a 40 nm amplification bandwidth, and is blocked by a first faraday isolator (ISO 1).

A second faraday isolator (ISO2), was inserted into the ring to ensure unidirectional counter clock-wise oscillation. Starting from the ISO2, the oscillation goes into a first stage, made by a polarizer followed by a polarization controller, able to optimize the oscillating polarization state for the SOA1. This stage is followed by ISO2 and by a second block made by a Fabry-Perot cavity (FP), realized with two optical mirrors whose reflectivity is equal to 99%, and a variable optical delay line (ODL1). A subsequent 1 nm optical tunable filter is used to select the wavelength needed. The -30 dBm clock signal, extracted via a 10% coupler (C2), goes into an EDFA, a 1 nm bandwidth optical filter able to erase the ASE contribution and then is subsequently processed by an external SOA (SOA2), an Opto-Speed 1550MRI/X model with 30 dB gain for small signal and 30 nm amplification bandwidth, to obtain amplification and to simultaneously erase the amplitude noise. This particular configuration permit to obtain a sensible enhancement of the pulses train extracted.

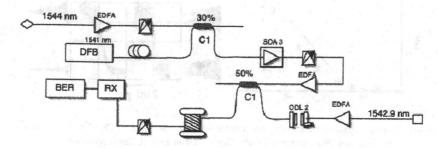

Figure 4. Re-Timing Unit set-up

Fig. 4 shows the experimental layout related to the Re-Sampling Stage. After SOA2, the recovered clock information at 1544 nm, is injected into a SOA (SOA3), an Alcatel M18 bulk model, with a CW signal through a 30% coupler. The cross gain modulation inside the SOA3 modulates the CW pump with the same information carried by the signal coming from the ring. In this manner we have a clock information with a narrower optical spectrum. At this point this signal at 1541 nm is coupled with the original 10 Gbit/s NRZ PRBS signal at 1542.9 nm, through a 50% coupler, and amplified in order to perform the retiming process inside a 6 km long DS fiber via a FWM process. An OPL2 is used to set the exact position of the data pulses in order to match it with the ones coming from the clock.

3.   DISCUSSION OF THE RESULTS

To explain the generation of a so high number of components, we could argue that the non linear optical Kerr effect determines a phase modulation due to the beating between the pump and the signal to be converted, near λ_0 (1539 nm in our case); this beating cause a modulation of the refractive index of the fiber with frequency equal to frequency difference between pump and signal. This effect produce a phase modulation of the *whole* incoming field that generates new frequencies spaced by the pump-to-signal detuning range (ω_P-ω_S) [3]. Then any other carrier present in the fiber is also phase modulated, and the phase-modulated carrier's spectrum presents a series of replicas typical of the phase modulation. The transfer function of these channels shows a non-zero threshold and a spread maximum that we can exploit for reshaper use. By adjusting the signal and pump power, at the DS fiber input, it is possible to choose the better working point in order to reshape the incoming corrupted signal, as reported in fig. 5.

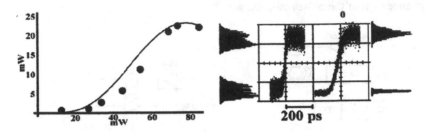

Figure 5. Comparison between experimental transfer function, dotted line, of Ch. 47 with simulation, continuous-line. Noise compression in a Non-Return-to-Zero 10 Gbit/s coded signal

The behavior of the converted signals depends both on the SNR, due to the different generation efficiency, and to the chromatic dispersion regime seen by each channel along the device. The non-linearity at the base of the process is able produce a phase modulation of the *whole* incoming field not only the optical beating field, that induce the non-linearity itself, but any kind of other optical carrier propagating with the beating field the fiber. Subsequently, the optical spectrum of the new carrier, at 1535.8 nm in this case, will present a series of components each one reproducing the high data rate amplitude modulation of the original incoming signal and each one with the same Bessel-like transfer function.

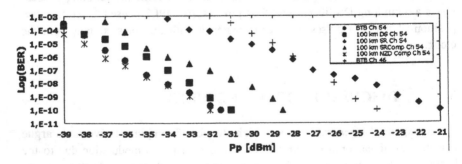

Figure 6. BTB and propagated BER measurements vs received power for ITU Ch 46 and 54

In Fig. 6 we report the BER results obtained at 10 Gbit/s, in the back-to-back configuration and in propagation, for the ITU-GRID channels 54 (1534.2 nm) and 46 (1540.5 nm). The propagation was performed using a cable deployed between the cities of Rome and Pomezia encompassing different kind of fibers, G.652, G.653 and G.655. The better performance on the Ch.54 depends on the better signal to noise ratio respect to the Ch. 46.

In Fig. 7a we report the clock information at the output of the SOA2 used to obtain amplification and to simultaneously erase the amplitude noise from the ring.

Regarding the CRU unit and in the case of NRZ signals, the Fabry-Perot cavity plays a key role since it simultaneously filters out a big amount of modes, that oscillate inside the cavity, and, furthermore, by tuning the mirrors with a piezo-driver control, forces the cavity to resonate at the repletion incoming data rate. Without such resonance the locking conditions, in the case of NRZ signals, can be extremely difficult. The reasons of such behavior stand on the large frequency band of resonance of the ring. If the FP is instead introduced in the loop and its band is tuned around the frequency of the data rate input, it will select the exact oscillating frequency giving raise to a pulse trains. The cavity fundamental frequency was 7.5 MHz and the locking bandwidth of the ring has been measured and it is equal to about 5 MHz.

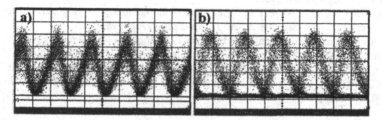

Figure 7. a) Clock recovered at the output of SOA2. b) Retiming of the incoming signal

In Fig. 7b we report a preliminary result concerning the converted replica at 1544.7 nm obtained at the end of the 6 km long DS fiber. The FWM is obtained by the beating of the incoming signal at 1542.9 with the clock extracted at the output of the SOA2 at 1541 nm. It is possible to see that the retiming process, due to width of the generates clock pulses, determines a format translation from NRZ to RZ format.

4. CONCLUSIONS

We have investigated three different stages able to implement the following functions: 2R regeneration and wavelength conversion, timing extraction and signal re-sampling. The results demonstrate the possibility to adopt a common cascade configuration in order to implement an all-optical 3R regeneration device.

ACKNOWLEDGMENTS

This work has been carried out in the framework of the Italian project FIRB RBAU01XEEM_005 "Photonics Networks" and in the cooperation between Fondazione Ugo Bordoni and Ministero delle Comunicazione under the project INTERN.IT

We thank ISCTI (Istituto Superiore delle Comunicazioni e Tecnologie dell'Informazione of Rome) supplying of instrumentation and the Rome-Pomezia experimental deployed cable.

REFERENCES

[1] F.Curti, F. Matera and Giorgio Maria Tosi-Beleffi, "Wavelength converter-reshaper based on multi-wavelength spectral components generation in optical fiber", *Opt Comm, 208, pp. 85-89, 2002.*

[2] M. Karlsson, Four-wave mixing in fiber with randomly varying zero-dispersion wavelength, *J. Opt. Soc. Am. B, vol. 15, pp. 2269-2275, 1998.*

[3] R. Thompson, R. Roy, "Nonlinear dynamics of multiple four-wave mixing processes in a single-mode fiber", *Physical Review A, vol. 43, No. 9, pp. 4987-4996, 1991.*

[4] J.P. Turkiewicz, E. Tangdiongga, All Optical 10 Ghz Clock Recovery from 160 Gbit/s OTDM signals using a mode-locked fiber ring laser, *IEEE/LEOS Benelux Chapter 2002, Amsterdam*

EXPERIMENTAL STUDY OF RESHAPING RETIMING GATES FOR 3R REGENERATION

M. Gay, L. Bramerie, G.Girault, V. Roncin, J-C. Simon
Laboratoire d'Optronique, CNRS UMR 6082 FOTON,
Groupement d'Intérêt Scientifique FOTON,
ENSSAT/Université de Rennes1, 6 rue Kerampont, BP.447 22305 Lannion Cedex, France,
e-mail : Simon@enssat.fr

Abstract: The linear degradation of the bit error rate as a function of the number of regenerators is experimentally observed, with an optoelectronic or with an original all-optical 3R repeater. We demonstrate that Q factor measurements are not suitable for a correct assessment of optical links incorporating 2R or 3R regenerators.

1.   INTRODUCTION

Some optoelectronic and all-optical regenerators enabling Re-amplification, Reshaping (2R) and Re-timing (3R) have shown their capability to ensure high bit-rate ultra-long haul transmission systems [1]. Noise distribution and Bit Error Rate (BER) evolution through this kind of device are of great interest to understand basic features of regeneration. In this paper, we show experimentally the BER evolution through different kinds of Non Linear Gates (NLG). We finally compare Q factor and BER measurements in optical transmission links including 3R regenerators.

2.   EXPERIMENTAL SET UP

Experiments were carried out with an optical or with an optoelectronic (O/E) regenerator in order to compare two types of 3R regenerators. One is ideal (the optoelectronic one) which presents a step-like shape of the transmission versus input power characteristics, while the other one (the optical regenerator) presents a smoother S-shape. Both regenerators have the same O/E retiming device to ensure 3R regeneration.

2.1   Transmission experiment

The 10 Gbit/s transmission experiment is carried out with a 100 km recirculating loop composed of two 50 km Non-Zero Dispersion Shifted Fibre (NZDSF) spans, with chromatic dispersion of 4ps/nm/km. The fibre link dispersion is compensated (DCF). Figure (1) shows the experimental set-up of the recirculating loop.

Losses are compensated by Erbium Doped Fibre Amplifiers (EDFA) and counterpropagating Raman pumping ensuring a low noise accumulation line. The transmitter consists of a 2^{15}-1 pseudo-random bit sequence combined with a logical gate which produces an RZ electrical signal. This signal modulates the optical 1552 nm source thanks to a LiNbO$_3$ modulator which produces a 50 ps full width at half maximum signal. The signal is injected into the recirculating loop thanks to Acousto-Optic Modulators (AOM).

Noise is artificially included in the loop using an Amplified Spontaneous Emission source (ASE) in order to degrade the Optical Signal to Noise Ratio (OSNR) in front of the regenerator. That is necessary to measure a BER in regenerated signal experiments [2].

A polarization scrambler (polarization modulation frequency ~ 1 MHz) is placed in front of (in the optical case) or behind (in the optoelectronic case) the regenerator in order to take polarization effects into account.

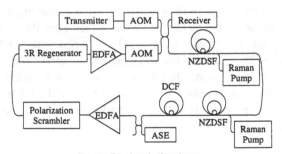

Fig.1: Recirculating loop.

2.2   Optical regenerator's architecture

The optical regenerator is made of two SOA-based wavelength converters. The first converter consists of a Non-Linear Optical Loop Mirror whose non-linear element is a SOA (SOA-NOLM) [3]. The second wavelength converter is a Dual Stage of SOA (DS-SOA) [4]. Figure (2a) represents the all-optical regenerator scheme.

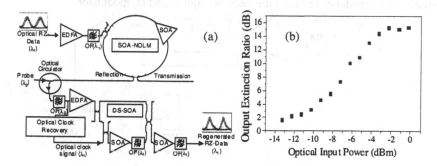

Fig. 2a: All-optical regenerator architecture.
Fig 2b: Output extinction ratio as a function of input power with an input extinction ratio of 15 dB.

The SOA-NOLM is based on a Sagnac interferometer which is intrinsically more stable than all-fibre Mach-Zehnder Interferometers (MZI) provided that fibre arms are short enough. In our case, polarization maintaining fibres are used in order to improve the stability.

Regeneration with NOLM has already been investigated [5] but never, to our knowledge, in a reflective configuration. This allows a better stability with regard to the phase effects and a data output inversion which reduces the converter's polarization dependence.

The DS-SOA as the second wavelength converter stage is an original architecture. In addition to converting the signal back to the initial signal wavelength and to creating a second data output inversion, it improves the output extinction ratio by more than 4 dB. Moreover, the DS-SOA is composed of low polarization sensitivity SOA (0.5 dB) from Alcatel. Consequently, combined with the SOA-NOLM, this results in a polarization insensitive reshaping gate.

The extinction ratio of the overall regenerator is 14 dB for a minimum input extinction ratio of 8 dB. Figure (2b) shows the output extinction ratio versus input power characteristics of the global regenerator that presents an S-shape required for reshaping [6].

2.3 Optoelectronic regenerator architecture

The optoelectronic regenerator has a classical architecture presented on figure (3). It is composed by a 10 GHz PhotoDiode (PD) feeding a Broadband Amplifier (BA) followed by a limiting amplifier. One output of the amplifier is used to recover the clock, the second one feeds the Decision Flip-Flop (DFF). After being amplified by a broadband amplifier, the reshaped and retimed electrical signal finally modulates a local DFB laser through a LiNbO$_3$ modulator.

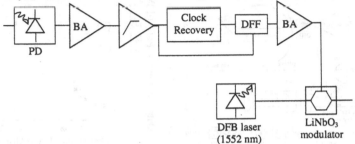

Fig. 3: Optoelectronic regenerator architecture.

3. BIT ERROR RATE EVOLUTION THROUGH A RESHAPING RETIMING GATE

We experimentally show BER evolution with the number of laps for the first time to our knowledge with the two regenerators described above. Results are presented on figure (4) with an OSNR of 17 dB (measured on 0.1 nm). Through the ideal gate (the step function), as initially theoretically reported in [7], the BER in a transmission line with regenerators, linearly increases with the number of concatenated regenerators:

$$BER \approx N \cdot \exp(-k \cdot OSNR) \qquad (1)$$

with N the number of laps, k a suitable constant and OSNR the Optical Signal to Noise Ratio at first lap.

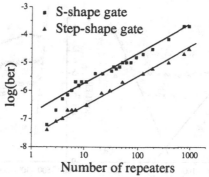

Fig. 4: Experimental BER evolution as the number of laps for an initial OSNR of 17 dB (measured on 0.1 nm).

Through the S-shape NLG, such a linear evolution is observed after about ten laps, this can be explained by the fact that when concatenating the S-shape gate ten times, it tends toward a step function as the transfer function is raised to the tenth power.

Consequently the BER is strongly dependent on the OSNR in front of the first regenerator. The key point will then be to locate the repeater at an early enough stage in order to match a targeted BER for a given link length.

4.   Q FACTOR AND BER THROUGH A NON LINEAR GATE

The BER is commonly expressed as a function of the Q factor as:

$$BER = \frac{1}{2} \cdot erfc\left(\frac{Q}{\sqrt{2}}\right) \qquad (2)$$

When the BER is not directly measurable (typically BER<10^{-10}), it is deduced from the Q factor measurement [8]. Pertinence of BER measurement deduced from Q factor measurement is studied in that part.

BER was studied as a function of decision threshold at different points of the transmission link. Experimental results are presented on figure (5).

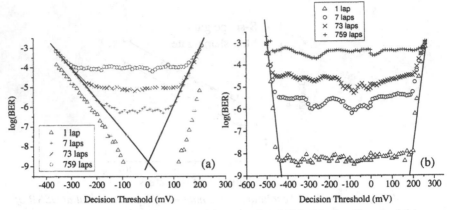

*Fig. 5: BER evolution as a function of decision threshold with the all-optical (a)
and the optoelectronic (b) regenerators.*

The first thing to be noticed is that the BER reaches a plateau, consequently the BER is almost independent of the decision threshold, and the decision is taken by the regenerator through a NLG.

Secondly, in the S-shape case, the plateau width becomes broader as the number of laps increases, namely as the gate tends to a step-like shape. This is the reason why in the step-like shape case, the plateau width remains identical.

On figure (5), extrapolation of the sides is plotted to deduce the Q factor. This measurement would have led to the same deduced BER value, whereas the direct BER measurement leads to an increase of one decade when the lap number is multiplied by ten. As a consequence, we can conclude that Q factor measurement is inadequate to deduce BER evolution as soon as non linear gates are introduced in the transmission line.

5.   CONCLUSION

Signal degradation through different non linear gates was investigated experimentally in this paper. The linear degradation of the BER as a function of the number of regenerators was observed, as predicted by the theory, with an optoelectronic or with an original all-optical 3R repeater. Then, to enhance transmission performance, repeaters must be located early enough in the line in order to reach a targeted BER at the link end-side. Also a BER versus decision threshold study leads to the conclusion that Q factor measurement is not an adequate assessment way for optical transmission links including 2R/3R regenerators.

ACKNOWLEDGMENTS

This work was supported by the "Ministère de la Recherche et des Nouvelles Technologies", the "Conseil Régional de Bretagne", and the European Commission (FEDER).

REFERENCES

[1] O. Leclerc et al., "Optical regeneration at 40 Gbit/s and betond", *J. Lightwave Technol.*, vol. 21, no.11, pp. 2779–90, nov. 2003.

[2] W. Kuebart et al., "40 Gbit/s transmission over 80 000 km dispersion shifted fibre using compact opto-electronic-3R regeneration", *proc. Europ. Conf. Optical. Comm.*, MO4.3.1, 2003.

[3] M. Eiselt et al., " SLALOM : Semiconductor Laser Amplifier in a Loop Mirror ", *J. Lightwave Technol.*, vol. 13, pp. 2099–2112, 1995.

[4] J-C. Simon et al., "Two stages wavelength converter with improved extinction ratio ", in *Proc Opt. Fiber Comm* San Jose, PD15-2, 1995.

[5] F. Seguineau et al., "Experimental demonstration of simple NOLM-based 2R regenerator for 42.66 Gbit/s WDM long-haul transmissions", in *Proc Opt. Fiber Comm*, WN4., 2004.

[6] J-C. Simon et al., "All-optical regeneration techniques", *Ann. Telecommun.*, vol. 58, no. 11-12, nov. 2003.

[7] P. Öhlen et al., "Noise accumulation and BER Estimates in Concatenated Nonlinear Optoelectronic Repeaters", *Photon. Technol. Lett.*, vol. 9, no. 7, pp. 1011, July 1997.

[8] N.S. Bergano, "Margin measurements in Optical Amplifier Systems", *IEEE Photon. Tech. Lett.*, vol. 5., no. 3, march 1993.

CHIRP-FREE TRANSMISSION THROUGH A NOLM BASED OPTICAL REGENERATOR

K.Sponsel[1], M.Meissner[1], K.Cvecek[1], B.Schmauss[2], G.Leuchs[1]

[1] *Institute of Optics, Information and Photonics, University Erlangen-Nuremberg*
Guenther Scharowski Str.1, 91058 Erlangen, Germany
[2] *Electrical Engineering Department, University of Applied Science Regensburg*
Prüfeninger Str.58, 93049 Regensburg, Germany

Abstract: We experimentally show that a strongly asymmetric nonlinear optical loop mirror (NOLM) based 2-R regenerator does not degrade the pulse quality in a fiber optic transmission line. The signal at the NOLM output port shows only negligible chirp if a fundamental soliton propagates in the strong arm of the NOLM. This behavior is measured for splitting ratios of 91:09 and 85:15. The NOLM does not insert penalties regarding the pulse shape and is therefore suitable for inline regeneration of optical transmission lines.

1.   INTRODUCTION

In optical communication at high bitrates optical regeneration [1,2] is a major issue of current reasearch to reduce costs and extend system reach. Though the focus presently lies on optical 3-R regenerators, 2-R regenerators become more and more interesting. They can offer a cost saving alternative as retiming is not always needed. An example of a 2-R type regenerator is the asymmetric NOLM first suggested by Doran and Smith [3]. It is capable of considerably reducing the amplitude noise introduced by amplifiers on the "1" bit and thus improving the BER and the bitrate distance product [4]. An amplifier noise reduction of 12dB on the "1" bit has already been shown [5]. But for the use as an inline regenerator in fiber optical transmission lines not only the reduction of amplitude jitter is of interest. At high bitrates of 40GBit/s and more, the pulse quality is of major importance. In particular, the low chirp tolerance of such systems will be a major problem. For the investigation of pulse quality Frequency Resolved Optical Gating (FROG) [6] is a powerful technique, as it provides a complete description of the electrical field envelope and phase of the pulses as a function of time. In this paper we use the FROG technique to analyse the pulse quality at the output port of a highly asymmetric NOLM with special emphasis on chirp.

2.   PRINCIPLE OF OPERATION OF THE NOLM

The NOLM represents a fiber Sagnac interferometer, consisting of a fiber coupler and a fiber loop connected to the two output ports of the coupler, as shown in Fig.2. An incoming signal pulse is split into two counter-propagating pulses with different peak powers, according to the splitting ratio. Because of the Kerr nonlinearity in the fiber the two pulses gather different nonlinear phase shifts, depending on their different peak powers, and interfere at the coupler after one roundtrip. As this interference is power dependent it leads to a nonlinear input output power transfer characteristic that depends on the splitting ratio, as displayed in Fig.1.

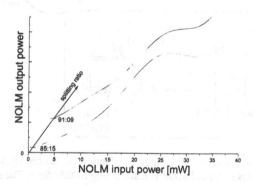

Figure 1.   NOLM power transfer characteristics for splitting ratios of 91:09 and 85:15. For more symmetric splitting ratios, higher input powers are needed to get into the plateau region.

With the used splitting ratios of 85:15 and 91:09 regions with low slopes are present in the the NOLM characteristics. This regions between the points of slope smaller than one are called plateau regions. For the given setup and a splitting ratio of 91:09 the plateau is located between input powers of 22.5mW and 27.5mW, corresponding to peak powers of 750W and 918W. Amplitude noise reduction is achieved if an incoming noise distribution is located on the plateau region, as its width is significantly reduced when transfered to the output port [5]. Best noise reduction is possible if the noise distribution is centered around the middle of the plateau. With decreasing optical signal to noise ratio (OSNR) the noise distribution broadens. Therefore the splitting ratio of the

NOLM has to be adapted to the input OSNR, because the width of the plateau region increases when the splitting ratio becomes more symmetric. For an OSNR better than 44dB a ratio of 91:09 is needed and for an OSNR of 28dB a ratio of 85:15 [7]. Optimal noise reduction performance is observed around an OSNR of 30dB. The splitting ratios of 91:09 and 85:15 in our experiment are chosen to investigate the OSNR interval, where the NOLM is operating as a regenerator. [8].

3. EXPERIMENTAL SETUP

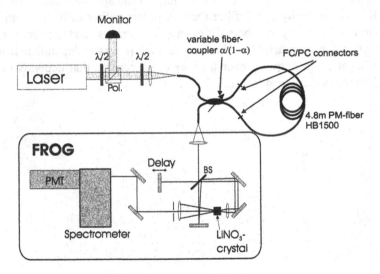

Figure 2. Experimental setup. The input polarization is matched by a half wave plate to one of the principle axis of the polarization maintaining NOLM fiber. The splitting ratio of the NOLM can be adjusted by the variable fiber coupler.

The experimental setup is shown in Fig.2. The laser source is a mode-locked $Cr^{4+} : YAG$ laser emitting pulses of sech-shape with 162fs pulse duration at 1495nm and a repetition rate of 163MHz. As we use soliton transmission within the NOLM, the results are transferable to optical communication systems where e.g. 1.5ps are used for 160GBit/s. Neglecting higher order effects as the stimulated Raman effect the properties of solitons such as self stabilization and pulse shaping are independent of the fiber type and pulse length [9]. The drawback is, that for longer pulse durations much longer fiber loops or specially designed photonic crystal fibers [10] are needed. To reach the plateau region the phase differences $\Delta\Phi$ due to the nonlinear Kerr effect of the two counterpropagating pulses in the NOLM must be about 1.5 π. For a fundamental soliton in the intense arm of the NOLM the required loop length for plateau formation is then given by equation 1. In units of dispersion length L_D the fiber loop length depends only on the splitting ratio α and not on the pulse duration [9].

$$L = L_D \frac{\Delta\Phi}{|1 - 2\alpha|} \tag{1}$$

In our experiment the optical power of the laser beam is adjusted with a half wave plate and a polarizing beam splitter at the NOLM input. The polarization of the signal is matched to one of the principal axis of the NOLM fiber by a second half wave plate. The NOLM comprises a tunable fiber coupler consisting of polarization maintaining HB1500 fiber and 4.9m fiber loop of the same type of fiber. The length of the loop is chosen to aquire the 1.5 π phase shift for the plateau region with a soliton in the intense arm at a splitting ratio of 91:09, according to equation 1. The NOLM output pulses are analysed with the FROG, consisting of a Michelson interferometer based SHG autocorrelator and a monochromator to spectrally resolve the SHG signal. Finally the signal is detected by a photomultiplier at the spectrometer output as a function of wavelength and time.

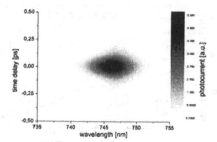

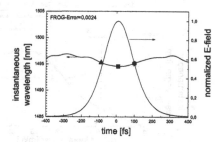

Figure 3.   a) Experimental FROG trace of the NOLM output pulse, taken at an input peak power of 835W and a splitting ratio of 91:09.

Figure 4.   b) Electrical Field (right axis) and instantaneous wavelength (left axis) derived from the FROG trace in Fig.3 The dip in the pulse center originates from a slight quadratic chirp of the $Cr^{4+} : YAG$ Laser

The FROG traces were taken for several input powers along the nonlinear transfer characteristic. These traces were recorded with a resolution of 128 steps in the time domain and 512 steps in the spectral domain and were corrected for the spectral response of the photo multiplier. The high resolution in the spectral domain allows for a simple method for noise reduction of the signal by averaging over 4 pixels in the spectrum. Therefore, for the signal retrieval FROG traces of 128 times 128 pixels were used, as shown in Fig.3. From each FROG trace the electrical field envelope and the phase were calculated as functions of time as shown in Fig.4. Differentiating the phase the instantaneous frequency can be calculated [6] and converted into an instantaneous wavelength, which is more convenient. To quantify the chirp of the transmitted pulses, we take the instantaneous wavelength at the points of half the maximum of the electrical field envelope and at the pulse center, shown in Fig.4.

4. EXPERIMENTAL RESULTS

To investigate the pulse quality behind the NOLM for lower OSNRs, like 28dB we chose a splitting ratio of 85:15. The results are presented in Fig.5. The input-output characteristic shows a flat plateau region which starts around an average input power of 25mW. The instantaneous wavelength at the FWHM and in the pulse center of the E-field are plotted versus the NOLM input power in the left hand side of Fig.5.

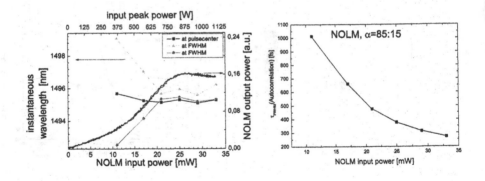

Figure 5. NOLM characteristic and development of the instantaneous wavelength at a splitting ratio of 85:15 (left), and autocorrelation width of the transmitted pulses versus input power (right)

Input powers below 20mW lead to dispersive pulse broadening and therefore a linear chirp, which can be seen in the autocorrelation and the instantaneous wavelength in Fig.5. The instantaneous wavelengths of the pulse edges are about 2nm away from the center wavelength. Increasing the input power leads to soliton formation in the strong arm of the interferometer and causes a reduction of chirp of the output pulse. The pulse length decreases. To match the power of the fundamental soliton in the intense arm an average NOLM input power of 26.5±1.5mW, corresponding to a peak power of about 885W is needed. In this case the maximum difference of the instantaneous wavelength within the FWHM of the E-field of the pulse is below 0.65nm. This is only a small fraction of the pulse spectral width of $\Delta\lambda_{FWHM}$=24.2nm in the electrical field, implying that the pulses after the NOLM are nearly free of chirp. The power range for chirp-free NOLM output pulses stretches from 20mW to about 30mW, covering the range of input powers in the plateau region of the nonlinear transfer characteristic which is used for noise reduction. Note that even the intrinsic chirp of the laser, resulting in a wavelength difference between the pulse edges and the center of each laser pulse of 2.7nm is reduced. The lasers quadratic chirp originates from its internal group velocity dispersion compensation. The end of the plateau region could not be reached due to the experimental limitation in the maximum available laser power.

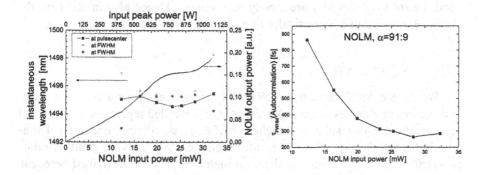

Figure 6. NOLM characteristic and development of the instantaneous wavelength at a splitting ratio of 91:09 (left), and autocorrelation width of the transmitted pulses versus input power (right)

The nonlinear input output characteristic for a splitting ratio of 91:09 is depicted in Fig.6. The fundamental soliton was measured at an average input power of 23±1.5mW. Accordingly at the given splitting ratio, an input power of 25±1.5mW is needed to launch a soliton into the intense arm of the NOLM, corresponing to a peak power of about 835W. For an average NOLM input power of 25mW and a splitting ratio of 91:09 electrical field and instantaneous wavelength are presented in Fig.4. As expected from a soliton the change of chirp is quite similar to the case with a splitting ratio of 85:15 shown in Fig.5. For input powers close to 25mW the output pulses are practically free of chirp. When the input power exceeds the fundamental soliton power in the strong arm above 28.5mW, the chirp of the pulse increases again. This is due to effects like soliton formation with emission of dispersive waves and changes in the interference as the weak pulse is stronger affected by the Kerr-effect. But here the NOLM input output characteristic has a slope larger than unity, thus this part is not suitable for noise reduction and is therefore of no practical interest.

5. DISCUSSION

For an optimal pulse transfer through the NOLM the propagation of a fundamental soliton in the intense arm of the interferometer is used. Fundamental soliton are self stabilising, have a constant phase and are thus free of chirp. In the weak arm of the interferometer the pulse is broadened in time by dispersion. At the fiber coupler the soliton in the strong arm interferes and recombines with the dispersive pulse of the weak arm. But as the weak pulse added has only 11% of the soliton power for a splitting ratio of 91:09, and 17.6% for a splitting ratio of 85:15 the perturbation of the soliton pulse shape is small and the soliton properties are mostly maintained. Hence also in this case the NOLM output pulse is practically free of chirp.

6. SUMMARY

We have demonstrated that a NOLM type 2-R regenerator does not introduce chirp or changes in the pulse shape in an extended region around its point of operation. This results from the usage of a self stabilizing fundamental soliton in the intense arm of the fiber interferometer. We analysed splitting ratios of 91:09 and 85:15. Therefore the asymmetric NOLM, as described here, can be used in high speed telecommunication systems as an optical regenerator, without introducing a penalty regarding to the pulse shape and chirp. Therefore no changes in the fiber link design are needed which makes the NOLM a promising alternative to 3-R regenerators, regarding performance and price.

ACKNOWLEDGMENTS

We would thank Arne Striegler from the Chair for Microwave Engineering and High Frequency Technology of the University Erlangen-Nuremberg for his support and the fruitful discussions. We would also like to greatfully acknowledge the financial support of the DFG.

REFERENCES

[1] E.S. Awad, P.S. Cho, C. Richardson, N. Moulton, and J. Goldhar,"Optical 3R regeneration with all-optical extraction and simultaneous wavelength conversion using a single Electro-Absorbtion Modulator", *Proc. ECOC*, Copenhagen, Denmark, Vol. 3, 6.3.2, 2002

[2] G. Gavioli and P. Bayvel,"Novel, High-Stability 3R All Optical Regenerator Based On Polarization Switching in a Semiconductor Optical Amplifier", *Proc. ECOC*, Copenhagen, Denmark, Vol. 3, 7.3.2, 2002

[3] N.J. Smith and N.J. Doran,"Picosecond soliton transmission using concatenated nonlinear optical loop-mirror intensity filters",J. Opt. Soc. Am. B12, 1117-1125 (1995)

[4] R. Ludwig, A. Sizmann, U. Feiste , C. Schubert, M. Kroh, C.M. Weiner and H.G. Weber,"Experimental Verification of Noise Squeezing by an Optical Intensity Filter in High-Speed Transmission", *Proc. ECOC*, Amsterdam, Netherlands, 2001

[5] M. Meissner, M. Roesch, B. Schmauss, G. Leuchs,"12dB of noise reduction by a NOLM based 2-R-regenerator",IEEE Photonics Technology Letters 15,1297-1299 (2003)

[6] R. Trebino, *Frequency-Resolved Optical Gating: The Measurement of Ultrashort Laser Pulses*(Kluwer Academic Publishers, 2002)

[7] M. Meissner, M. Roesch, B. Schmauss, N. Korolkova and G. Leuchs,"Optimum Splitting Ratio for amplifier noise reduction by an asymmetric nonlinear optical loop mirror", *Proc. ECOC*, Copenhagen, Denmark, Vol. 3, P3.08, 2002

[8] M. Meissner, M. Roesch, B. Schmauss and G. Leuchs,"Noise reduction performance of a NOLM based 2-R-regenerator in dependence on the OSNR", *Proc. ECOC*, Rimini, Italy, Vol. 3, We4P102, 2003

[9] G.P. Agrawal, *Nonlinear Fiber Optics*(Academic Press, 1995)

[10] P. Russell,"Photonic Crystal Fibers", *Science*, Vol. 299, p.358, 2003

PROPAGATION OF UNEQUAL OTDM DATA CHANNELS IN 2R REGENERATED SYSTEM

Zhijian Huang, Ashley Gray, Igor Khrushchev, Ian Bennion
Photonics Research Group, Electronic Engineering,
Aston University, Birmingham, B4 7ET, United Kingdom
E-mail: z.huang@aston.ac.uk

Abstract: Switching of an optical signal comprising individual OTDM channels of unequal amplitudes in a nonlinear optical loop mirror is investigated. The propagation dynamics of unequal-channel data streams in a switch-guided, dispersion-managed link are also studied

1.   INTRODUCTION

Optical time division multiplexing (OTDM) is currently the only way to form optical data streams at ultrahigh speeds of 80Gb/s and above. Due to the finite accuracy of channel equalisation, some variations of amplitude between the channels is inevitable in OTDM transmitters. Other fundamental effects, such as interferometric noise induced by finite pulse extinction ratio and phase noise from the laser source [1] also contribute to these variations. Moreover, it is sometimes advantageous, from the system point of view, to introduce the channel inequality deliberately either to reduce the pulse-to-pulse interaction[2] or to simplify the task of clock recovery [3].

All 2R regenerators, including nonlinear optical fibre loop mirrors, NOLMs [4-6], possess nonlinear transmission response characteristics. Therefore, OTDM channels of different amplitude will experience different transformations when switched in a 2R device. This is an important factor in a 2R-supported data transmission link. In this paper, we investigate how the channel inequality affects the nonlinear signal switching in a 2R regenerator and the data transmission in a nonlinear switch-guided system.

2. EXPERIMENTAL SETUP

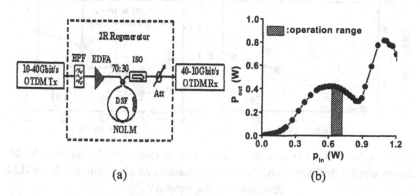

Fig.1 a) Experimental setup b) Switching curve of NOLM.

The NOLM shown in Fig. 1(a) comprised a 70:30 fibre coupler and 2.3km of dispersion shifted fibre with an anomalous dispersion of +2.8 ps/(nm·km) at the wavelength of 1550nm. A mode-locked fibre laser generated Gaussian-shaped 3.5-ps pulses with an extinction ratio higher than 33dB. The pulses were modulated at 10Gbit/s by a pseudo-random bit stream (PRBS) of length $2^{31}-1$ and subsequently optically multiplexed to form a single-polarization, 40Gbit/s OTDM data stream by using a two-stage fibre delay line. It was possible to vary the amplitudes of the four OTDM channels in the 40Gbit/s signal individually by adjusting the polarization controllers inside the delay line. At the receiver, the 40Gbit/s signal was optically de-multiplexed to 10Gbit/s using an electro-absorption modulator.

3. RESULTS AND DISCUSSION

Figure 1(b) shows the measured transmission of the NOLM as a function of average power of the input 40Gb/s signal. The shaded area indicates the input optical power range that usually provides the best switching and noise suppression performance [4,5].

The switching behaviour of the OTDM signal was investigated by using histogram analysis of the eye diagrams taken on a 50GHz digital sampling oscilloscope equipped with a 32GHz photo-detector. The signals were measured before and after the NOLM. We used the relative standard deviation of amplitude, ΔV, as a measure of inter-channel non-uniformity. The mean voltage value V_i (i=1,2,3,4) of marks in each individual eye determined the amplitude of a corresponding channel. We defined the inter-channel amplitude difference as $\Delta V=\sigma/\mu$, where σ is the standard deviation of V measured channel-to-channel and μ is the averaged value of V over the four channels.

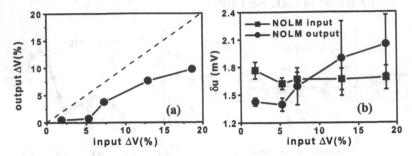

Fig.2 a) Inter-channel amplitude variation, ΔV, before and after the NOLM. b) Random amplitude fluctuations δu before (squares) and after (circles) the NOLM, depending on the input ΔV.

The input ΔV was varied by adjustment of the OTDM multiplexer. The average power at the NOLM input was optimised in order to minimize the ΔV of the signal at the NOLM output. The average power arriving at the photo-detector was maintained constant. Fig. 2(a) shows the output ΔV as a function of that at the NOLM input. One can see that the inter-channel amplitude variation is always reduced by the NOLM, but the effect becomes weaker as the input signal becomes less uniform.

We also studied the dynamics of another important parameter, the random fluctuation of amplitude. This was characterised by $\delta u = (\Sigma \rho_i)/4$, where ρ_i was the standard deviation of amplitude within each channel.

Fig. 2(b) shows δu as a function of the input amplitude difference, ΔV, measured before and after the NOLM. The height of the error bar indicates the difference between the minimum and maximum ρ_i measured in the most stable and in the noisiest of the four channels, respectively. One can see that the effect of the input channel non-uniformity on the amplitude noise suppression is significant. When the input amplitudes are relatively uniform (ΔV less than 5% in the experiment), the NOLM efficiently suppresses the amplitude fluctuations in all channels. However, when the input channels become considerably unequal (ΔV larger than 10%), the efficiency of the amplitude noise suppression strongly varies among the channels. As a result, the overall quality of the switched signal deteriorates. In fact, a signal comprising very non-uniform OTDM channels actually experiences an amplitude noise increase as a result of switching in the NOLM.

Overall, the inter-channel amplitude difference is usually reduced by the NOLM, whilst amplitude fluctuations of individual channels may be either suppressed or increased depending on the non-uniformity of the input signal.

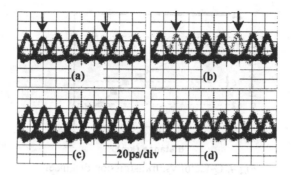

Fig.3. Eye diagrams. a) Initial ΔV = 15%, back-to-back. b) ΔV = 15%, measured after 2,500km. c) ΔV = 5%, back-to-back. d) ΔV = 5%, after 2,500km.

Non-uniformity of the input signal is likely to affect the data transmission in a system employing NOLMs as in-line 2R regenerators. If the format of the input signal is such that the amplitude fluctuations in some channels increase in the first NOLM, these channels will subsequently deteriorate during propagation through subsequent 2R elements, probably resulting in unstable propagation of corresponding channels.

We studied the effect of channel non-uniformity on the data transmission by monitoring the eye diagrams and Q-factor of a digital signal propagating in a re-circulating loop with the in-line NOLM acting as a 2R element [6]. The 2R regeneration was performed after every 200km of fibre. The Q-factor was calculated by using decision threshold measurements [7]. Figure 3 shows the eye diagrams measured back-to-back and after propagating over 2500km of SMF. The initial amplitude difference, ΔV, was set at a level of either 15% (Fig.3(a,b)) or 5% (Fig.3(c,d)) for this experiment.

With an initial ΔV of 15% (Fig.3a), the signal considerably deteriorated during transmission, showing drop-out of the selected channel after propagating over 2,500km (Fig.3b). The overall bit-error-rate of the 40Gb/s data stream increased dramatically during the first several hundred kilometres of transmission.

The higher quality signal with an initial ΔV of 5% (Fig.3c) propagated in a totally different manner. Channel non-uniformity was eliminated after multiple transmissions through the 2R element, resulting in a virtually perfectly uniform OTDM data stream observed after 2,500km of propagation (Fig.3d). The estimated Q-factor was as high as 19dB after 2500km of SMF (Fig.4) and the error-free propagation (Q~15.6dB) distance was estimated to be in excess of 5800 km.

564

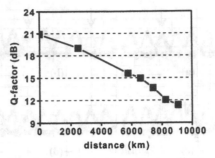

Fig.4. Q-factor vs. transmission distance.

4. CONCLUSION

We have shown that channel-to-channel amplitude differences in OTDM data streams have a strong impact on the switching behaviour of individual channels in a 2R-regenerator. Depending on the inter-channel amplitude difference, the optical pulses in different channels experience either suppression of the amplitude noise, or a noise increase. Appropriate control of the channel uniformity in the OTDM transmitters is necessary in order to support stable long haul transmission in 2R regenerated systems.

REFERENCES

[1] I. Y. Khrushchev, et al, "OTDM applications of dispersion-imbalanced fibre loop mirror", *Electron. Lett.,* , vol. 35, pp. 1183-1185, 1999

[2] G. P. Agrawal, Nonlinear Fiber Optics, Academic Press,1989), 130-133

[3] A. H. Gnauck, et al "1-Tb/s (6×170.6Gbit/s) transmission over 2000-km NZDF using OTDM and RZ-DPSK format", *IEEE Photon. Technol. Lett.*, vol. 15, pp. 1618-1620, Nov., 2003

[4] N. J. Smith et al, "Picosecond soliton transmission using concatenated nonlinear optical loop-mirror intensity filters", *J. Opt. Soc. Am. B*, vol. 12, pp. 1117-1125, 1995

[5] S. Boscolo, et al "Study of the operating regime for all-optical passive 2R regeneration of dispersion-managed RZ data at 40 Gb/s using in-line NOLMs", *IEEE Photon. Technol. Lett.* , vol.14, pp. 30-33, Jan., 2002

[6] Z. Huang, et al "40Gb/s transmission over 4000km of standard fibre using in-line nonlinear optical loop mirrors", in *Proceedings of the 29th European Conference on Optical Communication (ECOC/IOOC 2003)*, Rimini, Italy, paper Mo4.6.3, 21-25 September 2003

[7] N. S. Bergano, F. W. Kerfoot, and C. R. Davidson, "Margin measurements in optical amplifier systems," *IEEE Photonics Technol. Lett* , vol. 5 pp. 304-306, Mar. 1993

2-R REGENERATION EXPLOITING SELF-PHASE MODULATION IN A SEMICONDUCTOR OPTICAL AMPLIFIER

Gianluca Meloni,[1] Antonella Bogoni,[2] and Luca Poti[2]

[1] *Scuola Superiore Sant'Anna, P.zza dei Martiri della Libertà 33, 56100 Pisa*
[2] *CNIT, Via Cisanello 145, 56124 Pisa;*

Abstract: All-Optical Re-amplification and Re-shaping (2R) exploiting spectral shift induced by Self-Phase-Modulation (SPM) in a Semiconductor Optical Amplifier (SOA) is presented. Theoretical model taking into account Carrier Density (CD), Carrier Heating (CH) and Spectral Hole Burning (SHB) is experimentally validated using 3 ps optical pulses.

1.   INTRODUCTION

The increase of transmission bandwidth request in the optical communications and networks opens new research perspective towards ultra-fast all-optical systems. Optical Time Division Multiplexed (OTDM) systems together with all-optical packet switching networks represent the future. However, ultra short signals are strongly affected by different kind of noise sources as, for example, Polarization Mode Dispersion, Chromatic Dispersion, Nonlinear fiber effects, Amplified Spontaneous Emission, etc... In this contest opto electronic conversion, regeneration, and electro-optic conversion, are unthinkable. The electrical bottleneck must be overcame exploring all-optical regeneration techniques.

In this paper, we propose a new technique to perform 2R (Re-shaping and Re-amplification) all-optical regeneration using a Semiconductor Optical Amplifier (SOA) and a Tunable Optical Filter (TOF). The principle exploits the phenomenon of Self Phase Modulation into the SOA. SOA can be a key device for such applications in future optical networks because of its integrability, nonlinear efficiency, low cost and reliability.

The paper describes the model used to numerically investigate the phenomenon. An experimental validation of the results are also reported.

1.1   2R-Regenerator Basic Principle

2R Regeneration stands for Re-amplification and Re-shaping. Former functionality is easily obtained, in the optical domain, using Fiber, Semiconductor or Raman amplifiers. However, amplification is not always sufficient to guarantee good performance for high capacity transmission systems, where Re-shaping can be introduced. The principle of operation for the re-shaper is based on the ideal characteristic shown in Fig.1 where the output power is plotted as a function of the input power.

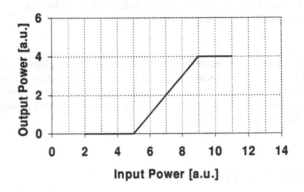

Figure 1.   Re-shaper ideal characteristic.

The input/output characteristic represents a typical soft limiter transfer function. In this way, if the re-shaper signal input has power fluctuations, due to any kind of noise accumulated during its propagation, they can be partially or completely eliminated by the limiter nonlinear characteristic. The effect can be exploited both on the space and mark levels. An ideal 2R-Regeneration, hence, is a nonlinear amplifier that, at the same time, properly shapes and amplifies the input signal. For this purpose, we exploit Self-Phase Modulation (SPM) in a SOA. When a powerful signal propagates into a SOA, it experiences a nonlinear phase modulation (SPM) induced by a refractive index variation due to carrier change in the conduction bandwidth. As the carrier change is governed by fast phenomena (Carrier Density - CD, Carrier Heating - CH, and Spectral Hole Burning - SHB), the phase modulation exactly follows the peak power time evolution. When a powerful pulsed or modulated Return-to-Zero (RZ) signal propagates through a SOA, its phase is modulated, thus giving an optical spectrum broadening and red shift. In other words, an high mark level at the SOA output is wavelength-shifted in respect to a space level. The wavelength shift is proportional to the input peak power. It is possible, than, to use an optical filter centered on the shifted signal to eliminate the noise on the space level and to limit the power fluctuation on the mark level.

1.2   Self-Phase Modulation model for SOAs

In [1] the nonlinear Kerr effect (refractive index modulation) is included in an accurate model and is described by the dependence of the effective refractive index n on intensity I:

$$n(\omega, I) = n_0(\omega) + n_2 I \tag{1}$$

where n_0 is the linear index, n_2 is the Kerr coefficient[1], and ω the pulsation. When the intensity $I(t)$ is modulated, the refractive index changes in time, and the propagation constant becomes:

$$\beta(\omega, I) = \frac{\omega n(\omega, I)}{c} \tag{2}$$

The study of the medium response to the optical field leads to the following set of three equations:

$$\frac{\delta P}{\delta z} = (g - \alpha_{int})P \tag{3}$$

$$\frac{\delta \Phi}{\delta z} = -\frac{1}{2}\alpha g \tag{4}$$

$$\frac{\delta g}{\delta \tau} = \frac{g_0 - g}{\tau_c} - \frac{gP}{E_{sat}} \tag{5}$$

where α_{int} is the internal loss, E_{sat} is the saturation energy of the amplifier, $P(z)$ and $\Phi(z)$ are the power and the phase, g_0 is the small signal gain, τ_c is the spontaneous carrier lifetime, α is the line-width enhancement factor ([2], [3]), and g is the saturated gain. Equation 4 shows the origin of SPM. Carrier population dynamics must considered in the model to properly account for ultra-fast intra- and inter- band phenomena:

- Carrier Density (inter-band phenomenon)
- Spectral Hole Burning (intra-band phenomenon)
- Carrier Heating (intra-band phenomenon).

Inter band means that carrier are exchanged between valence and conduction bandwidths, whereas intra band includes those phenomena that are generated by a carrier displacement in the same bandwidth. The former have time constants of about 100 ps, and the latter three order of magnitude lower.

---

[1]n_2 is approximately $3.2 \times 10^{-16} cm^2/W$

568

The intra/inter-band phenomena are included in the model by considering a gain:

$$g(t) = \overbrace{\underbrace{\frac{g_N(t)}{f(t)}}_{\text{SHB}}}^{\text{CD}} + \overbrace{\Delta g_T(t)}^{\text{CH}} \qquad (6)$$

where $g_N(t)$, $\Delta g_T(t)$, and $f(t)$ represent CD, CH, and SHB respectively.

2. NUMERICAL RESULTS

When an ultra-short optical pulse propagates through a SOA, the gain quickly saturates, due to the carrier consumption on the conduction bandwidth, and afterward it recovers with time constants that are strictly related to the intra/inter band phenomena. As an example, Fig.2 shows the SOA gain time saturation and recovery when a 1.2 ps long pulse crosses the device. Few carriers are moved within the conduction bandwidth very quickly by CH and SHB giving the first fast gain recovery, whereas most of the bandwidth is repopulated by the bias current (CD).

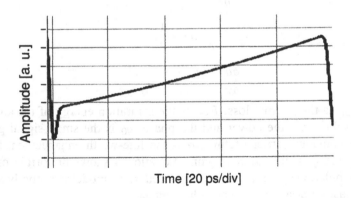

Figure 2. Gain dynamic.

The model is here used, than, to simulate for the spectrum red shift. In Fig.3 is plotted the optical spectrum of the signal at the input (line) and output (dashed-line) of the SOA and at the output (dotted-line) of the Tunable Optical Filter (TOF) after the amplifier.

First we simulate a 15 ps clock signal regeneration. A 10 GHz pulsed signal at 1548.7 nm is propagated through a SOA and the output is selected using a TOF centered around 1550 nm with a bandwidth of 1 nm. The time signals are shown in Figg.4(a) and 4(b)

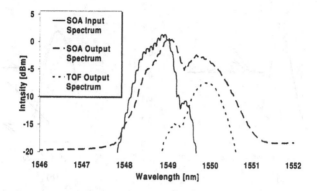

Figure 3. Simulated Spectra at the SOA input (line), output (dashed-ine), and after the TOF (dotted-line)

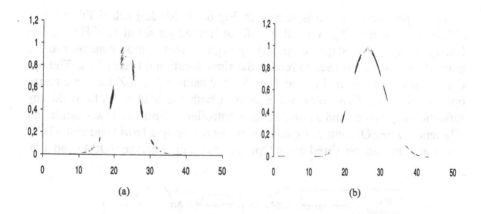

Figure 4. (a) Input pulses; (b) Output regenerated pulses

Numerical results shows a sensible noise reduction on the peak power. An input peak change of about ±10% around the mean level is reduced to less than ±2% at the TOF output.

As a second case study, a modulated signal, with same parameter as before, has been simulated. The considered sequence is ...10101010... where the optical modulator extinction ratio is 20 dB. Input and output eye diagrams are shown in Fig.5.

Also in this case we may appreciate an improve of about ±3% on the mark level, and a small noise reduction also on the space level.

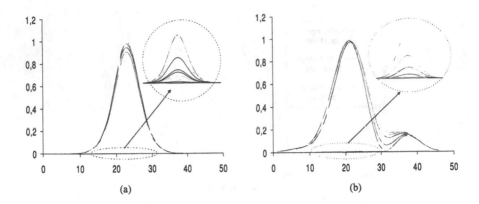

(a)   (b)

Figure 5.   (a) Input eye diagram; (b) Output eye diagram

3.   EXPERIMENTAL ACTIVITY

The experimental setup is shown in Fig.6. A Mode-Locked Fiber Laser (MLFL) generates 20 ps optical transform-limited pulses at 10 GHz. A following compression stage composed by higher order soliton generator and a pedestal suppressor is used to reduce the time duration down to 2 ps. The optical pulses are than amplified and modulated using a Mach-Zehender electro-optic modulator. Two isolators are used at both the SOA ports to avoid any reflection and lasing, and a polarization controller assures maximum nonlinear efficiency. The SOA output signal is than filtered using a $1 nm$ bandwidth TOF. The signal is than measured by an Optical Spectrum Analyzer (OSA) and a 50

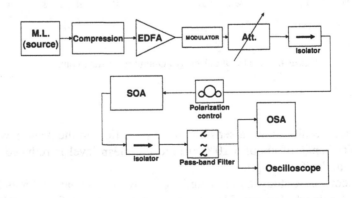

Figure 6.   Experimental Set-up.

GHz bandwidth sampling oscilloscope. Fig.7shows the measured specrta at the SOA input, output and after the TOF. The red shift is clear from the graph.

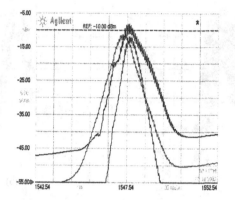

Figure 7. Measured spectra with an input power of $5dBm$

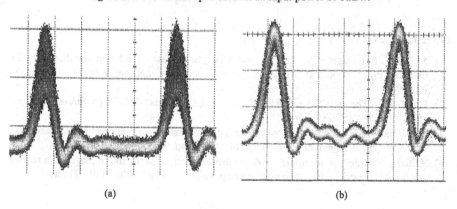

(a)   (b)

Figure 8. (a) Input distorted pulses; (b) Output regenerated pulses

The measured pulses time trace are shown in Fig.8 when an input mean power of 5 dBm is used. By increasing the compressor power it is possible to increase both the input SOA power and the pulse distortion. Fig.9 shows the distorted input (left) and regenerated output (right) pulses for an input power of about 8 dBm. The reshaping effect, in this case, is more evident.

4.   CONCLUSION

In this paper we have theoretically demonstrated and experimentally implemented a 2R regenerator. Nonlinear reshaping characteristic is obtained exploiting SPM in SOA and following spectral filtering. The scheme has been successfully implemented using 10 GHz optical transform-limited pulses with time duration lower than 3 ps.

572

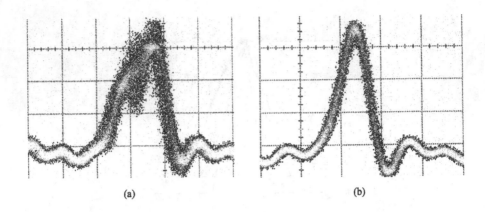

(a)   (b)

Figure 9.   (a) Input distorted pulses; (b) Output regenerated pulses

REFERENCES

[1] M.Stern, J.P.Heritage, R.N.Thurston, and S.Tu, "Self-phase modulation and dispersion in high data rate fiber-optic transmission systems," *Journal of Lightwave Technology*, vol. 8, july 1990.

[2] C.H.Henry, "Theory of the linewidth of semiconductor lasers," *IEEE J. Quantum Electron.*, vol. QE-18, pp. 259–264, 1982.

[3] M.Osinski and J.Buus, "Linewidth broadening factor in semiconductor lasers-an overview," *IEEE J. Quantum Electron.*, vol. QE-23, pp. 9–29, 1987.

[4] A.Bizzi, "Sviluppo di un simulatore di amplificatori ottici (soa) e suo impiego nella realizzazione di un modulatore ottico ultraveloce per sistemi otdm," 2000-2001. Thesis.

TWO PUMP OPA FOR OTDM PULSES AMPLIFICATION

Lucia Marazzi[1], Paola Parolari[1], Pierpaolo Boffi[1], Elisabetta Rognoni[1], Paolo Gaviraghi[1] and Mario Martinelli[1,2]
[1] CoreCom, via G. Colombo, 81, 20134 Milano, Italy marazzi@corecom.it
[2] Politecnico di Milano, piazza L. da Vinci 32, 20133 Milano, Italy

Abstract: Necessity to exploit new bandwidth has moved research interest towards new amplification solution, which support the exploitation of the fiber bandwidth and of short optical pulses. 2P-OPA seems quite promising as they offer high project flexibility, high gain uniformity and low accumulated dispersion. We present and discuss a 2P-OPA realization and its behavior with CW and picosecond pulses.

1. INTRODUCTION

Exploitation of available bandwidth has moved towards two directions: accommodation of many channels at different wavelengths in Dense Wavelength Division Multiplexed (DWDM) systems or very high bit rate Optical Time Division Multiplexed (OTDM) systems. Nowadays this distinction may not be very significant as each WDM channel may be in fact a high bit rate OTDM channel, which may employ picosecond optical pulses. In this scenario the development of new amplification techniques, which support the exploitation of the fiber bandwidth and of short optical pulses, has gained a great interest. In particular amplification solutions based on nonlinear effects show good potentialities as they avoid the constraint of ions fixed energy levels and lifetimes. Both Optical Parametric Amplifiers (OPA's), which are based on fiber Four Wave Mixing (FWM), and Raman amplifiers present high project flexibility. In principle their bandwidth can be arbitrarily chosen provided correct pump wavelength and power. In particular broadband significant gain [1] and low noise figure [2] OPA's

574

have been demonstrated. Moreover the Kerr effect is known to respond at less than 100 fs and seems thus very interesting for picosecond OTDM pulses amplification.

Aim of this contribution is to experimentally analyze the impact of OPA's exploiting Dispersion Shifted (DS) fiber on OTDM pulses. We will concentrate on two-pump OPA (2P-OPA) solutions, which seem particularly suitable for OTDM pulses amplification. As can be seen in Fig.1a, in the three waves scheme (1P-OPA), the pump, close to the zero dispersion wavelength, transfers energy toward signal and idler, which are symmetrically located with respect to the zero dispersion wavelength. On the contrary in the 2P-OPA (fig.1b) the two pumps are symmetric with respect to the DS fiber zero dispersion wavelength. Thus amplification region is not only very homogeneous, as can be seen in the inset of Fig.1, where simulation results are shown, but also shows very low dispersion. Both characteristics can better support picosecond pulse propagation, which is affected by dispersion and, due to broad picosecond pulse spectra, also by gain non-uniformities. Moreover total pump power in the two-pump scheme is divided into two pumps thus reducing Stimulated Brillouin Scattering (SBS) impact [3] with respect to the one pump solution.

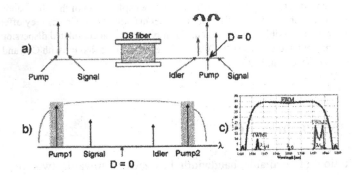

Figure 1. a)OPA e b)2P-OPA schemes. In c) 2P-OPA gain profile is shown together with the two TWM's gain regions. Simulation parameter are: fiber length 5 km, λ_0-$\lambda_c \cong 0.2$nm, λ_{pump1}=1532.25 nm, λ_{pump2}=1557.57 nm, P_{pump1}=P_{pump2} = 0.5 W.

2.   TWO PUMP-OPA PROJECT

The influence of 2P-OPA project parameters such as pump power and spectral allocation on the gain profile of 2P-OPA have been investigated with a simplified model, which takes into account also pump depletion. We focused on main process, i.e. non-degenerate FWM: two photons, one from each pump, combine to generate one signal photon and one idler photon satisfying the relation

$\omega_{p1}+\omega_{p2}=\omega_s+\omega_i$ where ω_i is idler pulsation. TWM1, which involves pump1, the signal and the respective idler1, and TWM2, which involves pump2, the signal and the respective idler2, are separately considered. The analysis of the complete equations relating TWM's [5,6], taking into account also pump depletion, has evidenced that the same actions which advantage 2P-OPA gain uniformity, e.g. pump power increase and pump distance reduction, also advantage extensions of the TWM gain bandwidths [4]. Actually in order to have 2P-OPA uniform gains it is necessary to look for a trade-off between these two effects. A sample of the performed simulations is presented in Fig.1c and shows that this can partly be accomplished by unbalancing pump powers, giving a first project hint. This hint together with limits due to available devices and nonlinear media has been taken into account in the realization of a 2P-OPA, which has been tested for OTDM pulse amplification and will be discussed in the next section.

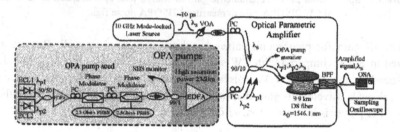

Figure 2. Set up.

3.   EXPERIMANTAL SET UP

Experimental setup is shown in Fig.2. Nonlinear medium is a 9.9 km long DS fiber with zero dispersion wavelength of 1546.1 nm and dispersion slope of 0.062 ps/(nm²km) and nonlinear coefficient $\gamma=1.5$ km^{-1}W^{-1}. Pumps are generated by two ECL lasers coupled together and subsequently amplified and further fed into the nonlinear fiber through a 90/10 coupler. Pumps wavelengths and powers are respectively λ_{pump1} = 1537.32 nm, λ_{pump2} = 1553.16 nm P_{pump1} = 14.11 dBm, P_{pump2} = 13.5 dBm. Probe signal is a 10 GHz 7 ps mode-locked pulsed source by Pritel. Due to high powers involved two phase modulators, operated with two different 2.5Gbit/s PRBS sequences, broaden pump optical spectrum to suppress SBS insurgence. Relative state of polarization of the three interacting fields is controlled. Fig.3 shows CW characterization of the projected and realized 2P-OPA. As spontaneous amplification (a) and small signal gain (b) show, with 14.84 nm pumps spacing, nearly 13 nm of uniform bandwidth are found. Uniformity is within 1dB. Signal power is -17.2 dBm and average on-off gain is 6.5 dB. NF values are all higher than 20 dB, which is an unacceptable value for a telecom amplifier, yet it has already been demonstrated [2] that by properly filtering pump noise contributions NF can be significantly reduced.

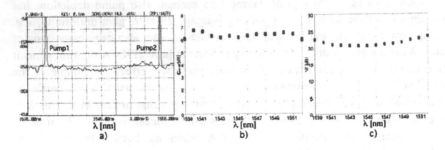

Figure 3. a) Spontaneous 2P-OPA parametric gain measured with optical spectrum analyzer, b) CW 2P-OPA gain c) measured 2P-OPA noise figure.

On-off gain for picosecond pulses has been measured as well. Results are plotted in Fig.4, compared with CW small signal gain. Slightly higher gain value are found for low power OTDM pulses than for CW signal: this is probably due to spectral peak wavelength gain measure. As expected this value saturates when peak pulse power is increased from −3 dBm to 11 dBm.

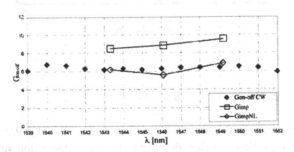

Figure 4. 2P-OPA on-off gain for CW, low power pulsed and high power pulsed signal.

Final OTDM pulse amplification has been evaluated by means of pulse autocorrelator traces, to measure pulse broadening which results for OTDM signal in power penalty at the receiver. Employed set up is shown in Fig.5: an EDFA is needed in front of the autocorrelator and polarization controllers maximize second harmonic generation. We measured pulse broadening in the fiber without and with OPA amplification along the amplifier bandwidth. Fig.6 shows an example of the performed measure. OTDM pulse wavelength is 1546.1 nm, which corresponds to the fiber zero dispersion wavelength. As expected, no broadening is observed between the pulse at the fiber input (Fig. 6a, first row) and at the fiber output (Fig.

6b, first row), while when amplification is taken into account pulse broadens from 7.5 ps to 7.79 ps (3.9%) (Fig. 6c, first row). All over the amplifier bandwidth we observed broadening ranging from 3.9% to 4.9 % (close to the two pumps). Input pulse peak power is −3dBm. When input power increases to 11 dBm observed broadening ranges from 2%, as expected in the anomalous dispersion region, to 4% as can be see on the second row of Fig.6.

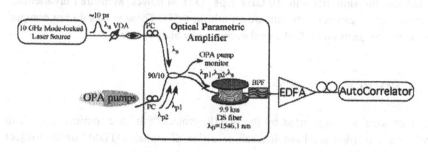

Figure 5. OTDM pulse broadening measure set up.

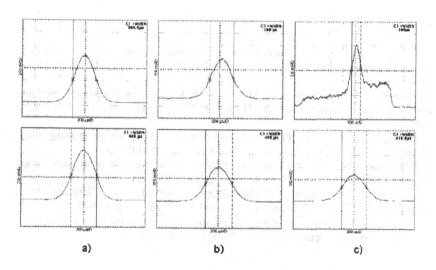

a)   b)   c)

Figure 6. Pulse autocorrelation traces at the fiber input (a), at the fiber output without parametric amplification (b), at the 2P-OPA output (c). Input signal peak power -3dBm (first row), 11 dBm (second row).

4. CONCLUSIONS

In conclusion we analyzed the 2P-OPA behavior as a possible solution to the necessity of exploiting new optical bandwidth when dealing both with WDM and high bit rate OTDM systems. OPA-2P has shown many advantages as high flexibility of project and gain uniformity over significant bandwidth. We projected and realized a 2P-OPA, which has been fully characterized for CW signals. We tested also the amplifier with 10 GHz 7 ps OTDM pulses. Measured broadenings over the amplification bandwidth are all within 4-5%. These values do not comport strong impairments on OTDM signal even at very high bit rate.

ACKNOWLEDGMENTS

This work was supported by the MIUR project "Studio e sperimentazione di architetture in fibra per l'amplificazione ottica di segnali OTDM" under Project 2001098217_004.

REFERENCES

[1] M.E. Marhic, Y. Park, "Broadband fiber-optical parametric amplifier and wavelength converters with low-ripple Chebyshev gain spectra", Opt. Lett., vol. 21, n. 17, pp. 1354-1356, 1996.

[2] L. Marazzi, P. Parolari, S. Seghizzi, M. Martinelli "Raman-generated pump impact on optical parametric amplification" IEEE Photon. Technol. Lett., 16, 1, pp.78 – 80, 2004.

[3] C.J. McKinstrie, S. Radic, "Parametric amplifiers driven by two pump wave with dissimilar frequencies", Opt. Lett., vol. 27, n. 13, pp. 1138-1140, 2002.

[4] C.J. McKinstrie, S. Radic, R. Chraplyvy, "Parametric Amplifier Driven by Two Pump Waves", J. Sel. Top. Quantum Electron., 8, 3, pp. 538-547, 2002.

[5] J. Hansryd, P.A. Andrekson, "Fiber-Based Optical Parametric Amplifiers and Their Applications", J. Sel. Top. Quantum Electron., vol. 8, n. 3, pp. 506-518, 2002.

[6] K. Inoue, T. Mukai, "Signal wavelength dependence of gain saturation in a fiber optical parametric amplifier", Optics Letters, vol. 26, n. 1, pp. 10-12, 2001.

APPLICATIONS OF FREE SPACE OPTICS FOR BROADBAND ACCESS

E. Leitgeb[1], M. Gebhart[1], U. Birnbacher[2], S. Sheikh Muhammad[1], Ch. Chlestil[1]

[1]Institute of Broadband Communications (at the Department of Communications and Wave Propagation, INW), TU Graz, Inffeldgasse 12, 8010 Graz, Austria, leitgeb@inw.tugraz.at
[2]Institute of Communication Networks and Satellite Communications, TU Graz, Inffeldgasse 12, 8010 Graz, Austria, birnbacher@inw.tugraz.at

Abstract: Free Space Optics (FSO) is an excellent supplement to conventional radio links and fibre optics. It is a broadband wireless solution for the "Last Mile" connectivity gap throughout metropolitan networks.
At the Department of Communications and Wave Propagation several FSO-systems have been developed within the past 5 years. Thereby the main effort was laid on using LEDs instead of laser diodes. One system is based on a modular concept using available standard components, used for distances of about 300 m at a specific power margin of 25 dB/km.
Additionally the research group "OptiKom" investigates the reliability and availability for different network-architectures (ring, mesh and star).

1 INTRODUCTION

Today, many people need high data rates for connecting to the Internet or for other access-services. Therefor Free Space Optics (FSO) is a well suited technology. The high bandwidth of the backbone (fibre network) is also available for the end user. The workgroup for Optical Communication ("OptiKom") at the Department of Communications and Wave Propagation, Technical University Graz, has carried out research in the field of Free Space Optics (FSO) over a period of about ten years. The work includes the development of equipment for research purposes and the evaluation of commercially available FSO systems for the climate in Graz (Austria) in co-operation with industrial partners. The main projects in this field have been funded by Telekom Austria AG, InfraServ Gendorf and the Government of Styria.

At the Department of Communications and Wave Propagation FSO-links are investigated and developed (in research projects and also in diploma and PhD theses). Additionally, existing systems are evaluated in cooperation with distributors, telephone companies and providers and optical links) are estimated in regard to range, bandwidth, traffic and weather (calculation of link budget and margin). In the work of "OptiKom" possibilities are investigated to increase the channel capacity and the reliability and availability in Optical Free Space Links.

2 FREE SPACE OPTICS-APPLICATIONS

Optical Wireless is the broadband solution (high data rates without any cabling) for connecting end users to the backbone (Last-Mile-Access). This technology is an excellent supplement to conventional radio links and fibre optics [3].

Applications for short range ("Last Mile", max. 1 km) can be used as quick installation of optical wireless links in urban areas (Point-to-Point / Point-to-Multipoint Systems) or as Broadband links for railway, highway or river crossings. Additionally FSO-links between buildings of companies or institutions (protection of ‚wiretapping') and rapid replacement of broken cable-links can be well suited solutions. To install FSO-systems for mobile / nomadic use (f.e. seminars, meetings, events) or links between different locations of dislocated events or for disaster management are also possibilities for using optical wireless technology.

In the following section different architectures are described and compared in regard to their advantages.

2.1 Optical Wireless in Ring Architecture

In the following schematic (figure 1) a FSO-network in ring architecture is shown. The distances between the buildings are up to 500 m. In the minimum configuration two Optical Receiver / Transmitter Units (OSE) are installed on the top of each building. In the event of a broken link (failure) between building 1 and building 2 (direct connection), the indirect link can be used. Instead, the information is sent in the other direction of the ring network (building 1, building 6, building 5, building 4, building 3, and building 2). Thereby, a partial security against failure can be achieved.

The installation of additional links increases the availability and the security against failure (redundant links). An Optical Repeater has to be used, if there is no line of sight between transmitter and receiver. In a field test four FSO-links for short ranges (up to 300 m) have been connected serially at the Department of Communications and Wave Propagation. In this experiment six of the connected FSO-units (three pairs) operated as repeaters.

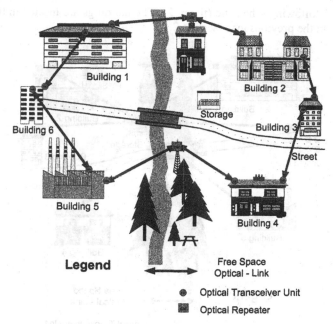

Figure 1. Optical Wireless (Ring Architecture)

2.2 Optical Wireless in Star Architecture

FSO networks with star architecture have been installed at the Department of Communications and Wave Propagation (project-group "OptiKom"). The coverage area of this FSO-network is about 300 m in diameter. An Optical Multipoint Unit (OZS) was mounted at the top of the department (Inffeldgasse 12). Five users (employees of the TU Graz) are permanently connected by their optical transceiver units to the Optical Multipoint hub-station. The five users are located inside the surrounding buildings (Inffeldgasse 10, 12, 18 and 16). Thereby, the user PCs are connected via the FSO-link to the OZS. In this configuration the OZS is realised with five FSO-Point-to-Point units, each of them directed to user FSO terminal. The Optical Multipoint Unit is interconnected by switches with the backbone network of TU Graz.

The advantage of this configuration are the shorter distances between any two FSO-units, because the Optical Multipoint Unit (OZS) is used as a repeater. In general, the Optical Multipoint Unit is in the centre of the area, but this architecture has the disadvantage of a single point of failure. If the Optical Multipoint Unit fails, a system breakdown of the whole installation is caused. To improve the reliability of this architecture, a redundant Multipoint Unit has to be installed. A second Optical Multipoint can also be implemented on moveable platforms (e.g. a broadcast van). For the installation of Multipoint Units on cars (for short term use) FSO-systems with "autotracking" are preferable. A network in star architecture is

582

shown in the following schematic (figure 2). The buildings are located in the same locations as in the previous map (section 2.1).

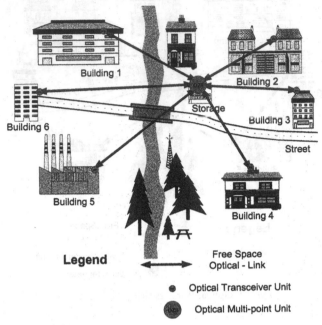

Figure 2. Optical Wireless (Star Architecture)

2.3 Optical Wireless in meshed architecture

For high reliability, the optimum network architecture is a meshed network, because it combines the advantages of the above described architectures. Different connections are possible, of which two examples (experimentally verified at the department) are described in this paper.

In figure 3 a mixture of a ring and a star network is shown. In each configuration, the central FSO-unit (OZS) can be connected to satellites, directional radio links, (mobile) telephone networks, or fibre optics. The Optical Multipoint Unit (OZS) can be connected with a switch or router to the backbone network. The clients in the buildings are linked with their FSO units to the OZS. This solution is similar to the FSO-network installed at the Department of Communications and Wave Propagation at the TU Graz.

Free Space Optics-systems with Light Emitting Diodes (LED) as source are developed [5] by the research-group „OptiKom" at the Department of Communications and Wave Propagation (section 3). Using special LEDs has the advantage that problems with Laser- and Eye Safety are minimised. The „OptiKom" group has developed low cost FSO systems for data rates of 10 and 100 Mbps by using LEDs instead of laser diodes. Most of this work was funded by the government of Styria and InfraServ Gendorf.

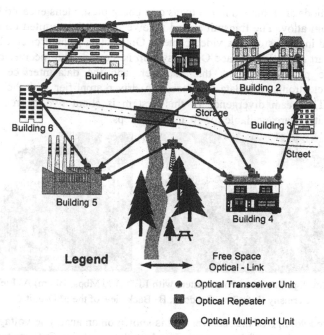

Figure 3. Optical Wireless (meshed architecture)

3 DEVELOPED FSO-SYSTEMS WITH LIGHT EMITTING DIODES

The developed systems combine available standard components to realise cost-effective solutions. The beam divergence usually covers values from about 8 to 60 mrad, allowing an easy alignment for the user without the need of a telescope. All elements including the optics do not require a precision as high as needed for other approaches, allowing higher tolerances at production, and the use of simple mounts.

Due to large divergence, the requirements for a stable underground are not very high allowing a quick installation. Suitable distances for high availability operation are limited by the wide beam angle and depending on local climate to up to 300 meters. Due to Laser Safety Regulations the use of sources with larger emitting area allows more output power in the same safety class, which improves the link budget.

3.1 Free Space Optics system for 100 m

The developed Free Space Optics system for a range of 100 m consists of two main parts, one transmitter (LED) and one receiver (photo-PIN-diode). The LED and the

584

photo-PIN-diode are mounted in the focus of a cheap plastic lens encased by a tube of Aluminium alloy. The transmitter and receiver units are mounted on a printed circuit board in a housing of a video camera. On a single printed circuit board the electronic part of the Free Space Optics-system is located. The receiver electronic includes the photo-PIN-diode, the amplifier, and the data-interface. For the transmitter electronic, the data-interface, the driver amplifier and the LED are necessary. The beam divergence of this system is about two degrees. At the moment solutions are available for 10 and 100 Mbps.

Figure 4. Developed Point-to-Point-system with LED (100 Mbps, 100 m) A) Field-test in Germany (InfraServ Gendorf), B) Back view of the FSO-unit

The measured voltage of the receiver unit is shown on an analogue voltage display (integrated at the back of the system). The installation of this FSO-system is very easy, because of the beam divergence of two degrees and the above mentioned voltage display.

3.2 Free Space Optics system for 300 m

The developed Free Space Optics system for a range of 300 m is mounted in a plastic housing. The Free Space Optics-unit consists of 8 transmitters (LED) and one receiver (photo-PIN-diode). The LEDs are also mounted in the focus of a cheap plastic lens encased by a tube of Aluminium alloy. The received light is focused by a large lens to the photo-PIN-diode.

Figure 5. Developed Point-to-Point-System A) with 3 transmitter units (for 100 m) and B) with 8 transmitter units (for 300 m)

The electronic part on the printed circuit board is similar to the small system for the 100 m range. By using VCSELs instead of LEDs, the range of the systems can be increased up to 800 m.

3.3 Folk-Festival "Aufsteirern" – Nomadic use of Free Space Optics

A real service demonstration for nomadic use of optical wireless networks took place at the folk-festival „Aufsteirern 2002" in the historical town of Graz.

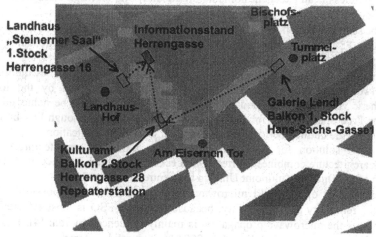

Figure 6. Location "Landhaushof" and „Tummelplatz" shown in the city map of Graz
(Event „Aufsteirern")

Folk-dancers and brass bands were located at different places and streets. The task of the project-group "OptiKom" was to transmit live video pictures (with a data rate of 10 Mbps) from two locations ("Landhaushof" and "Tummelplatz") to the central enquiry kiosk ("Herrengasse"). Thereby, the visitors of the festival could inform themselves what was going on at the other locations "Landhaushof" and "Tummelplatz". In figure 6 the setup for this demonstration is shown in the city map of Graz. For the FSO links, the described systems were used to transmit live video data streams from the yard inside of an historical building ("Landhaushof") and from place ("Tummelplatz") to the enquiry kiosk ("Herrengasse"). The transmission from the location "Tummelplatz" was realised with the use of an optical repeater, because there was no line of sight between "Tummelplatz" and the enquiry kiosk. For this purpose two systems (for 300 m range) have been (inter)connected.

In a similar scenario we installed Free Space Optics- and Satellite applications at a civil-military exercise in Spring 2004 in Styria (southern part of Austria). In this civil-military cooperation we use a mobile Satellite Earth Station (equipped with

Free Space Optics and Wireless LAN) for Videoconferencing between military and civil organisations (police, fire brigades and departments of the government). Additional Free Space Optics-links operate with high bit-rate for data-transfer between the military division and civil organisations (police etc.). This application is a new possibility of Broadband-Communication in Civil-Military-Cooperation (CIMIC).

4 CONCLUSIONS

Optical Wireless is an excellent nomadic broadband solution (high bandwidth) for connecting the end users to the backbone (Last-Mile-Access). This technology should be seen as supplement to conventional radio links and Fibre Optics. The use of low cost FSO-systems for short distances makes this technology interesting for private users.
At the moment the main work in this field is to increase reliability and availability. Those two parameters of the FSO-link are mainly determined by the local atmospheric conditions. So good reliability and availability can be achieved by using the Free Space Optics for short distances, by calculating enough link-budget and by using the optimal network architecture for each FSO application.
The optimal solution for FSO configurations is a meshed architecture. This network architecture combines shorter distances and high reliability, because of the location of the Optical Multipoint Unit (in the centre of the area).
The combination of FSO and microwave-links is also a further possibility for increasing reliability and availability, because terrestrial FSO is most effected by fog, whereas the microwave propagation is mainly influenced by rain [4]. Results of investigations show a reliability of 99.9991 % for hybrid systems.

REFERENCES

[1] Leitgeb E., Bregenzer J., Fasser P., Gebhart M., *Free Space Optics – Extension to Fibre-Networks for the „Last Mile",* Proceedings and Presentation at IEEE / LEOS 2002 Annual Meeting, 10th -14th of November 2002, Glasgow, Scotland

[2] Leitgeb E., Gebhart M., Fasser P., *Reliability of Free Space Laser Communications – Investigations at the TU Graz,* Proceedings. and Presentation at the 8th Annual WCA Techn. Symp., 14th - 16th of January 2002, San Jose, CA, USA

[3] Leitgeb E., Gebhart M., Fasser P., *Free Space Optical Access,* Proceedings. and Presentation at tcmc2001, 15th / 16th of October. 2001, Graz

[4] Schrotter P., Birnbacher U., Kogler W., Leitgeb E., Koudelka O., *Increased availability with hybrid optical / microwave networks,* Presentation at WG1-Meeting COST 270, 9th / 10th of April 2002, Graz

[5] Gebhart M., Leitgeb E., Fasser P., *Short-Range FSO-Systems - Increasing Network Flexibility and Reliability,* Presentation at WG1-Meeting COST 270, 9th / 10th of April 2002, Graz

[6] Leitgeb E., Gebhart M., Fasser P., Bregenzer J., Tanczos J., Impact of atmospheric effects in *Free Space Optics transmission systems,* Proceedings and Presentation at SPIE Photonics West, LASE 2003, 25th - 31st of January 2003, San Jose, CA, USA

POLARIZATION CONVERSION INDUCED IN A NON-CONVENTIONALLY BIASED CENTROSYMMETRIC PHOTOREFRACTIVE CRYSTAL

Claudio Crognale, Luigi Rosa
CNX S.P.A. Siemens, S.S.17, Località Boschetto, 67100, L'Aquila, Italy
claudio.crognale@siemens.com
luigi.rosa@siemens.com

Abstract: In an non-conventionally biased m3m centrosymmetric photorefractive crystal, the combination of the internal photorefractive field with the external static electric field can introduce a conversion between the transverse vector components of the read-out beam at infrared wavelengths.

In the last years, a great interest has been devoted to the photorefractive materials, with a particular attention to the class of m3m centrosymmetric crystals [1]. Recent works carried on biased Potassium Lithium Tantalate Niobate (KLTN) crystals in paraelectric phase have shown that, in these crystals, after the waveguide formation at the visible wavelengths, the internal charge field created by means of the photorefractively active light is still present even when the crystal is illuminated with an infrared beam. Then the quadratic response of the material nonlinearly combines the internal photorefractive field with any external static electric field, so that the index pattern generated by the active light can be strongly modified, up to produce an antiguiding effect on an infrared probe beam [2] (i.e. the read-out field). It is our opinion that even more attractive properties of such photorefractives could be highlighted, when non-conventional biasing configurations are taken into account. In this work, we introduce a non-conventional scheme, which allows to vary the orientation of the external biasing static field respect to that of the optical field vector, thus enhancing the tensorial features of the crystal. We found that, by imposing a proper non-conventional boundary to the crystal, after the waveguide formation, the combination of the internal photorefractive field with the external static electric field can properly modify the bulk refractive index pattern, introducing an attractive conversion mechanism between the transverse vector components of the probe beam at longer wavelengths.

1.1 THEORETICAL MODEL.

Let us consider a visible optical beam polarized in the x-y plane that propagates along the z axis of a centrosymmetric photorefractive crystal [3]:

$$\vec{A}(x,y,z) = \hat{x}A_X(x,y,z)e^{ik_Xz - i\omega t} + \hat{y}A_Y(x,y,z)e^{ik_Yz - i\omega t} + c.c. \quad (1)$$

with a Gaussian input field spatial distributions $A_{X,Y}(x,y,0)$ [2]. We assume that the x-y transverse polarization plane and the propagation direction of the optical field are according to an (x, y, z) coordinates system oriented along the principal dielectric axes of the crystalline medium as well. Moreover, we consider a centrosymmetric biased by means of a purely transverse (i.e. independent on the z axis) external static field distribution. At the steady-state, the propagation of the transverse Gaussian needle beam through the biased photorefractive can be described by means of the Kukhtarev's and Helmoltz's nonlinear coupled equations [3], [4], [5] (by assuming the validity of the slowly varying approximation for the field amplitudes $A_{X,Y}$):

$$\nabla \cdot \left[(I_0 + I_B)\nabla \vartheta_0 - \frac{K_B T}{q}\nabla(I_0 + I_B) \right] = 0 \quad (2)$$

$$\frac{\partial A_X}{\partial z} - \frac{i}{2k_X}\left(\frac{\partial^2 A_X}{\partial x^2} + \frac{\partial^2 A_X}{\partial y^2} \right) = \frac{ik_X}{n_X}\Delta n_{11} A_X + \frac{ik_X}{n_X}\Delta n_{12} A_Y \exp[i(k_X - k_Y)z] \quad (3a)$$

$$\frac{\partial A_Y}{\partial z} - \frac{i}{2k_Y}\left(\frac{\partial^2 A_Y}{\partial x^2} + \frac{\partial^2 A}{\partial y^2} \right) = \frac{ik_Y}{n_Y}\Delta n_{21} A_X \exp[-i(k_Y - k_Y)z] + \frac{ik_Y}{n_Y}\Delta n_{22} A_Y, \quad (3b)$$

In the expressions above, ∇ is the transverse gradient operator, ϑ_0 is the local electrostatic potential (according to the assigned boundary conditions), $I_0(x,y,z) = |A(x,y,z)|^2 = |A_X(x,y,z)|^2 + |A_Y(x,y,z)|^2$ the optical intensity of the propagating beam in nonlinear regime, I_B the artificial dark irradiance (constant), K_B the Boltzmann constant, T the absolute temperature, q the electron charge, n_x and n_y the zero-field principal refractive indices for light polarized along the x and y principal dielectric axes of the crystal, respectively, $k_x = kn_x$, $k_y = kn_y$, in which $k = \frac{2\pi}{\lambda}$ is the vacuum wave number (λ is the vacuum wavelength). We take in account the tensorial nature of the crystal by means of the terms Δn_{ij} in (3a), (3b), which are associated to the refractive index modulation due to the local space-charge field distribution, induced by the photorefractive effect. Without any lack of generality, let us consider for our investigation an m3m cubic centrosymmetric structure, such as the Potassium LithiumTantanate Niobate (KLTN, [2]). Thanks to its symmetry properties, in such a medium $n_x = n_y$, and the tensorial properties are completely described by one value of the relative dielectric constant ε_r, and by only three values of the components of the electro-optic quadratic tensor: $g_{11} = g_{22}$, $g_{12} = g_{21}$, and g_{66}, because the other terms $g_{16} = g_{61} = g_{62} = g_{26} = 0$. The nonlinear terms in (3a), (3b) can be written as [6]:

$$\Delta n_{11} = -\frac{1}{2}n_x^3\varepsilon_0^2(\varepsilon_r - 1)^2\left(g_{11}E_x^2 + 2g_{16}E_xE_y + g_{12}E_y^2\right) \tag{4a}$$

$$\Delta n_{12} = -\frac{1}{2}n_x^2 n_y\varepsilon_0^2(\varepsilon_r - 1)^2\left(g_{61}E_x^2 + 2g_{66}E_xE_y + g_{62}E_y^2\right) \tag{4b}$$

$$\Delta n_{21} = -\frac{1}{2}n_x n_y^2\varepsilon_0^2(\varepsilon_r - 1)^2\left(g_{61}E_x^2 + 2g_{66}E_xE_y + g_{62}E_y^2\right) \tag{4c}$$

$$\Delta n_{22} = -\frac{1}{2}n_y^3\varepsilon_0^2(\varepsilon_r - 1)^2\left(g_{21}E_x^2 + 2g_{26}E_xE_y + g_{22}E_y^2\right) \tag{4d}$$

In (4a)-(4d), ε_0 is the absolute dielectric constant, E_x and E_y the component of the local static space-charge electric field in the biased medium. We have performed our analysis by numerically solving the nonlinear propagation in (3a), (3b) (4a)-(4d) with the Beam Propagation Method (BPM), and the potential equation in (2) with a 5 points Finite Difference Method (FDM). The boundary conditions for the potential ϑ_0 were assigned by means of the external biasing voltage $V_B(x, y)$ applied to the crystal. For our analysis, we have introduced a particular theoretical model in which we assigned, in correspondence of the transverse section perpendicular to the z axis, the following boundary conditions for the potential ϑ_0 :

$$V_B\left(x = -\frac{L_X}{2}, y\right) = \left(\frac{V_2}{L_Y}\right)\left(y + \frac{L_Y}{2}\right) + V_{B0} \qquad -\frac{L_Y}{2} \leq y \leq \frac{L_Y}{2} \tag{5a}$$

$$V_B\left(x = \frac{L_X}{2}, y\right) = \left(\frac{V_2}{L_Y}\right)\left(y + \frac{L_Y}{2}\right) + V_{B0} + V_1 \qquad -\frac{L_Y}{2} \leq y \leq \frac{L_Y}{2} \tag{5b}$$

$$V_B\left(x, y = +\frac{L_Y}{2}\right) = \left(\frac{V_1}{L_X}\right)\left(x + \frac{L_X}{2}\right) + V_{B0} + V_2 \qquad -\frac{L_X}{2} \leq x \leq \frac{L_X}{2} \tag{5c}$$

$$V_B\left(x, y = -\frac{L_Y}{2}\right) = \left(\frac{V_1}{L_X}\right)\left(x + \frac{L_X}{2}\right) + V_{B0} \qquad -\frac{L_X}{2} \leq x \leq \frac{L_X}{2} \tag{5d}$$

In (5a)-(5d), L_X and L_Y are the transversal sizes of the boundary, and V_{B0}, V_1,

V_2 the boundary voltage values that are assumed independent on the ·x and y coordinates. From a practical point of view, this approach is equivalent to consider a crystal with transverse sizes much more larger than L_X and L_Y, that is biased by means of two independent, perpendicularly oriented external static fields, respectively given by $\vec{E}_{Xb} = -\hat{x}\dfrac{V_1}{L_X}$, $\vec{E}_{Yb} = -\hat{y}\dfrac{V_2}{L_Y}$, when the beam illumination is absent. Moreover, we also assume in this model that the transversal sizes of the boundary are much larger than the beam spot diameter, so that the presence of the optical beam field does not change the distribution of the electric static field at the boundary. According to this approach, we considered transverse boundary sizes L_x, L_y at least $20 \times (\Delta x, \Delta y)$ ($\Delta x, \Delta y$ are the Full Widths at Half Maximum of the optical field intensity along x and y directions, respectively).

1.2   NUMERICAL RESULTS AND DISCUSSION

Now consider a Gaussian read-out probe field at the input of the crystal after the waveguide associated to $\Delta n_{ij}(x, y)$ is formed. According to [2], in the numerical simulation, we study the propagation of a probe beam with the same wavelength of the active beam. This approach does not introduce any lack of generality, and all results reported below can be extended to other investigations in which longer wavelength optical probe beams are considered. The propagation of the read-out field through the biased photorefractive can be described by numerically solving the corresponding set of coupled equations (3a), (3b), with the (4a)-(4d), in which the effective read-out static electric field components are inserted:

$$E_{xeff} = E_x - E_{xb} + E_{xm},$$ (6a)

$$E_{yeff} = E_y - E_{yb} + E_{ym}.$$ (6b)

In (6a), (6b), E_{xm}, E_{ym} are the external biasing read-out static electric field vector components. We considered a $4mm$-long KLTN crystal, according to the conditions (reported in [2]): $\Delta x = \Delta y = 10 \mu m$ for both spatial distributions $A_{X,Y}(x, y, 0)$, $I_{PX}/I_B = I_{PY}/I_B = 1.3$ (I_{PX}, I_{PY} are the peak intensities of the orthogonal polarization states), $\lambda = 532nm$, $n_x = n_y = 2.4$, $\varepsilon_r(293K) = 9000$, $g_{11} = g_{22} = 0.12m^4 C^{-2}$, $g_{12} = g_{21} = 0.02m^4 C^{-2}$, $g_{66} = 0.05m^4 C^{-2}$. The value of term g_{66} for the KLTN is not available from the literature. Then, in our theoretical model, we have assumed this component to be of the order of the term g_{66} of the Barium Titanate ($BaTiO_3$) electro-optic quadratic tensor, because this crystal exhibits in the paraelectric phase the same structure of the KLTN [6]. The crystal was biased by means of two independent, perpendicularly oriented external static fields with the same magnitudes, equal to $E_{xb} = E_{yb} = 3 \cdot 10^5 V / m$. In the read-out phase, we imposed $E_{xm} = E_{xb} = 3 \cdot 10^5 V / m$, $E_{ym} = E_{yb} = 3 \cdot 10^5 V / m$, and $I_{POX}/I_B = 2$, $I_{POY}/I_B = 0$ (I_{POX}, I_{POY} are the peak intensities of the probe field orthogonal vector components). In this way, we simulated the propagation of a read-out Gaussian beam with the optical electric field vector lying just along the x axis. To analyze the result of our theoretical study, let us introduce the following parameters $P_{0X}(z)$, $P_{0Y}(z)$:

$$P_{0X,0Y}(z) = \frac{\iint I_{0X,0Y}(x, y, z) dx dy}{\iint (I_{0X}(x, y, 0) + I_{0Y}(x, y, 0)) dx dy} \qquad (7)$$

$P_{0X}(z)$, $P_{0Y}(z)$ are, respectively, the percentages of optical power contained in the x and y read-out optical field vector components at the crystal length z. The result is summarized in Fig.1.

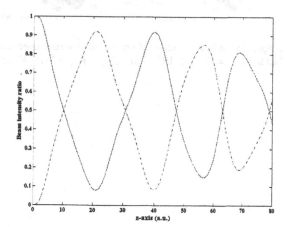

Figure 1. Power exchange between polarization vector components
($E_{xm} = E_{xb} = 3 \cdot 10^5 V / m$, $E_{ym} = E_{yb} = 3 \cdot 10^5 V / m$)

This figure reports the values of $P_{0X}(z)$ (continuous curve), $P_{0Y}(z)$ (dashed-

592

dotted curve) obtained according to the read-out launching conditions expressed above (in all figures reported here, *a.u.* stands for *arbitrary units*). It clearly appears a quasi-periodic exchange of power between the transverse vector components, with a quite good efficiency (higher than 90%) of the conversion process for $z = 21a.u.$ (corresponding to around $1mm$ propagation distance). Figs. 2(a), 2(b) show the spatial intensity distributions of both orthogonal polarization vectors at this distance (x-polarized component in Figs. 2(a), y-polarized component in Fig. 2(b)).

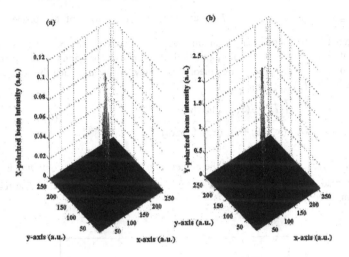

Figure 2. Spatial intensity distributions of orthogonal polarization vectors at $z = 21a.u.$ ($E_{xm} = E_{xb} = 3 \cdot 10^5 V/m$, $E_{ym} = E_{yb} = 3 \cdot 10^5 V/m$)

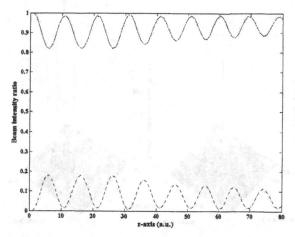

Figure 3. Power exchange between polarization vector components
$$(E_{xb} = E_{yb} = 3 \cdot 10^5 V / m, E_{xm} = 0, E_{ym} = 0)$$

The impact of the diffraction is quite low, at this distance, and the optical beam spot of the y-polarized component results well focused on the crystal transverse section.

Now let us consider the result in Fig.3 (the continuous curve is referred to $P_{0x}(z)$, the dashed-dotted curve to $P_{0Y}(z)$). We obtained this behavior by launching the x-polarized optical Gaussian probe beam at the input of the unbiased crystal ($E_{xb} = E_{yb} = 3 \cdot 10^5 V / m$, $E_{xm} = E_{ym} = 0$). This figure shows that, in absence of an external biasing field, a quasi-periodic mutual energy exchange between the orthogonal vector components of the linearly polarized read-out field occurs. Nevertheless, at the distance of $z = 21 a.u.$ considered above, the conversion process is extremely low (less than 5%) so that, in this case, the x-polarized component appears well focused on the crystal transverse section. This is shown in Figs. 4(a), 4(b), where the spatial intensity distributions of both orthogonal polarization vectors at this distance are reported (x-polarized component in Fig.4(a), y-polarized component in Fig. 4(b)).

594

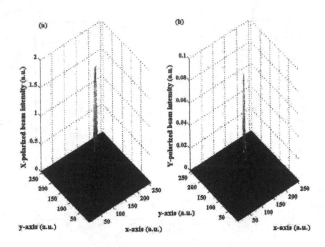

Figure 4. Spatial intensity distributions of orthogonal polarization vectors at
$$z = 21a.u. \,(E_{xb} = E_{yb} = 3 \cdot 10^5 V / m, E_{xm} = 0, E_{ym} = 0)$$

REFERENCES

[1] G. M. Tosi-Beleffi, F. Curti, D. Boschi, C. Palma and A. J. Agranat, "Soliton-based Y-branch in photorefractive crystals induced through dispersion-shifted optical fiber," *Opt. Lett.*, vol. 28, pp. 1561-1563, 2003

[2] E. DelRe, M. Tamburrini and A. J. Agranat, "Soliton electro-optic effects in paraelectrics," *Opt. Lett.*, vol. 25, pp. 963-965, 2000

[3] M.Segev, G. C. Valley, S. R. Singh, M. I. Carvalho, andD. N. Christodoulides, "Vector photorefractive spatial solitons" *Opt. Lett.*, vol. 20, pp. 1764-1766, 1995

[4] B. Crosignani, P. Di Porto, A. Degasperis, M. Segev, and S. Trillo, "Three-dimensional optical beam propagation and solitons in photorefractive crystals," *J. Opt. Soc. Am. B*, vol. 14, pp. 3078-3090, 1997

[5] B. Crosignani, P. Di Porto, M. Segev, G. Salamo, and A. Yariv, *Rivista del Nuovo Cimento*, vol. 21, pp.1-37 , 1998

[6] A. Yariv and P. Yeh, *Optical Waves in Crystals: Propagation & Control of Laser Radiation*, Ed. New York: John Wiley & Sons, 1983

TRANSIENT CONTROL BY FREE ASE LIGHT RE-CIRCULATION IN EDFA BASED WDM RING NETWORKS

Giovanni Sacchi[1], Simone Sugliani[1]Antonella Bogoni[1], Antonio D'Errico[1], Fabrizio Di Pasquale[2], R. Di Muro[3], R. Magri[4], G. Bruno[4], F. Cavaliere[4]
[1] *Photonic Networks National Laboratory ,CNIT, Via cisanello 145, 56124 Pisa, Italy*
[2] *Scuola Superiore Sant'Anna Via Cisanello 145, 56124 Pisa, Italy*
[3] *Marconi Corporations plc, New Century Park, CV3 1 HJ Coventry,*
[4] *UK Marconi Communications, via Negrone 1A , Genova, Italy*

Abstract: We demonstrate that a free ASE light re-circulation in EDFA based WDM ring networks provides an effective gain clamping technique. Proper network and amplifier design ensure signal power overshoots lower than 2.5 dB under 23/24 WDM channels add-drop operations.

1.   INTRODUCTION

Low cost and effective amplifier power transient controls are becoming essential for future WDM metro-core network development. Although new amplification technologies, such as Raman [1] and semiconductor optical amplifiers [2] can provide remarkable performances in terms of gain bandwidth and flexibility, standard EDFAs are still the most attractive solution as the best trade-off between cost end performances.   When EDFAs are used in add/drop networks, or in presence of traffic bursts, signal power transients due to the variable input signal load, can cause serious performance degradation, therefore gain control techniques must be used.

In WDM ring networks, in which EDFAs are used to compensate the losses of both fiber spans and network elements, closed optical paths can be formed, giving rise to uncontrolled lasing oscillations which can impair network performances under WDM channels add-drop operations [3]. Although it has been previously shown that closed cycle lasing can be made stable, and used to stabilize re-configurable WDM networks [4], no transient effects and network performance analysis has been reported, to the best of our knowledge.

In this paper we experimentally demonstrate an effective gain clamping technique for EDFA based WDM ring networks, which is based on free amplified spontaneous emission (ASE) light re-circulation along the ring. We show that proper network and amplifier design, can make the system robust to WDM

596

channels add-drop operations, providing, at the same time, a cost effective signal power transients control and acceptable optical signal to noise ratio (OSNR) performances.

2.   EXPERIMENTAL SET-UP

Fig. 1 shows the experimental set-up we have used to reproduce a worst case scenario in term of transient effects. Seven, high power, 100 GHz spaced DFB lasers (from 1552.5 nm to 1557.6 nm) are multiplexed and switched on and off at 100 Hz by an acouto-optic modulator, before being combined with a probe signal at 1551.7 nm (the modulation period is 5 ms, much longer than the network round trip time, which is about 950 µs). Note that in order to reproduce adding and dropping of 23/24 channels, the power/channel of the seven loading signals is always chosen about 5 dB higher than the probe power (10log10(23/7)=5.16).

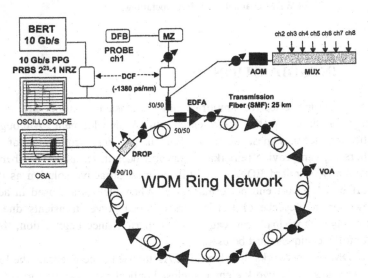

Fig. 1: Experimental set-up representing a worst case scenario.

All WDM signals are then inserted into the ring, at the first EDFA input, and then propagated along the network before being extracted at the last EDFA output, through a fixed 8 channels add-drop multiplexer which leaves the ASE light to freely circulate in the ring. All EDFAs in the network are operated at constant pump power (100 mW at 980 nm). The set-up shown in Fig. 1 allows us to measure the output spectrum, optical signal to noise ratio (OSNR), the probe power transients induced by add-drop operations and the probe bit-error-rate performances at 10 Gb/s. Note that two dispersion compensating fiber spools (-1380 ps/nm) are introduced respectively at the transmitter and receiver side, to compensate exactly for the accumulated chromatic dispersion. Variable optical attenuators (VOA) are used in each fiber span (25 km of standard SMF) and at the transmitter side in order to investigate network performances in different operation conditions such as varying the input power per channel and the span losses.

3.   WDM RING NETWORK PERFORMANCES

We have first investigated the probe power transient behaviour (see Fig. 2) at the last EDFA output, under 23/24 WDM channels add-drop operations at the first EDFA input; the input probe power is −17 dBm and the span loss is 20 dB, high enough to ensure stable gain peaking at around 1532 nm, that is far enough from the WDM signal band (from 1542 nm to 1561 nm with 24, 100 GHz spaced channels).

Note that, for a given EDFA structure and input power per channel, the span loss must be optimized in order to ensure the best compromise between good OSNR performances and lasing stability at around 1532 nm, under full WDM channels add-drop operations.

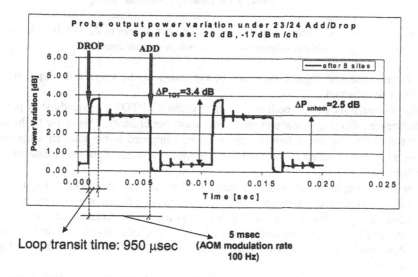

Fig. 2: Transient behaviour at the last EDFA output induced by 23/24 WDM channels add-drop at the first EDFA input.

598

From Fig. 2 we can notice a maximum probe power overshoot (PTOT) of about 3.5 dB, which is very small if compared with the strong power transient which would be expected in such a long EDFA chain without any gain control (≈13 dB). Also note that after each loop transit time (≈950 μs) the lasing ligth, re-circulating along the ring, clamps the gain with typical probe power transients induced by the lasing relaxation oscillations. The clamping mechanism, provided by the ASE light re-circulation, is only partially effective and the steady-state probe power level remains about 2.5 dB above the steady-state condition with full network load (ΔP_{UN-HOM}); this is due to spectral hole burning and un-homogeneous gain.

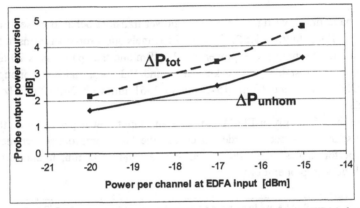

Fig. 3: Probe power excursion at the last EDFA output induced by 23/24 WDM channels add and drop versus input signal power per channel.

We have also investigated the probe power transients by varying the input signal power per channel.

Fig. 3 clearly shows that both maximum overshoot ΔP_{TOT} and steady-state power difference ΔP_{UN-HOM}, grow with the input power per channel. This is due to the fact that the more the lasing light is predominant, compared to the total signal power, the more the clamping mechanism, provided by ASE light re-circulation, is effective.

The probe output OSNR is greater than 24 dB (resolution bandwidth: 0.1 nm) and its maximum OSNR variation, induced by polarization dependent effects, has been measured to be less than about 0.7 dB. Also the probe relative intensity noise (RIN) has been measured and compared with open and closed ring, at the same OSNR value: no penalties have been observed due to RIN transfer from laser light to signals.

However, in order to exclude all possible potential transmission penalties, related to both probe power transients and noise transfer from lasing light to WDM channels, we have performed BER measurment at 10 Gb/s.

The probe signal has been externally modulated at 10 Gb/s (PRBS 223-1, NRZ format) and BER measurments versus OSNR have been performed under WDM channels add-drop operations.

Fig. 4 shows that no BER penalties have been observed with respect to back-to-back conditions (receiver characterization by noise loading).

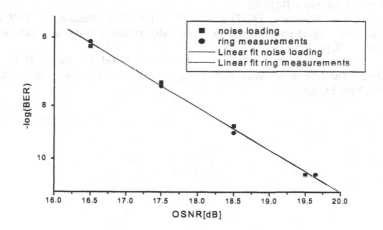

Fig. 3: BER versus OSNR with noise loading and with ASE light recirculation.

4.   CONCLUSIONS

We have presented and experimentally demonstrated an effective gain clamping technique, based on free ASE light re-circulation, in EDFA based WDM ring networks for metro applications.

Experimental results confirm that a proper EDFA and network design ensure robustness to add-drop operations and acceptable OSNR performances, without any transmission penalty related to the presence of ASE light re-circulation.

ACKNOWLEDGMENTS

We thank Mr. P. Ghiggino for helpful discussions and suggestions.

REFERENCES

[1] M.Islam, "Raman amplifiers for telecommunications 1, Physical Principles" and Raman amplifiers for telecommunications 2, Sub-Systems and Systems", Springer-Verlag, New York 2003.

[2] P.Iannone, K. Reichmann, "In-service up-grade of an amplified 130-km metro CWDM transmission systems using single LOA with 140 nm bandwidth", OFC 2003, Atlanta, Georgia, USA, paper ThQ3.

[3] Y. Sun, A.K. Srivastava, J.L. Zyskind, J.W. Sulhoff, T.A. Strasser, C. Wolf, J.R. Pedrazzani, "Signal power variations in optically amplified WDM ring networks", ECOC 1997, pp. 135-137.

[4] W. Xin, G.K. Chang, B.W. Meagher, S.J.B. Yoo, J.L. Jackel, J.C. Young, H. Dai, G. Ellinas, "The benefits of closed cycle lasing in transparent WDM networks", ECOC 1999, Nice, France.

NOVEL OPTICAL DIRECT DETECTION SCHEME FOR DPSK SIGNALS USING FIBRE BRAGG GRATINGS

P. Munoz[1], I. T. Monroy[2], R. Garcia[1], J.J. Vegas[2], F.M. Huijskens[2], S. Sales[1], A. Gonzalez[1], J. Capmany[1], A.M.J. Koonen[2]

[1] IMCO2, Universidad Politecnica de Valencia, Camino de Vera s/n, 46022 Valencia - SPAIN - ssales@dcom.upv.es

[2] COBRA Institute, Eindhoven University of Technology, The Netherlands, I.Tafur@tue.nl

Abstract: A novel scheme for direct detection of DPSK signals using FBGs is proposed. It alleviates the stability requirements of conventional one-bit-delay demodulators and it is suitable for very high data-rates.

1. INTRODUCTION

Differential Phase-Shift Keying (DPSK) has been proposed as an attractive alternative to On-Off Keying (OOK) in optical fibre communication systems since it is robust to the nonlinear transmission impairments [1, 2]. Balanced DPSK receivers using a one-bit-delay interferometer have been widely proposed because of its higher receiver sensitivity [1–3] and can profit from the advantage of integrated optics to realize stable and compact interferometers and balanced detectors. However, there are some impairments present in one bit- delay interferometers for balanced DPSK detection [3]: arising from amplitude imbalance, finite extinction ratio of the interferometer, phase imbalance, delay-tobit rate mismatch, frequency offset and polarisation dependent delay. Furthermore, the higher the data bit rate the more difficult it is to mitigate the aforementioned degradations.

We propose a novel DPSK receiver using Fibre Bragg Gratings (FBGs) which alleviates the stabilisation drawbacks of one-bit-delay balanced DPSK receivers. Two similar approaches have been proposed before [4, 5], but to the best of our knowledge, this is the first time that a correct recovery of the optical DPSK signal is demonstrated.

2. SYSTEM DESIGN

The phase shift keying modulation consists on encoding a binary data stream as phase shifts in a signal [6]. For an optical wave, this can be represented with the following expression:

$$E(t) = \sqrt{P_0}e^{j(\omega_0 t + \Delta\phi \sum_k a_k p(t - kT_b))} \tag{1}$$

where P_0 is the electrical field power (constant), ω_0 is the electrical field central frequency, a_k are the binary symbols, $p(t)$ is the electric pulse shape, $\Delta\phi$ is the phase shift corresponding to each binary transition, usually π and T_b the bit period. In particular, optical DPSK consists in encoding a logical change in the bit stream, represented by the summation term in Eq. 1 by a phase shift of the optical wave [6]. Hence, if we consider ideal rectangular pulses:

$$p(t) = \prod \left(\frac{t - T_b/2}{T_b}\right) \tag{2}$$

the instantaneous frequency of the electrical field is given by:

$$
\begin{aligned}
f_i &= \frac{1}{2\pi}\frac{\partial\phi(t)}{\partial t} \\
f_i &= f_0 + \frac{1}{2}\sum_k a_k \left[\delta(t - kT_b) - \delta(t - (k+1)T_b)\right]
\end{aligned} \tag{3}
$$

Hence, the phase shift can be also regarded as an instantaneous frequency shift, so a data transition from 1 to 0 corresponds to an instantaneous down frequency shift, while a 0 to 1 transition results in an instantaneous up frequency shift, which is represented by the delta functions in Eq. 3. Therefore, all the information from the original bit stream is encoded in these instantaneous frequency shifts.

Based on this properties, we propose a novel receiver scheme to recover data using direct detection and optical pre-filtering. The setup for the receiver is shown in Fig. 1.

Thus, to detect properly the optical signals, two FBG cantered in the upper (FBG+) and lower (FBG-) frequencies with respect to the central wavelength can be placed as shown in Fig. 1. This splits the positive and negative frequency shifts, which after direct detection are transformed in intensity peaks. After the photodetectors some simple electronic components can be employed to convert the detected transitions in the received pulses.

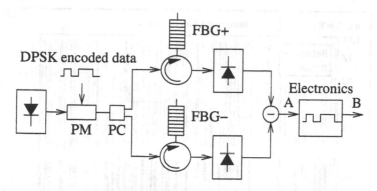

Figure 1. Schematic diagram of the proposed DPSK receiver(PM: Phase modulator, PC: Power Coupler, FBG: Fibre Bragg Grating).

3. SIMULATIONS

The receiver was simulated using available commercial software (VPItransmissionMaker), and the results are shown in Fig. 2, for a 10 Gbps bit stream. The figure shows the intensity peaks corresponding to the transitions, detected each time data symbols change from 0 to 1 and from 1 to 0 (dark line), and how using a comparator and a low-pass filter, the bit stream is reconstructed (light line). The simulations were carried out for data rates from 2.5 Gbps up to 40 Gbps, and in all the cases we found good agreement between the theory and the simulations.

4. EXPERIMENTAL RESULTS

We have experimentally implemented the set-up presented in Fig. 1, upto point A, to test the theoretical analysis and simulations. Only a single branch (one FBG and a single optical photodetector, followed by and oscilloscope) was used, replacing the FBG conveniently to measure the positive and negative frequency shifts respectively. We used a continuous wave tunable laser source with an output optical power of -2 dBm. An external optical phase modulator was driven by 10 Gb/s PRBS sequence. The bit sequence length was $2^7 - 1$ with a mark probability for the logical "1" of 7/8, in order to obtain a small number of logical "0" in the fixed pattern. The FBG had a reflectivity of 0.5 with a Full-Width Half Maximum of 6 GHz. The FBG was used in reflection, as depicted in Fig. 1.

The results are plotted in Fig. 3. In both plots, the electrical transmitted signal and the detected pulse edge transitions are shown, both for the negative and positive frequency shifts, from the laser nominal frequency. From the figure, there is a meaningful agreement between the experimental data and the results from the simulations shown in Fig. 2.

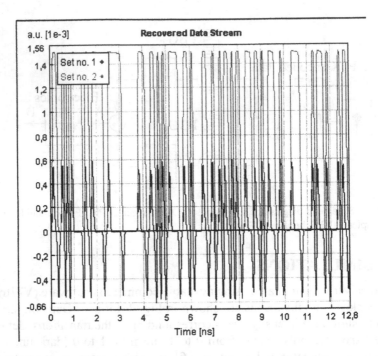

Figure 2. Simulation results intensity peaks at point A (dark line) of Fig. 1 and recovered bit stream (light line) at point B of Fig. 1.

The scale of the oscilloscope is 50 mV/div and 1 ns/div. The upper trace of each plot in Fig. 3 is the transmitted signals and the lower trace is the recovered data at point A in the setup shown in Fig. 1. Fig. 3-(a) shows the detected pulse leading edges when the FBG is placed at frequencies lower than the optical carrier, which is in good agreement with the theory and simulations. Fig. 3-(b) shows the other case, with similar results and conclusions. Also, some ripples are observed in the figures, which are attributed to the fact that each time the spectrum is shifted some of the signal energy falls within the FBG side lobes, and is detected by the photodiode. This can be alleviated by a proper design in the pulse shaping electronics stage after detection, from A to B in Fig. 1.

Fig. 4 shows the results for the case when the FBG is aligned with the transmitter wavelength. It can be seen that each time a logical "0" (a transition in the upper trace) is transmitted, the detected power decreases due to the shift of the spectrum. Thus, the receiver scheme of Fig. 1 can be simplified to use a single FBG placed at the central wavelength at the expenses of receiver sensitivity, as has been also reported by other authors [5].

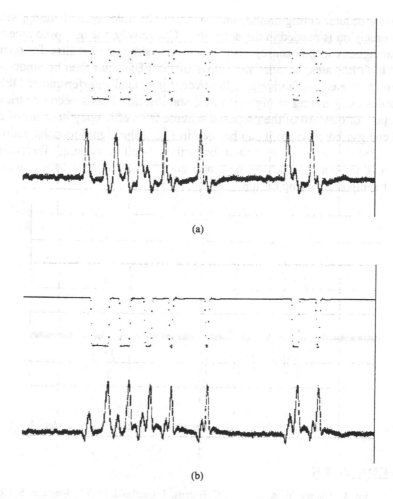

Figure 3. Transmitted electrical signal and measured pulse edge detection for (a) negative (FBG-) and (b) positive (FBG+) frequency shifts.

5. CONCLUSIONS

We have proposed a novel receiver scheme for DPSK signals which alleviates the impairments of stabilisation of the conventional one-bit-delay optical interferometer. The receiver is based in optical separation, using a filter as for example a fibre Bragg grating, of the lower and upper frequency of a optical PSK signal, which detected independently, yielding intensity peaks after the photodiodes at instants corresponding to the trailing and leading edges, respectively, of the transmitted pulses. Although the FBG requires thermal stabilisation, it is less complex than stabilisation of optical interferometers. In

606

the interferometer configuration, increasing the bit rate is challenging, since higher precision is needed in the delay line. Conversely, the proposed receiver is advantageous in this sense, since it becomes simpler to realize for operation at high bit-rates, because the slopes of the FBGs can then be smoother, and therefore easy to fabricate. Moreover, for optical interferometer DPSK demodulators operating at high bit rates, stabilisation issues become stricter. One aspect of concern of the proposed scheme is its efficiency in terms of detected energy; however, as it can be seen in Fig. 3 the signal-to-noise ratio of the detected signals is good enough for further signal processing. Techniques to improve the performance of the proposed scheme are under study such as pulse shaping and grating design.

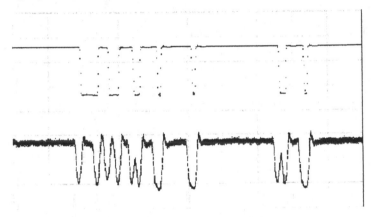

Figure 4. Detected signal when the FBG is aligned with the transmitter wavelength.

REFERENCES

[1] J. Sinsky, A. Adamiecki, A. Gnauck, C. Burrus, J. Leuthold, O. Wohlgemuth, S. Chandrasekhar, and A. Umbach, "RZ-DPSK transmission using a 42.7-Gb/s integrated balanced optical front end with record sensitivity," *J. Lightwave Technol.*, vol. 22, pp. 180–185, Jan. 2004.

[2] D. Gill, A. Gnauck, X. Liu, X. Wei, and Y. Su, "$\pi/2$ alternate-phase on-off keyed 42.7 Gb/s long-haul transmission over 1980 km of standard single-mode fiber," *IEEE Photon. Technol. Lett.*, vol. 16, pp. 906–908, Mar. 2004.

[3] P. Winzer and K. Hoon, "Degradations in balanced DPSK receivers," *IEEE Photon. Technol. Lett.*, vol. 15, pp. 1282–1284, Sept. 2003.

[4] A. Royset and D. Hjelme, "Novel dispersion tolerant optical duobinary transmitter using phase modulator and bragg grating filter," in *European Conference on Optical Communications*, vol. 1, pp. 225–226, Sept. 1998.

[5] D. Penninckx, H. Bissessur, P. Brindel, E. Gohin, and F. Bakhti, "Optical differential phase shift keying (DPSK) direct detection considered as a duobinary signal," in *European Conference on Optical Communications*, vol. 3, pp. 456–457, September 2001.

[6] B. Sklar, *Digital Communications.* Prentice Hall, 2nd ed., 2001.

Authors' Index

608

610